Lecture Notes in Computational Vision and Biomechanics

Volume 37

Research related to the analysis of living structures (Biomechanics) has been carried out extensively in several distinct areas of science, such as, for example, mathematics, mechanical, physics, informatics, medicine and sports. However, for its successful achievement, numerous research topics should be considered, such as image processing and analysis, geometric and numerical modelling, biomechanics, experimental analysis, mechanobiology and enhanced visualization, and their application on real cases must be developed and more investigation is needed. Additionally, enhanced hardware solutions and less invasive devices are demanded. On the other hand, Image Analysis (Computational Vision) aims to extract a high level of information from static images or dynamical image sequences. An example of applications involving Image Analysis can be found in the study of the motion of structures from image sequences, shape reconstruction from images and medical diagnosis. As a multidisciplinary area, Computational Vision considers techniques and methods from other disciplines, like from Artificial Intelligence, Signal Processing, mathematics, physics and informatics. Despite the work that has been done in this area, more robust and efficient methods of Computational Imaging are still demanded in many application domains, such as in medicine, and their validation in real scenarios needs to be examined urgently. Recently, these two branches of science have been increasingly seen as being strongly connected and related, but no book series or journal has contemplated this increasingly strong association. Hence, the main goal of this book series in Computational Vision and Biomechanics (LNCV&B) consists in the provision of a comprehensive forum for discussion on the current state-of-the-art in these fields by emphasizing their connection. The book series covers (but is not limited to):

- Applications of Computational Vision and Biomechanics
- Biometrics and Biomedical Pattern Analysis
- Cellular Imaging and Cellular Mechanics
- Clinical Biomechanics
- Computational Bioimaging and Visualization
- Computational Biology in Biomedical Imaging
- Development of Biomechanical Devices
- Device and Technique Development for Biomedical Imaging
- Experimental Biomechanics
- Gait & Posture Mechanics
- Grid and High Performance Computing on Computational Vision and Biomechanics
- Image Processing and Analysis
- Image processing and visualization in Biofluids
- Image Understanding
- Material Models
- Mechanobiology
- Medical Image Analysis
- Molecular Mechanics
- Multi-modal Image Systems
- Multiscale Biosensors in Biomedical Imaging
- Multiscale Devices and BioMEMS for Biomedical Imaging
- Musculoskeletal Biomechanics
- Multiscale Analysis in Biomechanics
- Neuromuscular Biomechanics
- Numerical Methods for Living Tissues
- Numerical Simulation
- Software Development on Computational Vision and Biomechanics
- Sport Biomechanics
- Virtual Reality in Biomechanics
- Vision Systems
- Image-based Geometric Modeling and Mesh Generation
- Digital Geometry Algorithms for Computational Vision and Visualization

In order to match the scope of the Book Series, each book has to include contents relating, or combining both Image Analysis and mechanics.

Indexed by Scopus.

More information about this series at https://link.springer.com/bookseries/8910

Mousumi Gupta · Sujata Ghatak · Amlan Gupta ·
Abir Lal Mukherjee
Editors

Artificial Intelligence on Medical Data

Proceedings of International Symposium, ISCMM 2021

Springer

Editors
Mousumi Gupta
Sikkim Manipal Institute of Technology
Majitar, India

Amlan Gupta
Sikkim Manipal Institute of Medical Sciences
Sikkim, India

Sujata Ghatak
Institute of Engineering and Management
Kolkata, India

Abir Lal Mukherjee
Henry Ford Hospital
Detroit, MI, USA

ISSN 2212-9391 ISSN 2212-9413 (electronic)
Lecture Notes in Computational Vision and Biomechanics
ISBN 978-981-19-0150-8 ISBN 978-981-19-0151-5 (eBook)
https://doi.org/10.1007/978-981-19-0151-5

This Springer imprint is published by the registered company Springer Nature Singapore Pte Ltd.
The registered company address is: 152 Beach Road, #21-01/04 Gateway East, Singapore 189721, Singapore

Preface

Medical information is generated ubiquitously in many settings ranging from personal physician offices and hospitals to government and non-government health agencies. Some of these data are individual data points of single patients, while others are conglomeration of health demographic data in large server databases. Then, there are data that are generated in clinical laboratories testing different healthy persons for screening for disease along with basic science research that are being translated to clinical trials. The analysis of such data may supply insights for decision making from the level of individual patient/health worker to large-scale policy making for the entire population of country, region or racial groups. Such diverse generation of data has given rise to many developments in the method to analyse data. One of the primary concerns of these platforms has been sieving and sifting proper analysable data.

Many computer scientists and health professionals have been working in tandem, but there are also many individual efforts that go undocumented. Thus, this second symposium of ISCMM 2021 was conducted to encourage such individuals and small groups to present their findings and observations. Many of them have innovative ideas or findings that get drowned by a cacophony of larger groups pushing their agendas.

The symposium gave emphasis to use of artificial intelligence in medical images. This could be microscopic or radiologic images and image obtained by signal processing of individual patient to images generated when bid population data visualization is done.

Presentations having application of artificial intelligence in some side of medical information or those pointing to areas of signal processing to create or analyse medical images were included have wider scope for interaction. Papers related to medical image processing and artificial intelligence include neuroimaging for Alzheimer's disease and brain tumour detection, mammography, cancer and tuberculosis detection in chest X-ray images, artificial ophthalmoscopic and retinoscopy images of eye, microscopic images of blood smear, melanoma cells, viral detection in electron microscopy CT images in COVID-19 patients a very contemporary phenomenon. Some applications of artificial intelligence in signal processing of medical or health-related issues have been covered in chapters dealing with wearable tachycardia monitoring system, walking assistant for vision-impaired. Some authors have looked at artificial intelligence in medical databases, especially those generated by COVID-19 pandemic. Some images not usually disease but particularly important in health care is wearing mask during the pandemic. There are two chapters on use of artificial intelligence to detect mask. It forms a very major issue in Preventive and Community Medicine Departments of Academic Hospitals. One major problem faced by medical professionals is transmission of images without compromising on confidentiality of patients, and there are chapters dedicated to use of encryption and decryption of medical image data.

The chapters are segregated into distinct aspects of intelligent computing in health care, biomedical signal processing; health monitoring; computational image processing; image informatics. This will enable the reader to develop a directed structured approach to research in widely different applications of artificial intelligence on medical science and medical data.

The book aims to create a bridge the postgraduate student and research communities working in the disciplines of Computer Science and Engineering with those medical science. Thus, some of the independent problem-solving exercises are converted to cross-disciplinary activity. However, the researchers working in the allied domain of pattern recognition and others will also find the book useful as it will have some newer approaches, computational intelligence methodologies and applications in medical science.

The editors believe that the selected papers will encourage and give opportunity to both academic researchers and industrial workers to project their research contribution to wider community. Each chapter inspires young scholars to learn newly created avenues of research at an international academic forum. Researchers and research organizations can refer the book for developmental works and incremental researches.

The editors acknowledge the funding received from ICMR project ID 2020-5638 for the proceeding and the institutional support received from Dr. Murlidhar Pai, Dean, Sikkim Manipal Institute of Medical Sciences, and Dr. Ashis Sharma, Director, Sikkim Manipal Institute of Technology. The editors thank Dr. Atul Kumar (former faculty of Gastroenterology and Applied Mathematics at Stony Brook University) for his inaugural address. We would like to thank Springer publishing house for valuable help in the preparation of this proceedings.

Majitar, India
Kolkata, India
Sikkim, India
Detroit, USA

Mousumi Gupta
Sujata Ghatak
Amlan Gupta
Abir Lal Mukherjee

Contents

About the Editors

Dr. Mousumi Gupta is an Associate professor of Computer Application at SMIT with more than a decade of teaching and research experience in the field of pattern recognition. She graduated with majoring in mathematics and did her masters in computer application. She further specialized in Information Technology with an M.Tech. from Sikkim Manipal Institute of Technology (SMIT) in 2007. She completed her Ph.D. from Sikkim Manipal University in 2015 during which she worked in the area of pattern recognition using digital image processing. She started her professional career after joining as a lecturer in the Department of Computer Science and Engineering in 2007 at SMIT. Her present research area include development of pattern recognition algorithm for digital pathology images. Presently, she has two research grants from ICMR and 26 peer reviewed publications.

Dr. Sujata Ghatak is an Assistant Professor of Computer Applications at Institute of Engineering and Management, India. She graduated from Calcutta University majoring in Computer Science following which she received Masters in Computer Science. She completed Master of Technology in Computer science from Sikkim Manipal Institute of Technology (SMIT), Majitar, Sikkim. She is one of the editor for Springer LNNS conference proceedings for IEMIS 2018. Her research interests include Image processing, Pattern Recognition, Applied Dynamical System in image processing, Natural Language Processing and Data Science. She is the member of several technical functional bodies such as The Society of Digital Information and Wireless Communications (SDIWC), Internet Society as a Global Member (ISOC), International Computer Science and Engineering Society (ICSES). She has chaired several sessions in various conferences in India. She has published several papers in reputed journals and conferences.

Dr. Amlan Gupta graduated in medical sciences from NMCH Patna in 1989 and Post-Graduated in Pathology from Calcutta University in 1995. After a brief interlude of private practice, he started his academic career in MCOMS Pokhara as an Assistant professor in 1997. He moved from there to SMIMS Gangtok and joined as an Associate professor in March of 2002. He became a professor in 2007. As a faculty he has been teaching, researching and taking care hospital diagnostics including surgical pathology and blood banking. His present area of research is wound healing and neural tissue healing. During his tenure he has 32 publications with 4 extramural research grants including one on quantitative histopathology.

Dr. Abir Lal Mukherjee is a board certified anatomic and clinical pathologist, working at Henry Ford Hospital, Detroit, Michigan as Chief of Clinical Neuropathology. He is a medical graduate from the oldest medical college in India, Calcutta Medical College. He post-graduated in pathology from JIPMER Pondicherry. He had further specialization in pathology at MD Anderson Cancer Center Houston Texas. He began his academic career in MCOMS Pokhara as assistant professor in 1996. He rose to the position of Associate Director, Pathology Residency Program and Director of Autopsy and Neuropathology, Department of Pathology and Laboratory Medicine, Temple University, School of Medicine, Philadelphia, PA. He is member of major scientific and educational committees, e.g., Graduate Medical Education Committee. He has 19 high impact publications. His primary research interests are in the area of Morphological spectrum of central nervous system malignancy and post mortem appearances of CNS in encephalitis and general incidence of undiagnosed malignancies in autopsy.

Biomedical Signal Processing

Comprehensive Study—A Deep Learning and Machine Learning Classification Methods for Cardiogram Images

D. Damodharan and Amit Kumar Goel

Abstract An echocardiogram is one of the heart testing techniques which is widely used in medical field for test and detecting the human heart illness, at present stage of heart functions, diseases, blood flow identification and any other issues to predict. In medicine, the resolution and quality of the pictures are most important. The problem of the medical images is not able to find issues as sooner and also various kinds noises mixed with it. This paper completely analysis the cardiovascular issue in all the multiple fields of technical views surveyed, with basic related issues to cardio diseases pattern identification and its identify the category of the diseases. At most challenging tasks are removing the noises, classifying and recovering the images in advanced technology with less cost. The machine learning techniques and models will help to improve the quality of the pictures and classify the images to practice for clinical and research purpose.

Keywords Classification · Cardiogram · Heart diseases and issues · Medical image

1 Introduction

Due to environmental changes, we see a lot of diseases are facing in the world. Due changing the environment and people food and habits are creating various new type of diseases are entering, so for in this situation only solution is identified early stage and control the diseases in earlier stage. One of the main things that human being needs to know is that cardiovascular disease is one of the foremost cause of deaths among men and women, not breast cancer and so on. Unhealthy food habits, genetic disorders, and age factors like old age people are affected more by these diseases. For Following some basic principle, we got good health and maintain every one body and heath, that is ABCD principles A-active, B-Blood pressure C-Cholesterol and D-Good balanced diet. It will care of diabetes and no smoke, working environment, adequate sleep daily.

D. Damodharan (✉) · A. K. Goel
School of Computing Science and Engineering, Galgotias University, Greater Noida, India
e-mail: itechdamu@gmail.com

M. Gupta et al. (eds.), *Artificial Intelligence on Medical Data*,
Lecture Notes in Computational Vision and Biomechanics 37,
https://doi.org/10.1007/978-981-19-0151-5_1

In this part delivers images of the correlated similar areas of this broadside such as machine learning, its methods with transitory descriptions, information pre-processing, concert evaluation system of measurement and a brief description of the utmost used sentiment illness dataset.

A. Supervised learning: Supervised learning is an approach to creating AI artificial intelligence model in which are trained using labeled training data and based on the output, it provided input data well correct output data to ML Machine learning model [1, 2].
B. Unsupervised learning: The approach of this method is not included in the dataset. Hear, there is no structure to give a proper format. So as a result, the algorithm attempts to recognize comparisons between input variables besides too classifies them accordingly. The clustering approach is part of the unsupervised learning strategy [3].
C. Reinforcement learning: This method is in managed and unaided realizing, where the model works on its exhibition as it collaborates with the climate. Consequently, it figures out how to address its slip-ups. It should get the right outcome through assessment and evaluate various conceivable outcomes [4].

The limit of the information are had structure and different Patten and different class of in the fields so for the identify impression method to work and practice strategies; especially the whole body various technique that is typically castoff the assessment. In this paper, we generally exertion on the idea that pre-owned association cycles to investigate cardiovascular sickness.

2 Classification ML Techniques

This section provides a brief overview of the furthermost extensively used organization methods used for cardio disease prediction. The methods described here are all examples of supervised mathematics techniques.

2.1 *Naive Bayes (NB)*

Classifiers based on naive Bayes theorem belong to the probabilistic classifier family. These classifiers assume that all features are independent, and this is essential in making predictions. They are easy to construct and perform well regardless of sample size. Hence, it is suitable for use in the medical science field as well as in diagnosing diseases [5, 6].

As far as naive Bayes (NB) classifiers go, the difference between them is mainly based on their assumptions about how they will model $P(x_i|y)$, where x is the posterior likelihood of class (x, target) given predictor (y, attributes). However oversimplified their assumptions may be, naive Bayes classifiers have proved useful in a number of real-world applications, including document classification and spam filtering. Detailed estimation of parameters is possible by using a small set of training data.

A classifier is based on naive Bayes method; it can be terribly fast, related to other methods based on additional complex statistical analysis. An extrication of the class restricted features leads to an estimation of each distribution as an independent one-dimensional distribution. Through improving the image dimensional related issues, the curse of dimensionality becomes less severe.

This calculation includes tracking down the *k*-nearest information focuses in the preparation set to the information point for which an objective worth is inaccessible and appointing the normal worth of the discovered information focuses to it.

2.2 Artificial Neural Network (ANN)

Algorithms have been established in order to mimic like the nerve cell in the brain. Each node or neuron is connected to another, and its output is the input of the next node. Despite the fact that each node receives multiple inputs, it only produces one output. The layer involves as an input beside concealed layer and a production layer. Multi-layer perceptions (MLP) are one of the most widely used types of ANN. Under different circumstances, each layer contains a different number of neurons.

A network consists of several components and various layers. Artificial neurons are arranged in large number layers consisting of neural network. Each layer involves an established of artificial nerve cell. Analyzing layers cutting-edge a neural network we will explore various types of layers. Programmers can input data in a variety of formats as indicated by the name. Input and output layers are separated by a hidden layer. By calculating all the hidden features and patterns, it finds hidden patterns and features.

They are equipped with numerical values that can accomplish more than one task simultaneously. Unlike traditional programming, which is stored on one machine, traditional programming uses the network as a storage medium. One or two data pieces disappearing from one place does not affect the network's ability to function. From Fig. 1, AI neural network is training is conducted on inadequate data, and most output may still be produced. It is the importance of missing data that determines the performance loss here. For an ANN to be able to adapt, examples must be determined and the network must be encouraged through these examples in order to achieve the desired output. When the network does not see the event in entirely of its pieces, it will produce false results, because sequence of the network is straight comparative to selected artworks. ANN can still generate output even when some cells have been removed, making it fault-tolerant.

According to Das et al. [7] an artificial neural network (ANN) ensemble technique was designed for the system in question. Through the combination of multiple models trained on the same task, we were able to achieve greater generalization. 89.01% accuracy was observed when model predictions were made using SAS enterprise miner 5.2 software.

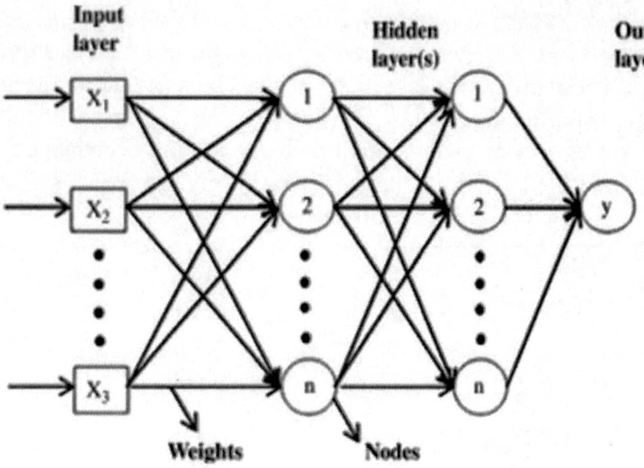

Fig. 1 AI neural networks (ANN)

Chen et al. [8] devised an algorithm that predicts heart disease using artificial neural networks (HDPS). Research performed in this study utilized learning vector quantization (LVQ), a type of ANN. Users can make predictions using the developed system by filling in the thirteen medical attributes. As a result, it displays the prediction result as either healthy or unhealthy, along with the ROC curve and various measurements of accuracy, sensitivity, specificity, and the time spent displaying the result. During development of the system, C programming language was used, as well as C# to implement the user interface. Eighty percentage, 85%, and 70%, respectively, were revealed to be the accuracy, sensitivity, and specificity of the model.

Dagare and Apte in [6] developed a predetermination system (HDPS) to predict whether patients would suffer from heart problems. Both datasets were attained from UCI's source and contain thirteen medicinal characteristics, including the Cleveland heart disease dataset. The precision of this model was improved by adding two additional attributes, obesity, and smoking, making fifteen in all. During the experiment, we used WEKA tool. When using the thirteen attributes, the prediction accuracy was 99.25%, while when using the fifteen attributes, the prediction accuracy was nearly 100%.

2.3 Decision Tree (DT)

This calculation has a DT technique likes flowchart and tree design construction. It comprises of twigs, plant leaves like, hubs besides a cause hub. The inside hubs hold the qualities though the divisions address the aftereffect of separately trial on every hub. DT is broadly utilized meant for grouping resolves since it need not bother with much information in the field or situation the boundaries for it to effort [9].

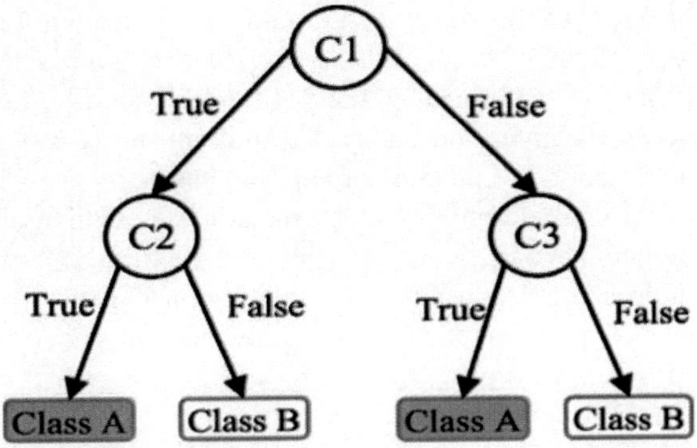

Fig. 2 Decision tree classification to predict

Among supervised machine learning procedures, the decision tree procedure is one of the greatest commonly methods to use for both classification and regression problems can be solved it. The decision tree conducts the tree illustration of the problem to solve the problem of predicting a target variable whose leaf node corresponds to a class label and whose attributes are placed on the internal node of the tree [3] (Fig. 2).

Our training set starts with the root, which we refer to as the entire set. Feature values with categorical attributes are preferred; however, continuous values will be converted to discrete values before a model can be built. Defining Entropy and Information Gain: Before we jump on to decision tree algorithms, we need to understand what is Entropy.

As a measure of disorder, impurity, or uncertainty, entropy measures the amount of impurity. Decision trees have their own way of handling data split according to entropy. As a result, decision trees have different boundary parameters. In a system with entropy values ranging from 0 to 1, less entropy is guaranteed to be trusted.

Equation for Entropy:

$$H(S) = -\text{probability of } \log_2(p+) - \\ - \text{probability of } \log_2(p-) \tag{1}$$

$(p+)$ % of positive class.
$(p-)$ % of negative class.

Formulating the values:

$$H(S) = -x/y \ \log_2(x/y) - x'/y' \log_2\left(x'/y'\right) \tag{2}$$

S refers to the number of positives in the calculation as form this Eq. 1 we are calculated probability *xy* represents the number of rows.

The same applies if we perform the entropy function, we get a bit, which is an attribute as there are 50% positives and 50% negatives. This splitting will not end, so we will not reach a pure subset. We will have either all positives or all negatives in this situation since we have a pure subset. In a tree, the steps to split based on which features to focus on at each level are guided by information gain. Simplicity is best when it comes to our tree, so we want it to be small. It is ideal to have the fewest daughter nodes at each step in this process. It is commonly understood that information is a measure of purity in business.

We can determine what information values tell us about a class based on its representation at a particular time point on the tree. First, the divided with the maximum information expansion will be engaged and a subsequent divided will be made till all of the derived nodes have a 0-information gain, or when all the nodes are pure.

Information gain:

$$\text{Gain}(S, A) = H(s) \cdot (|Sv|/|s|) * H(Sv) \tag{3}$$

Based on the algorithm, each split is analyzed for information gain; the split that results in the highest information gain is selected. The information gain corresponds to calculating the average value of the entropy following the split based on all the inputs.

Sv Total of sample after the split.
S Total of Sample.

Information gain plays an important role in determining which split receives selection.

Based on the distribution of classes in a dataset, Gini Impurity of new, random data indicates the likelihood of the class being incorrectly classified if it were assigned a random class label. Considering entropy has a maximum value of 1 and Gini impurity has a maximum value of 0.5, the Gini impurity requires less computational time, as it does not contain a logarithmic function.

A decision tree J48 algorithm was used Prerana and Shivaprakash at in [10] to select features and predict heart disease. For training purposes, 240 records were used, while for testing, 120 records were used. The dataset contained thirteen medical attributes/features. By selecting only certain features, the accuracy increased to 76.67%, which was better than the accuracy achieved by using all features. Additionally, improving accuracy to 85% can be achieved by restricting irrelevant features. The J48 algorithm is said to enable high prediction accuracy by selecting the minimum features.

Patel et al. [11] used several decision tree algorithms to assess the incidence or the presence of cardio disease using the WEKA tool on the UCI dataset. In addition to J48, logistic model trees and random forests were also tested. An accuracy of 56.76% was achieved by the J48 algorithm when compared to the rest.

2.4 K-Nearest Neighbor (KNN)

There are several kinds of machine learning algorithms, but the K-nearest neighbors (KNN) algorithm is capable of both regression and classification predictions. But this method is mostly used by the industrial sector to solve problems of classification and prediction. It is a good candidate for defining KNN because it is characterized by two characteristics.

KNN is a lazy learning method that routines all the data for physical activity training and classification, allowing lazy learning. The neural network takes into account no assumptions about what is underlying the data, which makes it a nonparametric learning algorithm (Fig. 3).

In the K-nearest neighbors (KNN) algorithm, the values of new data points are predicted by analyzing the characteristics of their features, which is in turn raised on how closely they match the parameters in working the traditional. Here are some steps that will help us understand how it works. The first step to implementing any algorithm is to acquire data. So, we need to load training and test data in the first phase of KNN.

In the next step, we must choose the nearest data points, where K is the value. K may be an integer of any size. The third step is to apply the following formula to each of the test points:

- Analyze the distance among each row of verification for the test information and every row of working out facts using any of the following methods: Manhattan/Hamming, Euclidean distance. The most commonly used technique to calculate distance is Euclidean.
- Categories the data to ascending order founded on the detachment value.

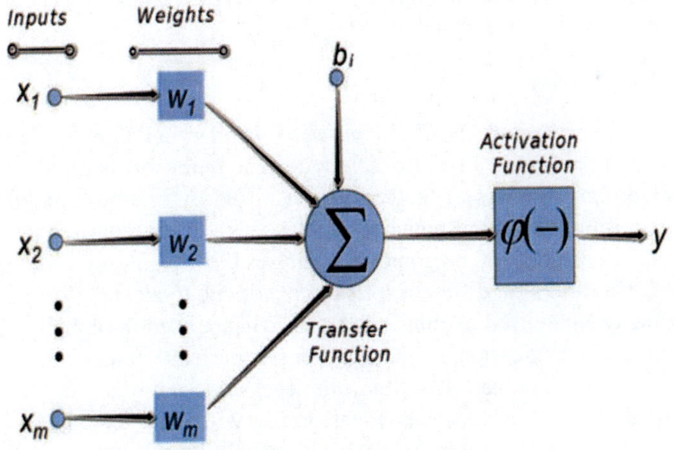

Fig. 3 KNN algorithm architecture diagram

- The following step will select the topmost K rows in the range based on the sort order.
- The class for the test point will now be determined by the most frequent class for the rows.

Based on the Cleveland dataset, Samu et al. in [12] practical K-Nearest Neighbor (KNN) to expect cardio disease. In the study, KNN applications alone and KNN applications with voting were compared. Data are divided into subsets by voting, and each subset applies the classifier. A cross-validation of tenfold is used for the evaluation. Various values for K showed an accuracy range of 94–97.4% without voting. A 97.4% accuracy was attained when $K = 7$. In spite of this, the accuracy did not improve using the voting technique. In the results, we noted a decrease in accuracy at $K = 7$–93%.

2.5 *Support Vector Machine (SVM)*

This calculation takes a valuable characterization precision. It is characterized by way of limited dimensional direction space which comprises of a measurement for each element/quality of an item.

A. Hereditary Algorithm

It is a developmental calculation that is constructed dependent on Darwin's hypothesis of advancement. It impersonates techniques in nature like transformation, hybrid, and normal determination. One of the for the most part utilized benefits of the hereditary calculation is its use to introduce loads of the neural organization standard [13]. That is the reason its utilization close by ANN is seen in numerous investigates to create a half and half forecast mode.

B. Troupe Learning

This technique joins various classifiers into one typical to build the precision. There are n kinds of collaborative learning technique. The main sort is deceiving, which is conglomerating classifiers of the comparative kind by casting a ballot method. Boosting is the subsequent kind, which resembles sacking, so far the new prototypical is influenced by past models' outcomes. Loading is the 3rd sorting which resources collecting AI classifiers for different sorts to deliver one model [14].

Among many supervised learning model, SVMs are used for equally classification and addition to regression. Primarily, it is second hand in machine learning to resolve classification glitches. This algorithm seeks to determine the best decision edge which, along with m-dimensional categories, will allow us to categorize new data points in the future easily. A hyperplane is a boundary that represents the best decision. This helps set up the hyperplane by selecting extreme points and vectors. Thus, support vector machine is called a method that considers these extreme cases as

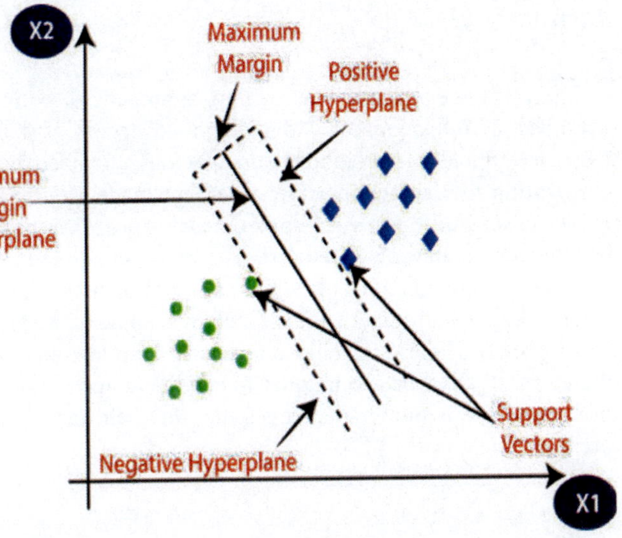

Fig. 4 SVM model architecture with maximum and vector with $x1,\ x\ 2$

support vectors. As shown in the following diagram, there are two different types of categorizations which are classified by defining a decision boundary or a hyperplane:

By using an example, one may understand how SVM algorithms work. Consider a dataset that has two tags (green and blue). Figure 4: The dataset also has two features $x1$ and $x2$, and these can be found here. A classifier should be able to distinguish green from blue coordinate combinations $(x1, x2)$.

Amelio et al. [15] examined SVM method types on the UCI database data to assess their accuracy in diagnosing heart disease. There were several types of SVMs studied various algorithms. An algorithm called Min–Max scaler was initially applied to the dataset. Using the above SVM algorithms, we then trained the algorithm on the dataset. With a 61.86% overall accuracy, BTSVM had improved performance as compared to the other algorithms.

RNNs are typically regarded as deep learning networks that are used for an unpredictability of tasks related to US image analysis in medicine field [16]. With RNNs, the complexity of the network is resolute by the quantity of sequences of input sample data. Xt and $ht1$, the last two states of a plain RNN, are mapped in a nonlinear fashion to give ht at time t. In this case, W-Weight and Weight-R are both collective over time, with b representing the bias parameter.

3 Conclusion

In our study, analysis reviewed machine learning approaches used in three popular tomography modes in ultrasonography, including an overview of standing machine learning techniques typically convolution neural networks founded that are premeditated for classification and segmenting diverse cardiac image patterns and structures. Furthermore, we discussed the fresh improvement of deep learning model-based segmentation approaches in each of the three segments, charted the possible of the deep learning-based segmentation procedures, and outlined any remaining limitations that could make them unsuitable for wide clinical application. With this review, we intend to provide a deeper understanding of the deep learning algorithms that have contributed many advanced techniques to cardiac image segmentation as well as highlight some of the common challenges that this field faces that will require future research.

References

1. Gárate-Escamila AK, El Hassani AH, Andrès E (2020) Classification models for heart disease prediction using feature selection and PCA. Inform Med Unlock 19:100330. ISSN 2352-9148. https://doi.org/10.1016/j.imu.2020.100330
2. Liu S, Wang Y, Yang X, Lei B, Liu L, Xiang Li S, Dong N, Wang T (2019) Deep learning in medical ultrasound analysis: a review. Engineering.https://doi.org/10.1016/j.eng.2018.11.020
3. Gavali P, Banu JS (2019) Deep convolutional neural network for image classification on CUDA platform. In: Deep learning and parallel computing environment for bioengineering systems, pp 99–122
4. Mohanasundaram R, Malhotra AS, Arun R, Periasamy PS (2019) Deep learning and semi-supervised and transfer learning algorithms for medical imaging. In: Deep learning and parallel computing environment for bioengineering systems, pp 139–151
5. Balaji K, Lavanya K (2019) Medical image analysis with deep neural networks. In: Deep learning and parallel computing environment for bioengineering systems, pp 75–97.https://doi.org/10.1016/b978-0-12-816718-2.00012-9
6. Dangare C, Apte S (2012) A data mining approach for prediction of heart disease using neural networks. Int J Comput Eng Technol 3(3):30–40
7. Das R, Turkoglu I, Sengur A (2009) Effective diagnosis of heart disease through neural networks ensembles. Expert Syst Appl 36(4):7675–7680
8. Chen C, Qin C, Qiu H, Tarroni G, Duan J, Bai W, Rueckert D (2020) Deep learning for cardiac image segmentation: a review. Front Cardiovasc Med 7:25. https://doi.org/10.3389/fcvm.2020.00025
9. Arivazhagan S, Deivalakshmi S, Kannan K, Gajbhiye BN, Muralidhar C, Lukose SN, Subramanian MP (2007) Performance analysis of wavelet filters for image denoising. Adv Comput Sci Technol 1(1):1–10. ISSN 0973-6107
10. Prerana THM, Shivaprakash NC et al. Prediction of heart disease using machine learning algorithms—Naïve Bayes, introduction to PAC algorithm. Compar Algorithms HDPS 3:90–99
11. Zhenya Q, Zhang Z (2021) A hybrid cost-sensitive ensemble for heart disease prediction. In: BMC medical informatics and decision making
12. Samundeeswari ES, Saranya PK, Manavalan R (2016) Segmentation of breast ultrasound image using regularized K-means (ReKM) clustering. In: International conference on wireless communications, signal processing and networking. IEEE, pp 1379–1383

13. Gonzalez RC, Woods RE (2002) Digital image processing, 2nd edn. Prentice-Hall, Englewood Cliffs, NJ
14. Damodharan D, Kumar Goel, A (2021) Novel approach on classification in machine learning model-based USG cardiogram images. In; 2021 international conference on advance computing and innovative technologies in engineering (ICACITE), pp 947–950. https://doi.org/10.1109/ICACITE51222.2021.9404752
15. Amelio L, Amelio A (2018) Classification methods in image analysis with a special focus on medical analytics. In: Intelligent systems reference library, pp 31–69.https://doi.org/10.1007/978-3-319-94030-4_3
16. Damodharan D, Suresh KN, Anandhan K, Jerald Nirmal KS (2019) J Innov Comput Sci Eng 9(1):29–33. Print ISSN: 2278-0947. Online ISSN: 2455-3506

Face Mask Recognition Using Modular, Wavelet and Modular-Wavelet PCA

Kavya Duvvuri, Kanisettypalli Harshitha, Vemula Srihitha, and Sarada Jayan

Abstract The world is currently under the grip of the COVID-19 pandemic. The only possible escape from this pandemic is wearing a mask. Mask checks are being done in most public places. This project brings out a good technique for recognizing a mask on a face. Principal component analysis (PCA) is a dimensionality reduction method which is used in image and signal processing. It uses only the principal components of a dataset and ignores rest of the components. In this paper, three different pre-processing methods (modular, wavelet and a combination of both) have been performed on an image dataset. The resultant data has been processed through PCA. Through comparison of the processed data vectors, the similarity between images has been established. Also, a comparison between different PCA techniques is developed through recognition of distorted faces. Initially, a comparison between different PCA techniques is developed through recognition of distorted faces. Then, a comparison is done between the Modular PCA, Wavelet PCA and Modular-Wavelet PCA techniques by observing the performance for mask recognition. Finally, a conclusion is drawn as to which method is the best in terms of mask recognition.

Keywords PCA · Modular PCA · Wavelet PCA · Modular-Wavelet PCA

K. Duvvuri (✉) · K. Harshitha · V. Srihitha
Department of Computer Science Engineering, Amrita School of Engineering, Bengaluru, Amrita Vishwa Vidyapeetham, Bengaluru, India
e-mail: krishnakavya.duvvuri@gmail.com

S. Jayan
Department of Mathematics, Amrita School of Engineering, Bengaluru, Amrita Vishwa Vidyapeetham, Bengaluru, India
e-mail: j_sarada@blr.amrita.edu

M. Gupta et al. (eds.), *Artificial Intelligence on Medical Data*,
Lecture Notes in Computational Vision and Biomechanics 37,
https://doi.org/10.1007/978-981-19-0151-5_2

1 Introduction

Most of the popular applications in image processing and signal processing have been possible due to various results from linear algebra [1–6].

Principal component analysis (PCA), that is developed using linear algebra, is a data processing method which ensures dimensionality reduction. It facilitates data storage and analysis. In this paper, PCA in combination with a few pre-processing techniques, is used for face mask recognition. The first section of this paper gives an introduction to the paper along with a brief literature survey. Section 2 describes the methods of image representation, and Sect. 3 details PCA and its three types that are used in this work—Modular PCA, Wavelet PCA and Modular-Wavelet PCA. Section 4 gives the experimental results and analysis, and the paper is concluded in Sect. 5.

According to the eigenface algorithm [2], any grayscale image with dimension $m \times n$ can be represented as a $1 \times (m \times n)$ row vector. A matrix can be formed with its rows as the row vector representations of every image of an image dataset. This matrix is processed using PCA to reduce its dimensions. A query image given by the user must be processed in the same way and compared with the row vectors of the matrix in order to select the most similar image from the given dataset.

The same method can be applied to pre-processed images. There are three different methods used in this paper: modular PCA, Wavelet PCA and Modular-Wavelet PCA [1].

In Modular PCA [3], the image that is to be processed is split into k equal sub-images. These sub-images are separately processed and converted into row vectors. The sub-image matrices are compared with the corresponding sub-image vector of the query image.

In Wavelet PCA [4], the image is first processed through discrete wavelet transformation (DWT). PCA is further applied to the LL sub-band of the obtained transformation. The same method is applied to the query image before comparison. DWT technique is useful in facial expression identification as demonstrated in [7]. In [8] Wavelet PCA has been combined with neuro-fuzzy fusion rule for medical image fusion.

In Modular-Wavelet PCA, a combination of both these methods is used to have the advantages of both. For comparing the similarity of the given query image with the images of the input dataset, Euclidean distance has been used. Interestingly, Euclidean distance can also be used in recognizing facial expressions as shown in [9].

In [10], a comparative study of several face recognition methods like PCA, convolutional neural network (CNN), linear discriminant analysis (LDA) and discrete cosine transformation (DCT) have been illustrated.

2 Image Representation

Image pixels can generally be represented in two different ways: RGB representation and grayscale representation. In RGB representation, the color of a pixel can be represented by a vector of three components which represent the ratio of red, green and blue colors that forms the color of the pixel. In the grayscale representation, the color of a pixel is represented by a single numeric value that represents the grayscale intensity.

In order to be able to represent a pixel using a single numeric value in the row vector, all the images in the image dataset are converted to grayscale before any further processing. After converting the images to grayscale, they are resized to ensure uniform image dimensions throughout the process.

3 PCA and Its Types

3.1 *Principal Component Analysis for Image Recognition*

An image of $m \times n$ dimension is converted into a row vector of length $m \times n$. All the k images of the given image dataset are converted into row vectors. These row vectors are populated as the rows of a $k \times (m \times n)$ matrix. Mean of each column of this matrix is found, and the means are subtracted from all the respective column elements. The matrix thus obtained is called the mean centralized matrix (X).

The covariance matrix of the mean centralized matrix is found using Eq. 1:

$$\text{Cov}(X) = \frac{X^T X}{k-1} \tag{1}$$

Now, the eigenvalues and eigenvectors of the covariance matrix are found, and the eigenvalues are arranged in decreasing order. The first i eigenvalues are considered as the most important, and the corresponding eigenvectors are arranged as the column vectors of a matrix. The matrix thus obtained is called the PCA matrix. The reduced matrix is obtained by multiplying the mean centralized matrix with the PCA matrix (P).

$$Y = PX \tag{2}$$

Equation 2 shows the formula to obtain the final reduced matrix using the mean centralized matrix and the PCA matrix (P). The query image given by the user is represented as a row vector after converting into grayscale. This row vector is centralized using the mean vector of the image dataset matrix. This centralized vector is multiplied with the PCA matrix obtained from the dataset matrix. The resultant vector is compared with each row vector of the reduced matrix to obtain the Euclidean

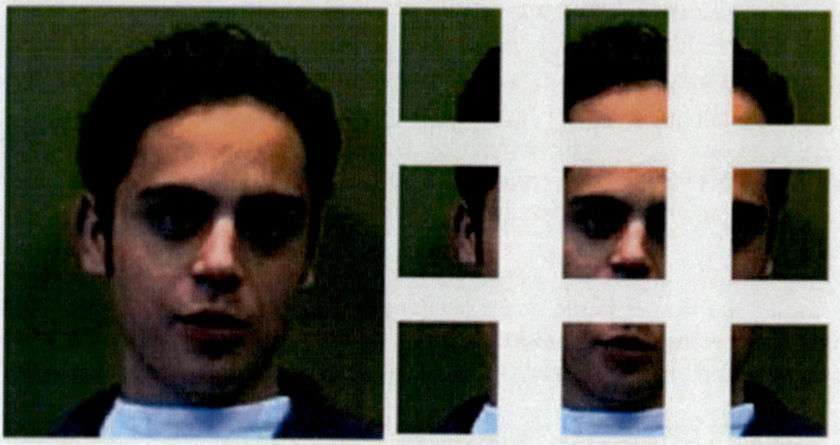

Fig. 1 Image before and after cropping into nine sub-images

distance vector. The image corresponding with the row vector with minimum distance from the centralized vector is said to be the recognized image.

3.2 Modular PCA

In Modular PCA all the images in the given image dataset are divided into k equal sub-images. Figure 1 gives an image before and after cropping into nine sub-images.

Corresponding PCA matrices and reduced matrices are obtained for each of these sub-images. The query image given by the user is also similarly divided into sub-images. A row vector is obtained for each of the sub-query image. The row vectors are compared with the rows of the matrices of the corresponding sub-images. Thus, each sub-query-image is matched with a recognized sub-image.

3.3 Wavelet PCA

In Wavelet PCA, each image of the given image dataset is processed through wavelet transformation. Wavelet transformation is a process of image compression. The first level Haar [11] decomposition of an image gives four sub-bands as the output: LL, LH, HL and HH. Figure 2 gives the images before and after Haar decomposition. LL sub-band [1] gives the approximation coefficients of the image and is approximately at half resolution of the original image. The LH sub-band gives the vertical detailed coefficients of the original image, while the HL sub-band gives the horizontal detailed

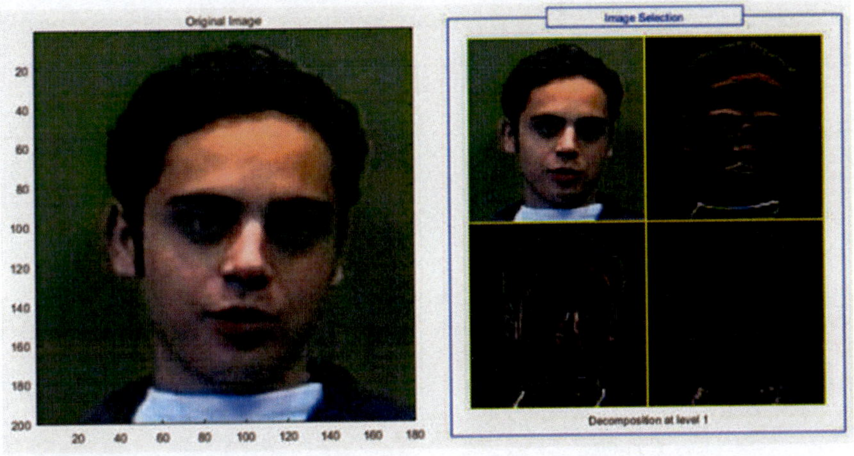

Fig. 2 Image before and after Haar decomposition

coefficients of the original image. The HH sub-band gives the diagonal detailed coefficients of the original image.

In this method, all the images in the given image dataset are decomposed using first level Haar transformation and each LL sub-band is processed through PCA. The given query image is also decomposed into the sub-bands, and the LL sub-band is processed through PCA. Finally, the query image vector is compared with the row vectors of the dataset matrix. The most similar image is identified and displayed.

3.4 Modular-Wavelet PCA

Modular-Wavelet PCA is a combination of Modular and Wavelet PCA. The images are first divided into k equal sub-images. These sub-images are decomposed into the four sub-bands using the first level Haar transformation. The LL sub-band of each sub-image is thus obtained and processed through PCA. The given query image is also similarly processed. The image vector is again compared with the row vectors of the reduced matrix, and then, the image with minimum distance is selected.

4 Experiments, Results and Analysis

In this work, two datasets have been used: Faces94 database [12] and Kaggle face mask dataset [13]. The Faces94 database has been used to compare plain PCA, Modular PCA, Wavelet PCA and Modular-Wavelet PCA. A set of 90 faces (three per individual) have been given as the image dataset. This dataset is loaded into the program. The process of PCA is common for all processes. The flowchart of the process followed is given in Fig. 3.

We distorted few images from the image dataset for testing the different PCA methods. Figures 4, 5, 6 and 7 give the results of plain PCA, Modular PCA, Wavelet

Fig. 3 Flowchart for PCA

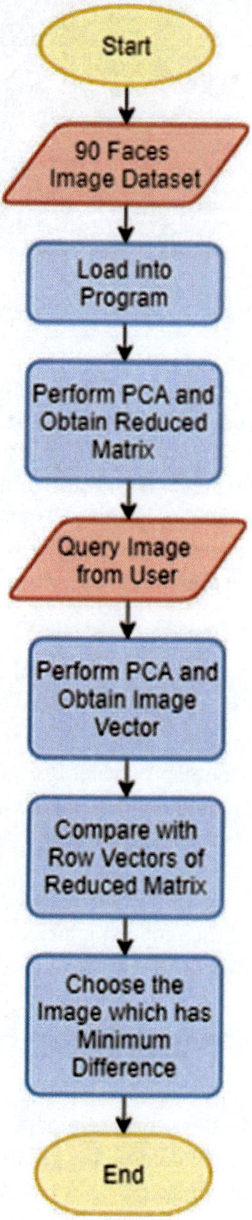

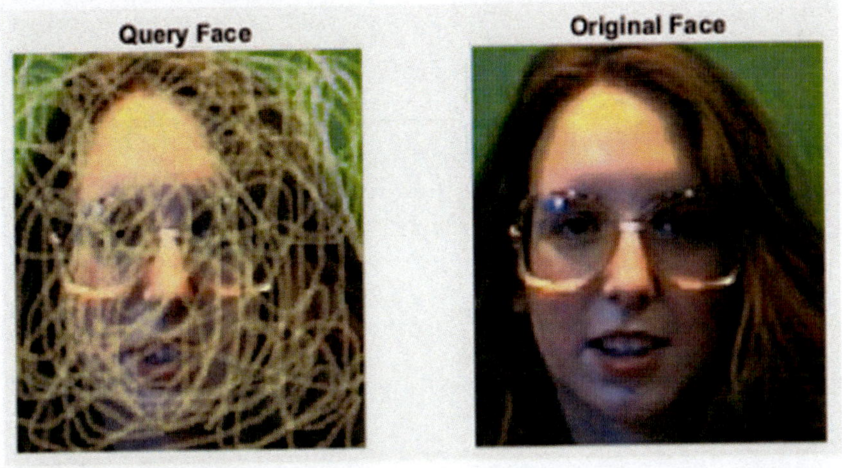

Fig. 4 Result for plain PCA

Fig. 5 Result for modular PCA

PCA and Modular-Wavelet PCA, respectively, in recognizing the same distorted image.

It can be observed from Fig. 4 that the distorted image is accurately recognized by using plain PCA.

It can be observed from Fig. 5 that the last two sub-images have been mismatched in case of Modular PCA.

It can be observed from Fig. 6 that the distorted image is accurately recognized in case of Wavelet PCA also.

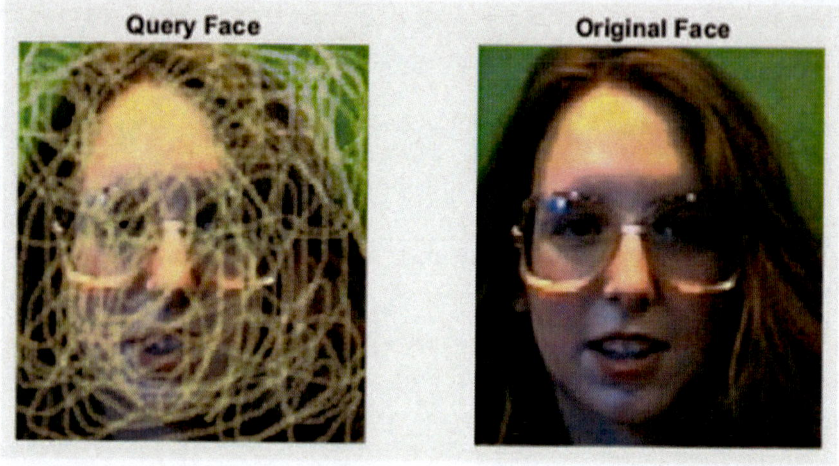

Fig. 6 Result for wavelet PCA

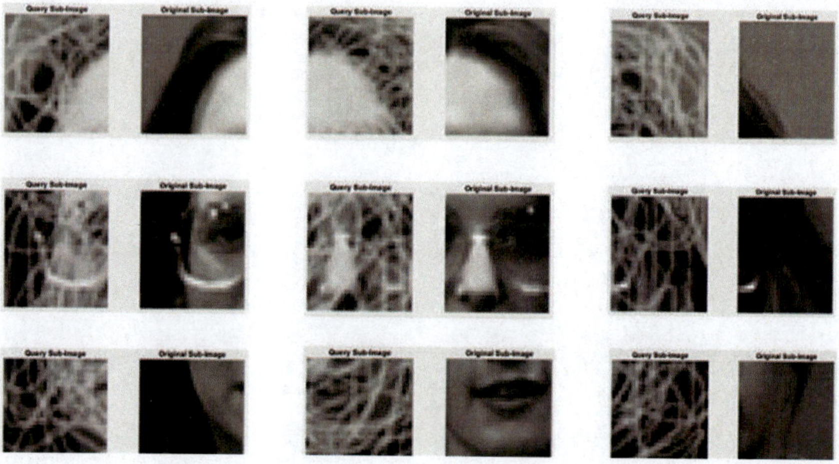

Fig. 7 Result for modular-wavelet PCA

It can be observed that the last two sub-images have been mismatched in case of Modular-Wavelet PCA also like in the case of Modular.

Execution time (seconds) is calculated and shown in Table 1 for different parts of the program for each of the four types of PCA. The simulation has been done on MATLAB in a system with 16 GB RAM.

It has been observed that Plain PCA takes longer duration for performance of PCA on the input dataset. Modular PCA takes moderate amount of time while Wavelet PCA takes lesser amount of time. Modular-Wavelet PCA takes the least amount of time for

Table 1 Execution time in seconds for different PCAs

Factor	PCA	MPCA	WPCA	MWPCA
Image reading time	0.421615	0.505515	0.550353	0.991048
PCA performance time	101.8981	0.800134	0.998963	0.058130
Image recognition time	0.716941	1.801003	0.884100	1.711521

PCA. For further analysis, only Modular PCA, Wavelet PCA and Modular-Wavelet PCA methods have been considered.

The Kaggle face mask dataset has been used for analyzing the use of the three PCA methods. One hundred ten masked face images and 110 unmasked face images have been loaded into the three different programs. For Modular PCA, the images have been split into two horizontal halves. Only the bottom sub-image has been considered for comparison. In case of Modular-Wavelet PCA, the bottom sub-image has been decomposed and its LL sub-band has been considered for further process.

It has been observed that the query images given from the input image dataset are recognized accurately in all the three cases. But in the case of query images outside the given dataset, the accuracy of Modular-Wavelet PCA and Wavelet PCA differs. A set of 100 masked images and 100 unmasked images have been given as query images to the Modular-Wavelet PCA and Wavelet PCA programs. These 100 images were the ones for which Modular PCA got 100% recognition. The number of accurate face mask recognitions that could be obtained by using Wavelet PCA and Modular-Wavelet PCA have been represented graphically in Fig. 10. Figures 8 and 9 show a sample output of mask recognition by Modular-Wavelet PCA and Wavelet PCA each.

It is observed that Modular-Wavelet PCA has greater accuracy than Wavelet PCA. Modular-Wavelet PCA also takes the least amount of time. The accuracy of Modular-Wavelet PCA may be further increased by increasing the number of principal components and the number of images in the input image dataset.

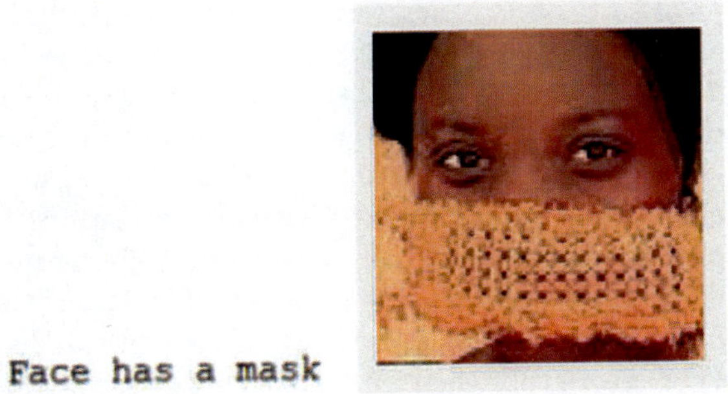

Fig. 8 Modular-wavelet PCA mask recognition

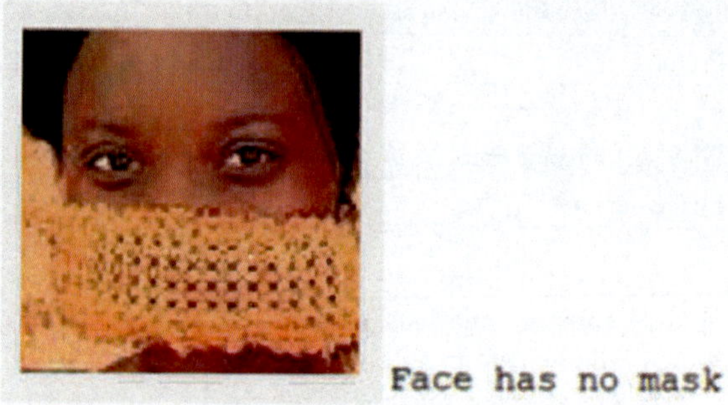

Fig. 9 Wavelet PCA mask recognition

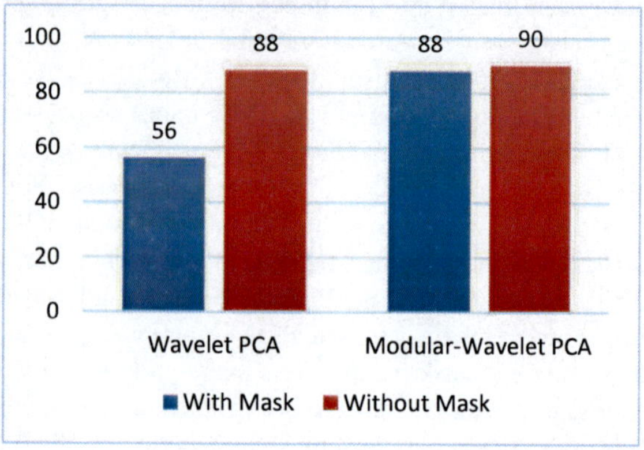

Fig. 10 Graph of number of accurate recognitions of wavelet PCA and modular-wavelet PCA out of 100 query images

5 Conclusions

From the observations made, it is understood that Plain PCA is a time-taking process. Modular PCA is a method which is useful in comparing certain specified regions of images. Wavelet PCA is used when general comparison has to be made between images. It is best preferable in cases where distorted images have to be identified. Modular-Wavelet PCA inherits the best properties of both Modular and Wavelet PCA. Modular-Wavelet PCA is more accurate and takes lesser amount of time. As observed in the masked image recognition test, only Modular and Modular-Wavelet PCA are recommendable in cases where only a part of the image must be focused upon during the comparison. But in the experiment, it is observed that in case of images that have distortions Wavelet PCA gives better results. The accuracy of all the discussed methods may be increased by increasing the number of images in the input dataset. Finally, it can be concluded that Modular-Wavelet PCA and Wavelet PCA are equally good methods of image recognition, and they must be chosen depending upon the type of application.

References

1. Puyati W, Walairacht A (2008) Efficiency improvement for unconstrained face recognition by weightening probability values of modular PCA and wavelet PCA. In: 2008 10th international conference on advanced communication technology, pp 1449–1453
2. Paul LC, Sumam AA (2012) Face recognition using principal component analysis method. Int J Adv Res Comput Eng Technol 1(9):135–139
3. Sankaran P, Asari V (2004) A multi-view approach on modular PCA for illumination and pose invariant face recognition. In: 33rd applied imagery pattern recognition workshop, pp 165–170
4. Feng GC, Yuen PC, Dai DQ (2000) Human face recognition using PCA on wavelet subband. J Electron Imaging 9:226–233
5. Tej Chinimilli B, Anjali T, Kotturi A, Reddy Kaipu V, Mandapati JV (2020) Face recognition based attendance system using Haar cascade and local binary pattern histogram algorithm. In: 2020 4th international conference on trends in electronics and informatics, pp 701–704
6. Pala S, Jayan S, Kurup DG (2020) An accurate UWB based localization system using modified leading edge detection algorithm. Ad-Hoc Netw 97:102017
7. Poorna SS, Anjana S, Varma P, Sajeev A, Arya KC, Renjith S, Nair GJ (2018) Facial emotion recognition using DWT based similarity and difference features. In: 2018 2nd international conference on I-SMAC, pp 524–527
8. Nair RR, Karumanchi SH, Singh T (2020) Neuro-fuzzy based multimodal medical image fusion. In: 2020 IEEE international conference on electronics, computing and communication technologies, pp 1–6
9. Divya MBS, Prajwala NB (2018) Facial expression recognition by calculating Euclidian distance for Eigen faces using PCA. In: Proceedings of the 2018 IEEE international conference on communication and signal processing, pp 244–248
10. Neha R, Nithin S (2018) Comparative analysis of image processing algorithms for face recognition. In: 2018 international conference on inventive research in computing applications
11. Porwik P, Lisowska A (2004) The Haar-wavelet transform in digital image processing: its status and achievements. Mach Graph Vis 13:79–98
12. Faces94 Face Recognition Database. https://cmp.felk.cvut.cz/~spacelib/faces/faces94.html
13. Kaggle Mask Identification Database. https://www.kaggle.com/ashishjangra27/face-mask-12k-images-dataset

CT Image Enhancement Using Variational Mode Decomposition for AI-Enabled COVID Classification

Rohith Ramakrishnan, Anirudh Vadakedath, Aakash Jignesh Modi, V. V. Sajith Variyar, V. Sowmya, E. A. Gopalakrishnan, and K. P. Soman

Abstract SARS-COV-2, also known as COVID-19 pandemic, has escalated calamity in the entire world. Due to its contagious properties, the disease spreads swiftly from person to person via direct contact. More than 210 million people got infected worldwide with more than 18 million active patients as of August 29, 2021. In numerous places, the test process for COVID-19 detection takes longer than 2 days. Once the patient is affected by COVID-19, the obstruction in lungs causes difficulty in analyzing the presence of other lung diseases, such as variants of pneumonia. In this paper, we propose an enhancement technique via the acclaimed signal processing method called variational mode decomposition (VMD) aiding any X-ray image classification method for the detection of pneumonia using convolutional neural networks (CNN). The experiments were conducted on VGG-16 model loaded with ImageNet weights followed by numerous configurations of dense layers.

Keywords Variational mode decomposition · X-ray classification · Signal processing · COVID pneumonia · VMD-feature extraction · Image classification

1 Introduction

Radiological images play a crucial role in the detection of COVID-19 [1, 2]. Chest X-ray (CXR) is an extremely useful tool that helps in decision making for further treatment and specific cases which include observing progression of COVID-19 in

R. Ramakrishnan · A. Vadakedath · A. J. Modi · V. V. Sajith Variyar · V. Sowmya (✉) · E. A. Gopalakrishnan · K. P. Soman
Center for Computational Engineering and Networking (CEN), Amrita School of Engineering, Amrita Vishwa Vidyapeetham, Coimbatore, India
e-mail: v_sowmya@cb.amrita.edu

M. Gupta et al. (eds.), *Artificial Intelligence on Medical Data*,
Lecture Notes in Computational Vision and Biomechanics 37,
https://doi.org/10.1007/978-981-19-0151-5_3

patients [3]. With the evolution of the infamous SARS-CoV-2 infection and the severity of lung aberrations, call for the latest methodologies in radiological evaluation to assist infected patients is a pressing priority. The above-mentioned problem falls under a field of AI, called Computer Vision. As the name suggests, computer vision deals with image-based classification, which as the name suggests is the process of binning images based on predefined classes or labels and object detection. Computer vision can be of great use and aid doctors in evaluating radiological images. One of the prevalent issues in the domain of computer vision is the prepossessing of images. Prepossessing steps for training any model in image classification include several tasks, namely feature selection and extraction, transformation, and structure identification. Deep learning-based classification is the primarily used method to identify and classify medical images [4]. Most of the time, the data can be messy and can be from different sources. The main aim of image processing is to improve the data by removing distortion, feature enhancement, and extracting important features relevant to certain applications like image segmentation, biomedical applications, to reduce the complexity and increase the accuracy of the used deep learning architecture. Numerous methodologies exist for image enhancement, such as spatial transformers to aid in the classification. Feature extraction is a major step in image classification and some of the methods include Pyramid Histogram of Oriented Gradients (PHOG), Contrast Features (CONTRAST), The Seven Hu Moments Features (HUGEO), Moments and Central Moments Features (MOMENTS), and Basic Intensity Features (BASICINT). Over the years, numerous signal and image processing algorithms based on frequency decompositions such as Fourier, wavelets, variational, and dynamic modes were used to enhance the quality of the input data [5, 6]. In recent years, deep learning algorithms are successfully applied over wide applications ranging from detection to classification. However, the performance of the deep learning algorithms depends on the quality of the input data and the feature extraction utilized. The objective of feature extraction is to discover the most informative set of features and distinct properties to improve the efficiency of the classifier [4]. Feature extraction plays a crucial role in biomedical signal classification because the model's performance might plunge if the necessary features are neglected. Therefore, in the present work, we used one of the decomposition algorithms called variational mode to enhance the input computed tomography (CT) data. The concept of variational mode decomposition (VMD) revolves around efficiently decomposing any given time series signal into modes with different center frequencies based on an non-recursive alternating direction multiplier method for optimization. Application of VMD is no stranger to the field of image processing, and it has been utilized for haze removal [7, 8], enhancement of images [5, 6], conversion of RGB images into gray maps [9], frequency-based dimensionality reduction technique [10], and infrared target detection [11]. The paper has been organized as follows. Section 2 gives a basic understanding on the theory, dataset, and preprocessing used in it and the model. Section 3 provides information on the metrics used to evaluate the model and the values obtained followed by conclusion.

2 Methodology

The dicom files are initially resized to 512px grayscale images. As portrayed in Fig. 1, the resized images are decomposed via variational mode decomposition and blended with the original images. After the creation of VMD dataset, a set of experiments are parallelly performed on the VMD and the original dataset to evaluate the efficiency of the VMD. The models utilized for these experiments are explored further in the upcoming sections.

2.1 Dataset Description

The data used for the study is SIIM-FISABIO-RSNA COVID-19 dataset [12] to identify abnormalities using chest radiographs. The dataset is divided into four class labels, negative for pneumonia, typical appearance, indeterminate appearance, and atypical appearance. The dataset contains a total of 6034 dicom files split into train and test sets. The data was unbalanced with 27% negative for pneumonia class, 47.4% typical appearance class, 18% indeterminate appearance class, and 7.6% atypical

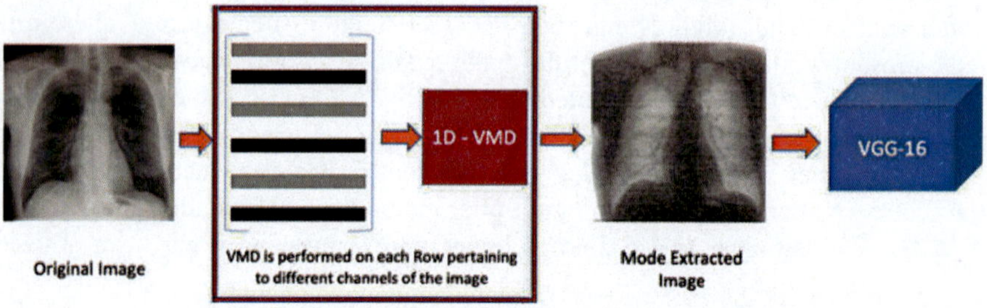

Fig. 1 Proposed architecture to enhance features via variational mode decomposition

Table 1 Datasets have been split for training and testing by 80–20 ratio, respectively

Dataset type	Train	Test
Original images	3617	903
VMD-images	3617	903

appearance class. The primary classes used in this experiments are negative for pneumonia and typical appearance consisting of 1656 and 2864 scans, respectively. The dataset was split into 80–20 for training and testing, respectively, as mentioned in Table 1. The original dimension of the DICOM Chest X-ray Scan file is 2100 × 1800 pixels. Variational mode decomposition was applied to the whole dataset and fed into the model for training. The dataset processed with VMD will be so-called as VMD-Images.

2.2 *Variational Mode Decomposition*

The objective of the novel variational mode decomposition (VMD) proposed by Dragomiretskiy Konstantin and Dominique Zosso [13] is the successor to the empirical mode decomposition (EMD), which aims at decomposing the signal f(t) into a finite sum of intrinsic mode functions [14]. Each mode k is required to be most compact around a center pulsation k determined along with the decomposition [13]. Let us consider f as the time series which is decomposed into a set of modes u_k around pulsation w_k according to the following constrained optimization problem

$$\min_{\{u_k\},\{\omega_k\}}\left\{\sum_k \left\|\partial_t\left[\left(\delta(t)+\frac{j}{\pi t}\right)*u_k(t)\right]e^{-j\omega_k t}\right\|_2^2\right\} \quad \text{s.t. } \sum_k u_k = f \tag{1}$$

where $\{u_k\} := \{u_1, .., u_k\}$ and $\{w_k\} := \{w_1, .., w_k\}$ are the shorthand notations for the set of all modes and their center frequencies, respectively. The implementation via alternating direction method of multipliers (ADMM) proposed by Dragomiretskiy Konstantin and Dominique Zosso [13]:

Algorithm 1: ADMM optimization for VMD [4]

initialise $\{u_k^1\}\,\{w_k^1\}\,\lambda^1, n < -0;$
while

$$convergence: \sum_k \left\|u_k^{n+1} - u_k^n\right\|_2^2 / \left\|u_k^n\right\|_2^2 < \epsilon \; or\, n \le N \tag{2}$$

do

n¡-n+1;
for k = *1:K* **do**
Update Equation for u_k:

$$u_k^{n+1} \leftarrow \arg\min_{u_k} \mathcal{L}\left(\left\{u_{i<k}^{n+1}\right\}, \left\{u_{i\ge k}^n\right\}, \left\{\omega_i^n\right\}, \lambda^n\right) \tag{3}$$

end
for k = *1:K* **do**
Update Equation for w_k:

$$\omega_k^{n+1} \leftarrow \arg\min_{\omega_k} \mathcal{L}\left(\left\{u_i^{n+1}\right\}, \left\{\omega_{i<k}^{n+1}\right\}, \left\{\omega_{i\ge k}^n\right\}, \lambda^n\right) \tag{4}$$

end
Dual Ascent Updation for λ:

$$\lambda^{n+1} \leftarrow \lambda^n + \tau\left(f - \sum_k u_k^{n+1}\right) \tag{5}$$

end

1D-VMD is often used to decompose the signal to attain useful information or denoise the signals via mode selection. In this study, the radiography data obtained from each channel will be considered as a collection of signals and hence decomposed into modes (u_k) via the parallel updation of central frequencies (w_k) and the Lagrangian multiplier (λ). The number of modes, as the name suggests decomposes the given signal into the desired number of modes (K), λ is the Lagrangian multipliers, which will render the problem unconstrained and are a common way of enforcing constraints strictly, τ is the penalty factor dealing with the updation of λ. The above-mentioned algorithm is iterated until convergence, as mentioned in (2) or until the maximum iteration count (N) is obtained.

2.3 Data Preprocessing

VMD is often used to decompose a signal to attain useful information or denoise the signal via mode selection. In this study, each image will be considered as a collection of signals and hence decomposed into modes. The VMD functions have multiple parameters such as number of modes (K), absolute tolerance (ϵ), maximum iterations (N), penalty factor (τ), initial Lagrangian multiplier (λ), etc. The K parameter set for the experiments is 3, and rest of them are set to their default values. Due to computation limitations, the image resolution was resized to 512×512 pixels with least possible loss. Every row of a given image is considered as a signal which is decomposed for the extraction of the first mode and blended with the original image in an attempt to suppress irrelevant details and enhance the rest.

The experiments were performed on a 2.4 GHz Quad-Core Intel Core i5 with 8 GB 2133 MHz MHz LPDDR3 machine. The proposed VMD processing for every image had an approximate runtime of 3 sec. Figure 2a, b corresponds to the output obtained for blended 1D-VMD for negative for pneumonia and typical appearance, respectively.

2.4 Model Description

Visual geometry group 16 (VGG-16) is a convolutional neural network (CNN) model first proposed in the paper "Very Deep Convolutional Networks for Large-Scale Image Recognition" [15] by Karen Simonyan and Andrew Zisserman. The model achieves an accuracy of 92.7% for ImageNet dataset of over 14 million images belonging to 1000 classes. For the proposed experiment, two models were used to analyze the performance with and without VMD, Model-1 consists of VGG-16 coupled to a dense layer of 512 neurons followed by a dropout of 0.3. Model-2 analogously consists of VGG-16 coupled to a dense layer of 128 neurons followed by a dropout of 0.5 as briefly visualized in Fig. 3a, b.

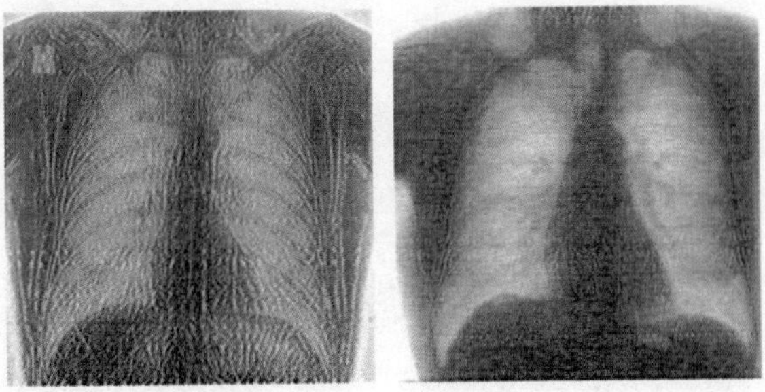

(a) Negative (b) Typical appearance

Fig. 2 Enhanced images obtained after 1D-VMD mode-blend

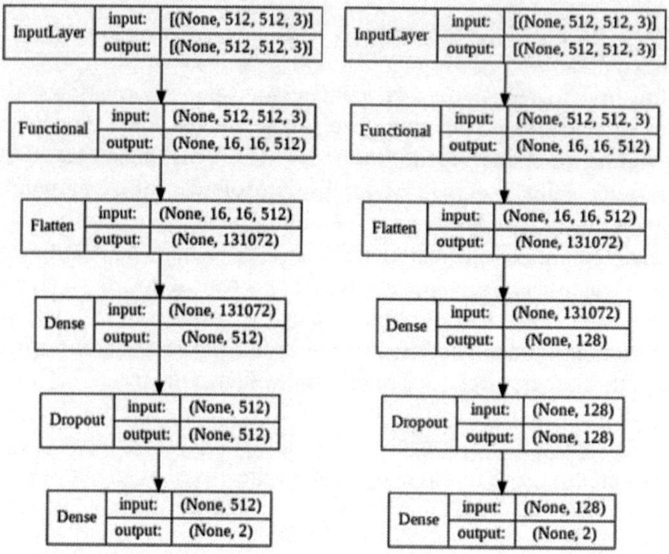

(a) 512 Dense Layer Model (b) 128 Dense Layer Model

Fig. 3 Models utilized for training

Two individual models were trained with the original and 1D-VMD data for analyzing the performance of feature enhancement via VMD. The model used for evaluating the performance was the VGG-16 architecture with pre-trained ImageNet weights.

3 Results and Discussion

Application of the VMD-based processing gave birth to two datasets, the original images and the VMD-enhanced image. The datasets were then employed to train 2 models, which resulted in 4 experiments. The metrics used to evaluate the performance of the models are accuracy, recall, precision, and $f1$-score. Model-1 gave an accuracy of 0.78 when trained with VMD-aided images and 0.63 when trained with the original dataset, while Model-2 gave an accuracy of 0.78 on training with VMD-aided images and 0.52 on training with the original images. From a general overview, it is observed that the models aided with VMD yield a better classification. Model-1 when trained with the original images could not classify the test data, whereas when assisted with VMD, a significant improvement of 23.8% and 50% for Model-1 and Model-2, respectively, in classification is noted. From the confusion matrix in Fig. 4,

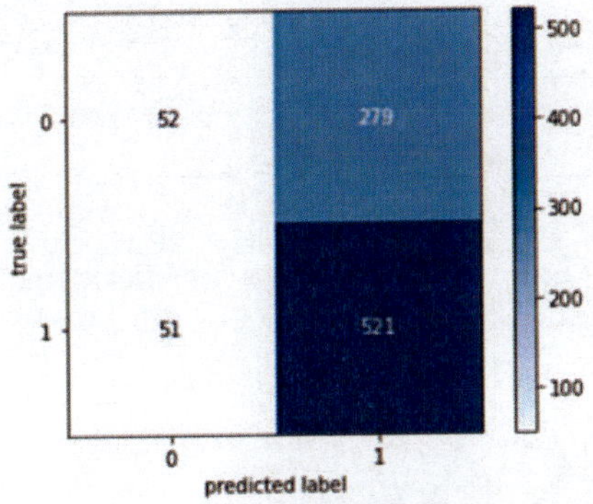

(a) Original Data with Model-1

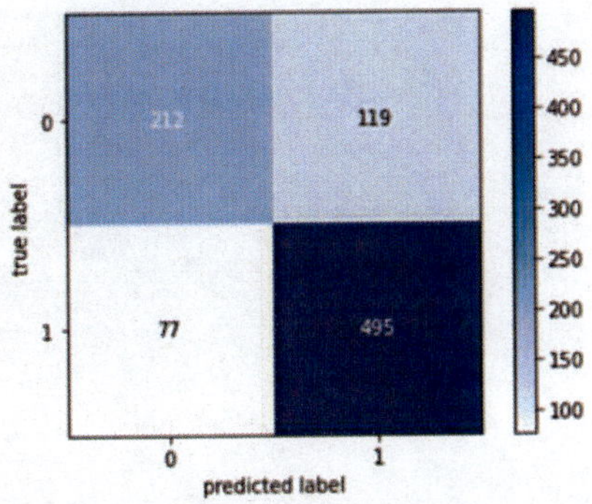

(b) 1D-VMD Data with Model-1

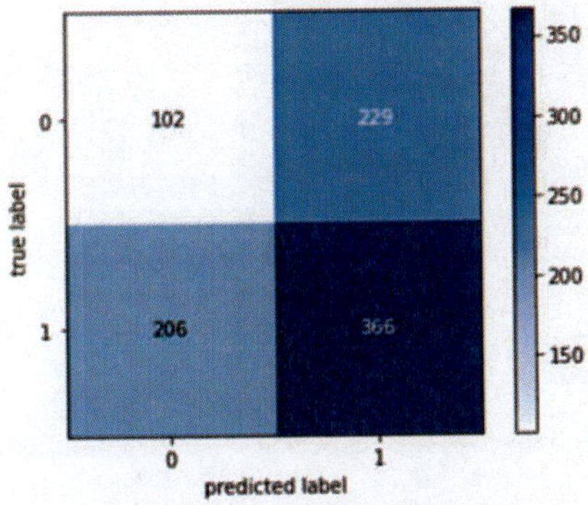

(c) Original Data with Model-2

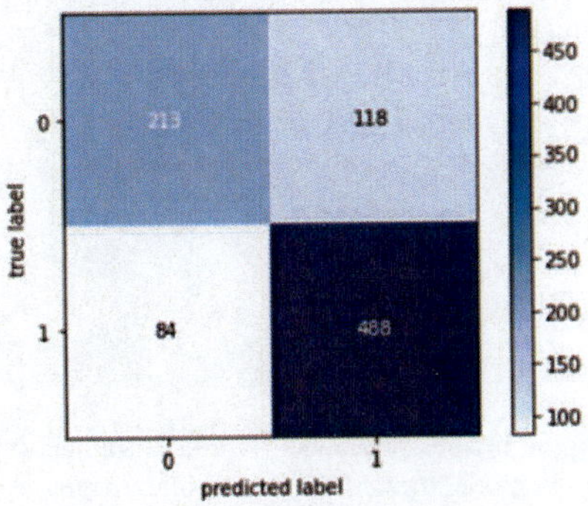

(d) 1D-VMD Data with Model-2

Fig. 4 Confusion matrices corresponding to four experiments conducted on two datasets

Table 2 Performance analysis of VMD and original datasets with the proposed models

Model no.	Class label	Accuracy		Precision		Recall		$F1$-score	
		Original	VMD	Original	VMD	Original	VMD	Original	VMD
1	Negative	0.63	**0.78**	0.5	**0.73**	0.16	**0.64**	0.24	**0.68**
	Typical			0.65	**0.81**	**0.91**	0.87	0.76	**0.83**
2	Negative	0.52	**0.78**	0.33	**0.72**	0.31	**0.64**	0.32	**0.68**
	Typical			0.62	**0.81**	0.64	**0.85**	0.63	**0.83**

Bold font indicates best results obtained on VMD vs. Original dataset

it can be observed that false positives and false negatives are significant for the model trained on the original dataset, meanwhile models trained on VMD dataset yielded much lower false categories.

From Table 2, we can note that in the proposed VMD-aided model, the recall and precision for each class are substantial compared to the original model. Another

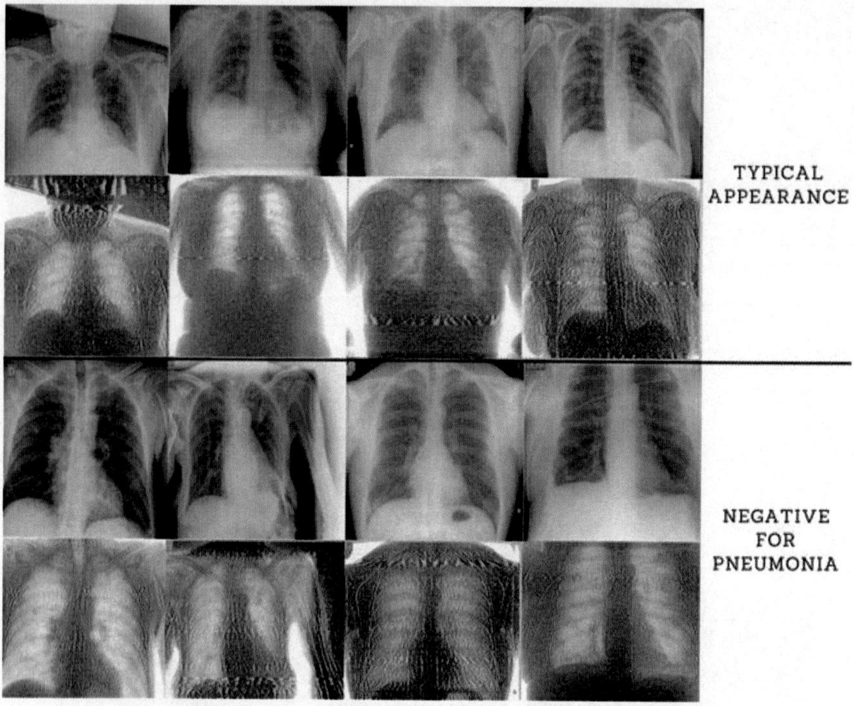

Fig. 5 Images predicted via model trained on VMD-enhanced images which were misclassified by the model trained on the original images

metric used for evaluation is recall. If the recall is high, it says that the false negative is less comparative to the true positives. Nearly a 40% increase in recall value is noted cumulatively for both models trained on VMD dataset compared to the original dataset. The recall values for the best-performing model, Model-2 are 0.64 and 0.85 for each class which is a tremendous increase from the recall obtained from the non-VMD variant of Model-2, 0.31 and 0.64. There is a notable difference in the recall values for Model-2, when trained on the original and VMD images. Another metric used to measure the performance was $f1$-score. Model-1 when trained on the original images yielded an $f1$-score of 0.24 and 0.76 while model-2 gave a score of 0.32 and 0.63 for the respective classes. On the other hand, model-1 when trained on VMD-aided images gave a score of 0.68 and 0.83 and the Model-2 gave a score of 0.68 and 0.86 for the classes negative for Pneumonia and typical appearance, respectively. This shows a 36% increase in $F1$-score, when trained on the VMD-aided images.

In Fig. 5, the images which were misclassified by the model trained on the original images but classified by the model trained on VMD-enhanced images are shown, the first two rows pertain to the images belonging to typical appearance of the first row contains the original images and the second row has the VMD-enhanced images. The same goes for negative for Pneumonia.

4 Conclusion

On comparing the performance of our set of experiments with the original chest X-ray scan and VMD extracted chest X-ray scan, it was noticed that the model trained with VMD chest X-ray scan outperformed the models trained with the original chest X-ray scans. In the field of biomedical classification, it is crucial that the models give a high precision and recall, which can be seen in our VMD-aided models. Due to the restrictions in computational power, a number of modes were set to 6 and the image resolutions were set to 512. With further tuning of the number of modes and an increase in image resolution, the model could hypothetically perform better in classifying the images. The future scope of this experiment involves exhaustive tuning of all parameters involved in the process of VMD with higher resolution of images and further replacement of 1D-VMD with 2D-VMD, which hypothetically could yield a better result. 2D-VMD would involve decomposing the whole image instead of individual rows of every image. The proposed experiment could also be tested with multi-class labels, which is beyond the scope of this paper.

References

1. Baltruschat IM, Nickisch H, Grass M, Knopp T, Saalbach A (2019) Comparison of deep learning approaches for multi-label chest x-ray classification. Sci Rep 9(1):1–10
2. Madaan V, Roy A, Gupta C, Agrawal P, Sharma A, Bologa C, Prodan R (2021) Xcovnet: Chest x-ray image classification for covid-19 early detection using convolutional neural networks. New Gener Comput:1–15
3. Xue Z, You D, Candemir S, Jaeger S, Antani S, Rodney Long L, Thoma GR (2015) Chest x-ray image view classification. In: 2015 IEEE 28th international symposium on computer-based medical systems. IEEE, pp 66–71
4. Anand R, Sowmya V, Menon V, Gopalakrishnan A, Soman KP et al (2021) Modified vgg deep-learning architecture for covid-19 classification using chest radiography images. Biomed Biotechnol Res J (BBRJ) 5(1):43
5. Fengji M, Junyi C, Hai W (2019) Two-dimensional compact variational mode decomposition-based low-light image enhancement. IEEE Access 7:136299–136309
6. Pankaj D, Sachin Kumar S, Mohan N, Soman KP (2016) Image fusion using variational mode decomposition. Indian J Sci Technol 9:45
7. Satapathy LM, Tripathy RK, Das P (2019) A combination of variational mode decomposition and histogram equalization for image enhancement. Nat Acad Sci Lett 42(4):333–336
8. Suseelan HT, Sowmya V, Soman KP (2017) Image dehazing using variational mode decomposition. In: 2017 international conference on wireless communications, signal processing and networking (WiSPNET). IEEE, pp 200–205
9. Yan H, Zhao P, Zhuang D, Yang X, Liu P (2021) Frequency division denoising algorithm based on vif adaptive 2d-vmd ultrasound image. Plos One 16(3):e0248146
10. Swarna M, Sowmya V, Soman KP (2018) Band selection using variational mode decomposition applied in sparsity-based hyperspectral unmixing algorithms. Signal Image Video Process 12(8):1463–1470
11. Wang X, Peng Z, Zhang P, He Y (2017) Infrared small target detection via nonnegativity-constrained variational mode decomposition. IEEE Geosci Remote Sens Lett 14(10):1700–1704

12. de la Iglesia Vayá M, Saborit JM, Montell JA, Pertusa A, Bustos A, Cazorla M, Galant J, Barber X, Orozco-Beltrán D, García-García F, Caparrós M, González G, Salinas JM (2020) Bimcv covid-19+: a large annotated dataset of RX and CT images from covid-19 patients
13. Konstantin D, Dominique Z (2013) Variational mode decomposition. IEEE Trans Signal Process 62(3):531–544
14. Zeiler A, Faltermeier R, Keck I, Tomé A, Puntonet C, Lang E (2010) Empirical mode decomposition—an introduction, pp 1–8
15. Simonyan K, Zisserman A (2014) Very deep convolutional networks for large-scale image recognition. arXiv preprint arXiv:1409.1556

Analysis of Machine Learning and Deep Learning Algorithms for Detection of Brain Disorders Using MRI Data

Deepthi Sudharsan, S. Isha Indhu, Kavya S. Kumar, Lakshaya Karthikeyan, L. Srividhya, V. Sowmya, E. A. Gopalakrishnan, and K. P. Soman

Abstract Brain diseases impact more than 1 billion people worldwide and include a wide spectrum of diseases and disorders such as stroke, Alzheimer's, Parkinson's, Epilepsy and other Seizure disorders. Most of these brain illnesses are subjected to misclassification, and early diagnosis increases the possibilities of preventing or delaying the development of these disorders. Magnetic Resonance Imaging (MRI) plays an important role in the diagnosis of patients with brain disorders and offers the potential of non-invasive longitudinal monitoring and bio-markers of disease progression. Our work focuses on using machine learning and deep learning techniques for the preemptive diagnosis of Schizophrenia using Kaggle data set and Alzheimer's using TADPOLE data set comprising of MRI features. Since the number of works using TADPOLE data set is minimum, we have chosen this for our study. Machine learning algorithms such as support vector machine (SVM), Decision Tree, Random Forest, Gaussian Naive Bayes, and 1D-CNN deep learning algorithm have been used for the classification of the disorders. It has been observed that Gaussian NB performed the best on Schizophrenia data, while Random Forest outperformed on Alzheimer's data compared to the other classifiers.

Keywords Magnetic Resonance Imaging (MRI) · Alzheimer's · Parkinson's · Brain disorders · 1D-CNN · Gaussian Naive Bayes

1 Introduction

Brain disorders affect individual's memory, speech, thought process, mobility, and different organ's functionalities. People affected with brain disorders are commonly observed to have structural and functional brain alterations. Brain functional imaging data, especially functional magnetic resonance imaging (fMRI) data, have been

D. Sudharsan (✉) · S. Isha Indhu · K. S. Kumar · L. Karthikeyan · L. Srividhya · V. Sowmya · E. A. Gopalakrishnan · K. P. Soman
Center for Computational Engineering and Networking, Amrita School of Engineering, Coimbatore, Amrita Vishwa Vidyapeetham, Coimbatore, India
e-mail: cb.en.u4aie19022@cb.students.amrita.edu

M. Gupta et al. (eds.), *Artificial Intelligence on Medical Data*, Lecture Notes in Computational Vision and Biomechanics 37, https://doi.org/10.1007/978-981-19-0151-5_4

employed to reflect functional integration of the brain. Alteration in brain functional connectivity (FC) is expected to provide potential bio-markers for classifying or predicting brain disorders.

Schizophrenia is a severely disorienting mental illness affecting more than 1.5 million people each year. The symptoms include delusions, hallucinations, disorganized thoughts which impairs the daily functioning of an individual. This illness has no well-established, non-invasive diagnosis bio-marker. Currently, due to the overlap of symptoms with other mental illnesses (like bipolar disorder), it can only be diagnosed subjectively, by process of elimination [5, 14]. Alzheimer's, which is a type of dementia, is a neurological disorder that progressively causes the brain to shrink. Dementia has approximately affected 50 million people worldwide, and around 60–70 of these people are estimated to have Alzheimer's disease. Early symptoms of Alzheimer's include cognitive, behavioral, psychological changes in activity [10]. With time, affected people will develop severe memory impairment and consequently struggle with everyday tasks. Therefore, early diagnosis and appropriate medication might temporarily improve or slow the progression of symptoms of Schizophrenia and Alzheimer's.

2 Related Works

Recently, several studies have been done on structural MRI data of Schizophrenia on the Kaggle data set [12], and accuracies of 45.23–81.07% have been achieved [9, 15]. In [15], the support vector machine (SVM) has been used to explore the potential utility for cortical thickness and surface area measurements in differentiating individual patients from healthy controls. An accuracy of 85% has been observed. In [8, 9], Gaussian Mixture Models were used to detect and extract the score vectors, based on partial least squares. In [4, 9], different data handling strategies are investigated, and the most important neuro-imaging features are studied for classification. We could obtain best accuracy of 85% for 101 features selected.

There are several data set sources available for the prediction of Alzheimer's disease, and most of the works have used the Open Access Series of Imaging Studies (OASIS) data set comprising of demented and non-demented classes [3]. In our work, we used the TADPOLE challenge data set to forecast the evolution of clinical diagnosis, cognitive decline, and neuro-degeneration (brain atrophy). In contrast to previous data sets, this is designed to inform clinical trials through identification of patients at the early stages of disease who are likely to progress over the short to medium term (1–5 years) [7]. In the case of OASIS data, the classification is carried out mainly based on Clinical Dementia Rating (CDR), Mini-Mental State Exam (MMSE) scores [6] of the MRI, and accuracy of up to 99.52% has been achieved by applying various supervised and ensemble learning techniques, whereas in TADPOLE data set (used in our work), features such as ADAS13, Ventricles, Hippocampus, WholeBrain, Entorhinal, MidTemp features (which are discussed in detail in Sect. 3.3) are used to determine the extent to which the patient is impacted and could achieve 97% accuracy using Random Forest classifier which is a benchmark for this data set.

The performance analysis of machine learning algorithms has not only been applied to MRI data but also to other types of microscopic images data [11]. In [15], Convolutional Neural Network (CNN) [2] has been applied on the ECG signal data for the diagnosis of diabetes. We drew similar motivation and applied 1D-CNN on Schizophrenia data set. In Sect. 3, we have included the proposed methodologies for the classification of the disorders. We pre-processed the data and applied feature selection to obtain optimum accuracy and then the data are passed to the machine learning classifier. The results are discussed in detail in Sect. 4.

3 Proposed Methodology

3.1 Data Set Description

We have used the data set from the Machine Learning for Signal Processing (MLSP) 2014 Schizophrenia classification Kaggle challenge [12]. The details of the medical attributes used in this data set are given in Table 1.

The data set includes 410 attributes, 32 SBM, and 378 FNC features. The two modalities of MRI scans, functional and structural, are used to obtain the features. The model trained using the optimal features that were selected by calculating ANOVA correlation coefficient was observed to have better accuracy than the machine learning models trained using all the features.

For the Alzheimer's disease, we have used the TADPOLE data set (Prediction Of Longitudinal Evolution Challenge) [1]. The features chosen for this study are the following: ADAS13, Ventricles, Hippocampus, WholeBrain, Entorhinal, MidTemp which are discussed in detail in Table 2.

Table 1 Medical attributes—Schizophrenia

Medical attributes	Description
Source-based morphometry (SBM)	Correspond to the weights of brain maps obtained from the application of independent component analysis (ICA) on the gray matter concentration maps of all subjects [12]
Functional network connectivity (FNC)	Measure of the temporal correlation between independent brain maps. The overall level of 'synchronicity' between brain areas of the subjects is indicated by FNC

Table 2 Medical attributes—Alzheimer

Medical attributes	Description
ADAS13	The Alzheimer's disease assessment scale (ADAS) is a neuro-psychological test to assess the level of cognitive dysfunction in Alzheimer's disease
Ventricles	Change in the volume of gray and white matter structures surrounding the ventricles indicate the presence of Alzheimer's disease
Hippocampus	There is a strong correlation between the atrophy (shrinkage) of the hippocampal areas with the presence of Alzheimer's disease which is initial areas of the brain that is affected by Alzheimer's disease
WholeBrain, Entorhinal, MidTemp	These are the other MRI features used in this study

3.2 Data Pre-processing

The training and testing data sets of the Schizophrenia have 86 and 119748 data points, respectively. Due to the absence of null entries, no rows were dropped. The training and testing data sets were merged based on the attribute 'Patient Id'. Based on the presence or absence of Schizophrenia, the patients are assigned labels '1' or '0'. In feature selection, the machine automatically selects those features that help in predicting the variable or output [6]. For the Schizophrenia data set, the features are selected based on univariate statistical test results. Since the response variable is continuous and the predictor variable is categorical, ANOVA-based feature selection is used. For varied number of features, such as 50, 101, 204, and 250, the results were analyzed to check the effect on the ROC scores. A grid search for the optimal parameters led to the conclusion that the default parameters for all the machine learning models were optimal. It was observed that selecting the top 50 features yielded the best accuracy.

In the Alzheimer data set, the label DX bl was numerically encoded and consequently had 8920 in the AD class and 3821 in the normal class. Due to the evident class imbalance, techniques like under-sampling and SMOTE oversampling were utilized to produce an equal number in both classes. Since the features were numerical, the features were normalized so as to remove any bias in the model. An unavoidable fraction of the data set had NaN values, and thus, imputation was necessary. Imputation with different statistical values was performed, and it significantly improved the performance of the model. Figure 1 shows the block diagram of the machine learning system for classifying Alzheimer data.

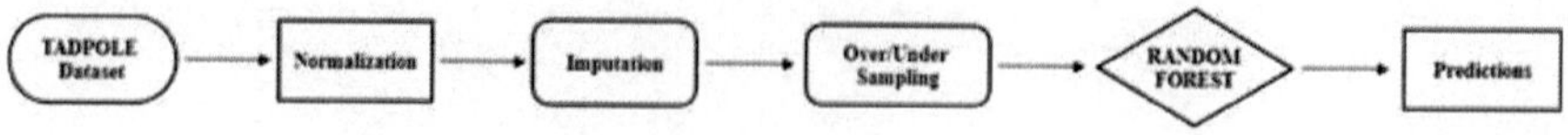

Fig. 1 Block diagram for Alzheimer data classification

3.3 *Training and Classification*

For Schizophrenia data, the features selected using ANOVA were used to train the various algorithms such as Naive Bayes, Random Forest, SVM, and Decision Tree. Since the numerical values of these features have similar ranges, normalization did not have much effect on the accuracies of the predictions, and hence, the features were not normalized for the final models. To find the optimal number of features for the model, different values were experimented with. Among the various algorithms, Gaussian Naive Bayes gave the best results on the training data with an accuracy of 1.00 with 50 components. Initially, the selected features were trained using the SVM and decision tree algorithms without normalization, and the resultant accuracies were recorded. Following this, the data were normalized, and the same algorithms were able to produce higher accuracies. Therefore, Naive Bayes and Random Forest algorithms were built to train the normalized data.

The Random Forest algorithm proved to be the most efficient, giving an accuracy of 0.71 on the Alzheimer data set. Due to this, the random forest model was chosen for further optimizations. The abnormal class had 8920 instances, whereas the normal class had 3821 instances. To combat this class imbalance problem, the under-sampling was first performed after dropping the instances with NaN values. This yielded an increased accuracy of 0.89 from 0.71. As an alternate method of dealing with the NaN values, imputation with median and mean values was tried before under-sampling. This increased the accuracy significantly from 0.89 to 0.95 and 0.96, respectively. Since the imputation with mean values produced better results, the same was tried with SMOTE oversampling as well. This further improved the accuracy to 0.97.

4 Results and Discussion

4.1 *Schizophrenia's Result*

Gaussian Naive Bayes performed the best, yielding an accuracy of 1 (with 50 components and 0.1 test split) for the training data and giving a maximum accuracy of 0.88717 for the testing data for 50 components (Private leader-board score calculated using 99% of the testing data). The results of the other classifiers are discussed in Table 3. A 1D-CNN was also trained using the data; however, for this model, no feature selection was performed. After performing a grid search for parameters such as number of epochs, batch size, and dropout rate, a ROC AUC score as high as 0.9 with a specificity of 1 was obtained on the training data. The performance of the various algorithms on the test data is discussed in Table 3. Figure 2 shows the training and validation accuracy of the 1D-CNN model plotted against five epochs. It is observed that as the training time increases, the model's validation accuracy

Table 3 ROC AUC scores of test data

N components	50	101	204	250
Models				
SVM	0.83076	0.85128	0.83589	0.83589
Gaussian NB	0.88717	0.85128	0.83589	0.83076
RF	0.84102	0.85641	0.77435	0.88205
DT	0.5641	0.64102	0.67948	0.68461

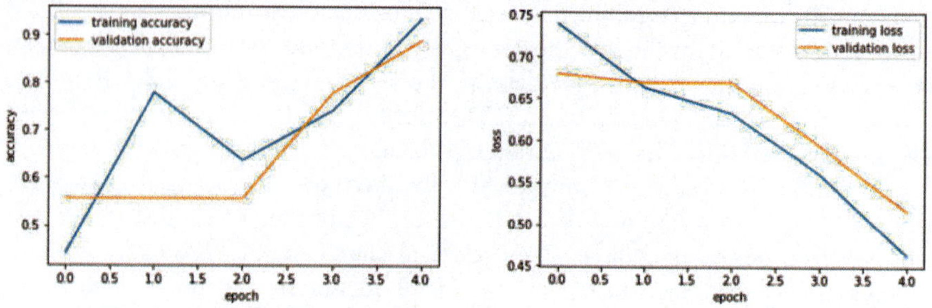

(a) Training and Validation Accuracy (b) Training and Validation Loss

Fig. 2 Accuracy and loss curves of the 1D-CNN model

increases and loss decreases. The model attains an accuracy of 93% on the training set and an accuracy of 88% on the validation set. Hence, the model has not over-fitted.

4.2 *Alzheimer's Result*

The best results have been yielded by the random forest model that was trained by oversampling the data and filling NaN values with the mean. From Table 4, it can be observed that both the specificity and sensitivity values are extremely high, showing that the model is able to recall both the positive and negative classes equally well.

Table 4 Evaluation metrics for random forest model

Metrics	Under-sampling			Oversampling
Imputation	–	Median	Mean	Mean
Accuracy	0.89	0.95	0.96	0.97
Specificity	0.914	0.97	0.963	0.973
Sensitivity	0.907	0.95	0.963	0.97
*F*1-score	0.91	0.96	0.96	0.97

Table 5 Confusion matrix for random forest model

	Predicted positive	Predicted negative
Actual positive	1746	48
Actual negative	47	1727

Generally, in medical diagnosis, the false negatives are not tolerable as they pose a greater risk to the patients. In the predictions, the false negative value is considerably low as observed from the confusion matrix given in Table 5.

5 Conclusion

In this work, machine learning and deep learning models have been proposed to predict the results of the classification of Schizophrenia and Alzheimer's using the Kaggle and TADPOLE data sets, respectively. Unlike data sets used in other works, both these data sources have significant MRI features to predict future outcome measurements of subjects at-risk of Schizophrenia and Alzheimer's disease. We have also applied feature selection on this data set which would help the physicians with the proper diagnosis of the disorder. Gaussian Naive Bayes classifier performed the best, yielding an accuracy of 1 (with 50 components and 0.1 test split) for the Schizophrenia data, whereas we could achieve benchmark accuracy of 97% using the Random Forest classifier for the Alzheimer data. The efficiency of the models is reported with respect to metrics like ROC score, accuracy, specificity, and sensitivity. This study can be further explored in the future by applying ML and DL techniques for multistage classification of brain disorders.

References

1. Alzheimer dataset-tadpole-grand challenge (n.d.). https://tadpole.grand-challenge.org/Data/#Data
2. Anupama M, Sowmya V, Soman K (2019) Breast cancer classification using capsule network with preprocessed histology images. Int Conf Commun Signal Process (ICCSP) 2019:0143–0147
3. Bansal D, Chhikara R, Khanna K, Gupta P (2018) Comparative analysis of various machine learning algorithms for detecting dementia. Proc Comput Sci 132:1497–1502
4. Chen Z, Yan T, Wang E, Jiang H, Tang Y, Yu X, Zhang J, Liu C (2020) Detecting abnormal brain regions in schizophrenia using structural MRI via machine learning. In: Computational intelligence and neuroscience
5. De Filippis R, Carbone EA, Gaetano R, Bruni A, Pugliese V, Segura-Garcia C, De Fazio P (2019) Machine learning techniques in a structural and functional mri diagnostic approach in schizophrenia: a systematic review. Neuropsychiatr Dis Treat 15:1605
6. Khan A, Zubair S (2020) An improved multi-modal based machine learning approach for the prognosis of Alzheimer's disease. J King Saud Univ Comput Inf Sci

7. Marinescu RV, Oxtoby NP, Young AL, Bron EE, Toga AW, Weiner MW, Barkhof F, Fox NC, Golland P, Klein S et al (2019) Tadpole challenge: accurate Alzheimer's disease prediction through crowdsourced forecasting of future data. In: International workshop on predictive intelligence in medicine, pp 1–10
8. Nair JJ, Mohan N (2017) Alzheimer's disease diagnosis in MR images using statistical methods. Int Conf Commun Signal Process (ICCSP) 2017:1232–1235
9. Nunes A, Schnack HG, Ching CR, Agartz I, Akudjedu TN, Alda M, Alnæs D, Alonso-Lana S, Bauer J, Baune BT et al (2020) Using structural MRI to identify bipolar disorders–13 site machine learning study in 3020 individuals from the enigma bipolar disorders working group. Mole Psychiatry 25(9):2130–2143
10. Priyanka G, Priya RT, Vasunthra S (2021) An effective dementia diagnosis system using machine learning techniques. J Phys Conf Ser 1916(1):012173
11. Saiprasath G, Naren Babu R, ArunPriyan J, Vinayakumar R, Sowmya V, Soman K (2019) Performance comparison of machine learning algorithms for malaria detection using microscopic images. IJRAR19RP014 Int J Res Anal Rev (IJRAR) 6(1)
12. Schizophrenia Dataset-Kaggle (2014). https://www.kaggle.com/c/mlsp-2014-mri
13. Suri GS, Kaur G, Moein S (2021) Machine learning in detecting schizophrenia: an Overview
14. Swapna G, Kp S, Vinayakumar R (2018) Automated detection of diabetes using CNN and CNN-LSTM network and heart rate signals. Proc Comput Sci 132:1253–1262
15. Xiao Y, Yan Z, Zhao Y, Tao B, Sun H, Li F, Yao L, Zhang W, Chandan S, Liu J et al (2019) Support vector machine-based classification of first episode drug-naive schizophrenia patients and healthy controls using structural MRI. Schizophr Res 214:11–17

EMD-Based Binary Classification of Mammograms

Anirban Ghosh, Pooja Ramakant, Priya Ranjan, Anuj Deshpande, and Rajiv Janardhanan

Abstract Mammography is an inexpensive and noninvasive imaging tool that is commonly used in detection of breast lesions. However, manual analysis of a mammogramic image can be both time intensive and prone to unwanted error. In recent times, there has been a lot of interest in using computer-aided techniques to classify medical images. The current study explores the efficacy of an Earth Mover's Distance (EMD)-based mammographic image classification technique to identify the benign and the malignant lumps in the images. We further present a novel leader recognition (LR) technique which aids in the classification process to identify the most benign and malignant images from their respective cohort in the training set. The effect of image diversity in training sets on classification efficacy is also studied by considering training sets of different sizes. The proposed classification technique is found to identify malignant images with up to 80% sensitivity and also provides a maximum $F1$ score of 72.73%.

Keywords Breast tumor · EMD · Mammograms · Image classification

A. Ghosh (✉) · A. Deshpande
SRM University AP, Neerukonda, Guntur, AP 522502, India
e-mail: anirban.g@srmap.edu.in

A. Deshpande
e-mail: deshpande.a@srmap.edu.in

P. Ramakant
King Georges' Medical University, Lucknow, Uttar Pradesh, India

P. Ranjan
Bhubaneswar Institute of Technology, Info Valley, Harapur, Odisha 752054, India
e-mail: priyaranjan@bit.edu.in

R. Janardhanan
Amity Institute of Public Health, Amity University, Sector 25, Noida, UP 201313, India
e-mail: rjanardhanan@amity.edu

M. Gupta et al. (eds.), *Artificial Intelligence on Medical Data*,
Lecture Notes in Computational Vision and Biomechanics 37,
https://doi.org/10.1007/978-981-19-0151-5_5

1 Introduction

The density of breast parenchyma has a well-established classification system based on mammographic findings. In younger women, the breast is denser, and as with age advancement, density reduces, and breast parenchyma tissue gets replaced by fat. Indian breast cancer spectrum is very much different from the western world breast cancer presentation. In India, we see higher incidence of breast cancer in younger population, denser breasts, more triple negative (estrogen, progesterone and HER2Neu all negative on immunohistochemistry), more aggressive disease biology, more locally advanced and metastatic stages, poorer survival outcomes and higher recurrences. Mammogram has been the standard imaging modality for detecting breast lesions. We add ultrasonography or MRI of breasts in selected subsets of patients. Ultrasound alone is not good enough to document each feature of breast cancer, and MRI is limited in use due to high cost. Hence, we need to add certain modifications in the mammogram technology to enhance its detection rates. In breast cancer, survival and outcomes are directly proportional to the stage of the disease. Hence, to achieve improved survival rates, early and timely detection of breast lesions is the only remedy. In western world, more than 90% patients with breast cancer survive more than 5–10 years, and in our country due to advanced stage presentation, less than 50% patients survive. Hence, we see more morbidity and mortality in our patients inspite of best possible clinical care and management which reemphasizes the requirement of an inexpensive technological tool to aid the process of early detection.

In recent times, several computer-aided diagnostic (CAD) tools have been developed, which are being used for various disease tagging to different degrees of accuracy [1–11]. It is expected that such computer-based tools would aid the healthcare professionals to make a more educated decision. In this paper, we have investigated the possibility of classifying breast lumps into benign and malignant by applying EMD-based algorithm to the mammogramic images of the patients which is based on the emulation of human perception to texture similarity. In [12], Bhaskar et al. showed that EMD-based implementation gives better result compared to other texture similarity measures when adopted for image classification based on texture which is seen as mixture of sinusoidal signals. A single texture can be defined as a distribution of signal energy in the frequency domain. EMD between images with many textures is obtained by considering single texture EMD as the ground distance in a space of textures [13]. A deep learning-based approach to the classification problem inherently suffers from instability and convergence issues which are never desirable in applications where human lives are at stake. We thus settle for a more stable approach using EMD. In younger Indian populace, detection of lesions from mammographic images is a major challenge which we eventually intend to address. However, such lump identification cannot be simply based on mammographic images due to increased breast density. Addition of multiple modalities such as clinical and epidemiological features of the subjects along with their mutation profiling may form the rationale toward the development of a multi-modal multi-fusion sensor technology for facilitating rapid and precision-oriented tagging of the lesions in dense breasts at not only tertiary care centers but also extended community. As prelude to the extensive beta testing of the proposed method, our current study is using the available dataset [14] to study the classification of lesions for the age group 40–50 to gain valuable insights on the etiopathogenesis of the breast cancer lesions on a spatiotemporal scale which will form the rationale for developing cutting edge, disruptive yet affordable and accessible mobile health applications aimed at evaluating the breast health of the Indian populace.

The rest of the paper is organized as follows: in Sect. 2, we elaborate and enlist the relevant works in literature. The detailed description of the proposed method is presented in Sect. 3, while Sect. 4 presents the efficacy of the proposed technique. Section 5 concludes the paper.

2 Related Works

Our explorations of automated disease tagging for breast cancer [15] using mammograms started with the understanding that there is lacunae in detection of breast lumps on mammogram in dense breasts. This work discovered that there was relatively more knowledge in the medical group; however, the skills of breast self examination (BSE) and attitude to prevention and early detection among rural and urban women were sub-optimal and do not differ significantly. Arti et al. [16], introduced an EMD-based approach to identify the tissue type from mammograms causing cancer. In addition, the authors also implemented a pixel-level classification of mammograms with a sensitivity of 92.64%. In this work, there was open question regarding selection of reference mammograms, which is resolved in current study using EMD in a very elegant fashion. Authors extract the content feature of the mammogram, histogram of oriented gradients, the edge direction histogram, the local binary pattern and the gray-level histogram and the image pair content similarity using the Earth Mover's Distance (EMD) with a precision of 0.83, but a sensitivity of 0.76 [17]. Another algorithm proposed in [18], leverages the same spatial features and an Earth Mover's Distance (EMD) flow vector between two corresponding regions in a pair of aligned mammograms. Their sensitivity is 0.82 with a modest dataset of 40 images and involves some complex data processing. In [19], authors developed partial signature matching using the EMD for identifying mammogram structures. They embedded histograms into a learned Euclidean space to accelerate computation but did not provide any quantitative results. Norlailah [20] observed that Chi-squared distance and EMD features are of statistical significance to be used in classification of normal and abnormal digital infrared thermographs, but the contrast measurement is computed to be statistically insignificant for future breast cancer classification work using digital infrared thermography. The final digital image obtained with a mammography system consists of several random processes, with diverse distributions. These precise distributions which form the given mammograms are subject of study in [21]. They also proposed new directions in the image acquisition process by introducing a simulation platform which contains models of scatter, image lag and electronic noise [22] but shed no light on the lesion classification problem. To summarize, a significant number of work in literature have explored an EMD-based approach in mammographic image classification to varied degree of success; however, they required either sophisticated image processing or involved a very small dataset or used approaches with relatively low sensitivity. To the best of our knowledge, this work is a novel attempt at EMD-based binary classification of mammograms with a reasonably large dataset producing a decent sensitivity without any requirement for image processing. The major contribution of this work can be summarized as

- Implementing LR scheme to identify the leader among the benign and malignant images from their respective training set.
- Proposing and realizing an EMD-based binary classification method to tag a given mammogram as benign or malignant.
- The tagging of images occurs with a decent $F1$ score and sensitivity using a reasonable sized data.

3 Methodology

After delineating the relevance of the current work in Sect. 2, we now describe the implementation of our proposed work. The section begins with the description of the data, extends the description to the training of the mentioned data elaborating on the LR scheme and finally concludes with the explanation of the validation process.

3.1 Gist of Dataset

The mammograms are obtained from the dataset in [14]. The database comprises of 3728 images from 1775 patients in different age brackets. The mammographic images contain biopsy confirmed tumors classified into benign and malignant. In our current study, we are interested to study mammographic images of women in the age group between 40 and 50 years where it has been statistically found to be more common. In addition, we consider only the right breast images for our proposed work. In the sequel, this is going to be considered as the *preset criteria* for our proposed model. We consider two training sets of different sizes—one consisting of images with benign tumor and the other with malignant ones. The validation set is different from the training set and contains a random mix of images with benign and malignant tumors which satisfies our preset criteria. The preset criteria and data used are summarized in Table 1.

Table 1 Preset criteria and summary of data used

Parameters	Values
Age bracket	40–50
Organ	Right breast
No. of training sets	2
Size of training set	10/20/50
Size of validation set	30

3.2 *Overview of Implementation*

An overview of the workflow of the proposed method is shown in Fig. 1. A certain number of right breast images of women in the cluster of 40–50 years are first segregated into two training sets corresponding to benign and malignant tags. In each set, the images are compared with every other image of the same set to build an EMD matrix. The EMD matrix is an aggregation of the texture dissimilarity scores between images of same type. A leader from each training set is then determined based on the LR scheme as explained in the next subsection which acts as the representative of its respective cohort. In the next phase, a validation set comprising of a random mix of images with benign and malignant lesions is curated. The EMD of each image from the testing set is determined with the leaders of the two cohorts. Since EMD essentially measures the perceptual dissimilarity between its argument, a higher value

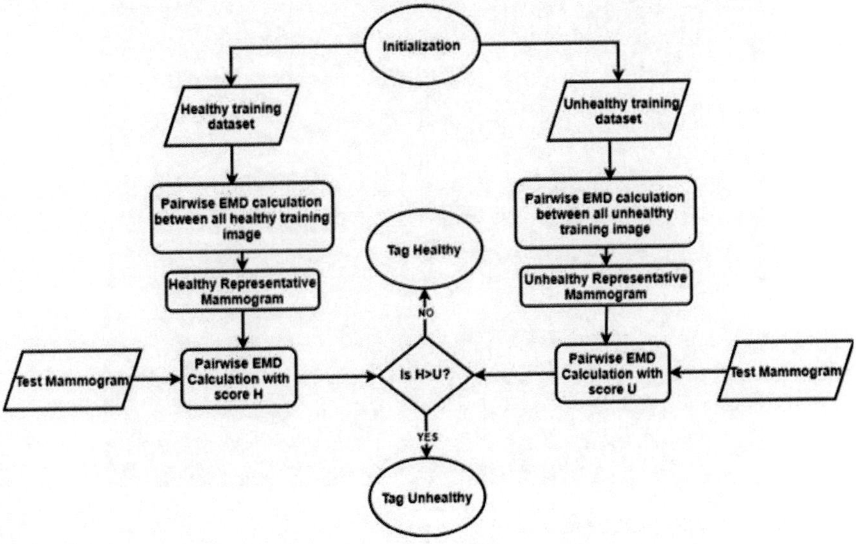

Fig. 1 Overview of implementation

indicates greater dissimilarity between arguments. Thus, as shown in the decision-making box of Fig. 1, if the EMD value of the testing image with the leader of the normal images is large, then the algorithm identifies the testing image to be closer to abnormal image and tags it as such and vice versa.

3.3 Training Set and LR Scheme

We have used three different training sets for implementation of the proposed work. The sets are differentiated based on their cardinality, and some samples from both the training sets are shown in Fig. 2. The motivation behind creating different training sets is to observe if increasing the diversity in the training sets has any effect in the identification efficiency. They are classified as

- **Case I**: The number of images in each training set is 10.
- **Case II**: The cardinality of each training set is 20.
- **Case III**: Each set in this case is of size 50.

Once the images are segregated into different training sets, we use the leader recognition (LR) schema to identify the most normal and most abnormal images from the respective groups. The LR schema is implemented as follows:

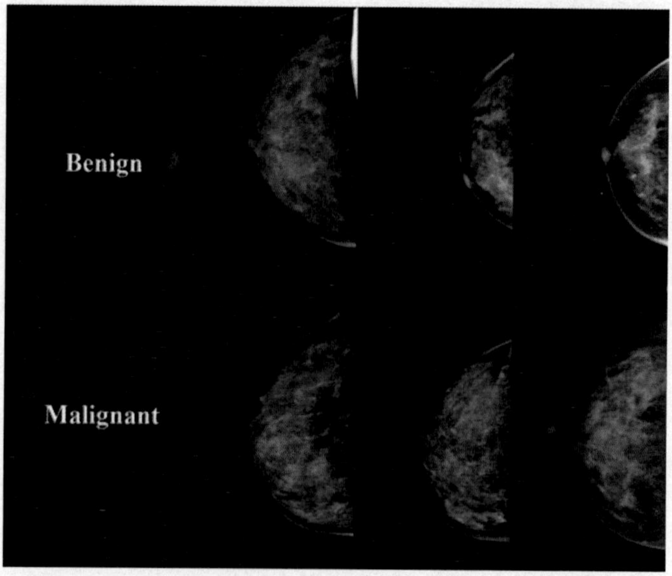

Fig. 2 Sample images from training sets

	n10	n11	n12	n13	n14	n15	n16	n17	n18	n19
n10	0	0.78731346	1.317723989	5.443656921	1.20466423	1.033955216	0.847761214	60.57052231	0.2791045	0.999067247
n11	0.787313461	0	8.843673706	15.4904995	12.75299072	38.55735779	13.98426723	26.46399307	15.708252	1.088808298
n12	1.317723989	8.84367371	0	18.57713127	11.03740406	24.79735947	3.472780943	34.32659912	18.5035744	1.42478478
n13	5.443656921	15.4904995	15.77086639	0	42.49219513	9.333973885	11.62497139	11.23469543	20.317461	62.00736237
n14	1.20466423	13.3121767	11.59926796	42.49219513	0	36.01356888	5.385020256	36.24471664	3.29869652	5.257815838
n15	1.033955216	38.7939796	24.79735947	9.333973885	36.01356888	0	26.30135918	4.245952129	43.3567009	4.054598808
n16	0.847761214	13.9842672	3.472780943	13.39538574	5.385020256	26.30135918	0	9.115343094	29.8572636	1.539080143
n17	60.57052231	24.365921	1.828373194	11.32490349	30.88265228	4.752351284	7.226522446	0	10.0343866	26.79938316
n18	0.279104501	15.708252	18.50357437	20.31746101	3.412604809	50.5657692	29.85726357	10.03438663	0	1.352514744
n19	0.999067247	1.0888083	1.42478478	62.00736237	5.257815838	4.054598808	1.539080143	26.79938316	1.35251474	0
Average	7.248376909	13.2374892	8.75584048	19.83825693	14.84389162	19.54102937	10.02390264	21.90355916	14.2707954	10.4523415
Rank	1	5	2	9	7	8	3	10	6	4

Fig. 3 EMD matrix of normal image cohort for *case I*

- Label the images in each group.
- Find the pairwise EMD between each image in a particular group to form the EMD matrix as shown in Fig. 3 for *Case I*.
- Next find the column-wise average of the pairwise EMD values calculated from the matrix as shown by the row labeled *Average* in Fig. 3.
- In the subsequent step, rank the averaged values in ascending order and select the image with the smallest average as the *leader*, seen highlighted in yellow in Fig. 3

The leader thus chosen using our novel LR scheme is the best representative of its particular training group. The leaders determined using the above scheme are found to be same for both the sets in *Case II* and *Case III*. This is not surprising, it only means that the extra images in *Case III* did not add any further diversity to the training sets to change the leader. The validation process is explained in the succeeding section.

3.4 Validation Process

Validation is implemented as a two step process—In the first step, we use the leaders from each cohort for the different cases to tag the rest of the images in the training set corresponding to that case. The motivation for this step is to ensure how well the choice of leader for each group is. This step however cannot be considered to compare the performances between the different scenarios; i.e., no inter-scenario performance comparison is possible since the training sets contain different images. In order to compare the performance between different scenarios, a random mix of 30 images with benign and malignant lesions is considered. In order to find the efficacy of the proposed method, pairwise EMD of the test image with the leaders is evaluated and then tagged in favor of the leader with which it has lower EMD value. We present the results of our proposed method in the following section.

4 Results

True positives (TP) and true negatives (TN) are considered as the outcomes, where the proposed method correctly identifies an image with malignant and benign lesions, respectively. On the other hand, a false positive (FP) is an outcome where a benign lesion in an image is wrongly classified as malignant, whereas in case of false negative (FN), a malignant lesion is wrongly tagged as benign. In this context, a model having higher false negatives is more dangerous as it entails wrong identification of malignant lesions as benign thereby undermining the criticality. In a similar vein, a model with higher TP is preferred as it aids in correctly identifying the critical cases requiring treatment. A confusion matrix aptly captures the proportion of TP, TN, FP and FN in a classification problem.

$$\text{Accuracy} = \frac{\text{TP} + \text{TN}}{\text{TP} + \text{TN} + \text{FP} + \text{FN}} \tag{1}$$

$$\text{Sensitivity} = \frac{\text{TP}}{\text{TP} + \text{FN}} \tag{2}$$

$$\text{Specificity} = \frac{\text{TN}}{\text{TN} + \text{FP}} \tag{3}$$

$$\text{Precision} = \frac{\text{TP}}{\text{TP} + \text{FP}} \tag{4}$$

$$F1\ \text{score} = \frac{2\text{TP}}{2\text{TP} + \text{FP} + \text{FN}} \tag{5}$$

Accuracy, sensitivity, specificity, precision and $F1$ scores are considered as some of the critical performance metrics in image classification models which can be calculated from a confusion matrix using Eqs. (1)–(5) as shown in [23]. Accuracy is a great comparison metric only when the dataset is symmetric; that is, it has equal number of false positives and negatives [24]. In case of uneven distribution between classes, $F1$ score, which is the harmonic mean of precision and recall, is the preferred metric. Precision is a measure of the confidence of a model in identifying true positives, whereas recall also known as sensitivity is a measure of the proportion of identifying actual positives. In a similar context, specificity is a measure of the percentage of identifying actual negatives. In the subsequent subsections, the performance of the proposed model is quantified.

4.1 *Results with Training Set*

In this subsection, the pairwise EMD values of the training set images with the leaders are evaluated. The image is tagged in favor of the leader with which it has lower EMD value. The process is repeated for all the three scenarios, and the results

Table 2 Confusion matrix with the training sets

	Actual labels		
Predicted labels	*Case I* (10 *images each*)		
		Benign	Malignant
	Benign	8	2
	Malignant	2	8
	Case II (20 *images each*)		
		Benign	Malignant
	Benign	15	6
	Malignant	5	14
	Case III (50 *images each*)		
		Benign	Malignant
	Benign	36	16
	Malignant	14	34

are captured in the form of a confusion matrix in Table 2. The observations from the results in Table 2 are summarized as follows:

- The intention of this exercise is to test how well the LR scheme has performed in selecting the leader of a cohort. It certainly is not a measure of the efficacy of our classification algorithm since the training sets are different.
- As the size of the training sets increases, the accuracy in correctly classifying the images seems to dip which can be attributed either to the increase in diversity of images or to the burden of identifying more images.
- The identification accuracy saturates to some reasonable value—it is 72.5% for *Case II* and 70% for *Case III*. In cognizance of the above trend, our understanding is —beyond a certain point just increasing the size of training set might not yield any better results.

4.2 *Results with Validation Set*

In this subsection, we quantify the performance of our proposed algorithm using definite metrics described earlier. Since the benign and malignant leaders for *Cases II and III* are same, so they are considered together in the presented results. As is evident from Table 3, for a validation set of 30 images, *Cases II and III* have only 3 false negatives as compared to 7 in case of *Case I*. Similarly, *Cases II and III* score over *Case I* in terms of true positives as well having 12 as against 8. So, it is clearly evident that *Cases II and III* give better result in identifying images with malignant lesions compared to *Case I*. The trend however reverses in identification of images

Table 3 Confusion matrix with validation set

	Actual labels		
Predicted labels	*Case I*		
		Benign	Malignant
	Benign	12	7
	Malignant	3	8
	Cases II and III		
		Benign	Malignant
	Benign	9	3
	Malignant	6	12

Table 4 Performance metric comparison

Parameters	Case I (%)	Cases II and III (%)
Accuracy	66.67	70
Sensitivity	53.33	80
Specificity	80	60
Precision	72.73	66.67
$F1$ score	61.54	72.73

with benign lesions. In the sequel, we evaluate the performance metrics to quantify which training scenario produces an overall better result.

The performance metrics of the presented model with the validation set are evaluated and shown in Table 4. The observations from the performance metrics can be summarized as follows

- Since in case of uneven distribution, $F1$ score is the true measure of the effectiveness of a model, so in our implementation, *Cases II and III* with larger and more diverse training sets appear to have a superior performance.
- The smaller and less diverse training sets used in *Case I* are more prone to correctly identifying images with benign lumps as expected from the confusion matrix and indicated by specificity value.
- As the training size increases (*Cases II and III*), the algorithm becomes more effective in identifying images with malignant lumps as is indicated by its higher sensitivity, but at the same time, it becomes quite conservative as reflected by its lower precision.

So, based on the two major observations, it can be concluded that a diverse and larger training set can improve the effectiveness of the proposed model which provides reason to test it across other age groups.

5 Conclusion

EMD appears to be a simple and easy to implement method for tagging of mammographic images of breast lumps into benign and malignant. It does not suffer from the inherent instability that is present in process-based approach, is computationally less demanding and yet classifies the images with a reasonable accuracy. The proposed method included images of a specific orientation containing lumps and from a particular age group to classify them as benign and malignant. In our future deliberation, we intend to improve on the LR algorithm since that forms the pivot of our classification scheme and apply the same to a dataset belonging to a younger subset where the problem is becoming more common and the disease biology is more aggressive. The final frontier in this domain would pertain to the integration of EMD-based classification of mammograms with clinical and epidemiological features of the subjects along with the molecular markers specific to the Indian population with a divergent genetic base. Such an approach will form the rationale for the development of a multi-modal multisensor technology aimed significantly at improving the precision of tagging the breast cancer lesions at the community in a skill independent manner. Such a strategy will also enable us to identify the etiopathological factors responsible for the onset and progression of breast cancer lesions. We do hope that the development of an indigenous mammometer scale (to be referred to as *Stan Arbudha Mapika*) would be deployed across the 735 districts of Indian subcontinent as a part of affordable and accessible healthcare option thereby significantly alleviating the menace of this unmet clinical need at a community level.

References

1. Ranjan P, Shrivastav KD, Vadlamani S, Janardhanan R (2020) HRIDAI: a tale of two categories of ECGs. International symposium on signal processing and intelligent recognition systems. Springer, Singapore, pp 243–263
2. Shrivastav KD, Arambam P, Das AM, Saeed S, Kaul U, Ranjan P, Janardhanan R (2020) Earth mover's distance-based automated geometric visualization/classification of electrocardiogram signals. Trends Commun Cloud Big Data 2020:75–85
3. Taneja A, Ujlayan A, Janardhanan R, Ranjan P (2021) Pancreatic cancer detection by an integrated level set-based deep learning model. In: Big data and artificial intelligence for healthcare applications. https://doi.org/10.1201/9781003093770-8
4. Bhatia V, Ranjan P, Taneja N, Singh H, Janardhanan R (2021) Early and precision-oriented detection of cervical cancer. In: Big data and artificial intelligence for healthcare applications. https://doi.org/10.1201/9781003093770-9
5. Khatri A, Jain R, Vashista H, Mittal N, Ranjan P, Janardhanan R (2020) Pneumonia identification in chest X-ray images using EMD. Trends Commun Cloud Big Data 2020:87–98
6. Goyal A, Tirumalasetty S, Bathla D, Arya MK, Agrawal R, Ranjan P, Hossain G, Challoo R (2019) A computational segmentation tool for processing patient brain MRI image data to automatically extract gray and white matter regions. Emerging research in computing, information, communication and applications. Springer, Singapore, pp 1–16

7. Shrivastav KD, Das AM, Singh H, Ranjan P, Janardhanan R (2018) Classification of colposcopic cervigrams using EMD in R. International symposium on signal processing and intelligent recognition systems. Springer, Singapore, pp 298–308
8. Taneja A, Ranjan P, Ujlayan A, (2016) Multi-cell nuclei segmentation in cervical cancer images by integrated feature vectors. Multimedia Tools Appl 77(8), 9271–9290; Taneja A, Ranjan P, Ujlayan A, (2016) An efficient SOM and EM-based intravascular ultrasound blood vessel image segmentation approach. Int J Syst Assur Eng Manage 7(4):442–449
9. Taneja A, Ranjan P, Ujlayan A (2016) Novel texture pattern based multi-level set segmentation in cervical cancer image analysis. In: Proceedings of the international conference on image processing, computer vision, and pattern recognition (IPCV). The steering committee of the world congress in computer science, computer engineering and applied computing (WorldComp), p 76
10. Taneja A, Ranjan P, Ujlayan A (2018) Automated cell nuclei segmentation in overlapping cervical images using deep learning model. In: Proceedings of the international conference on image processing, computer vision, and pattern recognition (IPCV). The steering committee of the world congress in computer science, computer engineering and applied computing (WorldComp), pp 165–172
11. Anju M, Shanu S, Sanjay K, Priya R, Amit U (2021) Effect of hand grip actions on object recognition process: a machine learning-based approach for improved motor rehabilitation. Neural Comput Appli 33(7):2339–2350
12. Purohit B, Bandyopadhyay T (2014) Beyond job security and money: driving factors of motivation for government doctors in India. Hum Resour Health 12:12. https://doi.org/10.1186/1478-4491-12-12
13. Rubner Y, Tomasi C, Guibas LJ (1998) A metric for distributions with applications to image databases. In: Sixth international conference on computer vision (IEEE cat. no. 98CH36271), pp 59–66. https://doi.org/10.1109/ICCV.1998.710701.
14. Cui C, Li L, Cai H, Fan Z, Zhang L, Dan T, Li J, Wang J (2021) The Chinese mammography database (CMMD): an online mammography database with biopsy confirmed types for machine diagnosis of breast. Cancer Imaging Archive. https://doi.org/10.7937/tcia.eqde-4b16
15. Ramakant P, Singh KR, Jaiswal S, Singh S, Ranjan P, Rana C, Jain V, Mishra AK (2018) A survey on breast cancer awareness among medical, paramedical, and general population in North India using self-designed questionnaire: a prospective study. Indian J Surg Oncol 9(3):323–327. https://doi.org/10.1007/s13193-017-0703-9
16. Taneja A, Ranjan P, Ujlayan A, Janardhanan R (2018) An earth mover's distance based density estimation for breast tissue type identification and grading of cancer. In: IEEE international symposium on biomedical imaging, poster presentation
17. Wang Z, Xin J, Huang Y, Li C, Xu L, Li Y, Zhang H, Gu H, Qian W (2018) A similarity measure method combining location feature for mammogram retrieval. J Xray Sci Technol 26(4):553–571. https://doi.org/10.3233/XST-18374
18. Erihov M, Alpert S, Kisilev P, Hashoul S (2015) A cross saliency approach to asymmetry-based tumor detection. In: Navab N, Hornegger J, Wells W, Frangi A (eds) medical image computing and computer-assisted intervention—MICCAI 2015. MICCAI 2015. Lecture notes in computer science, vol 9351. Springer, Cham. https://doi.org/10.1007/978-3-319-24574-4_76
19. Holmes AS, Rose CJ, Taylor CJ (2002) Transforming pixel signatures into an improved metric space. Image Vis Comput 20(9–10):701–707. ISSN 0262-8856. https://doi.org/10.1016/S0262-8856(02)00060-4
20. Norlailah L (2017) Digital infrared thermography analysis for breast cancer using image textures features, Doctoral dissertation, University of Malaya
21. Rosa RS (2019) Simulations and virtual clinical trials for the assessment of the added clinical value of angio-tomosynthesis over angio-mammography, Doctoral dissertation, Universite Paris-Sachlay

22. Rosa RSDl, Carton AK, Carvalho PMD, Bloch I, Muller S (2019) Analysis of cedbt and CESM performance using a realistic X-ray simulation platform. In: 2019 IEEE 16th international symposium on biomedical imaging (ISBI 2019), pp 1070–1073, https://doi.org/10.1109/ISBI.2019.8759527
23. Tufail AB, Ma YK, Zhang QN (2020) Binary classification of Alzheimer's disease using sMRI imaging modality and deep learning. J Digit Imaging 33:1073–1090. https://doi.org/10.1007/s10278-019-00265-5
24. Ghoneim S. Accuracy, recall, precision, *F*-score and specificity, which to optimize on? https://towardsdatascience.com/accuracy-recall-precision-f-score-specificity-which-to-optimize-on-867d3f11124

Comparative Analysis of Machine Learning Algorithms with Ensemble Techniques and Forecasting COVID-19 Cases in India

Nidhi Kumari Chauhan, Chhavi Goel, and Pooja Singh

Abstract Unanticipated information in December 2019 changed the world around us. A relatively contagious disease unfolded through the SARS-CoV-2 virus that travelled throughout the globe and was declared an epidemic by WHO in March 2020. The need of examining the scenario became the inducement behind this research. The assessment of COVID-19 in India is performed from 1 April 2020 to 20 May 2021 which amassed a total of 415 instances. Further, preprocessing of the dataset is executed with the use of normalization. The experimentation is executed through the use of four ensemble strategies which are bagging, boosting, stacking and voting with four distinct machine learning algorithms linear regression, sequential minimal optimizer for regression, multilayer perceptron and Gaussian process. The splitting of the dataset is completed at 75%, and machine learning algorithms with ensemble techniques are applied. Linear regression with the bagging ensemble method gives satisfactory outcomes with the correlation coefficient of 0.935 and 0.919 for confirmed cases and recovered cases, respectively, and Gaussian process presented the best results for deceased cases. In the case of ensemble strategies, bagging indicates the best correlation coefficient in each case. Therefore, with the help of the three best algorithms, confirmed cases, recovered cases and deceased cases predictions are performed. The paper has potential implementations that can foresee the COVID-19 confirmed cases, recovered cases and deceased cases based on historic data and subsequently structure the plan for the future.

Keywords COVID-19 · Machine learning · Ensemble technique · Regression · Prediction

N. K. Chauhan (✉) · C. Goel
Department of Computer Science and Engineering, Amity School of Engineering and Technology, Guru Gobind Singh Indraprastha University, New Delhi, India
e-mail: nidhikchauhan0411@gmail.com

P. Singh
Postdoctoral Fellow, Graduate Program in Telecommunication Engineering, Federal Institute of Education, Science, and Technology of Ceara (IFCE), Fortaleza, Ceara, Brazil

Associate Professor, Department of Computer Science & Engineering, GL Bajaj Institute of Technology & Management, Knowledge Park-3, Greater Noida (U.P.) 201306, India

M. Gupta et al. (eds.), *Artificial Intelligence on Medical Data*,
Lecture Notes in Computational Vision and Biomechanics 37,
https://doi.org/10.1007/978-981-19-0151-5_6

1 Introduction

The year 2020–2021 could be remembered by the world as one of the toughest years for humanity. A sudden outbreak of highly infectious and communicable disease unfolds all across the globe when COVID-19 was declared a pandemic by World Health Organization in March 2020 [1]. The disease is transmitted by a virus named SARS-CoV-2 [2]. According to WHO, the first cases were reported in December 2019 [2]. The origin of the virus was suspected in the city of Wuhan, China [3]. The preliminary signs and symptoms of the sickness had been pronounced to be much like influenza i.e. cough, cold and fever; however, a few reviews advocate gastrointestinal signs and symptoms related to this sickness, others describe asymptomatic infections, especially among younger children [4]. The virus severely impacted the world; in India, the virus paved its way and started affecting people drastically, leaving authorities all around the nation with the responsibility of saving their citizens and hence the world from the outbreak. To contain the unfold of virus in a very country with an outsized population, the Government of the Republic of India obligatory restrictions on travel, gatherings and projected 1st lockdown on 25 March 2020. It has not only impacted the health and well-being of the citizens but also has deeply affected the economy and development of the country. Since the virus is contagious, multiple lockdowns were imposed in 2020 with a few relaxations in every subsequent lockdown. The result of lockdown was observed as the cases were under control during the first wave, and it was observed that situation was improving by the end of the year [5]. The country was recovering from the primary wave withinside the early months of 2021, while the second wave hit the state in April 2021 [6]. The second wave confirmed an exponential growth in cases. The cumulative confirmed cases from 1 April 2020 to 20 May 2021 are proven in Fig. 1, respectively, [7]. A drastic growth withinside the instances was discovered in April–May 2021. The unexpected and extreme growth in the 2nd wave made a big strain on health services. It was reported on 23 April 2021, there have been approximate 15.9 million cases and 185,000 deaths throughout the nation [6]. Figure 2 signifies the cumulative deceased cases from 1 April 2020 to 20 May 2021 [7]. As shown in Fig. 2, the second wave resulted in more deaths than the first. According to many clinical portions of research and studies, the cause behind the unexpected growth is that the mutant virus has a better rate of transmission [6]. The unexpected boom in the COVID-19 cases results in the deficiency of health supplies and human resources which made the state of affairs worse implying a more difficult recovery [5]. In February 2021, successful vaccination drives started in India. India took the challenge of vaccinating a nation with a 1.38 billion population [8] to vaccinate 300 million people by August 2021 [9]. The consequences of COVID-19 have impacted people from all walks of life including workers, daily wage earners, students, freshers, businesses, families and children.

Therefore, the motivation at the back of this study is to examine the scenario in India using machine learning techniques, and in addition, to predict, confirmed cases, recovered cases and deceased cases so that required preparations and measures may

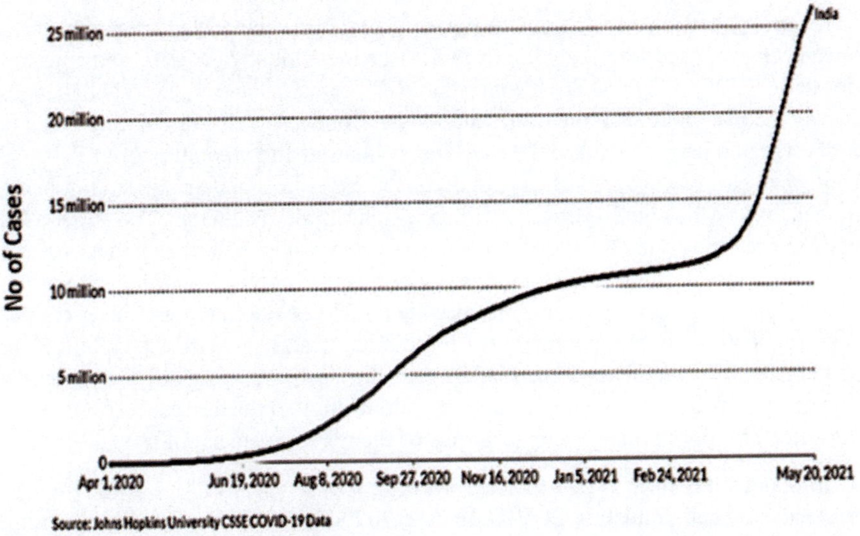

Fig. 1 Cumulative confirmed COVID-19 cases from 1 April 2020 to 20 May 2021. *Source* Adapted from [7]

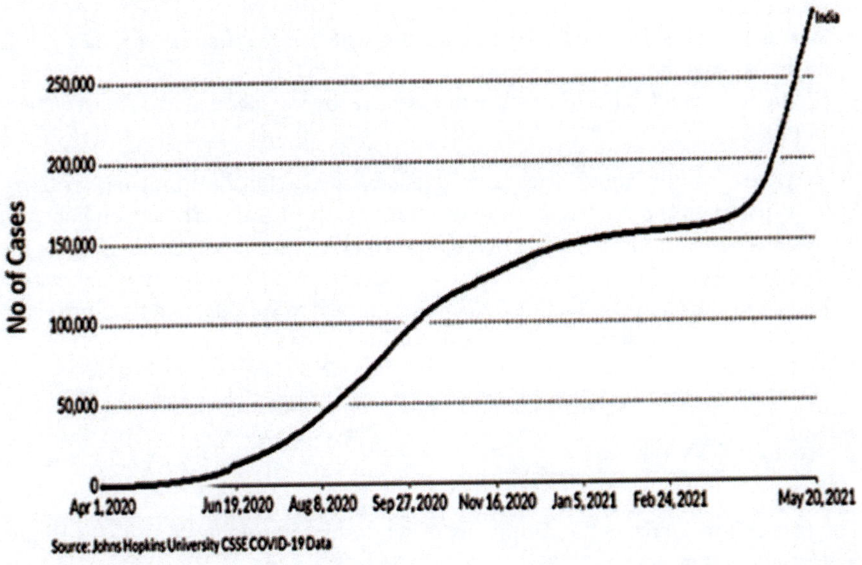

Fig. 2 Cumulative deceased COVID-19 cases from 1 April 2020 to 20 May 2021. *Source* Adapted from [7]

be taken ahead to keep away from the outbreak of future waves. Many research works have been proposed for predicting COVID-19 cases globally in 2020, to the best of our knowledge a very few are for the year 2021. India has faced several challenges during the pandemic, and therefore, the focus of our research is predicting cases in India with the help of updated dataset. The evaluation and prediction of COVID-19 confirmed cases, recovered cases and deceased cases in India are carried out using the supervised machine learning technique—ensemble technique. The supervised machine learning method specializes in analysing more than one model. We are featuring four ensemble strategies, namely boosting, bagging, stacking and voting, and the machine learning algorithms are taken into consideration are linear regression, sequential minimal optimizer for regression, Gaussian process and multilayer perceptron. The predictions are made using the best and most efficient machine learning algorithms obtained after training and testing all the models.

Therefore, the significant contributions of the research are as follows:

- In-depth study of COVID-19 pandemic in India.
- Analysing and evaluating COVID-19 cases in India by applying machine learning algorithms.
- Applying various ensemble techniques over machine learning algorithms.
- Comparative assessment of ensemble techniques over different machine learning algorithms.
- Assessing and selecting the best algorithm with the maximum efficiency in the ensemble method.
- Predicting the COVID-19 confirmed cases, recovered cases and deceased cases in India.

Structure of the Paper: The paper is established with the following sections; Sect. 2 discusses the related work that describes all of the work accomplished in the respective field. Section 3 is a perception right into a method that highlights the process followed to hold the research work. Section 4 explains the implementation and result of the analysis. Section 5 discusses threats to validity, and finally, Sect. 6 consists of the conclusion and future scope of the research work.

2 Related Work

The research task covers three domains of work, first unfolding of COVID-19 and its impact, 2nd in which the evaluation has been executed using machine learning algorithms and third on implementation of ensemble techniques.

2.1 COVID-19 Studies

The remarkable work done by fellow researchers gave insights into the topics and guided our research insignificantly. Majhi [10] observed rising COVID-19 cases in India and has build predictive models that can forecast number of positive cases with higher accuracy. Zu [11] has focused the research on aetiology, epidemiology and clinical symptoms of COVID-19 while highlighting the role of chest CT in prevention and disease control. The conclusion of this work highlighted the importance of CT scans by studying cases from Wuhan and also states that RT-PCR, however remains the primary method of detection of coronavirus. Nishiura [12] talks about the spread of the coronavirus by examining 28 infected pairs of people. The conclusions of the study indicate that the transmission period of the virus is short-lived and it can spread even before the disease affects the host.

2.2 Machine Learning Approach on COVID-19

Apart from evaluating the COVID-19 statistics, many previous studies have implemented machine learning techniques and algorithms on COVID-19 to forecast the future situation. Rajan [13] has studies the impact of the pandemic on India. Predictions have been accomplished in exponential and polynomial regression model. Also, death rate case predictions achieved through two-class classification models with an accuracy of 60%. Kwekha-Rashid [14] performed an overall analysis to recognize all previous research work that has emphasized using machine learning for studying COVID-19 cases. The paper covered different algorithms used in different countries used by authors based COVID-19 outbreak. The study shows the supervised learning algorithms were 92% accurate and showed better results than unsupervised learning algorithms. The research work by Barstugan [15] is on the classification of CT scan images using machine learning. The feature extraction in this study is applied by using algorithms, namely grey-level co-occurrence matrix (GLCM), local directional pattern (LDP), grey-level run length matrix (GLRLM), grey-level size zone matrix (GLSZM) and discrete wavelet transform (DWT). Support vectors were implemented for classification with twofold, fivefold and tenfold cross-validation. The authors claim to obtain 99.68% accuracy tenfold cross-validation and GLSZM feature extraction method. Tiwari, Dimple, et al. [16] presented an analysis to predict trend of pandemic COVID-19 using machine leaning algorithms, namely support vector machine (SVM), naive Bayes and linear regression, which are applied on real-time series dataset. Out of three, naive Bayes better results with less mean absolute error and less mean squared error.

2.3 Ensemble Techniques Approach

Ensemble techniques have grabbed researcher's attention in recent years. Ahmed et al. [17] focused on ensemble classifiers to detect network intrusion. The research proposed using more than one classifier and concluded the boom inaccuracy of consequences in evaluation to their earlier work and study. Tiwari [18] studied different ensemble paradigms on the J48 classifier. The conclusion drawn in the study highlighted the advantage of ensembles over individual model comparison, and they also found that stacking, blending worked best of all. Kotsiantis [19] used two ensemble methods bagging and boosting, and a comparison was performed on 25 sub-classifiers, evaluated on standard benchmarks. The research proved that there was a 9–16% increase in accuracy by using ensembles than individual classifiers. Abdar et al. [20] developed a two-layer nested ensemble classifier and performed comparative study with single classifiers Wisconsin Diagnostic Breast Cancer (WDBC) dataset. Two-layer ensemble models performed better, and also both SV-BayesNet-3-MetaClassifier and SV-naïve Bayes-3-MetaClassifier achieved accuracy 98.07%.

3 Methodology

COVID-19 enormously impacted the world directly or indirectly. Consequently, it becomes vital to acquire every bit of knowledge and foresee for the future to elevate, strategize and fight with the pandemic. The purpose of this study is to examine the statistics during the last twelve months to advantage a perception into the rise and fall of confirmed cases, recovered cases and deceased cases and consequently carry out predictions for the future. Machine learning is a well-known approach that gives the machine power to learn and predict, based on past inputs. Choosing the right algorithm for optimized outcomes is a difficult task. Supervised machine learning offers a method known as the ensemble technique that focuses on estimating multiple analytical models instead of using just one. Our research work is primarily based on the ensemble approach and consequently acquiring satisfactory models for predictions. The proposed method of the research work is illustrated in Fig. 3.

The initial section of the analysis involves the gathering of knowledge on COVID-19 cases in India. The subsequent step focuses on the optimization of the accessible dataset. Once the raw data is extracted, the data preprocessing techniques are performed. The data preprocessing techniques embody the cleansing of the raw dataset and feature scaling that is a necessary step for reducing the errors and complexity of the data. On obtaining the ultimate dataset, machine learning algorithms and ensembles techniques are applied to coach the data and to perform additional predictions. A comparative assessment is carried out between different machine learning algorithms and ensemble techniques to obtain the best, and most efficient machine learning algorithm which is also used for the prediction. The

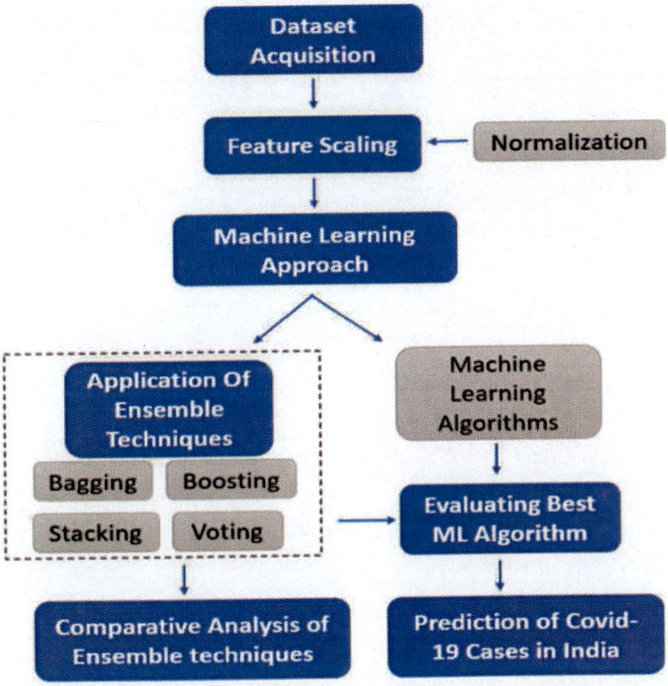

Fig. 3 Overview of research work

following sections cover insight of different phases of our research work as follows: Sect. 3.1 highlights the data acquisition process and details about the dataset, and Sect. 3.2 explains feature scaling. Section 3.3 focuses on the machine learning approach.

3.1 Data Acquisition

Dataset is the foundation and backbone of a research study. The fundamental requirement of machine learning-based research work is to obtain a consistent and accountable dataset. The COVID-19 dataset has been acquired from COVID-19 data hub [21]. The raw dataset contains several parameters; therefore, the selection of appropriate parameters is simply based on the relationship and dependency between the parameters. The COVID-19 data is continuous and cumulative; i.e. daily active cases have been added to new confirmed cases. Therefore, regression machine learning algorithms are used. The dataset comprises 415 instances starting from 1 April 2020 to 20 May 2021. The dataset includes five variables, two independent variables that are days and tests, and the other three are the targets or dependent variables. The

Table 1 Dataset variables and their description

Variables	Description
Days	Number of day starting from 1 April 2020
Tests	Cumulative number of tests
Confirmed cases	Cumulative number of confirmed cases
Recovered cases	Cumulative number of patients released from hospitals or reported as recovered
Deceased cases	Cumulative number of deaths or deceased cases

Table 2 Statistical analysis of dataset

Variables	Mean	Std	Min	Max
Days	208	119.94	1	415
Tests	114,393,100	99,055,340	27,688	322,356,200
Confirmed cases	7,186,683	6,048,042	2059	26,030,560
Recovered cases	6,460,989	5,432,575	169	22,705,840
Deceased cases	99,974.64	71,525.46	53	290,775

dataset variables are mentioned in Table 1, and their statistical insights are depicted in Table 2.

3.2 *Feature Scaling*

Feature scaling is a method to unify self-variables or feature ranges in data. Feature scaling is performed when the dataset consists of a wide range of differences between the variables. Therefore, while preparing the dataset, it is crucial to scale the predictor's attributes or the variables of the dataset so that each one of the parameters is proportional and in a narrowed range [22]. In the case of one attribute having a notably large magnitude, it can dominate over the other attributes, affecting the general accuracy of the model. The maximum used and common strategies for feature scaling are standardization and normalization. Normalization is to scale down the attributes so that they represent an equivalent range whereas in standardization the values are centered around mean with unit variance [23]. Thus, scaling brings all the parameters on the same page without any dominant factor involved. This balances the dataset for suitable and appropriate results. Moreover, this increases the coverage and hence accuracy of the model. In this research, normalization is carried out under the feature scaling technique.

Normalization: Normalization is a scaling technique within which values are shifted and scaled so they fall into the limit of [0,1]. This kind of normalization method is called MIN–MAX normalization [24]. In this research, to implement the

MIN–MAX normalization at the unbiased variable, the following formula is used [25]:

$$X' = \frac{X - X_{\min}}{X_{\max} - X_{\min}} \tag{1}$$

Here, $X_{\min}$ and $X_{\max}$ are the minimum and the maximum values of the attribute of the dataset being normalized. When X is minimum, then X' will be 0, and when X is maximum, then X' will be 1.

3.3 Machine Learning Approach

Our research work is focused on distinct machine learning algorithms and ensemble techniques. Ensemble techniques lie under the supervised category of machine learning algorithms. Therefore, the foundation of this research is based on a well-labelled training dataset. Ensemble technique envelopes analysis of multiple models, and different models cover different aspects of the input dataset. For achieving the motive of this research, four different ensemble techniques are used, namely boosting, bagging, stacking and voting. The models have been developed with the use of four different machine learning algorithms which are linear regression (LR), sequential minimal optimizer for regression (SMOReg), Gaussian process (GP) and multilayer perceptron (MLP). This implies four algorithms are executed that highlights various aspects of the data. The result analysis is performed in twofolds:

(1) First is the assessment between considered ensemble techniques and evaluating the best ensemble technique among the four techniques used.
(2) The second emphasis is on the overall performance of all of the machine learning algorithms implemented and therefore selecting the satisfactory machine learning algorithm; this is in addition used for the prediction. The information about the ensemble approach and algorithms is included withinside the subsequent section.

The tool used for this analysis is an open-source data mining software known as Weka[1] (Version 3.9.4). The Weka workbench is a set of machine learning algorithms, and data preprocessing tools that consists of all of the algorithms, virtually [26]. It offers a variety of options that help in the preparation and representation of a wide range of data. The visualization made comparative analysis efficient. For this study train–test, the splitting technique is used that simply divides the dataset into two subsets. The first set is the training dataset that is used to train the model, and also, the second subset is employed to check the model that is testing dataset. First, the model learns from the training dataset and builds the connection between dependent and independent variables, and then, the model predicts the variable for the testing dataset.

[1] https://www.cs.waikato.ac.nz/ml/weka.

The operating of numerous machine learning algorithms is analysed by evaluating various performance measures. The COVID-19 dataset has 415 data points that are split as 75% for training and 25% for testing.

3.3.1 Ensemble Techniques

As mentioned, the focal point of the research paper is the ensemble method which is a supervised machine learning approach. Ensemble learning emphases the concept of the opinion of many more than one. By combining different learning and analysis of the result tends to be less biased. The ensemble technique uses multiple machine learning algorithms for the same task, which provides a better prediction for an individual. This method increases the accuracy and consistency of the outcome and offers less biased results and fewer variance errors. Ensemble technique will be enforced in numerous ways, and during this research work, different machine learning algorithms are trained and tested using four best-known ensemble strategies as bootstrap aggregation or bagging, boosting, stacking and voting.

Bootstrap Aggregation or Bagging: Bagging is an ensemble method of combining multiple models of the same learning algorithm which is trained with a subset of the training dataset. These models are trained independently in parallel, and their result is combined by averaging process. Bags, i.e. a subset of the training dataset, is randomly picked. It first generates different samples of a dataset and then applies the base learning algorithm to train the model, and this process is performed on all the subsets to obtain multiple models. Then, each model is tested by using the same test dataset, and instead of using any one model for predictions, it averages the values predicted for test compounds over the whole ensemble of models [27]. Base Model with high variance an d low bias are well fitted for bagging. Additive regression implements the classic boosting algorithms in Weka. The meta classifier complements the overall performance of a regression-based classifier. Each iteration fits a base model to the residuals left through the classifier at the preceding iteration. Prediction is achieved by including the predictions of every classifier [27].

Boosting: Boosting is an iterative model which implies at the given instance the training model depends on previous model for input. Boosting ensemble methods offer a little variation to bagging; instead of training multiple models on subset simultaneously, in boosting, the data points that give wrong predictions or induces errors are selected. In boosting, a model is trained by employing a base learner on a subset of the training dataset. Then, testing is performed and the data points that gave wrong predictions are identified and a new subset or bag is formed using those data points along with some randomly chosen data points from the training dataset, and again testing is performed using the testing dataset. Then, similarly points of failure are isolated for consequent iteration. The method is repeated. Every iteration fits a brand new base model to the residuals left on the previous iteration. Prediction is accomplished by summing up the predictions of each base model [27]. This reduces error rates significantly hence increases the accuracy of the model. Base model with high bias and low variance are well fitted for boosting. Weka provides the boosting techniques as a classifier under a meta classifier named boosting. Along with this, the attribute classifier is used to select the learner classifier on which boosting is performed.

Stacked Generalization or Stacking: Stacking relies on using multiple learning algorithms and so employing a meta learner that learns to best mix the results of the training algorithms. In this method, the meta learner decides if the predictions made by the base learner are to be trusted or not. In stacking, the dataset is split into two folds, N weak learners make train on first split and then each N learner make prediction on second split. We fit the meta learner on second split using predictions of N weak learners. Only the base-level classifiers that are predicted to be correct are taken, and their predictions are combined by summing up the probability distributions [28]. The implementation of the stacking technique is done using a Weka meta classifier named stacking. In stacking, the meta classifier and base classifier are selected. This research uses meta classifier as Gaussian processes, multilayer perceptron and SMOReg, whereas linear regression is chosen as the base classifier.

Voting: Voting is a rather simplified version of stacking. In voting, multiple learners are used; however, rather than employing a meta learner, the predictions of the models are combined by either taking mean or mode of the result. It allows to give fair chance to all the models to contribute in the result. In the voting method, no result obtained is dropped rather every learner gets a chance to vote to predict the result, mean decides which value is to be thought of as result and which is marked as wrong. The voting ensemble can boost the robust error over each model [29]. Voting is recommended to be used when all the models are good performers and tend to agree on the result. The voting is implemented in Weka using the Weka meta classifier voting. The machine learning algorithms selected for voting are linear regression, Gaussian processes, multilayer perceptron and SMOReg.

3.3.2 Machine Learning Algorithms

Linear Regression (LR): Linear regression supports regression-based models. The dependence of attributes plays a key role in deciding the connection and therefore the coefficient on the dataset. It works by estimating the coefficient for a line or a plane that best fits the model. In linear regression, the relationship between dependent and independent variables is linear; i.e. y is proportional to α (where α is coefficient of regression). The equation of linear regression for one independent variable is given by [26]:

$$y = mx + c \tag{2}$$

where y represents the dependent variable, m is the slope, x is the independent variable and c is for the intercept for a given line.

SMOReg or Sequential Minimal Optimizer for regression (SMOReg): SMOreg implements the sequential minimal optimization algorithm for learning a support vector regression model (Smola and SchÃűlkopf 1998) [26]. The support vector regression model fairly finds hyperplane and the maximum margin that classifies the data points permitting them to either include or exclude the data point, and in the case of regression, the hyperplane modifies according to the points present for the same.

Gaussian Process (GP): Gaussian process is a strong statistical approach that gives a distribution of variables over functions that belongs to a continuous domain. When the input is training dataset X and testing dataset Y, the output of Gaussian process regression is the value and confidence at each point in testing dataset Y. The dataset is centered to have a mean zero, and covariance matrix is calculated on a combined dataset using a kernel function to find similarities. The joint distribution based on the covariance matrix needs to be calculated. Further, the conditioning matrix and marginalization is performed, which is to probability distribution on any given point based on other given data points. On having mean and standard deviation, a confidence interval is calculated from the standard deviation to the desired level. The mean is the prediction value of the Gaussian process [26].

Multilayer Perceptron (MLP): Multilayer perceptron is a neural network that is trained using backpropagation [26]. It consists of three types of layers—the input layer, output layer and hidden layer and a feed-forward network; in an MLP, the data flows in the forward direction from input to output layer. The neurons in the MLP are trained with the backpropagation learning algorithm. MLPs are designed to approximate any non-stop characteristic and may resolve problems that are not linearly separable [26].

4 Implementation and Result Analysis

The implementation process is based on a sequential approach which after acquiring and cleaning the dataset, goes for analysing the data, preparing training models and then performing predictions. For analyses and assessment of the model, certain performance measures are needed to be studied. The performance measures are also known as probability estimates which help in finding the ability of the algorithm or a model. Further, to decide which model or algorithm performs as per expectations, numeric measures are considered in machine learning that not only helps in evaluation but gives an insight into how the model is behaving on a given dataset. The details about the performance measures are as follows [24, 30]:

- Correlation Coefficient: It indicates how well the predictions are correlated or change with the actual output value. A value of 0 is the worst whereas a value of 1 is an accurately correlated set of predictions.

$$r = \frac{\sum(x_i - \overline{x})(y_i - \overline{y})}{\sqrt{\sum(x_i - \overline{x})^2 \sum(y_i - \overline{y})^2}} \tag{3}$$

where r is the correlation coefficient, x_i is the values of the x-variable in a sample, $\overline{x}$ is the mean of the values of the x-variable, y_i is the values of the y-variable in a sample and $\overline{y}$ is the mean of the values of the y-variable.

- Mean absolute error (MAE): This error measures the average of the absolute value of the difference between the predicted and actual values. The MAE is denoted by the following formula.

$$\text{MAE} = \frac{\sum_{i=1}^{n} |y_i - x_i|}{n} \tag{4}$$

where y_i is the prediction,x_i is the true value and n is the total number of data points.

- Root mean squared error (RMSE): Root mean square error (RMSE) is the standard deviation of the residuals or the prediction errors. This error is used to measure the difference between the predicted and actual value. It is denoted by the following formula:

$$\text{RMSE} = \sqrt{\frac{\sum_{i=1}^{N} \left(X_i - \widehat{X_i}\right)^2}{N}} \tag{5}$$

where i is the variable, N is the number of data points, X_i is the actual observation time series and $\widehat{X_i}$ is the estimated time series.

- Root relative squared error (RRSE): Mathematically, the root relative squared error $RRSEi$ of an individual model i is evaluated by the equation:

$$\text{RREi} = \sqrt{\frac{\sum_{j=1}^{n} \left(P_{ij} - T_i\right)^2}{\sum_{j=1}^{n} \left(T_j - \overline{T}\right)^2}} \tag{6}$$

where P_{ij} is the value predicted by the individual model i for record j (out of n records); T_j is the target value for record j and $\overline{T}$ is given by the formula:

$$\overline{T} = \frac{1}{n} \sum_{j=1}^{n} T_j \tag{7}$$

For a perfect fit, the numerator is equal to 0 and $Ei = 0$. So, the Ei index ranges from 0 to infinity, with 0 corresponding to the ideal.

- Relative absolute error (RAE): It is the ratio of dividing the absolute error by the magnitude of the exact value. The absolute error is the difference between the actual value and the individual measured value.

$$\text{RAE} = \frac{\sum_{i=1}^{n} |p_i - a_i|}{\sum_{i=n}^{n} |\overline{a} - a_i|} \quad (8)$$

where p_i is the predicted value, a_i is the actual value and $\overline{a}$ is the mean of actual value.

The study and machine learning approaches are individually executed for the confirmed cases, recovered cases and deceased cases. supported ensemble technique, the four eminent techniques are used bagging, boosting, stacking and voting, and also the learners or machine learning algorithms used are linear regression, SMOReg, multilayer perceptron, and Gaussian processes. Bagging and boosting are single learner and multiple model-based techniques whereas stacking and voting are multi-learners and multiple model-based techniques. Therefore, bagging is applied based on all four different machine learning algorithms, whereas stacking and voting are applied with a combination of the four algorithms mentioned above. The combination made with four ensemble techniques and four machine learning algorithms with their abbreviation used are depicted in Table 3.

The final result analysis is performed on all mentioned machine learning algorithms given in Table 3. The fourteen machine learning algorithms performance measures on 415 instances at a seventy-five per cent splitting magnitude relation is evaluated to forecast the COVID-19 cases for the future. The computed performance measures are displayed in Table 4 for confirmed cases, Table 5 for recovered cases and Table 6 for deceased cases. The bagging (LR) has the highest correlation coefficient for the confirmed cases and recovered cases whereas GP achieved maximum correlation coefficient for the deceased cases out of fourteen considered machine learning

Table 3 Machine learning algorithms with ensemble techniques considered with their abbreviation

Machine learning algorithms with ensemble techniques considered	Abbreviation used
Linear regression	LR
SMORegression	SMOReg
Multilayer perceptron	MLP
Gaussian processes	GP
Bagging with linear regression	Bagging (LR)
Bagging with SMORegression	Bagging (SMOReg)
Bagging with multilayer perceptron	Bagging (MLP)
Bagging with Gaussian processes	Bagging (GP)
Additive Regression with linear regression	AR (LR)
Additive regression with SMORegression	AR (SMOReg)
Additive regression with multilayer perceptron	AR (MLP)
Additive regression with Gaussian processes	AR (GP)
Stacking	Stacking
Voting	Voting

Table 4 Performances measures for confirmed cases

Performance measures		Correlation coefficient	RAE (in %)	RMSE (in %)
ML Algorithms	LR	0.9346	26.7179	27.7724
	SMOReg	0.9317	24.4405	29.1301
	GP	0.9306	43.3841	52.2524
	MLP	0.683	38.7354	54.853
Boosting	AR (LR)	0.9346	26.7179	27.7724
	AR (SMOReg)	0.9324	27.1425	28.2472
	AR (MLP)	0.6918	26.7593	28.385
	AR (GP)	0.9307	36.6108	53.6098
Bagging	**Bagging (LR)**	**0.935**	**26.8623**	**27.7637**
	Bagging (SMOReg)	0.9319	24.5127	29.0826
	Bagging (GP)	0.9304	43.1978	52.1221
	Bagging (MLP)	0.6792	39.0011	55.013
Stacking	Stacking	0.8598	37.9882	54.0218
Voting	Voting	0.9312	24.3315	37.7995

Table 5 Performance measures for recovered cases

Performance measures		Correlation coefficient	RAE (in %)	RRSE (in %)
ML algorithms	LR	0.9189	24.332	24.1871
	SMOReg	0.9081	19.1876	21.3047
	GP	0.9033	38.7099	44.954
	MLP	0.8088	24.5644	40.2967
Boosting	AR (LR)	0.9189	24.332	24.1871
	AR (SMOReg)	0.9149	24.5808	24.4571
	AR (MLP)	0.9058	23.1401	23.485
	AR (GP)	0.7659	22.6538	39.0271
Bagging	**Bagging (LR)**	**0.9191**	**24.6214**	**24.4468**
	Bagging (SMOReg)	0.908	19.468	21.4212
	Bagging (GP)	0.9031	38.5368	44.8262
	Bagging (MLP)	0.8089	24.6834	40.3546
Stacking	Stacking	0.8196	24.8108	40.3612
Voting	Voting	0.9076	15.403	26.0987

algorithms. Subsequently, the lowest coefficient correlation is recorded for bagging (MLP) in case of confirmed cases, AR (MLP) in cases of recovered cases, and AR (MLP) for deceased cases. The bold values signify the result from the respective table.

Table 6 Performance measures for deceased cases

Performance measures		Correlation coefficient	RAE (in %)	RRSE (in %)
ML algorithms	LR	0.8765	32.0263	32.3235
	SMOReg	0.8863	27.0231	27.1934
	GP	**0.8904**	**28.5749**	**35.4708**
	MLP	0.6385	23.7246	39.2046
Boosting	AR (LR)	0.8765	32.0263	32.3235
	AR (SMOReg)	0.8853	34.9649	35.3185
	AR (MLP)	0.8873	32.753	33.0696
	AR (GP)	0.5068	23.0704	38.96
Bagging	Bagging (LR)	0.8773	32.4311	32.7377
	Bagging (SMOReg)	0.8863	27.3764	27.5465
	Bagging (GP)	0.8902	28.3198	35.2786
	Bagging (MLP)	0.6372	24.1403	39.4287
Stacking	Stacking	0.7364	23.0592	38.5934
Voting	Voting	0.8811	17.2638	22.1684

The graphical representation of the other two performance measures that are mean absolute error (MAE) and root mean squared error (RMSE) for all fourteen machine learning algorithms evaluated on confirmed cases is shown in Fig. 4. Similarly, the

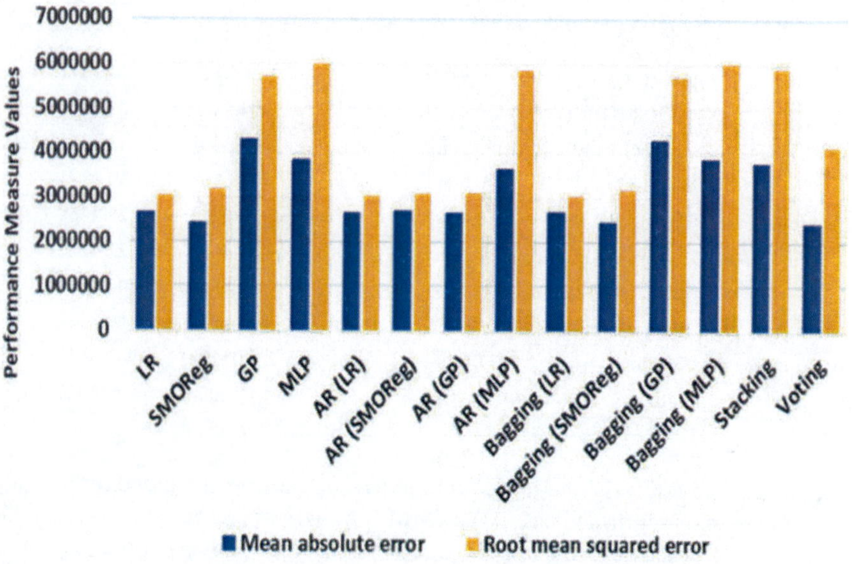

Fig. 4 Performance measure for different algorithms—confirmed cases

MAE and RMSE for recovered cases are depicted in Figs. 5 and 6 which represents deceased cases.

Based on the correlation coefficient, the top five algorithms for each case are illustrated in Table 7. *Hence, the bagging ensemble with linear regression classifier*

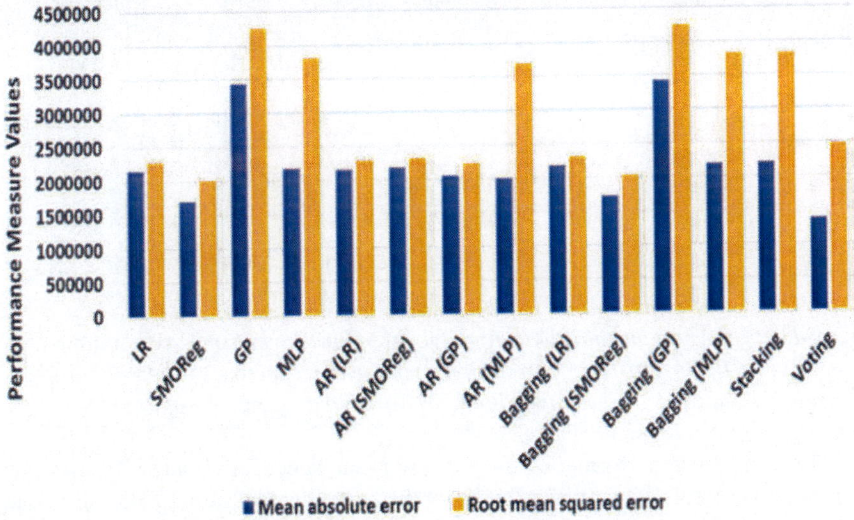

Fig. 5 Performance measure for different algorithms—recovered cases

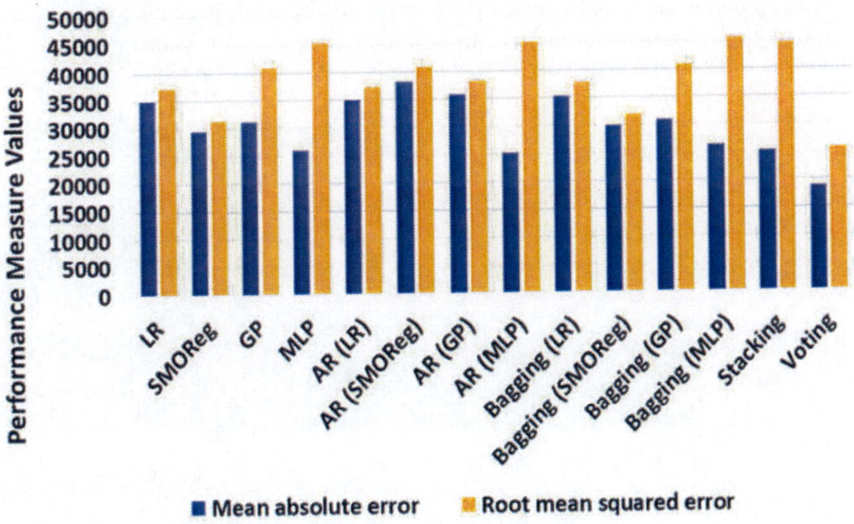

Fig. 6 Performance measure for different algorithms—deceased cases

Table 7 Top 5 algorithms based on correlation coefficient

Ranking	Confirmed cases		Recovered cases		Deceased cases	
	Machine learning algorithms	Correlation coefficient	Machine learning algorithms	Correlation coefficient	Machine learning algorithms	Correlation coefficient
1	Bagging (LR)	0.935	Bagging (LR)	0.9191	GP	0.8904
2	LR	0.9346	LR	0.9189	Bagging (GP)	0.8902
3	AR (LR)	0.9346	AR (LR)	0.9189	AR (GP)	0.8873
4	AR (SMOReg)	0.9324	AR (SMOReg)	0.9149	SMOReg	0.8863
5	Bagging (SMOReg)	0.9319	SMOReg	0.9081	Bagging (SMOReg)	0.8863

proved to be the best machine learning algorithm for the confirmed cases and recovered cases at 0.935 and 0.919 correlation coefficient, respectively, whereas Gaussian process proved to be the best machine learning algorithm in case of deceased cases with 0.8904 correlation coefficient.

The prediction is performed on the testing dataset; hence Figs. 7 and 8 illustrate the actual versus predicted confirmed cases by the best machine learning algorithm that is bagging (LR) for confirmed cases and recovered cases, respectively. Similarly, Fig. 9 illustrates the actual versus predicted deceased cases evaluated by the GP machine learning algorithm.

The second analysis is based on the comparison between ensemble techniques. The four ensemble techniques perform differently; hence, a comparative analysis is

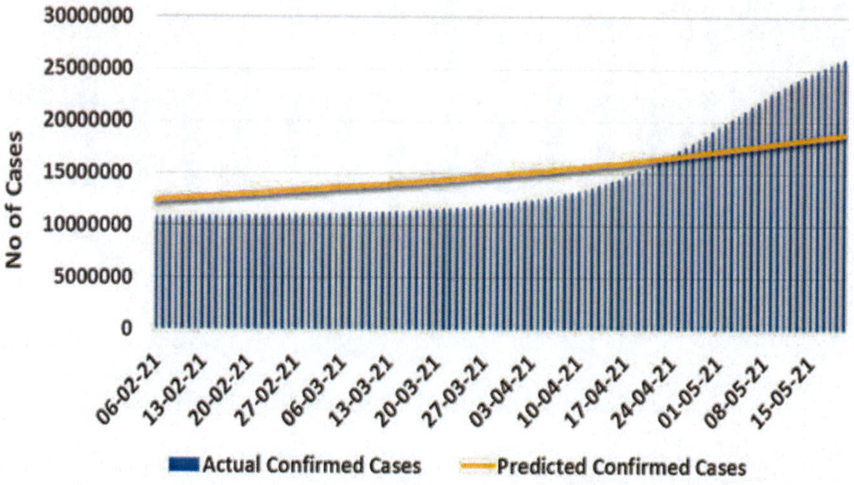

Fig. 7 Prediction analysis for testing data for Covid-19 confirmed cases—bagging (LR)

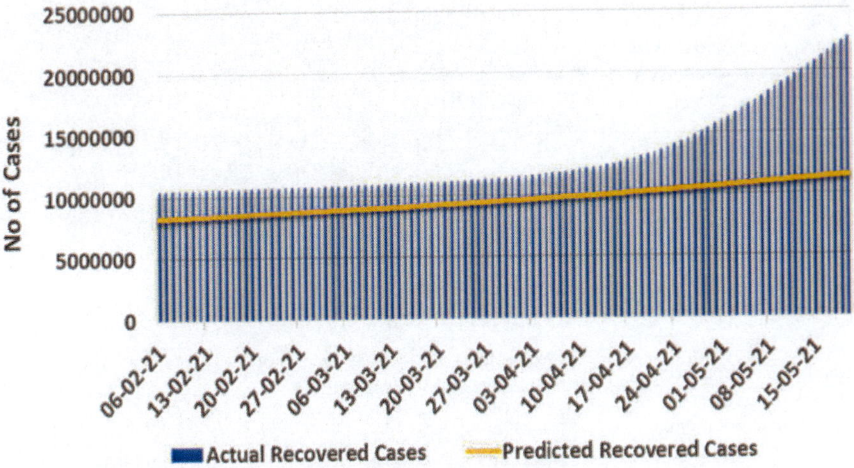

Fig. 8 Prediction analysis for testing data for Covid-19 recovered cases—bagging (LR)

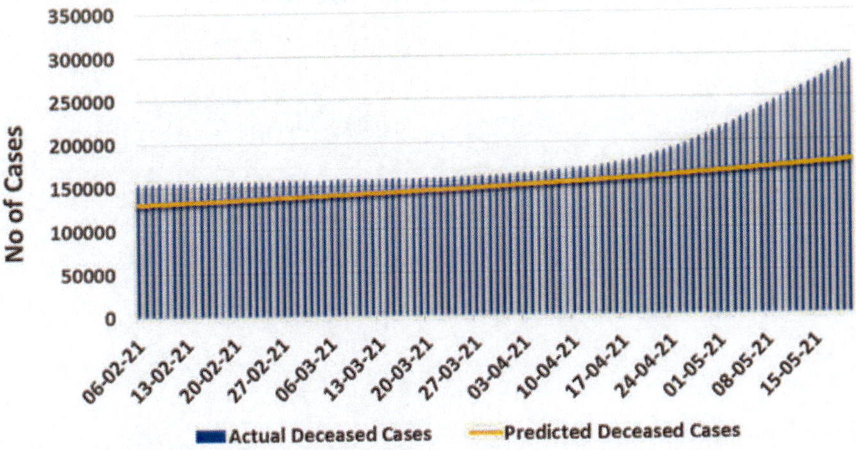

Fig. 9 Prediction analysis for testing data for Covid-19 recovered cases—bagging (LR)

accomplished to evaluate which ensemble method with which learner (in case of bagging and boosting) gave most optimized results for confirmed, recovered and deceased cases, respectively. Table 8 depicts the ensemble technique with their base learner algorithm in the case of confirmed cases with corresponding correlation coefficients. Similarly, Tables 9 and 10 determine for the recovered cases and deceased, respectively.

The bagging ensemble technique proves out to be the best ensemble technique for all the three cases. The boosting turns out to be the second-best ensemble technique in all the three cases, whereas stacking and voting stand out at last.

Table 8 Ensemble techniques comparison for confirmed cases based on correlation coefficient

Ensemble techniques	Base or learner algorithm	Correlation coefficient
Bagging (LR)	LR	0.9191
Boosting (LR)	LR	0.9189
Voting	LR, SMOReg, GP, MLP	0.9076
Stacking	LR, SMOReg, GP, MLP	0.8196

Table 9 Ensemble techniques comparison for recovered cases based on correlation coefficient

Ensemble techniques	Base or learner algorithm	Correlation coefficient
Bagging (LR)	LR	0.935
Boosting (LR)	LR	0.9346
Voting	LR, SMOReg, GP, MLP	0.9312
Stacking	LR, SMOReg, GP, MLP	0.8598

Table 10 Ensemble techniques comparison for deceased cases based on correlation coefficient

Ensemble techniques	Base or learner algorithm	Correlation coefficient
Bagging (LR)	GP	0.8902
Voting	LR, SMOReg, GP, MLP	0.8811
Boosting (LR)	GP	0.8873
Stacking	LR, SMOReg, GP, MLP	0.7364

5 Threats to Validity

This section discusses the threats to the validity of this research work. The research work used the ensemble method for analysing and predicting COVID-19 cases for the following ten days. The dataset considered is from 1 April 2020 to 20 May 2021. The prediction is done on assumption that data is accurate and patterns studied in past will continue to apply. However, the situation is unpredictable and uncertain; yet we believe predictive analysis of the situation using well-known methods might give an idea. Also, the analysis has been performed on a dataset of India, and therefore, it cannot be generalized as patterns may differ for different locations. Another factor is that the result analysis is drawn based on performance measures and more machine learning algorithms can be used for the prediction which might deflect the results. Prediction of COVID-19 cases may be performed using different machine learning algorithms or methodologies. The observations may vary as the base model of research changes, and for our research, we have adopted ensemble techniques for analysis and predictions.

6 Conclusion and Future Scope

An unprecedented situation created by COVID-19 has affected life and weaken the progress of the world. Therefore, forecasting would help in preparation to face the outbreak with efficient as well as sufficient resources. A deep analysis is performed on the pandemic situation rooted in India. The research work is performed and analysed on the COVID-19 situation from 1 April 2020 to 20 May 2021. Further, four different ensemble techniques, namely bagging, boosting, stacking and voting along with four distinct machine learning algorithms linear regression, SMOReg, Gaussian processes, and MultiLayer perceptron, are used. Thenceforth, the pre-processed dataset accumulated includes 415 instances on which 4 machine learning algorithms with their ensemble techniques are applied. The result from fourteen different combinations of techniques found out linear regression with bagging ensemble model showed the highest correlation coefficient of 0.935 and 0.919 for confirmed and recovered cases, respectively, and Gaussian process showed the highest correlation coefficient of 0.8904 in case of deceased cases. In addition to this, among different ensembles techniques, bagging proved out to be the best ensemble technique. Overall analysis showed ensemble techniques helps the machine learning algorithms perform better as they cover the input dataset efficiently by using multiple models. The COVID-19 outbreak has been uncertain, and moreover, the discovery of mutant virus further impacted the ongoing research and development, adding several challenges. Despite the uncertainty offered by the nature of the virus, deep analysis and predictions using effective models and approaches can assist in establishing one level of precautions and preventive measures that can eventually degrade the impact of the virus. One of the preventive measures is a vaccination that the Indian government also initiated on a large scale in February 2021. Our research work offers the massive potential in analysing as well as predicting future cases so that precautions and measures can be taken in advance.

Future Scope: More parameters for the prediction of COVID-19 cases can be included based on which the result may vary. For instance, the parameter of vaccinations done per day can be included to analyse the effects of vaccination on confirmed, recovered and deceased cases. The medical history of a patient is another parameter that plays a crucial role in determining recovery rate and the number of deaths, as COVID-19 affects a patient with medical history differently than a healthy individual. By efficiently considering more real-time parameters in the dataset, more accurate results can be achieved, and hence, model will be able to predict numbers close to real-world numbers. More robust machine learning techniques and ensembles prove to be more powerful. Therefore, with different base learners, machine learning algorithms and ensemble methods can be used in the future for the prediction of the pandemic disease COVID-19.

References

1. Sohrabi C et al (2020) World Health Organization declares global emergency: a review of the 2019 novel coronavirus (COVID-19). Int J Surg 76:71–76
2. World Health Organization (2020) Origin of SARS-CoV-2, 26 March 2020. No. WHO/2019-nCoV/FAQ/Virus_origin/2020.1. World Health Organization
3. Huang C et al (2020) Clinical features of patients infected with 2019 novel coronavirus in Wuhan, China. Lancet 395:10223:497–506
4. Chan JF-W et al (2020) A familial cluster of pneumonia associated with the 2019 novel coronavirus indicating person-to-person transmission: a study of a family cluster. Lancet 395(10223):514–523
5. Kar SK et al (2021) Second wave of COVID-19 pandemic in India: barriers to the effective governmental response. EClinicalMedicine 36
6. Jain VK, Iyengar KP, Vaishya R (2021) Differences between first wave and second wave of COVID-19 in India. Diabetes Metabolic Syndr
7. Dong E, Du H, Gardner L. An interactive web-based dashboard to track COVID-19 in real-time. Lancet Inf Dis 20(5):533–534. https://doi.org/10.1016/S1473-3099(20)30120-1
8. O'Neill A (2021) India—total population 2016–2026. Statista, 5 May 2021. www.statista.com/statistics/263766/total-population-of-india/
9. Bagcchi S (2021) The world's largest COVID-19 vaccination campaign. Lancet Infect Dis 21(3):323
10. Majhi R et al (2020) Analysis and prediction of COVID-19 trajectory: a machine learning approach. J Public Aff 21:e2537
11. Zu ZY et al (2020) Coronavirus disease 2019 (COVID-19): a perspective from China. Radiology 296(2):E15–E25
12. Nishiura H, Linton NM, Akhmetzhanov AR (2020) Serial interval of novel coronavirus (COVID-19) infections. Int J Infect Dis 93:284–286
13. Gupta R, Pal SK, Pandey G (2020) A comprehensive analysis of COVID-19 outbreak situation in India. MedRxiv
14. Kwekha-Rashid AS, Abduljabbar HN, Alhayani B (2021) Coronavirus disease (COVID-19) cases analysis using machine-learning applications. Appl Nanosci 2021:1–13
15. Barstugan M, Ozkaya U, Ozturk S (2020) Coronavirus (covid-19) classification using CT images by machine learning methods. arXiv:2003.09424
16. Tiwari D et al (2021) Pandemic coronavirus disease (Covid-19): world effects analysis and prediction using machine-learning techniques. Expert Syst
17. Mahfouz A et al (2020) Ensemble classifiers for network intrusion detection using a novel network attack dataset. Future Internet 12(11):180
18. Tiwari A, Prakash A (2014) Improving classification of J48 algorithm using bagging, boosting and blending ensemble methods on SONAR dataset using WEKA. Int J Eng Tech Res 2(9):207–209
19. Kotsiantis S, Pintelas P (2004) Combining bagging and boosting. Int J Comput Intell 1(4):324–333
20. Abdar M et al (2020) A new nested ensemble technique for automated diagnosis of breast cancer. Pattern Recogn Lett 132:123–131
21. Guidotti E, Ardia D (2020) COVID-19 data hub. J Open Sour Softw 5(51):2376. https://doi.org/10.21105/joss.02376

22. Wan X (2019) Influence of feature scaling on the convergence of gradient iterative algorithm. J Phys Conf Ser 1213(3)
23. Bhandari A (2020) Feature scaling: standardization vs normalization. Analytics Vidhya, 31 July 2020. www.analyticsvidhya.com/blog/2020/04/feature-scaling-machine-learning-normalization-standardization/
24. Smith TC, Frank E (2016) Introducing machine learning concepts with WEKA.Statistical genomics. Humana Press, New York, NY, pp 353–378
25. Jo J-M (2019) Effectiveness of normalization pre-processing of big data to the machine learning performance. J Korea Inst Electron Commun Sci 14(3):547–552
26. Frank E, Hall MA, Witten IH (2016) The WEKA workbench. Online appendix for "data mining: practical machine learning tools and techniques, 4th edn. Morgan Kaufmann
27. Baskin II et al (2017) Bagging and boosting of regression models. In: Tutorials in chemoinformatics, pp 249–255
28. Džeroski S, Ženko B (2004) Is combining classifiers with stacking better than selecting the best one? Mach Learn 54(3):255–273
29. Cheng M, Hsieh C-J, Dhillon I (2020) Voting based ensemble improves the robustness of defensive models. arXiv:2011.14031
30. Almarabeh H (2017) Analysis of students' performance by using different data mining classifiers. Int J Mod Educ Comput Sci 9(8):9

Imaging Radiomic-Driven Framework for Automated Cancer Investigation

M. Sreekrishna and T. Prem Jacob

Abstract Gliomas are the most ceaseless fundamental tumors in adults that cause genuine damage to the central tactile framework. The less forceful types of malady (e.g., second rate) in the clinical populace accompany a future of quite a long, while the more forceful variations (e.g., high evaluation) accompany a future of at most two years. For the two gatherings, magnetic resonance imaging (MRI) can give point-by-point pictures of the brain and is a standout among the most widely recognized neuroimaging conventions utilized when treated to uncover pieces of information about the infection qualities. In this undertaking division of brain tumors from MRI, datasets are of extraordinary significance for the enhanced conclusion, development rate expectation, and treatment arranging. The investigation in determining the healthy tissue from the tumor is a great challenge in medical image processing in terms of quantitative analysis. The biomedical tools are not suitable for extracting the quantitative data. When the gliomas tumor is in images, then radiomics can be used to extract valuable quantitative information for diagnosis and prediction of gliomas tumor. Automated feature learning from medical images can be extracted with the support of PyRadiomics an open-source platform. The quantitative features of radiomics are extracted from the MRI images that support in the detection of gliomas tumor automatically since it provides an excess of quantitative data. The generated data can be analyzed for both detection and prediction of gliomas cancer. In the proposed method, boundary extraction methodology is to perform segmentation on MRI images of brain cancer. This provides the visualization of data to find its correlation. The classifiers are trained and annotated with the clinical data for prediction. The statistics and the quantitative data that are obtained are processed to exhibit the relationship between the tumor and the normal one. The outcome supports in forecasting the tumor patients with the characteristic feature extracted from MRI images.

Keywords Image analysis · Quantification · Trained classifier · Detection · Tumor

M. Sreekrishna (✉) · T. Prem Jacob
Department of Computer Science and Engineering, Sathyabama Institute of Science and Technology, Chennai, India
e-mail: krishsree1212@gmail.com

M. Gupta et al. (eds.), *Artificial Intelligence on Medical Data*,
Lecture Notes in Computational Vision and Biomechanics 37,
https://doi.org/10.1007/978-981-19-0151-5_7

1 Introduction

Gliomas are a frequently occurring brain tumor that are found to be outrageous. A forthcoming report from the Tata Memorial Hospital gave one-year segment information and important tumor-related data on all patients enlisted in the neuro-oncology clinic. Gliomas established 38.7% (254 instances) of CNS tumors with 59.5% from high-grade gliomas including with 151cases, and 33.1% gliomas have a place with the framework of second rate which is surmised from79 cases. After dissecting oligodendrogliomas for 19 cases, anaplastic was induced for 7 cases, and 12 cases were grade-II. A large portion of these tumors were found in moderately aged guys. No investigation in regards to the drawn out clinical result of poor-quality gliomas has been distributed from an Indian place. Glioblastoma multiforme, the antagonistic type of malignancy, reports about the end of patients. As set up from western information, the foundation of the administration of these tumors has been a careful treatment, supported by radiotherapy. Around 13,000 passings and 18,000 new instances of essential dangerous mind and CNS tumors happen every year in the US expressed by the Central Brain Tumor Registry of the United States. The harmful cerebrum and tumor in central nervous system (CNS) among grown-ups matured 20 years and over in the US was 9.0 per 100,000 man a very long time during the year from 1998 to 2002 because of common age-changed incidence [1–5]. The 2016 World Health Organization (WHO) classification of tumors from the central nervous system (CNS) is both theoretical and commonsense. The classification depends on their minute similitudes with divergent putative cells of inference. The term gliomas shows to the tumor emerging from the gluey or strong tissue of the mind and addresses 27% of all cerebrum tumors and 80% of every single-threatening tumor. About 15.1% describe glioblastomas of all most significant mind tumors, and 55% represent gliomas. The American Brain Tumor Association anticipated that 12,120 new cases were assessed to have the most elevated number of cases with glioblastoma in 2016. It was identified that the average count of cancer was expected to be 14.5 lakhs and likely to reach 17.3 lakhs during 2020 by the Indian Council of Medical Research (ICMR). Globally, glioma, a particular brain cancer, was predicted mostly to mobile users [6, 7]. The analysis status at the International Agency for Research on Cancer (ICAR) with 31 scientists from various countries is discussed for long duration health concerns due to electromagnetic radiation discharged from mobile phones.

To improve the diagnosis of gliomas cancer, it is the forecasting of growth, and to manage the further treatment, segmentation of magnetic resonance images is of great importance [8, 9]. Due to tumor structure unpredictability and its imaging appearance of gliomas, automatic segmentation was challenging. To achieve this, a novel methodology was proposed to perform MRI images of gliomas segmentation automatically. The results achieved from the combination of random forest and active contour model were found to be more accurate than other segmentation techniques [10, 11].

Gliomas remain carefully testing cases because of complex life systems, including encompassing vasculature and the relationship to useful structures [12, 13]. To characterize the cancer feature related with forceful separate glioma evacuation just as its effect on long haul results and surveyed its part in deciding infection movement, dangerous change, and, at last, patient endurance. In any case, acknowledgment of the danger for gliomas at the underlying stage will help the anticipation and the board of cancer [14]. Early discovery and preventive consideration, for example, free affectation or improved well-being can incredibly decrease the danger of malignancy. The features of gliomas were found to be qualitative and quantitative. The non-statistical data was derived from qualitative data which cannot be generalized [15]. But, quantitative data generates sensitive random samples of data that are descriptive. Statistical analysis with systematic and standardization comparisons provides a more accurate result [16, 17].

2 Related Work

In 2011, the Lancet Oncology inferred intraoperative MRI is increasingly used in neurosurgery, although there is little evidence for its use. In 2015, National Center for Biotechnology Information infers that division of pre-usable second rate gliomas (LGGs) from attractive reverberation imaging is a significant advance for considering imaging biomarkers. Gliomas have sporadic tumor shape, heterogeneous arrangement, badly characterized tumor limits, and a set number of picture types. To get the improved question, division with the semi-mechanized cycle was set up which centers around T2-weighted (T2W) pictures. Pictures containing an enormous amount of information consume more measure of room. Quick preparing of those pictures sets aside more calculation effort for certifiable applications. Notwithstanding, ID and evaluation of the expected danger for gliomas at the early or pre-clinical stage itself would surely help the avoidance and the executives of disease. Consequently, it is fundamental to foster a non-steady representation-based framework, which is helpful to survey the lower limit issue. Also, early identification and preventive consideration, for example, free induction or improved well-being can extraordinarily diminish the danger of malignancy [18–20].

The machine learning algorithm focuses on the unsupervised learning that performs anomaly detection by searching the similar datasets available. The intensities of the image variance are taken into consideration. With characterization, profound learning can set up connections between, say, pixels in a picture and the name of concern. This can be a static forecast. The sign produced was stripped to examination among the correct information, where profound learning can set up connections between the current occasions and the future occasions.

Figure 1 shows the general processing of data. Example framework dependents on vector machine classifiers and quantitative highlights separated from MRI pictures, the utilization of this example involves in order techniques that is used to recognize various sorts of cerebrum tumors, like essential. This takes various PyRadiomics

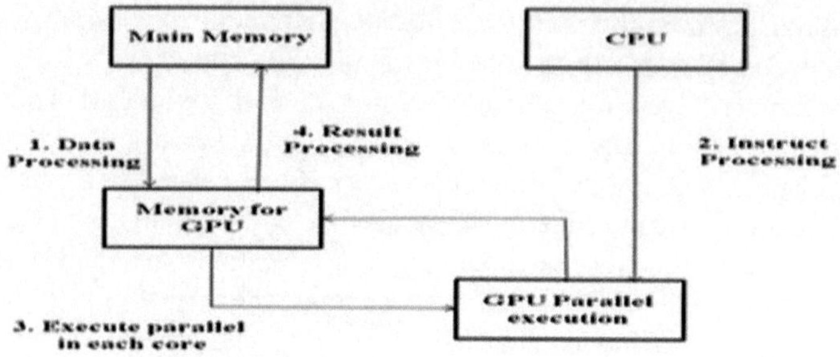

Fig. 1 Processing of generated data

properties that can take a million images to perform clustering based on the existing properties. Automatic extraction enables us to obtain many input layers producing an output.

3 Proposed System

The most common gliomas brain tumor diagnosis ability can be improved by performing segmentation automatically. The radiomic feature like shape and texture feature is extracted. This supports in prognosis identification of gliomas that support in applications of the medical system. Preprocessing and automatic segmentation of an image and extraction of radiomic features help in feature selection that improves the investigation in analysis of quantitative data.

3.1 Preprocessing of Gliomas Image

Image preprocessing involves converting the original image into a grayscale component which retains the grayscale color map as shown in Fig. 1. Median filters are used to perform the preprocessing generally based on average pixel value. It supports in preserving the edge. The unwanted distortions are eliminated that increase the brightness of the pixel. Upon increasing the brightness, the detection of pixel intensity features, their interrelationships, and their distribution which are changing abruptly can be identified (Fig. 2).

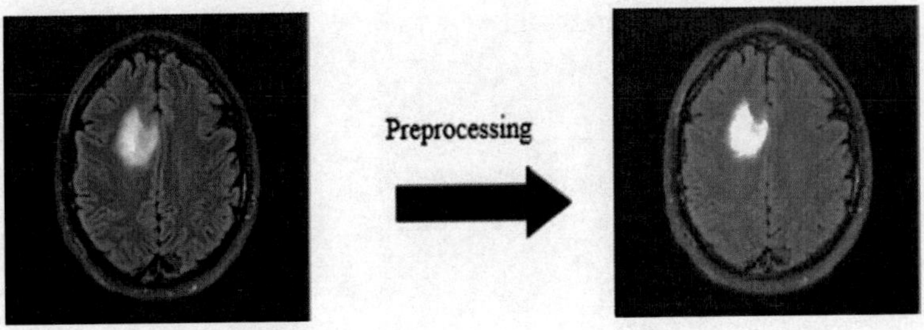

Fig. 2 Preprocessing using the median filter

3.2 Image Segmentation

Extracting the similar feature from different part of an image is processed in the segmentation. To improve the visualization of the medical image and to make the efficient image structure measurement quantitatively, segmentation plays a major role. This is done consequently that includes dividing an advanced picture into numerous portions. These sections are additionally called a bunch of pixels or picture objects. These progressions give the yield picture which is more significant and simple to dissect. Segmentation is applied here in order to view a clear picture of brain areas so as to detect the tumors if it is present inside portion of the brain (Fig. 3).

Using watershed segmentation and Sobel edge detection, the final output with tumor portion is extracted for process the radiomic feature of the gliomas. Figure 4 shows the histogram of various gliomas cancer image. This is mainly applicable for grayscale images that have brightness which is uniformly placed, and it includes histogram features. The left half and the right half of image's histogram are calculated (Table 1).

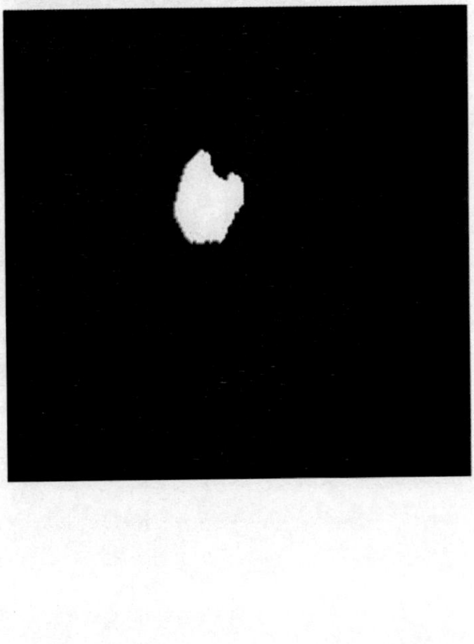

Fig. 3 Automatic segmentation

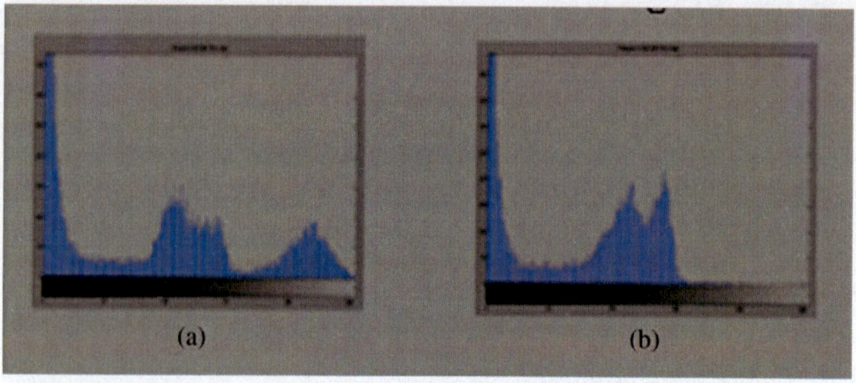

Fig. 4 Images **a** and **b** show the histogram of partitioned image

Table 1 Sampling the histogram of various gliomas cancer

	Original image	Left half image of Brain	Right half image of Brain	Left half image histrogram (Pixel count Vs Pixel value Curve)	Right half image histrogram (Pixel count Vs Pixel value Curve)
1	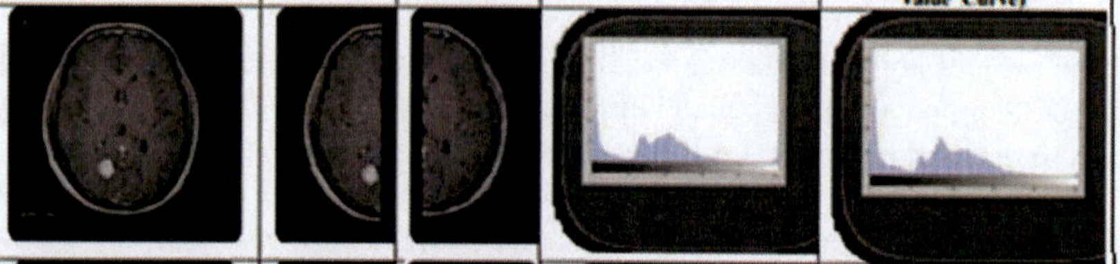				
2	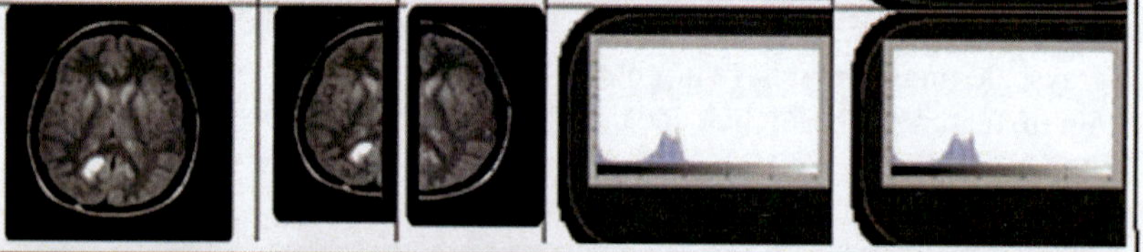				
3	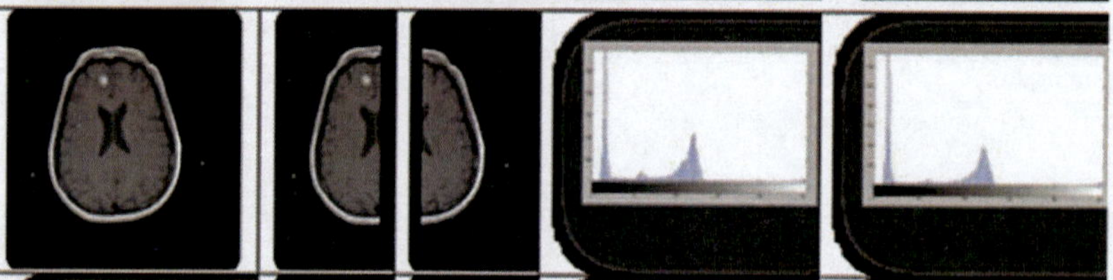				
4	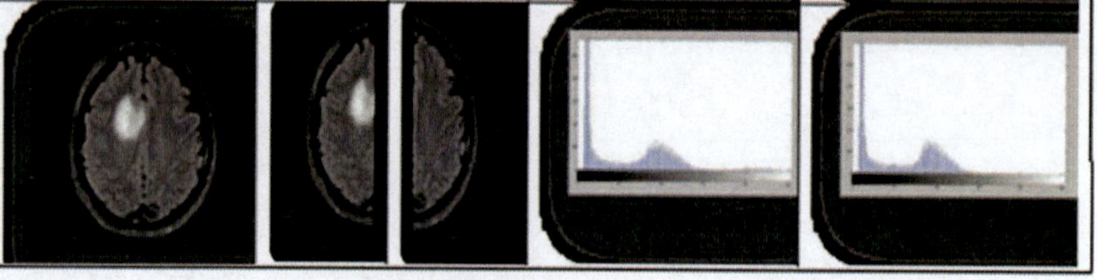				

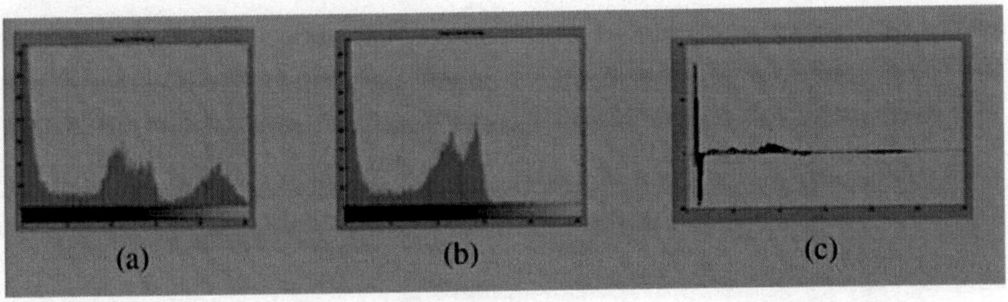

Figs. 5 **a** and **b** show the histogram for the partitioned image, and Image **c** shows the histogram curve having pixel count and pixel value curve

3.3 Radiomic Feature Extraction

Ahead of time with clinical imaging, radiomic includes extraction, and assume an essential part in getting the subtleties of power and surface qualities. This features support in generating the predictive model for diagnosis. They measure the heterogeneity in the region of interest from the segmented image. Now, the image is divided into two equal parts after segmentation. This supports in identifying the tumor region side that has a maximum size of gliomas tumor area calculated as 2143. This partitioning detects the brain side that is highly infected. The threshold value can be set at once the histogram is calculated.

3.4 Histogram Threshold Value Processing

The pixel and the intensity of the pixel are taken under consideration to plot the histogram. The plotting of the histogram is done, and the comparison of both the histogram output is processed in order to set the threshold limit. The pixel intensities are set between 0 and 255 as shown in Fig. 5 (Table 2).

3.5 Predicting the Radiomic Framework

The above histogram results are an important feature technique to analyze the tumor with segmentation and thresholding. The area of the tumor and its measurement play a vital role in analyzing the tumor. Since the brain is divided into two equal half, the left and right side analyses can be made accurate and support in finding the grade of the tumor. In the segmentation, the related affected area similar to the infected area alone can be measured without taking into account the normal area. The statistical analysis can be made accurate in each pixel of the brain, and variation prediction is compared. Upon checking with the histogram threshold graph, medical investigators

Table 2 Histogram threshold curve for the chosen sampling dataset, all the pixels are read, and when it exceeds the pixel intensity value, the incrementing of specific value is performed

	Original image	Histogram thresholding	Histogram thresholding Curve (Pixel count Vs Pixel value Curve)
1.	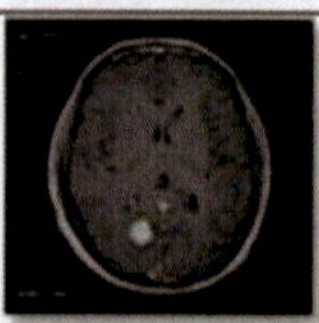	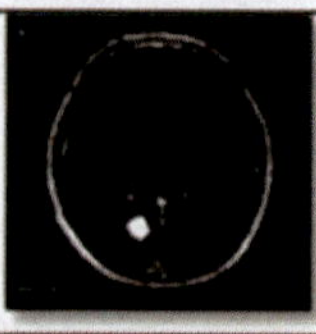	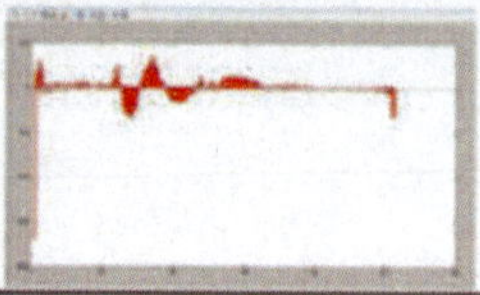
2.	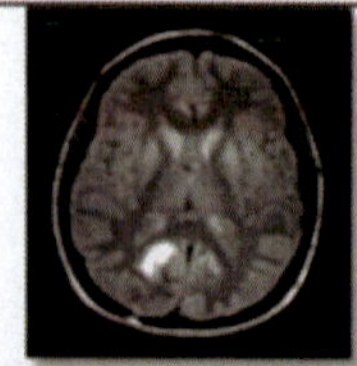	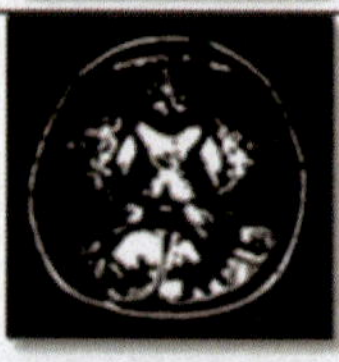	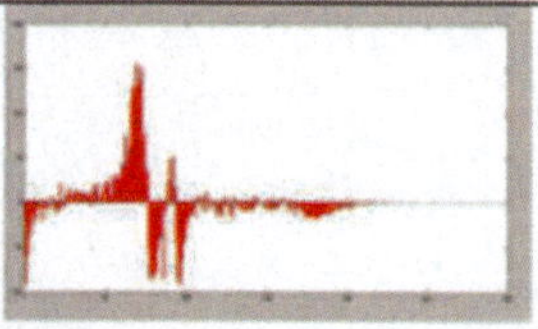
3.	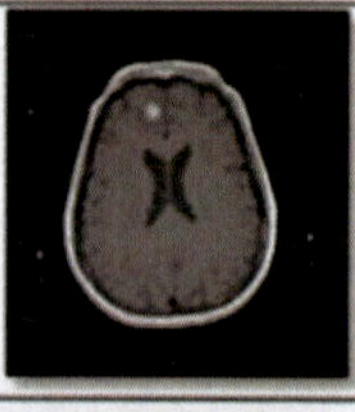	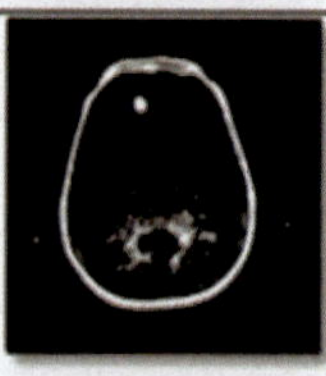	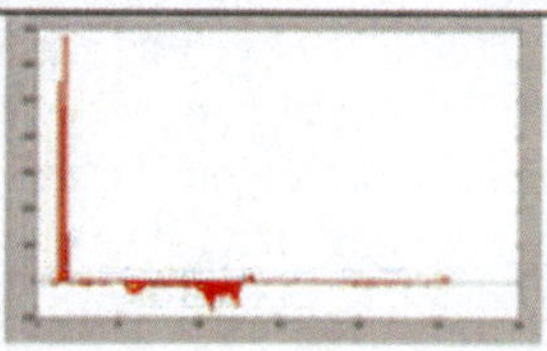
4.	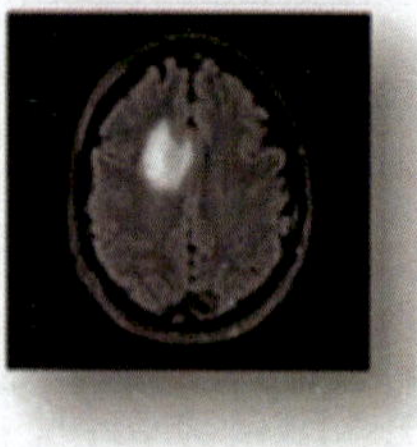	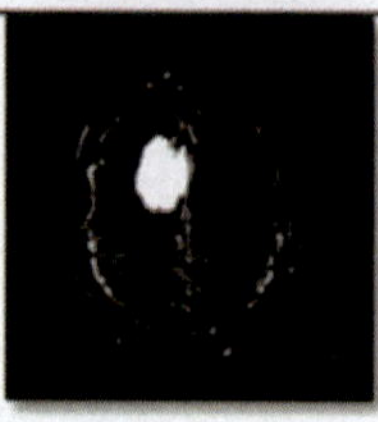	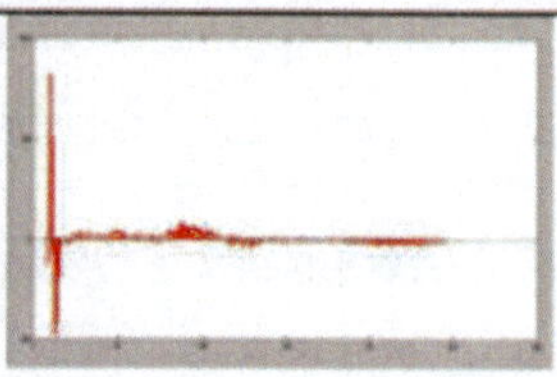

can analyze the value of the near future with the present tumor characteristics, and treatment can be made accordingly.

4 Conclusion

Thus, the extracted results are classified based on the quantitative data generated from the histogram which can be analyzed to find the next step of the comparison in the development of the known small-scale brain tumor. The currently obtained result involves calculating the tumor cell in its growth rate and its removal with further surgical studies. The tumor doubling time can be identified in the sensitivity of different locations that can be concluded. Thus, in the future, the obtained results can be compared with the other image dataset to perform classification. Maintaining those sensitive data as a whole for various patients is difficult. Further, the data obtained needs to be maintained for precision medicine in the future. This can be done with the support of any automation tool that can support the doctor and patient for ease of evaluation. The large dataset analysis can be processed with high-performance computing.

References

1. Islam A, Reza SMS, Iftekharuddin KM (2013) Multi-fractal texture estimation for detection and segmentation of brain tumors. IEEE
2. Shi X et al (2010) Detection and classification ofmasses in breast ultrasound images. In: Digital signal process. Elsevier, pp 1–12
3. Lin C, Wang S (2004) Training algorithms for fuzzy support vector machines with noisy data. Pattern Recogn Lett 25:1647–1656
4. Srimani PK, Mahesh S (2012) Knowledge discovery process in the image-segmentation data. IJKE 3(2):188–192
5. Canny J (1986) A computational approach to edge detection. IEEE Trans Pattern Anal Mach Intell 6:679–698
6. Beucher S, Meyer F (1992) The morphological approach to segmentation: the watershed transformation. Opt Eng 34:433
7. Borsotti M, Campadelli P, Schettini R (1998) Quantitative evaluation of color image segmentation results. Pattern Recogn Lett 19(8):741–747
8. Zhang YJ (1996) A survey on evaluation methods for image segmentation. Pattern Recogn 29(8):1335–1346
9. Agn M, Puonti O, Law I, Rosenschöld PMA, van Leemput K (2015) Brain tumor segmentation by a generative model with a prior on tumor shape. In: Proceedings of the multimodal brain tumor image segmentation chall, Munich, Germany, 5–9 Oct 2015, pp 1–4
10. Mohana Prasad K, Sabitha R, Oviya (2016) An efficient clustering formulation from resemblance in extant algorithms. ARPN J Eng Appl Sci 11(13):8278–8283. ISSN 1819-6608
11. Sethuraman R, Anand CDN, Sumanth G (2016) Multi-channel communication system for healthcare domai. Indian J Sci Technol 9:44
12. Posonia AM, Jyothi VL (2013) Structural-based clustering technique of XML documents. In: 2013 international conference on circuits, power and computing technologies (ICCPCT). IEEE, pp 1239–1242
13. Rajkumar V, Vikram S, Kamalesh (2016) Simplified interest classification using social media. ARPN J Eng Appl Sci 11(15):9379–9381
14. Reddy GM et al (2018) User based efficient video recommendation system intelligent data communication technologies and internet of things (ICICI 2018), vol 26. Springer, pp 1353–1362
15. Pradeep K, Jacob TP (2018) GHSA: task scheduling in heterogeneous cloud. In: International conference on intelligent data communication technologies and internet of things. Springer, Cham, pp 1202–1207
16. Rani AJM, Pravin A (2019) Multi-objective hybrid fuzzified PSO and fuzzy C-means algorithm for clustering CDR data. In: 2019 international conference on communication and signal processing (ICCSP). IEEE, pp 0094–0098
17. Govidan N, Rajasekaran Indra M (2018) Smart fuzzy-based energy-saving photovoltaic burp charging system. Int J Ambient Energy 39(7):671–677
18. Indra, MR, Govindan N, Satya RKDN, Thanasingh SJSD (2021) Fuzzy rule based ontology reasoning. J Ambient Intell Human Comput 12(6):6029–6035

19. Nagarajan G, Minu RI, Jayanthiladevi A (2019) Brain computer interface for smart hardware device. Int J RF Technol 10(3–4):131–139
20. Nirmalraj S, Nagarajan G (2021) Biomedical image compression using fuzzy transform and deterministic binary compressive sensing matrix. J Ambient Intell Human Comput 12(6):5733–5741

Neoplastic—Nuclei Cell Labelling Using Mask R-CNN

R. I. Minu, G. Nagarajan, Ankit Prabh, and Jenendar Bishnoi

Abstract Neoplastic cells are tumorous cells that damage the cells around them and are the prologue for cancer development in organs. However, identifying these cells poses a bottle-neck in the research of cancer cure as it is an extremely tedious job to manually isolate these from the rest of the cells in the tissue. Hence, the automation of this process using deep learning (DL)-based object detection and segmentation techniques such as Mask R-CNN will allow researchers and pathologists to save valuable time otherwise consumed in manually identifying these nuclei. The main objective of this research paper is to provide an instance segmentation technique to label and segment neoplastic cell nuclei from multiple instances of whole-slide images (WSI). For this process, a contemporary neural network architecture called the mask region-based convolutional neural network (Mask R-CNN) was used. This proposed technique generates a pixel-wise binary mask. These masks are capable of segmenting these instances and facilitating the advancement of intelligent systems in medical imaging and computational pathology. This time can instead be devoted to developing better cures by conducting more research. The paper also highlights the best techniques and practices that can be employed while training a model for a task of such complexity. The results of these techniques provide a mean average precision (mAP) score of 0.756 and a binary panoptic quality (bPQ) score of 0.675.

Keywords Medical imaging · Image processing · Neoplastic cell · Deep learning · Computer vision · Segmentation · Computational pathology · Mask R-CNN · Cancer research

R. I. Minu (✉) · A. Prabh · J. Bishnoi
Department of Computing Technology, SRM Institute of Science and Technology, Kattankulathur, India
e-mail: minur@srmist.edu.in

G. Nagarajan · A. Prabh
Department of Computer Science and Engineering, Sathyabama Institute of Science and Technology, Chennai, India

University of Pennsylvania, Philadelphia, PA, USA

M. Gupta et al. (eds.), *Artificial Intelligence on Medical Data*,
Lecture Notes in Computational Vision and Biomechanics 37,
https://doi.org/10.1007/978-981-19-0151-5_8

1 Introduction

According to the 2018 Data Science Bowl [1] competition, cell nucleus identification is one of the first steps in the long procedure of synthesizing cures for diseases. This is because DNA of each cell is present in the nucleus and analysis of that determines how quickly and effectively a cure can be made. Researchers and pathologists spent hours trying to identify the required cell nuclei to gauge the level of progress made. Each time, a new experiment is conducted. If this process can be automated, it can allow researchers to track the cells and measure the level of treatment on them, significantly improving the number of experimentations that can be conducted in a particular time span, and consequently, new and better cures for diseases such as cancer can be developed at a rapid pace.

With that being said, the process of cell identification and labelling is a complex task to automate, as stated by [2, 3]. Firstly, the shape and size of each type of cell vary greatly. Secondly, sub-types of certain cells such as red blood cells and white blood cells are present which induce an added layer of complexity in the process of cell identification. Thirdly, the various staining techniques that are used to visualize cells in computational pathology (CPath) can generate varied results when coupled with different algorithms. Lastly, due to the complex structure and arrangement of cells, a lot of the substantial ones can be hidden or overlapped, resulting in outliers.

Deep learning (DL), in recent years, has achieved impressive results in the field of computer vision (CV) and image processing where certain models have outclassed even humans. The ILSVRC [4] has facilitated significant contributions in the development of convolutional neural networks (CNNs) which have completely changed the paradigm of CV from the ages of hand-engineered algorithms to complete end-to-end trainable DL models. Szegedy et al. [5] proposed the CNN micro-architecture called as the "inception module" which allowed CNNs to be trained to greater depths yielding greater accuracies on complex problems. In this, they branched the inputs to the module into different-sized kernels and later concatenated the feature maps depth-wise. Gaining inspiration from the micro-architecture models [6] and later [7] were designed by Kaiming He et al. These worked on the concepts of residual blocks and identity mappings where the input to a series of convolution operations was added to the result of those operations. This solved the problem of vanishing gradients and allowed CNNs to be trained to depths of 100 or even 1000 layers. Today, CNNs have massively increased in strength and are able to tackle complex applications such as medical image diagnosis of novel diseases [8].

Mask R-CNN [9] is one such modern deep learning architecture which is able to perform object detection and image segmentation, allowing detected objects to not only be localized in images using a bounding-box but also be segment able by generating a binary mask of the particular instance of that object. It combines novel region of interest (ROI) generation networks such as the region proposal network (RPN) with a convolutional neural network (CNN) feature extractor and region detector by using shared parameters across the convolutional operations. This makes the entire model end-to-end trainable and reduces the time for inference while still giving robust detection results. A model trained on this state-of-the-art (SOTA) paradigm would be:

1. Capable of generalizing to different tissue types and detect neoplastic cells in WSI all on its own.
2. Robust and accurate
3. Able to localize the neoplastic cells with a binary mask which can further be used to extract cell data for post-processing
4. Able to save a lot of time for histopathology researchers and speed up cure development.

This paper gives a detailed description of the dataset used to train our model in Sect. 3. This section will also describe the different modules of our implementations and highlight our model training techniques. Section 4 will describe the results obtained, and Sect. 5 will present our conclusion and remarks on this project.

2 Related Work

Before DL [10] made its way through to medical imaging and cell segmentation, hand-engineered algorithms were used. Sankaran and Asari [11] used an adaptive thresholding technique to distinguish cell boundaries and applying a low-pass pre-processing filter, while [12] proposed morphological techniques to enhance the existing water-shed methods to detect cells. GrabCut [13] was also a popular choice for foreground segmentation.

As DL advanced, it was applied to more complex applications such as [14]. Ciresan et al. [15] used a deep CNN model that was able to classify each pixel of an image based on a pixel-patch from its neighbours. Long et al. [16] employed the use of fully convolutional networks to train semantic segmentation model which combined the abstract features from deeper layers with the local pixel-level features of the initial layers. This model gave better results than [13] and made it operational towards arbitrary dimensions. Drozdzal et al. [17] further improved the accuracy and lowered overfitting of the U-Net model combining short skips similar to that of ResNet [7] with the long skips. Chen et al. [18] proposed a hybrid model for faster and robust segmentation of small lung cancer. This model combined 3D convolutions which learned long-ranged 3D features along with the 2D convolutions which learned the short-range local features. A similar approach was used by [19] where they used a set of mixed convolutional blocks to better generalize on the dataset. It also made sure that their model adjusted quickly to the varying scenarios without compromising on accuracy.

The SOTA Mask R-CNN was built after a series of improvement to the CNN-based object detection system. In the year 2014, Girshick et al. published a paper [20] which was ground-breaking in the field of object detection. This model was called R-CNN and used selective search techniques to extract region proposals to detect objects. The proposed regions would then be converted into feature maps and classified using a SVM-based approach. The drawback of this method was its speed, so they proposed another faster system called as the fast R-CNN [21]. This model was end-to-end trainable and used a region of interest (ROI) pooling layer to filter out the large number of ROIs generated by selective search. Later, Girshick et al. published another improvement to the fast R-CNN model which was named faster R-CNN [22]. In this, they came up with two innovations which were baking the region proposal network (RPN) into the architecture itself which eliminated the use of selective search and used anchors to generate proposal regions. The RPN used a set of two convolutional kernels to calculate and separate the foreground from the background and generate a bounding-box for the image. This work was further improved to generate the Mask R-CNN model.

In recent years, Mask R-CNN has become a highly potent choice for segmentation models. Kurnianingsih et al. [23] used this to segment cervical cancer cells. Their model used a shallow base network for faster inference speeds. Ma et al. [24] used a Mask R-CNN to segment abnormal cells by adding attention-based mechanisms to their model to generate fixed size ROIs. These ROIs were then combined with the original ROIs to generate more accuracy masks. A similar effort was made by [25] to utilize attention-based activation maps to train on partially labelled cell datasets without losing out on accuracy.

3 Method

This section describes the methodology used in the paper. It gives a detailed description of the dataset, model architecture, training techniques and hyperparameter values in the following subsections.

3.1 Dataset Explanation

Powerful CNN architectures almost always manage to overfit the data given to them due to surface statistical irregularities [26] and subsequently do not generalize well in real-life scenarios. Hence, training a DL model for the task of cell nuclei segmentation should be facilitated by a generalized dataset of cells which enables the model to perform with the same efficiency regardless of the cell type and tissue area while not localizing to a single region of the body. Looking at and considering all the challenges associated with training a model for nuclei segmentation, a conclusion that the dataset must be just as comprehensive as the model was reached. Hence, this paper works with the PanNuke dataset [27] to train the model. This dataset as briefly illustrated by Fig. 1 consists of 200,000 labelled nuclei across five categories: neoplastic, non-neoplastic, epithelial, inflammatory, connective and dead. The speciality of this dataset is that it does not localize the data to one part or organ of the body but rather encompasses 19 different body tissues types. This pervasiveness allows the model

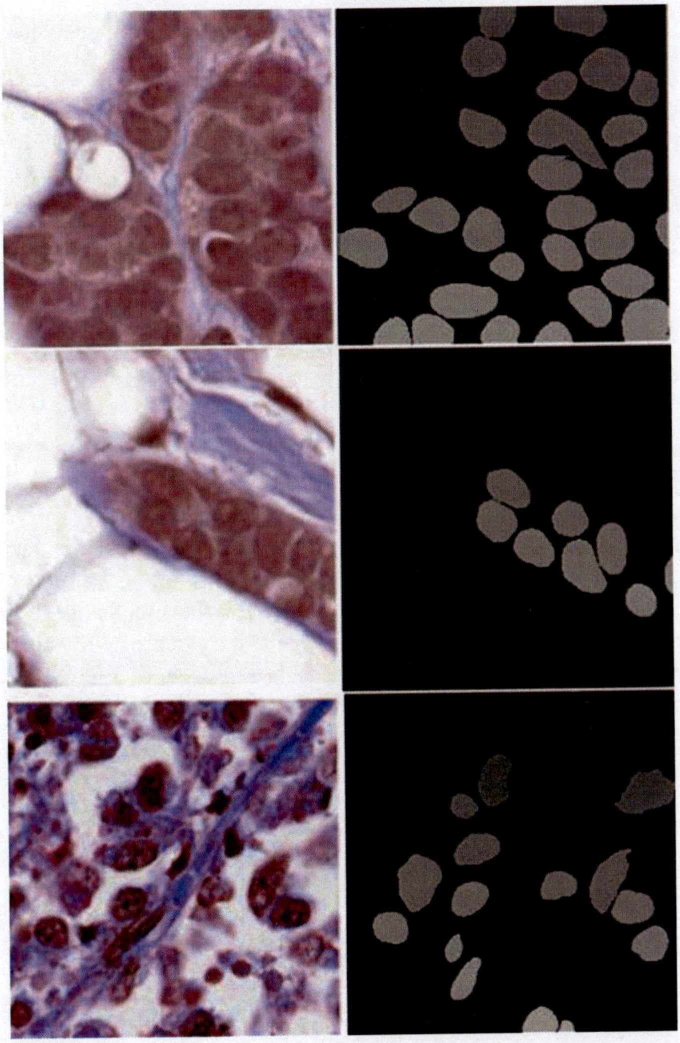

Fig. 1 Collage showing the cell images (left) and instance-wise segmented ground truth mask or neoplastic nuclei (right)

to combat overfitting. Furthermore, since the dataset takes into account different body tissues, it saves the tedious process of training multiple models for different tissue types. According to [28], nuclei detection done in is this manner also facilitates tissue phenotyping in an effective manner. For the purpose and scope of our research, we decided to work with only labelled neoplastic nuclei. After performing data pre-processing and extracting the neoplastic cells, we obtained a substantial number of labelled nuclei to train our model with. We decided to split 80% of the

Table 1 Describes the number of images and their splits used for training, validation and testing

	Training	Validation	Testing	Total
Images	2950	750	490	4190
Labelled cell nuclei	67,171	14,434	11,258	92,863

data for training and 20% for validation after randomly sampling 490 images as the test set. Detailed descriptions of our splits are given in Table 1.

3.2 *Model Architecture*

In this subsection, a brief description of the architecture of the Mask R-CNN model (Fig. 2) and highlights of some of its features that enable it to perform the segmentation task are given.

Backbone Network: The base network is responsible for feature extraction from images, so it should be a fully convolutional neural network of any architype such as ResNet or GoogLeNet. At the moment, looking at the complexity of the task, a ResNet 101 backbone is used. While it is not mandatory to use a fully convolutional network as at any time in the forward propagation process of a CNN one can just stop at a certain convolutional layer and retrieve the feature maps, it comes with the added advantage of accepting images of any spatial dimensions.

Region Proposal Network (RPN): The RPN is a small fully convolutional neural network that acts on the feature maps and generates proposals to feed into the deeper

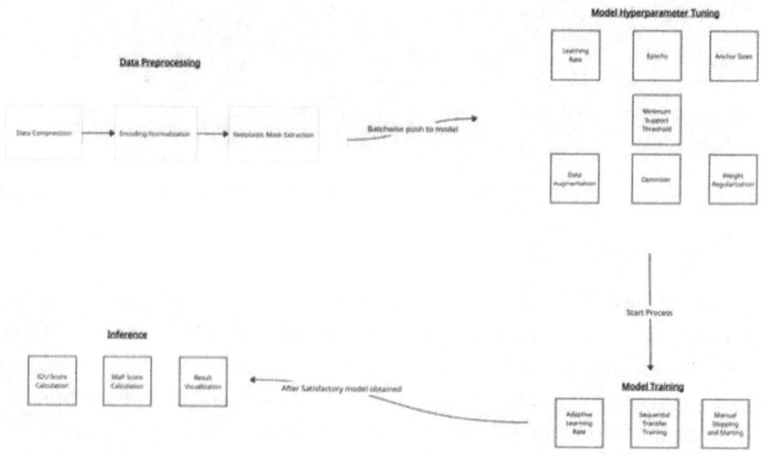

Fig. 2 Flow diagram describing the various modules of the Mask R-CNN model used

layers of the model. The RPN uses a set of anchors to generate these proposals, and based on the scale and aspect ratio, a total of nine anchor points can be generated to crop proposal regions from the feature maps. This works similar to and thus alleviates the needs of having a complex feature pyramid to detect objects at varying scales. The RPN also generates an objectiveness score for each region and classifies it as either positive, negative or neutral. Based on this score, top n proposals are selected and propagated to the next layer.

Region of Interest Align Layer: This layer is an improvement over the ROI pooling layer used in the faster R-CNN structure. The job of this layer is to further reduce the ROIs from the RPN by cropping out fixed size feature vectors from the feature maps. This layer also facilitates the binary mask generation by aligning the proposals with the ground truth masks. After this, the feature vectors propagate through three parallel layers.

Fully Connected Branch (FC): This branch includes two out of three parallel propagation paths. These are dense, fully connected layers which provide two outputs at the end. One of them of size $n + 1$ where n is the number of classes (one is added for background) provides the class labels for the dataset. The second FC output is of size $4 \times n$ and provides bounding-box coordinates, namely $(x1, y1, x2, y2)$.

Convolutional Branch: This branch was proposed as an overhead improvement to the faster R-CNN model by Girshick et al. in [9]. This is a parallel path from the ROI align layer which passes through a set of convolutional layers and outputs a binary mask at the end of it. The mask is then interpolated using nearest neighbour technique and overlayed on the input image.

3.3 Implementation

To simplify the implementation and reduce complexity and time, Matterport's Mask R-CNN implementation [29] is used, and the base classes are morphed as needed. This subsection highlights all the training techniques and hyperparameter choices made to optimize the model. Note that the loss metrics used to train all aspects of the model (bounding-box regressors, class probabilities and mask generation) are used exactly as stated in [9] and are not mentioned explicitly. Stochastic gradient descent [30] optimizer with a momentum value of 0.9 was used. To train the model, transfer learning [31] was used in the backbone network. To facilitate this, a pre-trained Mask R-CNN model which was trained on the MS COCO [32] dataset was used. While this dataset has no affiliation with cells, it still helps the CNN backbone network learn the features faster as the initial layers regardless of the dataset learn similar kinds of rudimentary low-level features such as lines and shapes. While using transfer learning, first only the heads of the network encompassing the feature pyramid network, RPN and the Mask R-CNN bounding-box and mask heads were trained for 20 epochs. This facilitates smoother overall training and lets the whole model in the later stages to make better updates. Later, the entire model along with the backbone was trained for 50 + epochs till the training plateaued.

Traditionally, while training model of such complexity, the learning rate (LR) is dropped by a factor of 10 each time the losses start to plateau. While this is very effective, a custom learning rate decay function was used in this paper which showed improved results in our experiments against using a sharp learning rate decay. A simple linear decay function which decayed the learning rate by a factor of approximately 16 between two sets of epochs was used. While training the heads, the LR was dropped linearly from $1e-3$ to $9.9e-5$ over 20 epochs. With the whole network training, the LR was dropped from $1e-4$ to $3.33e-6$ over 30 epochs. The formula of the learning rate decay is shown in Eq. 1.

$$\text{end}_{\text{lr}} = \text{start}_{\text{lr}} * \left(1 - \left(\frac{\text{current}_{\text{epoch}}}{\text{final}_{\text{epoch}}}\right)\right)^{\text{power}} \tag{1}$$

In Eq. 1, end_{lr} is final decayed LR, start_{lr} is current LR at epoch, $\text{current}_{\text{epoch}}$ is currently running epoch, $\text{final}_{\text{epoch}}$ is max epoch to train model to, power is 1.0 (for decay to be linear in nature). Since a ResNet101 backbone was used, the model started to overfit after the commencement of the second stage of training, irrespective of the size of the dataset. This is where some strong regularization techniques were used to train the model. A weight decay [33] value of $1e-4$ and a gradient clipping normalization value of 5 were employed. Furthermore, data augmentation was used to improve validation accuracy at the expense of training accuracy. An image augmenter randomly flipped the training data along the horizontal and vertical axis and also rotated it by 10°. This level of augmentation is possible to use in the Mask R-CNN model because of its translation invariance property [9]. Also, since the masks are aligned by the ROI align layer at the end of the model, we decided to use a mini-mask function which resizes the mask to a smaller dimension to reduce computational power.

4 Results

Initial experiments: This subsection highlights the initial experiments conducted with the Mask R-CNN model. At the beginning, the model was trained in three parts heads, ResNet 4 + layer modules and the full model. Later, it was found that training the model in two parts only gave better accuracy. Nevertheless, these results are included to show the effects of using augmentation even though the training data is sufficiently large. Figure 3 shows two stages of the training process. The training process without augmentation (Fig. 3a) overfits by epoch 37, while the process using augmentation (Fig. 3b) does not gives considerably good results. This experiment proved the potency of data augmentation, which was used in the final experiment as well.

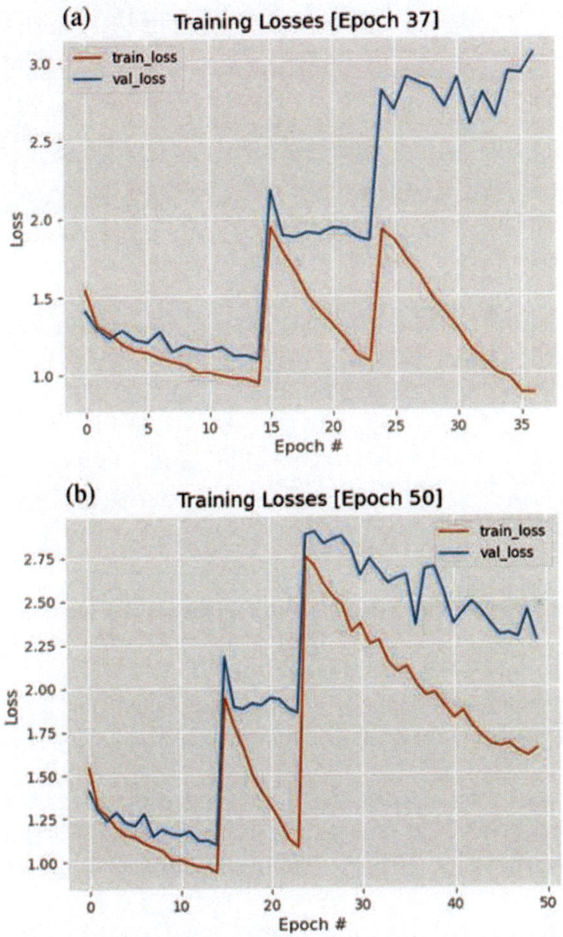

Fig. 3 **a** Graph showing training results without the use of augmentation techniques. **b** Graph showing training results with the use of augmentation techniques

Final results: Fig. 4 shows the final experiment conducted which bores the best results. As shown, raining the model in a two-stage process generates better accuracy towards the end, and the model remains relatively stable throughout the training process.

In order to evaluate the effectiveness of the model, two inference metrics, namely mean average precision (mAP) [34] and binary Panoptic Quality (bPQ) (a version of the PQ score that assumes all the cells belonging to one class and differentiating only between cells and background; which is sufficient for our case, having only two classes) [35] are used in this paper. The paper by Graham et al. [36] explains the merits of using bPQ as being one of the modern metrics in evaluating cell nucleus segmentation tasks. For both of these metric calculations, only those predictions

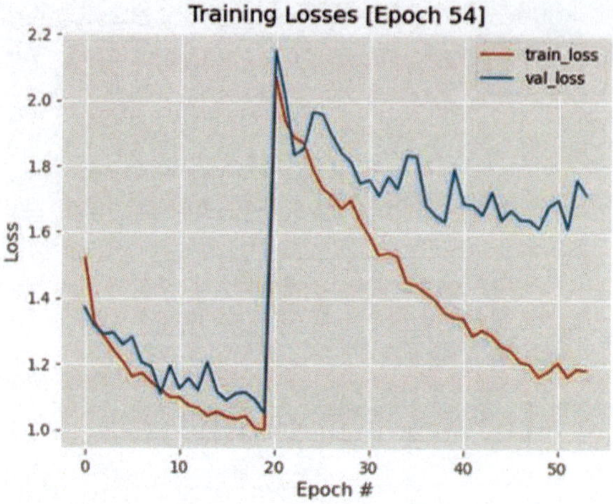

Fig. 4 Final experiment results where all mentioned training techniques are used. Losses start to plateau after 47 epochs

whose intersection over union (IoU) score was above 0.5 were considered. Table 2 describes the bPQ value across the various tissues and compares it to the Mask R-CNN baselines given in the PanNuke paper [27].

We also compared the average PQ (aPQ) score of our model and the PanNuke Mask R-CNN model only for the task of neoplastic cell detection. Results are shown in Table 3.

For the purpose of training this model and for inference, we used Nvidia's RTX 2070 Super Max-Q GPU. The time taken to perform inference and segment the cell nuclei was 0.44 s/per image. Figure 5 shows a collage of some of the results obtained by our model.

5 Conclusion

As seen in Tables 2 and 3, we think that our model shows improvement over the base Mask-CNN model for the following stipulated reasons:

- Due to the differences in the inference technique and the data splits used
- Our model only tackles neoplastic cell nuclei, while the base paper tackles four other classes
- The backbone used by us is more comprehensive than used in the base paper
- Best training practices used by us have improved the efficiency of the model.

Table 2 Comparison table of bPQ scores of our Mask R-CNN model and the model mentioned in the PanNuke paper

Tissue types	Our model bPQ	PanNuke mask R-CNN bPQ
Adrenal gland	0.717	0.555
Bile duct	0.652	0.557
Bladder	0.795	0.605
Breast	0.679	0.557
Cervix	0.673	0.548
Colon	0.586	0.460
Oesophagus	0.740	0.569
Head and neck	0.640	0.546
Kidney	0.657	0.509
Liver	0.742	0.609
Lung	0.612	0.513
Ovarian	0.744	0.578
Pancreatic	0.572	0.546
Prostate	0.735	0.579
Skin	0.667	0.502
Stomach	0.771	0.598
Testis	0.701	0.542
Thyroid	0.657	0.571
Uterus	0.585	0.559

Table 3 Comparison of the neoplastic cell detection average aPQ score between the two models

Cell type	Our model aPQ	PanNuke mask R-CNN aPQ
Neoplastic PQ	0.675	0.472

To make this research more thorough, we plan on using better and more efficient training strategies in future. We would also test our model on more external datasets similar to PanNuke and fine-tune the model using these datasets. Finally, we believe that the current model presented by us is very robust in its detection and should at least work as an assistive intelligent agent for researchers when they tackle the problem of neoplastic cell nuclei segmentation. The model being extremely fast in its inference time satisfies all of its proposed objectives and is a step forward in the use of intelligent systems for computational pathology.

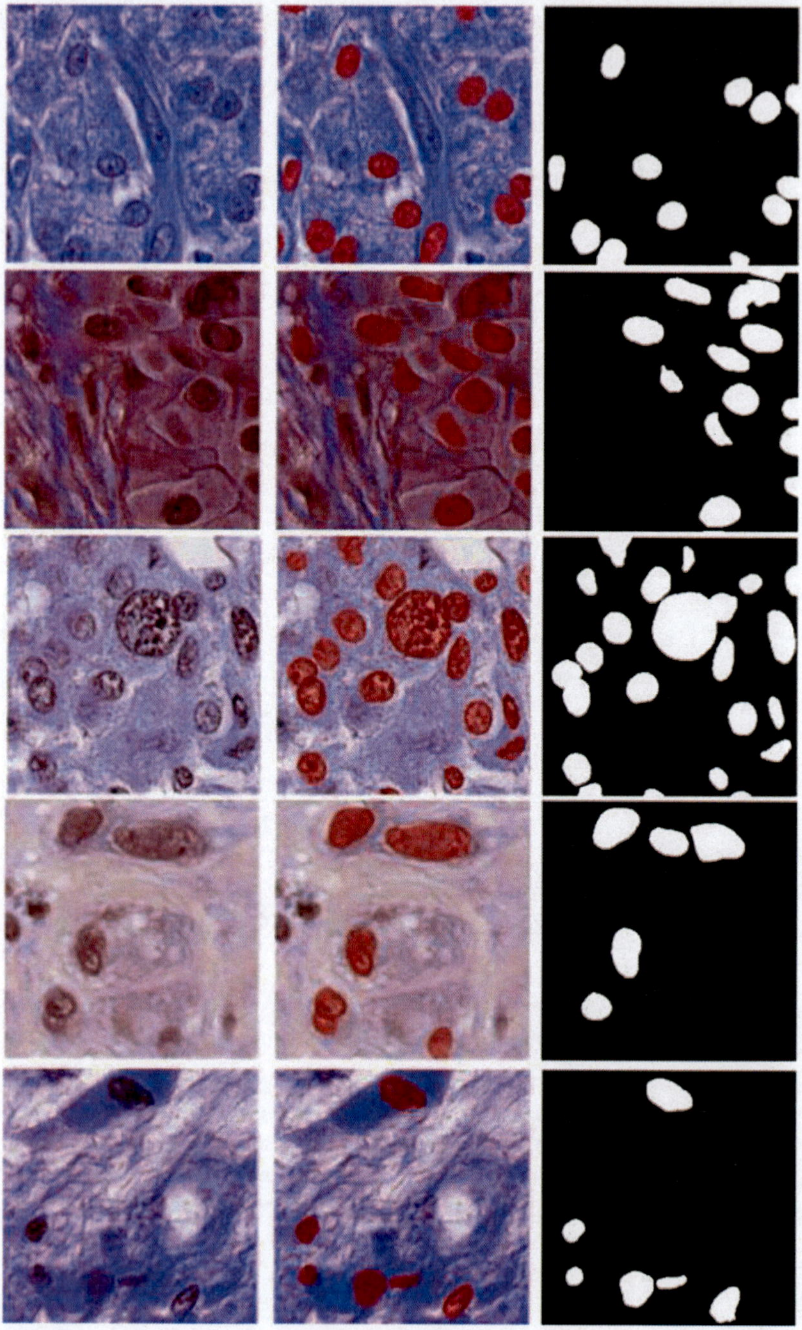

Fig. 5 Collage showing output from model inference. (First column) original cell nuclei image. (Second column) detected and segmented neoplastic cells with an alpha mask outputted from the model. (Third column) ground truth neoplastic nuclei binary mask thresholded to max pixel intensity for visibility

References

1. https://www.kaggle.com/c/data-science-bowl-2018
2. Deshmukh BS, Mankar VH (2014) Segmentation of microscopic images: a survey. In: International conference on electronic systems, signal processing and computing technologies, pp 362–364
3. Chen Y, Biddell K, Sun A, Relue PA, Johnson JD (1999) An automatic cell counting method for optical images. In: Proceedings of the first joint BMES/EMBS conference. In: 1999 IEEE engineering in medicine and biology 21st annual conference and the 1999 annual fall meeting of the biomedical engineering society, vol 2, p 819
4. Russakovsky O, Deng J, Su H et al (2015) ImageNet large scale visual recognition challenge. Int J Comput Vis:211–252
5. Szegedy C, Liu W, Jia Y, Sermanet P, Reed S, Anguelov D, Erhan D, Vanhoucke V, Rabinovich A (2015) Going deeper with convolutions. In: Proceedings of the IEEE conference on computer vision and pattern recognition (CVPR)
6. He K, Zhang X, Ren S, Sun J (2016) Deep residual learning for image recognition. In: Proceedings of the IEEE conference on computer vision and pattern recognition (CVPR)
7. He K, Zhang X, Ren S, Sun J (2016) Identity mappings in deep residual networks. In: Leibe B, Matas J, Sebe N, Welling M (eds) Computer vision—ECCV 2016. ECCV 2016. Lecture notes in computer science, vol 9908
8. Soufi SM, Kafieh R, Sonka M, Yazdani S, Jamalipour G (2020) Deep-COVID: predicting COVID-19 from chest X-ray images using deep transfer learning. Med Image Anal 65
9. He K, Gkioxari G, Dollar P, Girshick R (2017) Mask R-CNN. In: Proceedings of the IEEE international conference on computer vision (ICCV)
10. Sánchez GL, Kooi T, Bejnordi BE, Setio AAA, Ciompi F, Ghafoorian M, van der Laak JAWM, van Ginneken B, Clara I (2017) A survey on deep learning in medical image analysis. Med Image Anal 42:60–88
11. Sankaran P, Asari VK (2006) Adaptive thresholding based cell segmentation for cell-destruction activity verification. In: 35th IEEE applied imagery and pattern recognition workshop (AIPR'06), Washington, DC, USA, p 14. https://doi.org/10.1109/AIPR.2006.9
12. Di Rubeto C, Dempster A, Khan S, Jarra B (2000) Segmentation of blood images using morphological operators. In: Proceedings 15th international conference on pattern recognition. ICPR-2000, vol 3, pp 397–400
13. Bensch R, Ronneberger O (2015) Cell segmentation and tracking in phase contrast images using graph cut with asymmetric boundary costs. In: IEEE 12th international symposium on biomedical imaging (ISBI), pp 1220–1223
14. Porwal P et al (2020) IDRiD: diabetic retinopathy—segmentation and grading challenge. Med Image Anal 59
15. Ciresan DC, Giusti A, Gambardella LM, Schmidhuber J (2013) Mitosis detection in breast cancer histology images with deep neural networks. In: Medical image computing and computer-assisted intervention—MICCAI 2013, pp 411–418
16. Long J, Shelhamer E, Darrell T (2015) Fully convolutional networks for semantic segmentation. In: Proceedings of the IEEE conference on computer vision and pattern recognition (CVPR)
17. Drozdzal M, Vorontsov E, Chartrand G, Kadoury S, Pal C (2016) The importance of skip connections in biomedical image segmentation, deep learning and data labeling for medical applications, pp 179–187
18. Chen W, Wei H, Peng S, Sun J, Qiao X, Liu B (2019) HSN: hybrid segmentation network for small cell lung cancer segmentation. IEEE Access 7:75591–75603
19. Huang C, Ding H, Liu C (2020) Segmentation of cell images based on improved deep learning approach. IEEE Access 8:110189–110202
20. Girshick R, Donahue J, Darrell T, Malik J (2014) Rich feature hierarchies for accurate object detection and semantic segmentation. In: Proceedings of the IEEE conference on computer vision and pattern recognition (CVPR)

21. Girshick R (2015) Fast R-CNN. In: Proceedings of the IEEE international conference on computer vision (ICCV)
22. Ren S, He K, Girshick R, Sun J (2015) Faster R-CNN: towards real-time object detection with region proposal networks. Adv Neural Inf Process Syst 28:91–99
23. Kurnianingsih et al (2019) Segmentation and classification of cervical cells using deep learning. IEEE Access 7:116925–116941
24. Ma B, Zhang J, Cao F, He Y (2020) MACD R-CNN: an abnormal cell nucleus detection method. IEEE Access 8:166658–166669
25. Feng L, Song JH, Kim J, Jeong S, Park JS, Kim J (2019) Robust nucleus detection with partially labeled exemplars. IEEE Access 7:162169–162178
26. Jo J, Bengio Y (2017) Measuring the tendency of CNNs to learn surface statistical regularities. CoRR. abs/1711.11561
27. Gamper J, Koohbanani NA, Graham S, Jahanifar M, Khurram SA, Azam A, Hewitt K, Rajpoot N (2020) PanNuke dataset extension, insights and baselines. arXiv:2003.10778
28. Rajpoot SJ, Mahmood A, Fraz MM, Koohbanani NA, Benes K, Tsang Y-W, Hewitt K, Epstein D, Snead D, Rajpoot N (2020) Cellular community detection for tissue phenotyping in colorectal cancer histology images. Med Image Anal 63
29. Abdulla W (2017) Mask R-CNN for object detection and instance segmentation on Keras and tensor flow. GitHub repository
30. Kiefer J, Wolfowitz J (1952) Stochastic estimation of the maximum of a regression function. Ann Math Statist 23(3):462–466. https://doi.org/10.1214/aoms/1177729392
31. Torrey L, Shavlik J (2010) Transfer learning. In: Olivas ES et al (eds) Handbook of research on machine learning applications and trends: algorithms, methods, and techniques. IGI Global, pp 242–264. https://doi.org/10.4018/978-1-60566-766-9.ch011
32. Lin TY et al (2014) Microsoft COCO: common objects in context. In: Fleet D, Pajdla T, Schiele B, Tuytelaars T (eds) Computer vision—ECCV 2014. ECCV 2014. Lecture notes in computer science, vol 8693. Springer, Cham. https://doi.org/10.1007/978-3-319-10602-1_48
33. Krogh A, Hert JA (1992) A simple weight decay can improve generalization. In: Moody JE, Hanson SJ, Lippmann RP (eds) Advances in neural information processing systems, vol 4. Morgan Kauffmann Publishers, San Mateo CA
34. Beitzel SM, Jensen EC, Frieder O (2009) MAP. In: Liu L, Özsu MT (eds) Encyclopedia of database systems. Springer, Boston, MA. https://doi.org/10.1007/978-0-387-39940-9_492
35. Kirillov A, He K, Girshick R, Rother C, Dollár P (2019) Panoptic segmentation. In: Proceedings of the IEEE conference on computer vision and pattern recognition, pp 9404–9413
36. Graham S, Vu QD, Raza SEA, Azam A, Tsang YW, Kwak JT, Rajpoot N (2019) Hover-net: simultaneous segmentation and classification of nuclei in multi-tissue histology images. Med Image Anal 58:101563

Computational Medical Image Processing and Rehabilitation Engineering

A Study on Various Medical Imaging Modalities and Image Fusion Methods

B. Ashok Kumar, A. Sivaranjani, S. Senthilrani, and A. Senthil Murugan

Abstract Multimodal medical image fusion is the effective assimilation of different imaging techniques for enhancing the capability to analyze, to guide therapy, treatment, or predict end results. The accuracy of the image obtained from various medical imaging modalities has a significant impact on successful diagnosis of a disease, as image fusion provides supplementary valuable information. Complete and accurate information cannot be acquired from single medical imaging modality. Predominantly, the modalities can be classified into anatomy and metabolism; the former one gives the structural information of the body parts, and the latter one gives the functional information of the cells in the organ. Hence, the integration of various modality images will provide supplementary valuable information. In medical field, multimodal image fusion has become the focusing area of research and development of medical field. This survey paper portrays the background of multimodal medical image process, different multimodal imaging modalities, and various state-of-art methods for multimodal fusion methods. This paper also discusses the importance, opportunities, challenges, and advantages of multimodal medical image processing.

Keywords Multimodal · Image fusion · Various modalities · X-ray · CT · PET · MRI · SPECT · Fusion methods

B. Ashok Kumar (✉)
Department of Electrical and Electronics Engineering, Thiagarajar College of Engineering, Madurai, India
e-mail: ashokudt@tce.edu

A. Sivaranjani
Department of Computer Science and Engineering, University College of Engineering, Panruti, India

S. Senthilrani
Department of Electrical and Electronics Engineering, Velammal College of Engineering and Technology, Madurai, India
e-mail: ssr@vcet.ac.in

A. Senthil Murugan
Faculty of Information and Communication Engineering, Anna University, Chennai, India

M. Gupta et al. (eds.), *Artificial Intelligence on Medical Data*,
Lecture Notes in Computational Vision and Biomechanics 37,
https://doi.org/10.1007/978-981-19-0151-5_9

1 Introduction

In recent modern medical diagnosis system, multimodal medical image processing plays a key function for many clinical applications. Simply, it can be termed as more desirable interpretation of disease biology. In clinical applications, no single imaging modality can give complete information on the underlying technologies. A multimodal medical image is combination of several images from different imaging modalities [1–3]. It is also a technique of extracting harmonious information from a variety of resource images and binds them into a resulting image. The goal of multi-modal medical image fusion is to improve imaging quality while preserving specific features in order to increase the clinical usefulness of images for medical diagnosis and evaluation [4, 5]. Figure 1 shows the role of multimodal medical imaging in various clinical applications. Medical image fusion technologies are applicable to a wide range of fields. The areas include image processing, pattern recognition, computer Vision, machine learning, artificial intelligence, and deep learning [6].

In the areas of medical diagnostics, scrutiny, and historical documentation, there is a growing interest in utilizing medicinal imaging improvements [1]. The quantitative examinations of the images are empowered by computer-aided imaging techniques in assessment. It also improves the ability of medical specialists to make a fair and objective decision in a time when resources are limited. The enormous quantity of scholarly publications published since 2000 demonstrates the growing interest in this subject area [1]. Figure 2 schematically presents the milestones achieved in multimodal medical image fusion, showing the rising significance that the field has arisen in recent times.

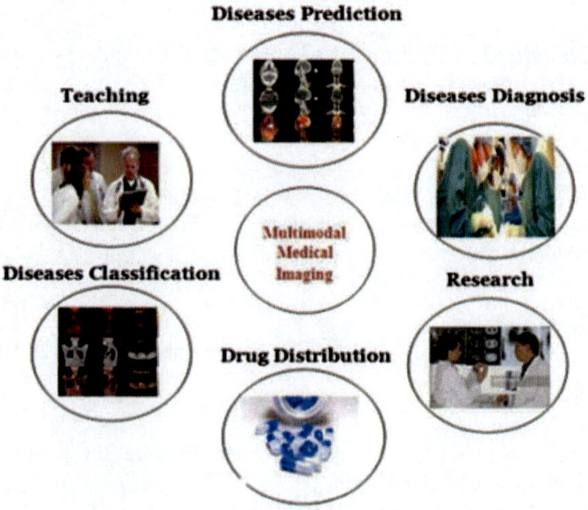

Fig. 1 Role of multimodal medical image fusion in various health diagnosis applications

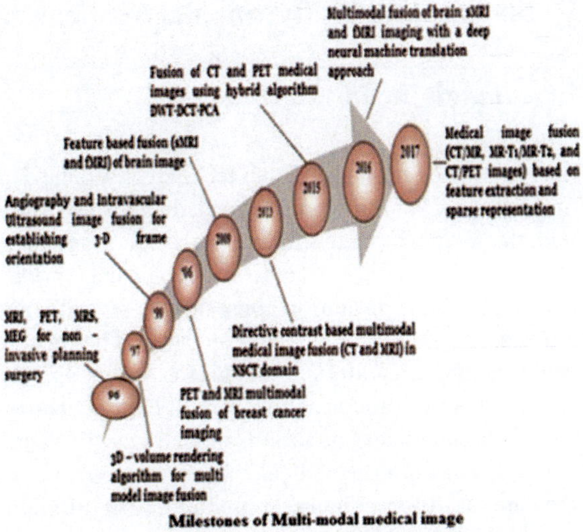

Fig. 2 Milestones of multimodal medical image

2 Related Surveys

The recent surveys [1–3, 7–9] have provided the comprehensive study of various multimodal image fusion techniques. James et al. [1] have presented a survey of the state-of-art methods for various medical image fusions. The measuring parameters for image quality assessment like standard deviation (STD), mean difference (MD), entropy, root mean square error (RMSE), average difference (DA), peak signal-to-noise ratio (PSNR) are evaluated, analyzed, and compared with different methods for good fusion result [10]. Though there are many advanced technical development in the medical imaging community, still, there are broad scientific challenges, issues, and many opportunities are interpreted in many research papers [11–13]. A detailed overview done on various fusion techniques was presented in [1–3, 6–9, 14–17] showed that there is a lot of work to be done to progress the image fusion approach.

This paper is organized as follows. In Sect. 2, the related surveys about the multimodal fusion image have been discussed. In Sect. 3, a brief explanation about the background of multimodal medical imaging is given. In this section, the types of image fusion levels and types of multimodal images have been presented. Section 4 gives state-of-art of various medical image fusion methods.

3 Background of Multimodal Medical Image Fusion

3.1 Image Fusion Architecture

The importance of image fusion techniques will be demonstrated in order to develop images capable of improving clinical diagnosis in to-day's medical world. In medical field, the medical images are classified according to their resolution, acquisition, or based on the imaging system to acquire image (various modality) [4]. The structural images give high-resolution apparent images with outstanding anatomical information and pinpoint accuracy in localization. The functional images afford low-spatial resolution images with functional information, useful for detecting cell activity, and related metabolic abnormalities [4, 5]. Consequently, the integration of anatomical and functional images of various modalities will afford auxiliary expensive information. The multimodal medical image fusion includes three main areas to consider. They are (a) different imaging modalities (b) different fusion algorithms (c) various organs [1]. Figure 3 shows the foremost attentive factors in multimodal medical image fusion.

In general, the image fusion is achieved by performing four essential steps. Figure 4 shows the process caught up in the image fusion. They are (i) registration (ii) feature extraction (iii) mapping the fusion rule (iv) quality assessment. The primary task in image fusion is image registration [18–21]. The process of coordinating the input images with the help of reference image is called image registration. Decomposition rules are used to break down the images to succession of sub-images. Secondly, the feature extraction is done to classify the feature for further process. To obtain the fused image, the third step is to map the features coordinates according to the fusion rule [22]. Final step is to carry out the quality evaluation using the quantitative metrics.

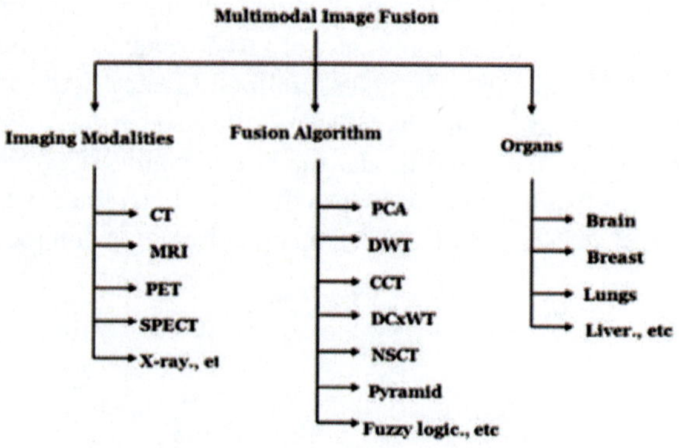

Fig. 3 Three main characters of multimodal medical image fusion

Fig. 4 Steps involved in image fusion

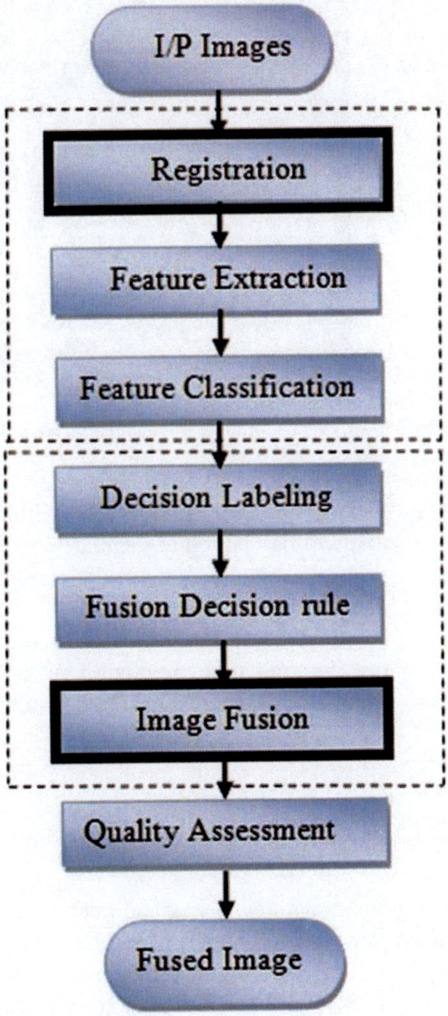

3.2 *Types of Image Fusion Levels*

The image fusion can be performed at three levels [2, 4, 8, 23–25]. They are pixel level [26], feature Level [27], and decision level [22]. Figure 5 shows the three levels of image fusion.

(1) Pixel Level: The image fusion is performed on pixel-by-pixel basis. It deals with the information associated with each pixel. The advantages of pixel-level image fusion include ease of implementation, measured the original amount, and efficient computing.

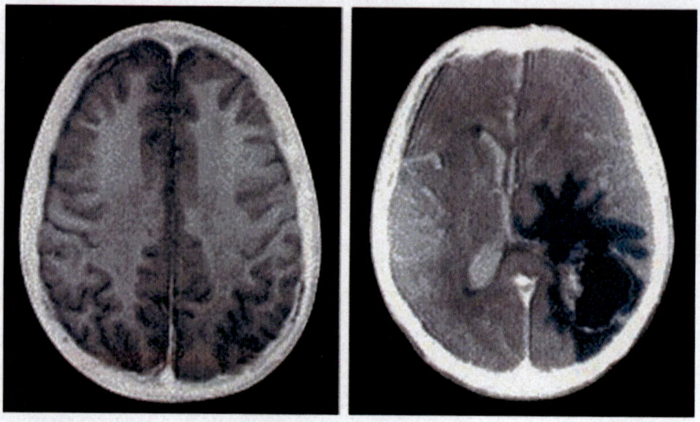

Fig. 5 X-ray image of unaffected and affected brain [29]

(2) Feature Level: It is concerned with extracting features from various images that will be merged to generate a new image. For fusion, the source image is split into regions, with attributes such as pixel intensities, edges, and texture being utilized.

(3) Decision Level: The decision-level fusion is a high-level fusion that uses statistics, determination, prediction, fuzzy logic, and other techniques. It is effective for intricate systems that are not suitable for widespread applications because it contains concise data.

3.3 Various Imaging Modalities

Different modalities of medical imaging represented different types of information about human organs and tissues; also, their applicability ranges differed. [1, 4, 7–9]. For instance, structural images like computed tomography (CT), Ultrasonography (USG), and magnetic resonance imaging (MRI) give high-resolution images with outstanding anatomical information and accurate localization ability. The functional images such as position emission tomography (PET), single-photon emission computed tomography (SPECT), and functional MRI (fMRI) provide low-spatial resolution images with functional information, which can be used to detect cell activity and metabolic problems. These abovementioned modalities and other modalities stumble on a wide range of clinical applications in making diagnosis, analysis, and assessments of medical condition effecting brain, lungs, bone marrow, breast, stomach, mouth, teeth, liver, intestines, bones, and soft tissues [4, 8, 28]. Table 1 shows the comparison for various multimodal medical images.

(1) **X-ray**: X-ray technology is one of the very old and mainly often utilized forms of medical imaging. It provides the photograph or images of the internal structure of a human organ by transient X-ray beam all the way through the human

Table 1 Comparison for various multimodal medical images

S. No.	Modalities and its nature	Type of information obtained	Type of imaging	Radiation source	Spatial resolution	Duration	Advantages	Disadvantages
1	X-ray—(ionizing)	Structural	Radiographic imaging	X-rays	Good	Short	Provides mostly anatomic information's	Manifestation to ionizing radiation
2	CT (ionizing)	Structural	X-ray computed tomography	Multiple X-rays	Good	Short	Provides cross-sectional layers	Uses high doses of radiation
3	MRI (non-ionizing)	Structural and functional	Magnetic resonance imaging	It makes use of magnetic fields as well as radio waves	Best	Long	Can get some chemical information	Slight movement can ruin the image, hence requiring retesting
4	PET (ionizing)	Functional	Nuclear imaging	Gamma rays	Poor	Long	It gives specific functional information, tissue functioning image	Can't be undertaken in some situations leads to insufficient information
5	SPECT (ionizing)	Functional	Nuclear imaging	Photons	Poor	Long	It depicts the flow of blood to various tissues and organs	Radiation exposure and very expensive

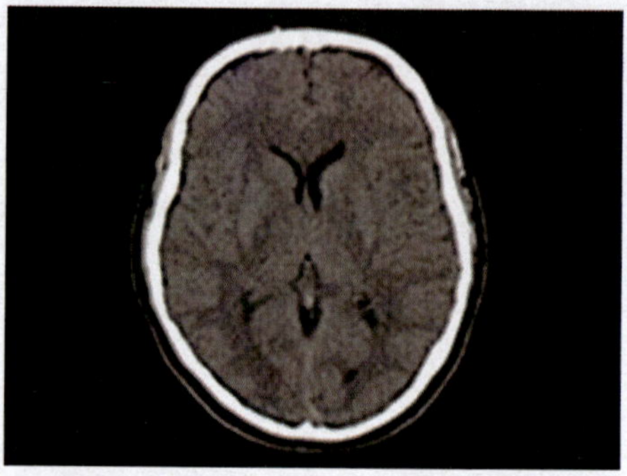

Fig. 6 CT image of brain [8]

body. X-ray images are normally used to estimate broken bones, swallowed objects, cavities, blood vessels, lungs, breast, etc. It won't provide the detailed information, and the ionizing rays are harmful to human skin [28]. Figure 5 shows the X-ray image of unaffected and affected brain.

(2) **CT**: Computed tomography (CT) is generally referred to as a computer-aided tomography. It is an ionizing technique. It uses X-ray and computer to construct the cross-sectional view of an organ. It provides the structural information of the body, including internal organs, blood vessels, bones, and tumors. CT images enable specialists to get very meticulous, 3D perspectives of specific parts of the body, such as soft tissues, the pelvis, blood vessels, the lungs, the brain, the heart, abdomen, and bones. CT is likewise often the favored method for diagnosing numerous malignancies, for example, liver, lung, and pancreatic cancers [28]. Figure 6 shows the CT image of brain.

(3) **MRI:** Magnetic resonance imaging (MRI) is a non-ionizing technique with complete 3D capabilities. It is excellent soft tissue contrast. It uses radio waves and a magnetic field to generate exhaustive images of organs and tissues. MRI has confirmed to be extremely effectual in diagnosing a number of situations by presenting the difference between normal and diseased soft tissues of the body. There are two types of MRI images, and they are structural MRI (sMRI) and functional MRI (fMRI) which provide the structural and functional information of the organ, cell, respectively. MRI is usually used to assess abnormal tissue, breasts, bones and joints, organs in the pelvis, chest and abdomen, spinal injuries, and ligament tear [28]. Figure 7 shows the MRI image of brain.

(4) **PET**: Positron emission tomography (PET) is a nuclear imaging technique. Usually, it presents the specialists with the functional information of cell, tissues, and organs. It shows chemical and other changes in the brain. It is used to monitor the metabolic activities of the body. It is frequently used in combination with CT imaging, to help out in creating exhaustive, computerized images of internal body parts. PET is often used to assess neurological diseases, cancer, effectiveness of treatments, and heart conditions [28]. Figure 8 shows the PET image of brain.

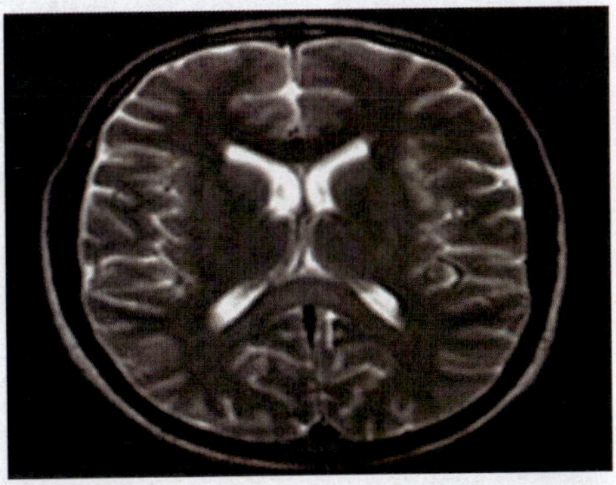

Fig. 7 MRI image of brain [8]

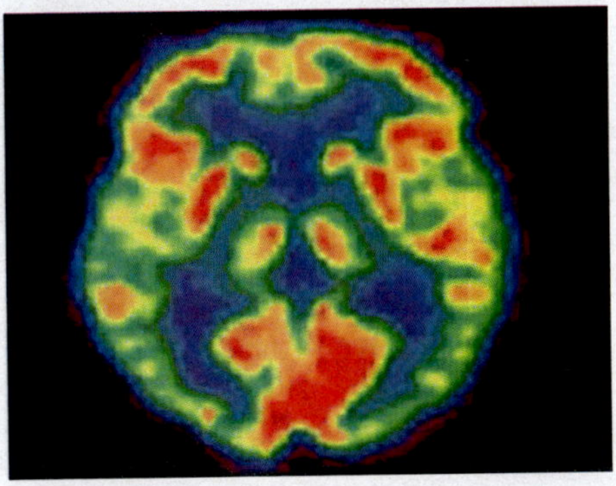

Fig. 8 PET image of brain [8]

(5) **SPECT**: A single-photon emission computerized tomography (SPECT) is also a nuclear imaging technique which uses gamma rays to emit the functional information of internal organs. It utilizes the radioactive substance and a special camera to create pictures which gets collected and translated into two-dimensional cross-sections by the computer. A SPECT is mostly used to assess how blood flows through arteries and veins in the brain. It is also used to detect reduced blood flow to injured sites [28]. Figure 9 shows the SPECT image of brain.

4 State-of-Art of Various Medical Image Fusion Methods

- **Color overlay, Color mixing, Interlacing**: Baum et al. [30] discussed the opportunity of using the three techniques for combining the PET and CT of breast image. The techniques are color overlay, color mixing, interlacing. One image is presented semi-transparently on peak of the former image using the color overlay technique. It is done by averaging the source images. Color mixing is a technique

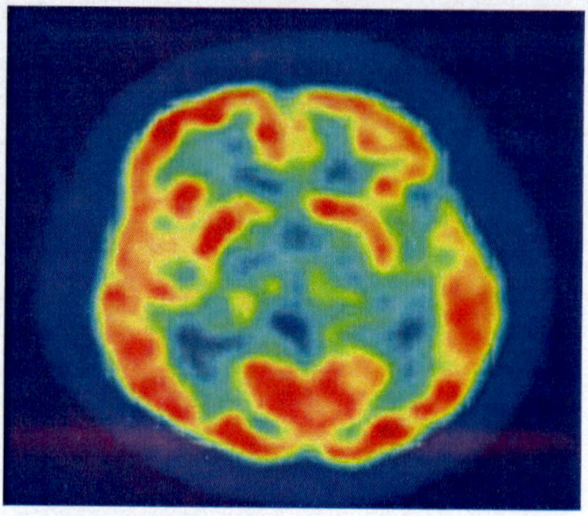

Fig. 9 SPECT image of brain [8]

for combining multiple one-channel images to generate a fused RGB image. Interlacing is a very common image fusion technique which interleaves the columns of pixels, rows, or even every pixel in the source images.

- **DWT and IHS**: Bhavana et al. [31] have done multimodal image fusion (MRI and PET) by discrete wavelet transform (DWT) to enhance the quality of the input images for brain disease images. Since to separate low/high activity region by "hue angle," here, the PET image is decomposed into intensity hue saturation (IHS) transform. Gaussian filter is used to eliminate the noise and headed for improve the image. By getting the low/high-frequency coefficients, the low/high activity region is decomposed using a four-level DWT transform to extract anatomical and spectral information. By combining the coefficients of MR image and PET image using averaging method, the fused frequency output is obtained. Then fused image is acquired using the inverse DWT.
- **SVM-based Adaptive Forward–Backward Greedy Algorithm (SVM-Foba)**: Jie et al. [32] developed a novel approach called SVM-FoBa, in order to overcome the limitation of both forward and backward strategies. Bipolar disorder (BD) and major depressive disorder (MDD) are distinguished using sMRI and fMRI of the brain (MDD). This study backs up the benefits of multimodal joint analysis and the efficacy of SVM-FoBa, which might be used to find biomarkers for a variety of mental illnesses.
- **DCxWT and SD**: Swathi et al. [33] have proposed a new approach by using Daubechies complex wavelet transform (DC $\times$ WT) and standard deviation (SD) to combine the multimodal (CT and MRI) medical image. Daubechies complex wavelet transforms that decomposes the input images into different levels to acquire the low/high-frequency coefficients. Each coefficient is divided into K blocks, and standard deviation for each block was computed. Value with the larger SD is selected for the fused result coefficients. Finally, the inverse transform is taken to acquire the combined image.
- **Hybrid Algorithm DWT-DCT-PCA:** Guruprasad et al. [10] have proposed a novel hybrid fusion algorithm for multimodal medical images (CT and PET) to

progress the image quality. For efficient image fusion, they integrated conventional and advanced fusion algorithm like DWT, DCT, and PCA. Initially, the input images are preprocessed; secondly, the resizing of image and conversion of RGB to gray level is done. Third step is to choose the hybrid fusion methods from methods like PCA, DCT-PCA, DWT-PCA, DWT, DCT, and DWT-DCT-PCA.

- **Wiener Filter in Shearlet Transform (WFST) fusion algorithm**: Biswas et al. [34] utilized the WFST fusion technique to merge the CT and MR images of the spine, providing additional information to aid diagnostic and treatment decisions. The source images (CT and MR) are decomposed by ST. The WF will estimate the best lowpass ST subbands. Highpass ST subbands are selected and combined by computing new coefficients and fused coefficients. CT and MR images with the leading directional ST subbands are chosen. The inverse ST transform is done to obtain the combined image.
- **Sparse Representation and Decision Map (SR and DM)**: Fei et al. [35] have proposed sparse representation and decision map-based combination mechanism for multimodal medical images to enhance the contrast and reserving more structure. Structure information map (SM), energy information map (EM), and structure and energy map (SEM) are decision maps used to produce results with more energy and edge information. Primarily, the resource images are resized into patches through sliding window and arranged into vectors. Separately group the vectors and construct the map based on their characteristics. The decision map is used to find out the input of the sparse representation system. The SR method is used for fusion, and result is generated according to the decision map.
- **MULTI-Nonsubsampled Direction Filter Bank (NSDFB)**: Geng et al. [36] have presented a novel algorithm and also proposed multi-NSDFB transform by combining multiwavelet with the nonsubsampled direction filter bank (NSDFB). Here, they combined CT and PET medical images using multi-NSDFB transform also found better approximation in image edge. Here, the image is decomposed by 2-level of multiwavelet, and every subband of multiwavelet is decomposed by 4-level of NSDFB. The spatial frequency is calculated to evaluate the salience feature in multi-NSDFB.
- **Deep Neural Machine Translation**: Amin et al. [37] have proposed a unique method for linking multimodal brain imaging data, such as structural MRI (sMRI) and functional MRI (fMRI). For the translation model, they used independent component analysis (ICA) characteristics. The purpose of the attention mechanism is to discover the alignment/linkage between the dynamic functional connectivity (dFNC) state and the anatomical components of the brain. The series predictor module predicts dFNC state for taking its alignments, and the attention network module used its learned parameters to estimate it at each time step.

The various multimodal imaging modality and its fusion methods have been compared and shown in Table 2.

Table 2 Various multimodal imaging modalities and its fusion methods

S. No.	Author and year	Modality	Methods and its motivation	Applications	Advantages
1	Baum et al. [30]	PET and MRI	Color overlay and color mixing are utilized to express the spatial links between the two modalities, improving diagnosis sensitivity, specificity, and accuracy	Breast cancer imaging	Simple to use and precise depiction of the original images
2	Bhavana et al. [31]	PET and MRI	Discrete wavelet transform (DWT) is used to improve the quality of the input images	Brain disease images	Accurate results with compact color distortion and with no loss of any anatomical information
3	Jie et al. [32]	sMRI and fMRI	SVM-FoBa is used to discriminate between BD and MDD	Bipolar disorder in brain disease image	It has been used to look for potential biomarkers for a variety of mental illnesses
4	Sui et al. [22]	sMRI-fMRI-EEG	Multi-set canonical correlation analysis (MCCA) is utilized to understand the anomalies that underpin schizophrenia patients, and it varies across multiple modalities	Discrimination of schizophrenia patients in brain image	In testing data, it has the maximum classification accuracy (91%) and prediction rate (100%)
5	Swathi et al. [33]	CT &MRI	Daubechies wavelet transform coefficient is used to increase the high-resolution images with structural and anatomical information	Any fusing image	Obtained better fused output and more information

(continued)

Table 2 (continued)

S. No.	Author and year	Modality	Methods and its motivation	Applications	Advantages
6	Amin et al. [37]	sMRI and fMRI	ICA and deep neural machine translation is used to illustrate considerable grouping differences among healthy controls and patients with schizophrenia in the learned alignments	Brain imaging data	Learns the alignment between the two modalities
7	Bhatnagar et al. [26]	CT and MRI	The NSCT method is used to compile the most relevant data from various sources into a single output	Alzheimer, subacute stroke, and recurrent tumor-related images	Provides a better fusion outcome
8	Guruprasad et al. [10]	CT and PET	Hybrid algorithm DWT-DCT-PCA is used to collect relevant, diverse, and complimentary data in a logical arrangement in order to improve the information visible in the images	Any fusion image	Provides accurate data and better utility
9	Biswas et al. [34]	CT and MRI	Wiener filter is used to provide the complementary information to take decision	Spine medical image	Gives good performance
10	Geng et al. [36]	CT and PET	Multiwavelet and nonsubsampled direction filter bank is used to provide the better approximation edge	Any fusion image	Effectively avoid the artificial information

5 Conclusion

A single modality of medical image cannot provide complete and accurate information. Image registration is another issue in multimodal medical image. The registration of the images needs a method to check and rectify the spatial misalignment among the different image datasets that regularly involve reward of variability resulting from scale changes, rotations, and translations. The problem of registration becomes complex in the occurrence of inter-image noise, missing features, and outliers in the images. The main challenge is to select the image fusion algorithms to make sure in the clinical significance and assist for better medical outcome. Another trivial task and challenge is to make right combination of the imaging modalities, feature selection, processing, extraction, decision-making which focus the particular medical condition. Another challenge of imaging is the existence of various nonmedical factors, which eventually may influence which modalities are used.

In conclusion, multimodal medical imaging, various modality imaging, and various techniques have been discussed in this survey. This review of the literature indicated that most researcher have worked on various fusion methods for the combination of PET/CT, MRI/CT, sMRI/fMRI. Many inventions are there, only for fusion of structural and anatomical information; only, few are there for functional and anatomical. In future, the main objective is to compose a new combination of both functional and anatomical information by developing a new image fusion method which is different from the usual. Hence, there is an opportunity for the researchers to find the solutions for the abovementioned challenges. We look ahead to that this survey on multimodal medical image processing, and its various fusion methods with affluent speculative information all along with the contents will offer valuable part to research works on multimodal image fusion and encourage new research.

References

1. James A, Belur V, Dasarathy (2014) Medical image fusion: a survey of the state of the art. Info Fusion 19:4–19
2. Constantinos SP, Pattichis MS, Micheli-Tzanakou E (2001) Medical imaging fusion applications: an overview. In: Signals, systems and computers, conference record of the thirty-fifth asilomar conference. IEEE, pp 1263–1269
3. Bhavana V, Krishnappa HK (2016) A survey on multi-modality medical image fusion. In: Wireless communications, signal processing and networking (WiSPNET), IEEE conference, pp 1326–1329
4. Deserno MT (2011) Fundamentals of bio-medical image processing. In: Deserno TM (ed) Biomedical image processing, biological and medical physics, biomedical engineering. Springer-Verlag Berlin Heidelberg, pp 1–52
5. Santarelli FM, Positano V, Landini L (1997) Real-time multimodal medical image processing: a dynamic volume-rendering application. IEEE Trans Inf Technol Biomed 1(3):171–178
6. Du J, Li W, Lu K, Xiao B (2016) An overview of multi-modal medical image fusion. Neurocomputing 215:3–20
7. Devanna H, Satish Kumar GAE, Giri Prasad MN (2017) A survey on multimodal medical image fusion. IOSR J Comput Eng (IOSR-JCE) 19(2):07–14

8. Swathi PS, Sheethal MS, Paul V (2016) Survey on multimodal medical image fusion techniques. Int J Comput Sci Eng Technol 6(1):33–39
9. Shingadiya RP, Joshi R (2015) Review on multimodality medical image fusion. Int J Eng Sci Res Technol 4(1):628–631
10. Guruprasad S, Kurian MZ, Suma HN (2015) Fusion of CT and PET medical images using hybrid algorithm DWT-DCT-PCA. In: 2nd international conference on information science and security (ICISS), pp 1–5
11. Weissman NJ, Soman P, Shah DJ (2013) 'Multimodality imaging: opportunities and challenges. JACC Cardio Cascular Imag 6(9):1022–1023
12. Daftary A (2010) PET-MRI: challenges and new directions. Indian J Nucl Med 25(1):3–5
13. Pichler B, Wehrl HF, Kolb A, Martin S, Judenhofer MS (2008) PET/MRI: the next generation of multi-modality imaging? Semin Nucl Med 38(3):199–208
14. Rani K, Sharma R (2013) Study of different image fusion algorithm. Int J Emerg Technol Adv Eng 3(5):288–291
15. Tamilselvan KS, Murugesan G (2014) Survey and analysis of various image fusion techniques for clinical CT and MRI images. Int J Image Syst Technol 24(2):193–202
16. Sharma A, Saroliya A (2013) A brief review of different image fusion algorithm. Int J Sci Res:2650–2652
17. Dogra A, Goyal B, Agrawal S (2020) From multi-scale decomposition to non-multi-scale decomposition methods: a comprehensive survey of image fusion techniques and its applications', digital object identifier. IEEE Access 5:16040–16067; Andrew RM (2020) Timely estimates of India's annual and monthly fossil CO_2 emissions. Earth Syst Sci Data 12:2411–2421. https://doi.org/10.5194/essd-12-2411-2020
18. El-Zahraa F, El-Gamal A, Elmogy M, Atwan A (2016) Current trends in medical image registration and fusion. Egypt Inform J 17:99–124
19. Miles B, Ayed IB, Law MWK, Garv-in G, Fenster A, Shuo L (2013) Spine image fusion via graph cuts. IEEE Trans Biomed Eng 60(7):1841–1850
20. Costin H, Rotariu C (2009) Registration of multimodal medical images. Comput Sci J Moldova 17(3,51):231–254
21. Orchard J (2007) Efficient least squares multimodal registration with a globally exhaustive alignment search. IEEE Trans Image Process 16(10):2526–2534
22. Sui J, Castro E, He H, Bridwell D, Du Y, Pearlson GD, Jiang T, Calhoun VD (2014) Combination of FMRI-SMRI-EEG data improves discrimination of schizophrenia patients by ensemble feature selection. In: Engineering in medicine and biology society (EMBC), 36th annual international conference of the IEEE, pp 3889–3892
23. Deshmukh C (2016) A Comparative study of different image fusion techniques for toned-mapped images. Int J Sci Eng Res 7(2):474–478
24. Sinija TS, Karthik M (2015) Three level multimodal medical image fusion. Int J Comput Sci Commun Netw 5(4):262–266
25. James AP, Dasarathy BV (2015) A review of feature and data fusion with medical images' book multisensor data fusion, chapter 9, pp 491–507
26. Bhatnagar G, JonathanWu QM, Liu Z (2013) Directive contrast based multimodal medical image fusion in NSCT domain. IEEE Trans Multimedia 15(5):1014–1024
27. Calhoun VD, Adalı T (2009) Feature-based fusion of medical imaging data. IEEE Trans Inf Technol Biomed 13(5):711–720
28. Wang Y (2017) Introduction to medicine for engineers', medical imaging techniques, Cornell MRI research lab summer immersion, p 1. http://weill.cornell.edu/mri/pages/book/chapter_4.html. Accessed 24 Apr 2017
29. Das A, Bhattacharya M (2015) Effective image fusion method to study Alzheimer's disease using MR, PET images. In: IEEE international conference on bioinformatics and biomedicine (BIBM), pp 1603–1607
30. Baum KG, Helguera M, Hornak PJ, Kerekes PJ, Montag ED, Unlu ZM, Feiglin HD, Krol A (2006) Techniques for fusion of multimodal images: application to breast imaging' image processing. In: IEEE international conference, pp 2521–2524

31. Bhavana V, Krishnappa HK (2015) Multi-modality medical image fusion using discrete wavelet transform In: 4th international conference on eco-friendly computing and communication systems, ICECCS, pp 625–631
32. Nan-Feng J, Mao-Hu Z, Xiao-Ying M, Osuch ME, Wammes M, Théberge J, Huan-Dong L, Zhang Y, Tian-Zi J, Sui J, Calhoun DV (2015) Discriminating bi-polar disorder from major depression based on SVM-Foba: efficient feature selection with multimodal brain imaging data. IEEE Trans Autonom Mental Dev 7(4):320–331
33. Swathi KV, Bindu CH (2013) Modified approach of multimodal medical image fusion using daubechies wavelet transform. Int J Adv Res Comput Commun Eng 2(11):4199–4201
34. Biswas B, Chakrabarti A, Dey KN (2015) Spine medical image fusion using wiener filter in shearlet domain. In: IEEE 2nd international conference on recent trends in information systems, pp 387–392
35. Fei Y, Wei G, Zongxi S (2017) Medical image fusion based on feature extraction and sparse representation. Int J Biomed Imag:1–12
36. Geng P, Su X, Xu T, Liu J (2015) Multi-modal medical image fusion based on the multiwavelet and nonsubsampled direction filter bank. Int J Signal Process Image Process Pattern Recogn 8(11):75–84
37. Amin MF, Plis SM, Damaraju E, Hjelm D, Choy K, Calhoun VD (2016) Multimodal fusion of brain structural and functional imaging with a deep neural machine translation approach. In: Southwest symposium on image analysis and interpretation. IEEE, pp 1–4

IoT Based Real-Time Wearable Tachycardia Monitoring System Using Machine Learning

Reuben O. Jacob and H. S. Niranjana Murthy

Abstract This paper proposes the design of a wearable IoT based, battery-powered real-time heart rate monitoring and alert system for tachycardia. The device uses an optical heart rate and a temperature sensor connected to a microcontroller with a GPS module to constantly send real-time medical and location data to a central server using a GSM module via AT commands. The central server will store the data periodically in a database based on MongoDB and NodeJS. In the database, the heart rate and temperature data are evaluated via an unsupervised machine learning algorithm using a K-means classification method whose centroids have been previously tuned. If the data is registered as dangerous, an SMS is sent to a selected guardian. The SMS received by the guardian will have the triggered location of the system in a link that displays the location on google maps. The time-stamped location and medical data from whenever an alert was sent is accessible to the user and is visualized on the Google Maps API via an Angular based website, hosted on an AWS server. The data is displayed on the website by the database using JavaScript upon user request, communicated by SQL queries.

Keywords K-Means classification · Unsupervised machine learning · SQL queries · MongoDB · NodeJS · Angular · AWS · Google Maps API · Central server · AT commands · Microcontroller · GSM · GPS

1 Introduction

Tachycardia is a form of arrhythmia where the heartbeat of an individual is abnormally fast. There are several forms of this condition whose symptoms are not usually

R. O. Jacob (✉) · H. S. Niranjana Murthy
Department of EIE, Ramaiah Institute of Technology, Bangalore, India
e-mail: reubenjacob16@gmail.com

H. S. Niranjana Murthy
e-mail: hasnimurthy@msrit.edu

M. Gupta et al. (eds.), *Artificial Intelligence on Medical Data*,
Lecture Notes in Computational Vision and Biomechanics 37,
https://doi.org/10.1007/978-981-19-0151-5_10

noticeable, such as atrial fibrillation, ventricular tachycardia, and ventricular fibrillation. Tachycardia can lead to multiple forms of medical emergencies such as stroke, heart attacks, and heart failure [1]. Hence when diagnosed with this disorder, patients need to be constantly monitored for possible imminent attacks. Therefore, patients usually employ home nursing or other forms of continuous medical observation to stay on top of this. Such a form of care is usually expensive.

This paper proposes a device that can continuously keep a check on tachycardia via real-time data transmission and machine learning. It aims to provide an alternative cheaper solution to constant medical observation. The operation of this device unlike commercially available options requires no technical dexterity and can be used to monitor individuals who need constant medical attention such as the elderly and high-risk adults and children. Onboard GPS, heart rate and temperature sensors continuously ping patient data with location to an online database. The UI of the hardware consists of only a simple push button to turn it on. A machine learning algorithm is employed on the server to assess patient status. If considered potentially harmful, a registered guardian is alerted via an SMS. For privacy concerns, the location data of the user can only be accessed by a registered account with a username and password.

The organization of the paper includes multiple sections. Section 2 is on the literature survey related to existing systems and research. Section 3 describes the methodology and software with block diagrams. Section 4 describes the results of the project. Section 5 is the conclusion followed by the references.

2 Literature Survey

The advancement in the Internet of Things (IoT) technologies has made it possible for internet-connected devices to talk together [2]. Body-worn sensors are the most among different gadgets that screen individual ailments. There has been an expanding interest in wearable sensors lately, and an arising set of new items are available for individual wellbeing checking and wellness. The two driving variables of this innovation are the IoT-based information collection and data evaluation by machine learning [3]. The advancement of Internet and wireless sensor networks [4] have prompted the birth of wearable ECG observing frameworks. A methodical survey of different portable medical care approaches was completed by [5].

Ventricular tachycardia causes a quick pulse and possible mortality without even a trace of prompt clinical mediation. As most unexpected heart deaths happen due to tachycardia [6], early forecast of VTA is critical to saving patients' lives. Various methods have been developed for the early prognosis of tachycardia.

IoT-based health monitoring with location tracking has been developed before. Personal tracking was implemented by using the GPS and GSM modules present in mobile devices [7]. The location history was stored in a database. A vehicle tracker was implemented, and a GSM module is used to transmit and update the vehicle location to a database [8]. A machine learning-based health monitoring system was developed in which five sets of parameters were measured using wearable sensors. The patient wore a transmitting circuit, and the nurses were alerted using a receiving circuit [9].

A real-time health monitoring system with location tracking was implemented, and a wearable system was developed to monitor the health and location of soldiers. Both K-means and hierarchical clustering methods were used to classify the data to assess the situation of the soldier [10]. A real-time machine-health monitoring system was developed using various sensors to measure machine details and was sent to a central server. The overall effectiveness of the machine was increased by analyzing the data using machine learning [11]. An IoT-based wearable health prediction device was implemented involving onboard sensors, and machine learning was used to detect and analyze impending health problems of the patient [12].

A smart 5G IoT-based healthcare monitoring system was built for patients by implanting wireless sensors on the body which collect different vital physiological parameters such as heart rate and blood pressure. Various machine learning models were used for efficient prediction [13].

A real-time cardiovascular health monitoring system was implemented in which IoT devices were used to transmit and store the data to a server. The KNN and random forest algorithms were used to predict various cardiovascular health issues [14]. A marine tracking system was implemented using a microcontroller and commercial GPS/GSM modules [15]. An app-based vehicle tracking device was implemented with data displayed on Google Maps [16]. The K-means clustering algorithm was adopted with data tunability. Various parameters were collected in this study from which heart rate and temperature were used [17]. However, there is a great scope for the design of a wearable IoT based, battery-powered real-time heart rate monitoring and alert system for tachycardia. Also, the use of Google Maps API via an Angular-based Web site, hosted on an AWS server will help to track the afflicted person and to provide emergency medical care. The key objective of this research is to create a cheap wearable health monitoring system that can be used by people of all ages. Existing systems are expensive and do not have an easy-to-use UI.

3 Methodology

The entire wearable system consists of a microcontroller and peripheral modules powered by an 18650 LiPo battery. Locally available GPS, GSM, and medical sensors were used for the monitoring system, and the device is turned on using a push button. The system design is shown in Fig. 1.

- A push button is used to switch on the device.
- Upon activation, the microcontroller compiles data from the heart rate and temperature sensors along with location data from the GPS and sends this data to the GSM.

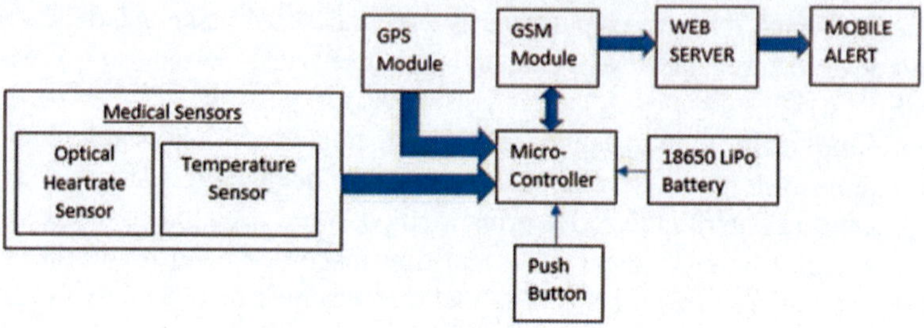

Fig. 1 Basic block diagram of the device

- The GSM module periodically transmits the data to the web server using AT commands.
- Once the database receives this data, the heart rate and temperature are entered into a pre-tuned K-means clustering algorithm.
- If the Euclidian distance of the new value falls closest to a centroid that suggests a possible heart attack, the guardian is alerted via an SMS.
- The guardians of the patients are alerted with an SMS containing a Google Maps link to where the device was triggered.
- An 18650 LiPo battery is used to power up the device.

3.1 *Software Implementation*

The guardian can log into the database using preset credentials to check on all the time-stamped locations and medical data of where the device was triggered. It is displayed on a map via the Google Maps API. The flowchart for web server operation is displayed in Fig. 2.

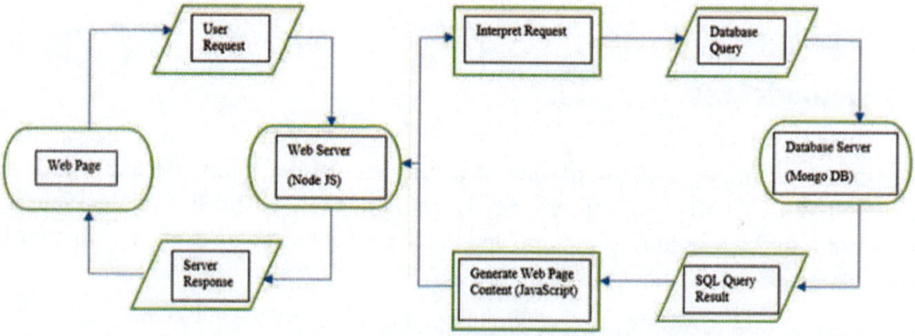

Fig. 2 Flow chart describing the flow of communication

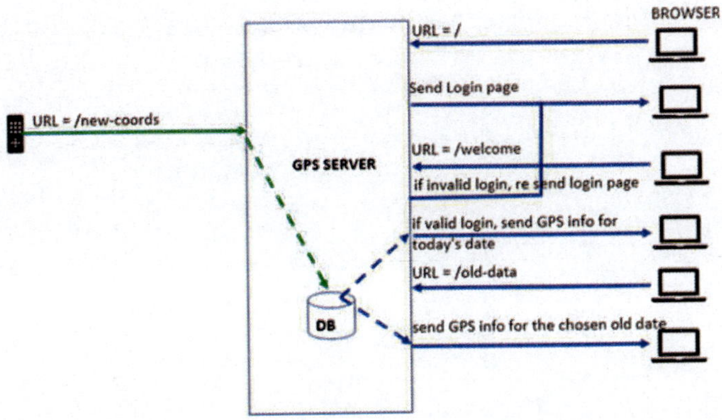

Fig. 3 Server sequence diagram

Table 1 Behavior of the web server

URL	Web server response
http://<ip address>>8080/	Displays the login page
http://<ip address>>8080/welcome	Displays the home page, containing GPS, and medical data of the current date
http://<ip address>>8080/old data	Displays the historic GPS and medical information of chosen date

3.2 *Sequence Diagram at the Server*

Figure 3 displays the sequence diagram at the server displaying what data the server produces based on user requests through the Web site (Table 1).

3.3 *K-means Clustering Algorithm*

- A K-means clustering algorithm was run on the dataset adopted from [17] after scaling it using the MinMaxScaler using Matplotlib.
- Four centroids were created differentiating the data into safe, exercise, and danger levels.
- Whenever a new value is entered into the database, the Euclidian distance from this point to the four centroid values are measured.
- The clustering algorithm sorts this data as safe, exercise, or dangerous conditions based on which centroid the point is closest to.
- The data is defined as safe when the HR is low despite the temperature. When both HR and temperature are high, the data is registered as the patient exercising.

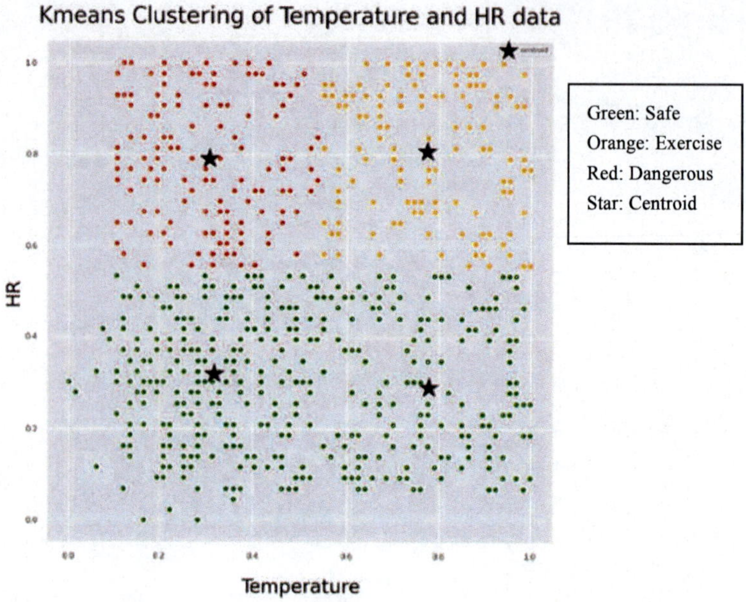

Fig. 4 K-means scatterplot

- When HR levels are high with low temperatures, the data is registered as impending danger.
- Based on this information, an SMS is sent to a registered guardian with the GPS and medical data whenever it is recognized as dangerous.
- The scatterplot of the clustering algorithm is shown below whose data points have been color coded (Fig. 4).

4 Results

The model was tested with patients in a medical facility who are constantly under observation for heart afflictions. The graph given below is a scatter plot with the times the patient recorded an actual tachycardia attack and the times when the device predicted an attack.

Figure 5 shows a scatter plot of the collected data from real-life observations which show an accuracy of about 90%.

The interface to the database is displayed by a web server on an Angular-based Web site, run on a Linux host on AWS servers. The user can log into this database only using registered credentials which ensures privacy. Once logged, Fig. 6 displays the web page where the guardian can enter past dates to see previous trigger data suggesting tachycardia by submitting a query.

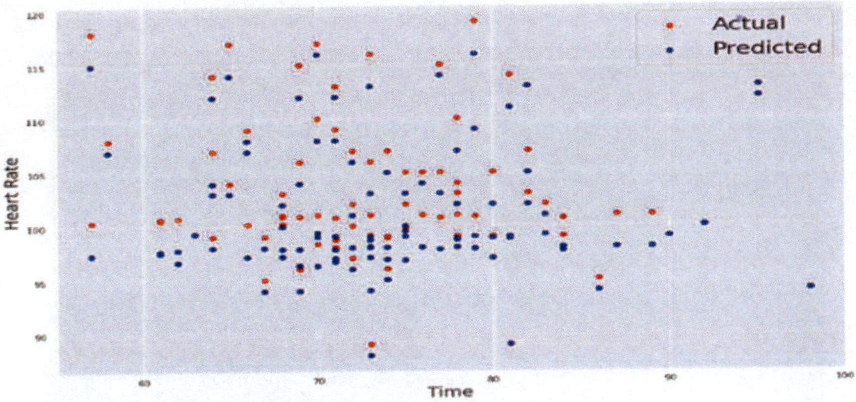

Fig. 5 Accuracy graph

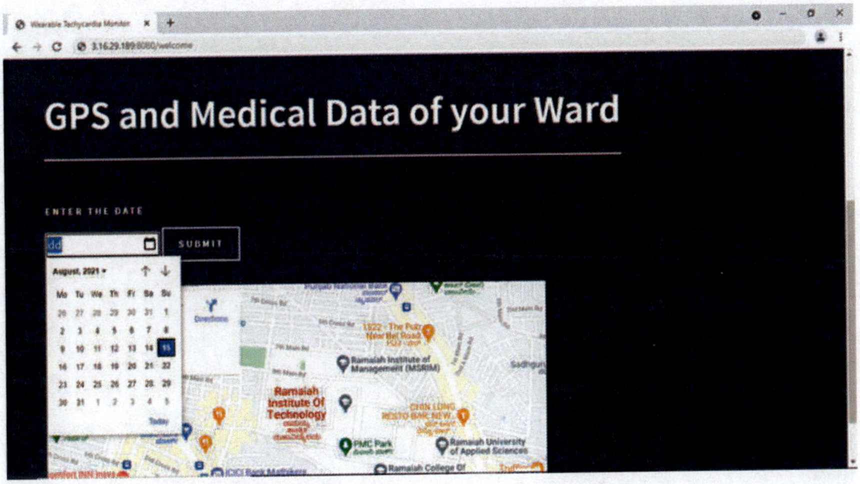

Fig. 6 Submit date query

Figure 7 shows the location data of the selected date where the system was triggered, displayed on the Google Maps API.

Upon clicking on each marked location, the Web site displays the heart rate and temperature during the time the data was registered. This is shown in Fig. 8.

Figure 9 shows the text messages received by the guardians. It contains the heart rate and temperature which was registered as dangerous. This message contains the Google Maps link as well.

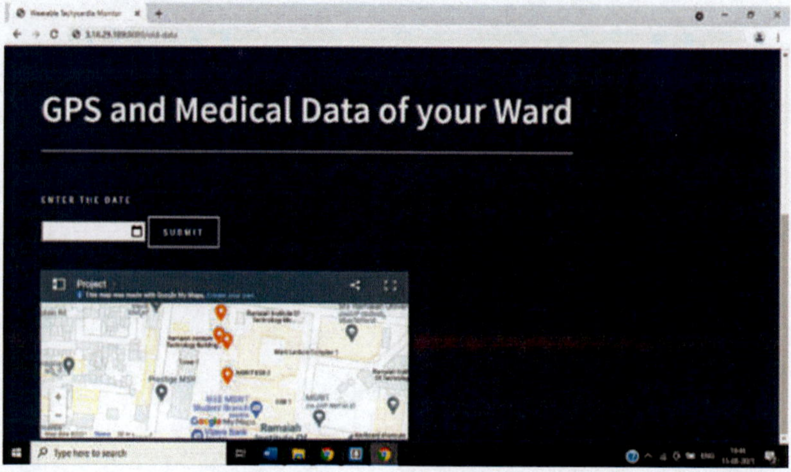

Fig. 7 Location history

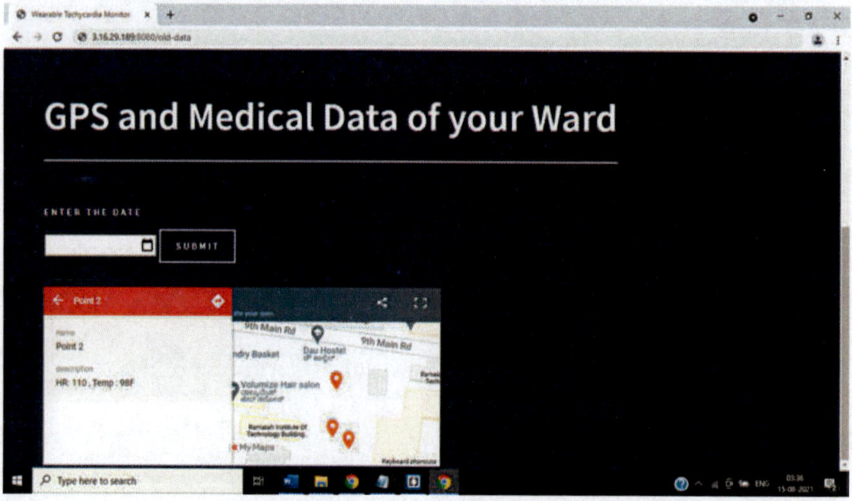

Fig. 8 Medical data of previous incidents

5 Conclusion

This paper exhibited the design and development of a wearable IoT based, battery-powered real-time heart rate monitoring and alert system for tachycardia. It involved an optical heart rate sensor, and a temperature sensor connected to a microcontroller, and a GPS to constantly send real-time medical and location data to a central server using a GSM module via AT commands. In the database, the heart rate and

Fig. 9 Message alert

temperature data are evaluated via machine learning using a K-means classification algorithm. When HR levels are high with low temperatures, the data was registered as impending danger and an SMS was sent to a registered guardian with the GPS and medical data indicating the dangerous situation. The Google Maps API on the Web site displays the heart rate and temperature during the time the data was registered. The results confirmed that the device was successful in detecting the afflicted person with 90% of accuracy. This device will help reduce the medical expenses of the patient by solving the need for constant care. It can also be employed in hospitals to ensure better care for its patients by providing care before an emergency occurs.

Further improvements can be made to reduce the footprint device, efficiency of the UI, and employ better machine learning algorithms.

References

1. Rajendra Acharya U, Fujita H, Lih OS, Hagiwara Y, Tan JH, Adam M (2017) Automated detection of arrhythmias using different intervals of tachycardia ECG segments with convolutional neural network. Inf Sci 405:81–90
2. Fernandez F, Pallis GC. Opportunities and challenges of the internet of things for healthcare: systems engineering perspective. In: International conference on wireless mobile communication and healthcare, pp 263–266
3. Hassanalieragh M, Page A, Soyata T, Sharma G, Aktas M, Mateos G (2015) Health monitoring and management using internet-of-things (IoT) sensing with cloud-based processing: opportunities and challenges services computing (SCC). In: 2015 IEEE international conference, pp 285–292

4. He S, Chen J et al (2014) Mobility and intruder prior information improving the barrier coverage of sparse sensor networks. IEEE Trans Mob Comput 13:1268–1282
5. Jersak LC, da Costa AC, Callegari DA (2013) A systematic review on mobile health care. Technical report 073, Faculdade de Informática PUCRS—Brazil, May 2013
6. Lee H, Shin SY, Seo M, Nam GB, Joo S (2016) Prediction of ventricular tachycardia one hour before occurrence using artificial neural networks. Sci Rep 6:1–7. https://doi.org/10.1038/srep32390
7. Kracht M, German Aerospace Centre (DLR) (2004) Tracking and interviewing individuals with GPS and GSM technology on mobile electronic devices. In: Conference paper submitted to the seventh international conference on travel survey methods, Costa Rica
8. Lee SJ, Tewolde G, Kwon J (2014) Design and implementation of a vehicle tracking system using GPS/GSM/GPRS technology and smartphone application. In: Conference paper submitted to the 2014 IEEE world forum on internet of things (WF-IoT)
9. Gnana Sheela K, Varghese AR, Department of ECE, APJ Abdul Kalam Technological University, Kerala, India (2019) Machine learning based health monitoring system. In: International multi-conference on computing, communication, electrical and nanotechnology, I2CN-2K19, 25th and 26th Apr 2019
10. Vineetha Y, Misra Y, Krishna Kishore K, ECE Department, GMRIT, Rajam, Affiliated to JNTUK University, AP, India (2020) A real-time IoT based patient health monitoring system using machine learning algorithms. Euro J Mole Clin Med 7(4):2912–2925
11. Wong TK, Mun HK, Phang SK, Lum KL, Tan WQ (2020) Real-time machine health monitoring system using machine learning with IoT technology. In: Conference paper submitted to the MATEC web of conferences 335, 14th EURECA 2020
12. SrivardhanReddy K, Sidaarth R, Reddy SA, Shettar R (2019) IoT based health monitoring system using machine learning. Int J Adv Res Innov Ideas Educ
13. Paramita S, Bebartta HND, Pattanayak P. IoT based healthcare monitoring system using 5G communication and machine learning models. In: Health informatics: a computational perspective in healthcare, pp 159–182
14. Vairam T, Sarathambekai S, Umamaheswari K, Jothibanu R, Manojkumar D (2020) Real-time cardiovascular health monitoring system using IoT and machine learning algorithms. In: Securing IoT and big data, 17 Dec 2020
15. Murthy SVSN, Satyanarayana BVV, Srinivas CVVS (2021) Location tracking and warning system of a ship using Arduino. Published in 2021 5th international conference on computing methodologies and communication (ICCMC)
16. Patil U, Mathad SN, Patil SR (2018) Vehicle tracking system using GPS and GSM using mobile applications. Int J Innov Sci Res Technol 3(5)
17. Mackowiak PA, Wasserman SS, Levine MM (1992) A critical appraisal of 98.6 °F, the upper limit of the normal body temperature, and other legacies of Carl Reinhold August Wunderlich. J Am Med Assoc 268:1578–1580

Performance Evaluation of EfficientNet Model Towards Malaria Parasite Detection in Segmented Blood Cells from Thin-Blood Smear Images

Swati Aggarwal, Akhil Vaid, Paras Kaushik, Anirudh Goel, and Aniket Kamboj

Abstract Malaria is a potentially fatal disease caused by infected Anopheles mosquitoes. Traditional diagnosis of malaria involves examination of thin blood smear slides under a microscope by trained microscopists to detect infected blood cells. This process is expensive, and results depend both on the quality of smear and on the expertise of the microscopist. Thus, work done in this field is focussed on automating this detection of infected cells. Early work for this task included using hand-engineered features and machine learning algorithms. This approach was taken over by the advent of CNNs which provided an end-to-end solution, right from feature extraction to classification. Work done in this field thus shifted towards using state-of-the-art CNNs. The authors of this paper found that most of the existing models that had been used for this problem had good classification accuracy but had big sizes that could not be run on a normal computing device. They identified that a new set of state-of-the-art CNNs, called EfficientNets which promised similar performance at much smaller sizes, had not been used for this task. So, they worked to evaluate the performance of EfficientNet models for this medical image classification task. The authors trained the EfficientNet-B0 model on the malaria images dataset taken from the National Library of Medicine (NLM) (Dataset used: https://lhncbc.nlm.nih.gov/LHC-publications/pubs/MalariaDatasets.html). The authors then computed the key performance parameters of the model and compared them against the existing models for this problem. The work done showed that using Efficient-Net models for this task achieved similar accuracy to existing models and significantly reduced the number of parameters.

Keywords Malaria · CNN · EfficientNet · Blood smear slides

S. Aggarwal · A. Vaid (✉) · P. Kaushik · A. Goel · A. Kamboj
Department of Computer Engineering, Netaji Subhas University of Technology, New Delhi, India
e-mail: akhilvaid21@gmail.com

S. Aggarwal
e-mail: swati.aggarwal@nsut.ac.in

P. Kaushik
e-mail: parask.co.17@nsit.net.in

M. Gupta et al. (eds.), *Artificial Intelligence on Medical Data*,
Lecture Notes in Computational Vision and Biomechanics 37,
https://doi.org/10.1007/978-981-19-0151-5_11

1 Introduction

Malaria is caused by a parasite that enters our body when bitten by infected female Anopheles-mosquitoes. In places like India, the transmission is seasonal. The peak time of transmission in India is during and just after the rainy season. Few symptoms of this disease are fever, headache, chills which are very difficult to diagnose as malaria. Malaria can turn out to be fatal if proper treatment is not given to the patient on time. It can be diagnosed by examining the blood smear under a microscope. Unfortunately, the diagnosis is dependent on human expertise, and sometimes, it is not very accurate, and it is frequently not available, especially in under-developed areas where malaria is preponderance.

According to the report released by WHO [1], there were 219 million cases of malaria in 2019 and 409,000 deaths. Children less than 5 years old are the most vulnerable amongst all groups affected by malaria. They accounted for 274,000 (67%) of all malaria deaths worldwide in 2019. According to that report, the union of India accounted for 3% of all malaria cases in the world and 6% of the deaths. The report estimated the total cases in India. The number estimated was 6.57 million (5.4–8.3 million) infected and 24,540 deaths (1600–46,500). The biggest stress of malaria in India is endured by people living in remote and backward areas because > 90–95% cases are reported from rural areas whereas mere <5 to 10% cases are from urban areas (Fig. 1).

Blood Smear Test [2]—Thick blood smears in combination with thin blood smears are needed by the doctors to figure out if you are infected by a malarial parasite. On

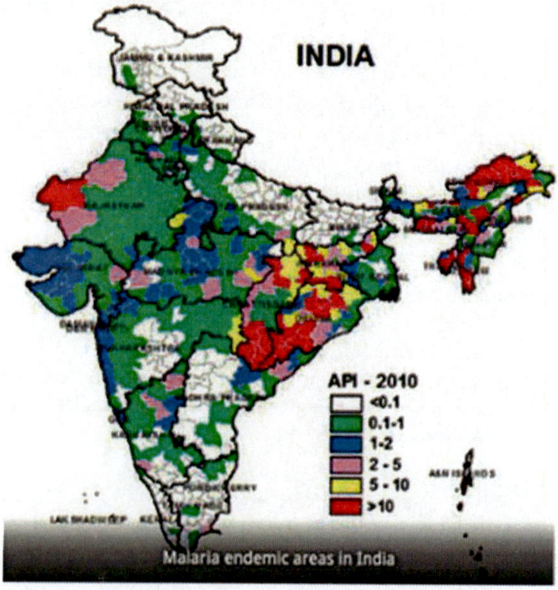

Fig. 1 Depiction of number of malaria cases in India

getting the test negative, doctors will repeat the test by taking blood smears every 8 h for 2 days to check for malaria.

Usually finger pricks are used to take the blood smear. The use of doing both thick and thin blood smear test is as follows:

- A thick blood smear is prepared by keeping a blood drop on a glass slide. Since we are seeing a large blood sample under a microscope, thick blood smears are dominantly used to determine the presence of malaria.
- Thin blood smear helps doctors to determine what species of malaria is present in the blood.

Of all the tests present for malaria diagnosis, blood smears are the medically accepted tests for malaria detection.

1.1 *Technological Background and Prior Work*

After getting the blood smear images, we can classify each image as parasitized or non-parasitized using machine learning techniques.

Generally, machine learning models used in this problem use CNNs.

In this research, we have focused on studying the following models for our problem statement:

(1) AlexNet (2) VGG-16 (3) ResNet-50 (4) Xception.

We shall discuss and elaborate each of them one by one.

A. *AlexNet*

The architecture of AlexNet [3] consists of the total layer, which is as follows according to Krizhevsky (2012) (Fig. 2):

An RGB input image is sent to the architecture and the final layer (FC3) contains the probabilities of the image belonging to each class. Out of those probabilities, maximum can be selected, and we can say that the image is predicted to be of that class.

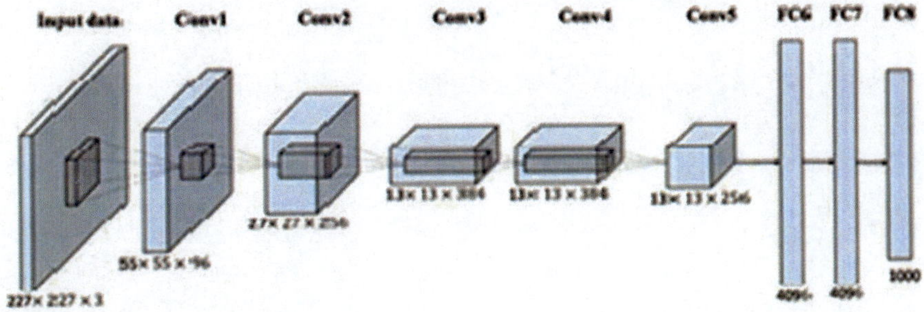

Fig. 2 Architecture of AlexNet

Some of the features of the AlexNet which makes AlexNet different from others are as follows:

1. ReLU nonlinearity
2. Dropout
3. Multiple GPUs.
4. Overlapping pooling
5. Data augmentation.

The number of parameters in AlexNet is **61 million.**

B. *VGG-16*

The architecture of VGG-16 [4] is divided into five blocks and three fully connected layers, which are as follows (Fig. 3):

The number of parameters in VGG-16 is **138 million** [4]

C. *ResNet-50*

The architecture of ResNet-50 consists of 50 layers [5] (Fig. 4):

The number of parameters in ResNet-50 [5] is **23 million.**

D. *Xception*

The architecture of Xception [6] consists of 36 layers divided into separate layers and blocks (Fig. 5):

The number of parameters in Xception [6] is **24 million.**

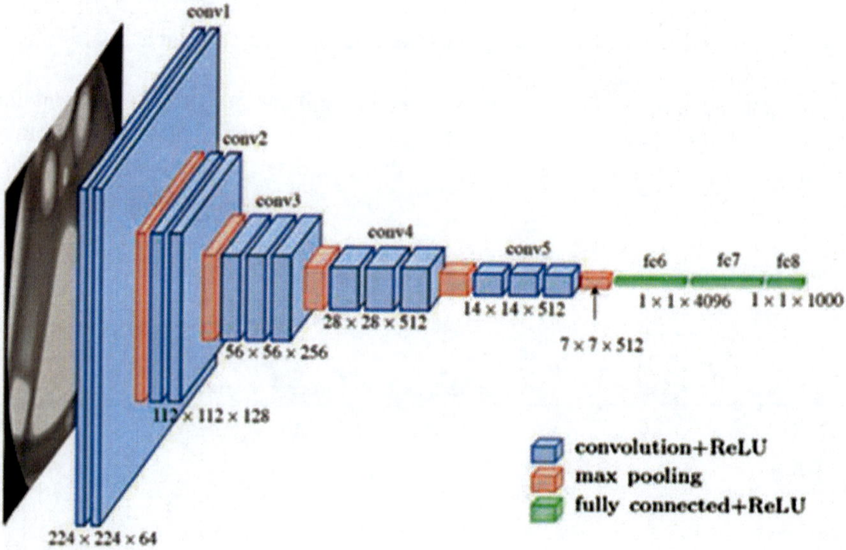

Fig. 3 Architecture of VGG16

layer name	output size	18-layer	34-layer	50-layer	101-layer	152-layer
conv1	112×112	7×7, 64, stride 2				
		3×3 max pool, stride 2				
conv2_x	56×56	[3×3, 64; 3×3, 64]×2	[3×3, 64; 3×3, 64]×3	[1×1, 64; 3×3, 64; 1×1, 256]×3	[1×1, 64; 3×3, 64; 1×1, 256]×3	[1×1, 64; 3×3, 64; 1×1, 256]×3
conv3_x	28×28	[3×3, 128; 3×3, 128]×2	[3×3, 128; 3×3, 128]×4	[1×1, 128; 3×3, 128; 1×1, 512]×4	[1×1, 128; 3×3, 128; 1×1, 512]×4	[1×1, 128; 3×3, 128; 1×1, 512]×8
conv4_x	14×14	[3×3, 256; 3×3, 256]×2	[3×3, 256; 3×3, 256]×6	[1×1, 256; 3×3, 256; 1×1, 1024]×6	[1×1, 256; 3×3, 256; 1×1, 1024]×23	[1×1, 256; 3×3, 256; 1×1, 1024]×36
conv5_x	7×7	[3×3, 512; 3×3, 512]×2	[3×3, 512; 3×3, 512]×3	[1×1, 512; 3×3, 512; 1×1, 2048]×3	[1×1, 512; 3×3, 512; 1×1, 2048]×3	[1×1, 512; 3×3, 512; 1×1, 2048]×3
	1×1	average pool, 1000-d fc, softmax				
FLOPs		1.8×10^9	3.6×10^9	3.8×10^9	7.6×10^9	11.3×10^9

Fig. 4 Architecture of ResNet50

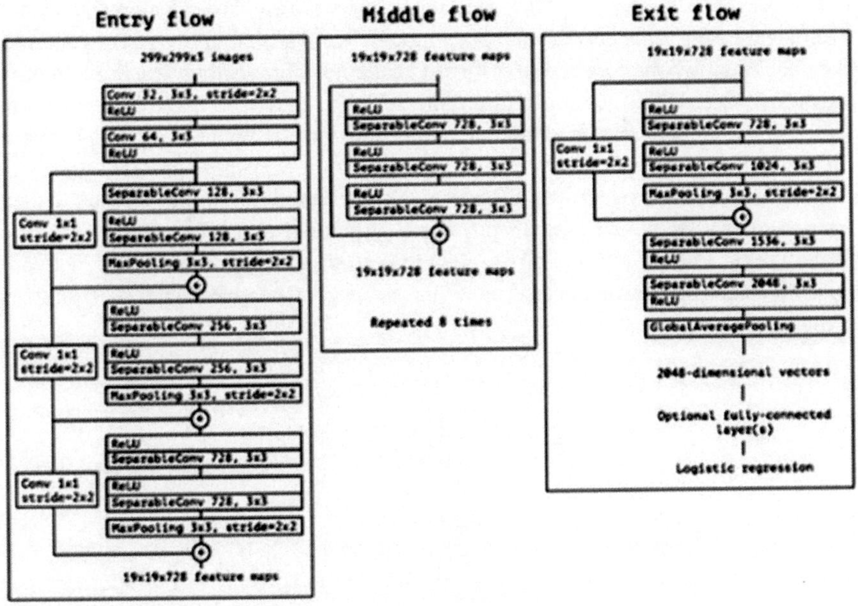

Fig. 5 Architecture of Xception

1.2 *Shortcomings and Challenges in the Prior Work*

Most of the work done in using computer programmes for the task of classifying blood cell images were based on techniques with manually engineered features for making decisions. This process of hand-engineered features requires priori expertise. This is scalable given the prevalence of malaria particularly in lower-income regions. Recent work in this field, starting from the base paper of 2018 [7], uses CNN as automatic feature extractors and then fits a classifier on top of the extracted features. Most of them use the CNNs which were state-of-the-art at their time as pretrained feature extractors. In all previous work, the practice is to scale one of the three dimensions. This arbitrary scaling follows no proper technique, and no method exists to improve CNNs.

1.3 *Efficient-Network*

A. *Introduction to Efficient-Net*

There are three dimensions of a CNN: depth, width and resolution. Depth is defined as the count of layers in the network, width means the number of filters in one layer and resolution is the input image resolution.

The prevalent methodology used by researchers was to first develop CNNs at a given level of resource cost. Then, those models were scaled up in order to achieve more accuracy when more resources were available. This scaling was done in only one of the three above-mentioned dimensions. This increased accuracy up to a point, but the performance saturated after that point. Also increasing the depth alone makes training a model difficult due to challenges such as the vanishing gradient problem.

The EfficientNet paper [8] proposes a novel model scaling method called compound scaling which aims to provide a structured and efficient way to scale up networks. The novelty lies in the idea that all three dimensions of the network must be scaled uniformly to maximize the accuracy for given level of resources available (Fig. 6).

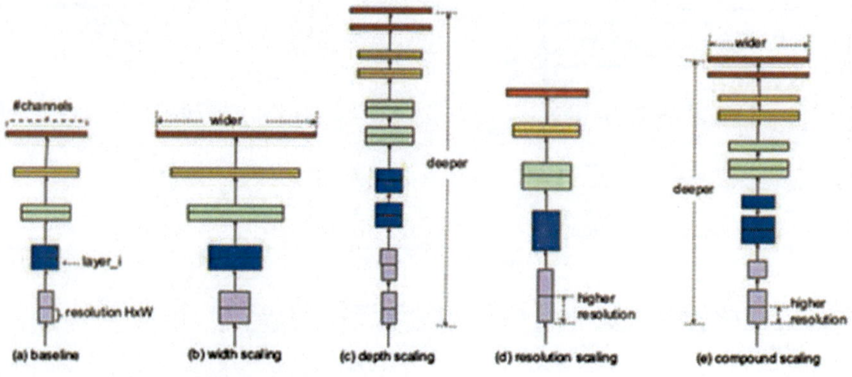

Fig. 6 Various dimensions for scaling CNNs

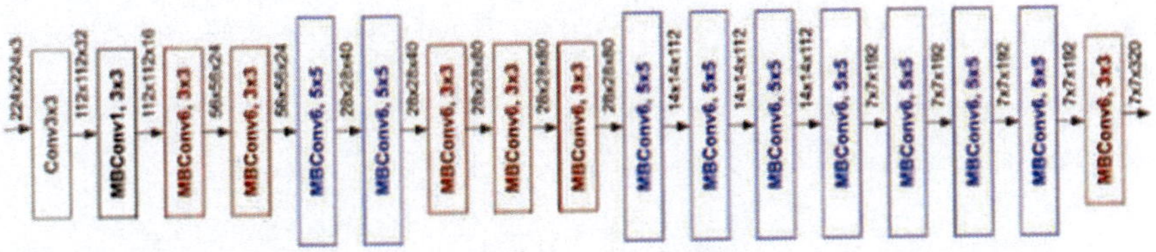

Fig. 7 Efficient B0-baseline network

B. *Compound Scaling*

The EfficientNet paper [8] proposed a simple scaling technique called compound scaling. It uses ϕ, a compound coefficient. This coefficient scales the three above-mentioned dimensions equally.

The compound coefficient is a user-specified coefficient which tells how many computational resources are available. Three constants α, β, γ were obtained using grid-searching, and these constants determine the increase in depth, width and resolution, respectively.

These equations determine the scaling:

Depth	$d = \alpha\ \phi$
Width	$w = \beta\ \phi$
Resolution	$r = \gamma\ \phi$
	s.t. $\alpha \cdot \beta \cdot \gamma \approx 2$
	$\alpha \geq 1, \beta \geq 1, \gamma \geq 1$

The paper sets a constraint (2.4.4) to quantify the relation that for any new ϕ the total number of FLOPS increase approximately by 2^{ϕ}.

C. *EfficientNet Architecture*

The paper [8] proposed a baseline B-0 model on which compound scaling was applied iteratively to scale up the baseline model. The parameter for compound scaling (ϕ) was iteratively increased to scale up the model giving the series of models from EfficientNet-B1 to EfficientNet-B7 (Fig. 7).

D. *Improvements over existing previous models*

The paper [8] compared EfficientNet with other existing CNNs on ImageNet dataset. The EfficientNet models achieve both higher accuracy and lesser number of parameters over existing CNNs (Fig. 8).

2 Dataset

The dataset [9] we used in our research was the Malaria Dataset from LHNCBC, which is a part of NLM.

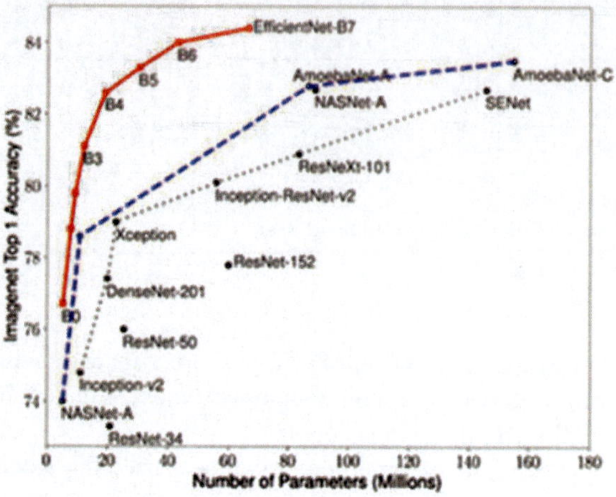

Fig. 8 Number of parameters v/s accuracy of various existing models

Some important points for the dataset are:

1. Giemsa-stained thin blood smear slides from 50 healthy patients and 150 P. falciparum-infected were photographed and collected at Chittagong Medical College Hospital, Bangladesh.
2. A smartphone's camera was used to acquire images of slides for the microscopic field of view.
3. Then, the de-identified images and annotations were archived at NLM.
4. Then, [7] the images were manually annotated by an expert slide reader at Mahidol-Oxford Tropical Medicine Research Unit in Bangkok, Thailand.
5. Then, the red blood cells were detected and segmented using a level-set-based algorithm.

The dataset [9] contains a total 27,558 cell images, out of which half of them are parasitized and half are uninfected cells (Fig. 9).

3 Methodology

Our approach consists of the following steps:

1. Importing required libraries
2. Importing data
3. Constructing model
4. Training model
5. Recording statistics and saving model
6. Plotting results.

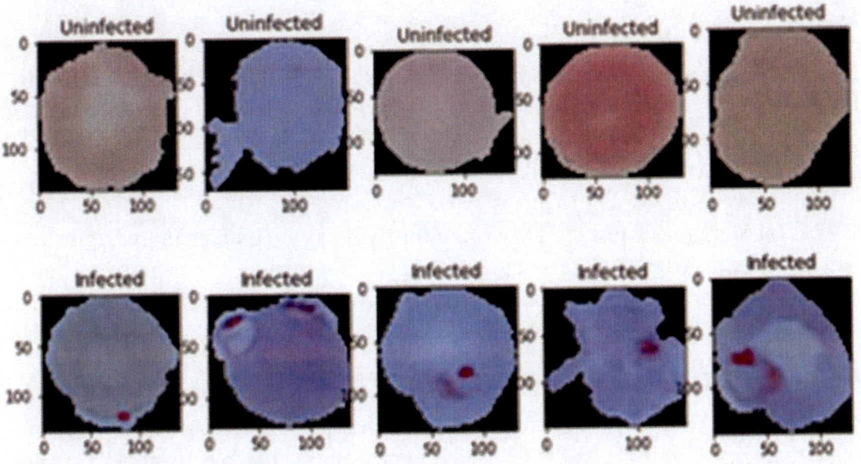

Fig. 9 Comparison of parasitised and uninfected cells

A. *Importing required libraries*

Code has been written in Python language. The following libraries have been imported subsequently:

Numpy

Numpy is a Python library used for manipulating large, multi-dimensional arrays and matrices and has a large collection of mathematical functions to be used on arrays. Version used in our code is: 1.19.4.

Pandas

Pandas is library available for Python which provides data manipulation and analysis. It offers DataFrame Objects, data alignment, tools for reading/writing data, slicing, splicing, merging, joining, etc. Version used in our code is: 1.1.5.

Os

Os is a core Python module allowing to carry out tasks with OS-specific routines. We have used Os to manipulate directories where data is being stored.

Pickle

Pickle is a module written for Python which implements protocols for serializing and de-serializing Python objects into binary form. The whole process is called 'Pickling', and we have used it to save our models securely. Version used in our code is: 4.0.

Google.colab

Google.colab is a library written for Python which we have called to unzip our dataset from Google Drive. Subsequently, it mounts our dataset into content subfolder. Version used in our code is: 1.19.4.

CV2

CV2 is a computer vision library written for Python which helps in image manipulation. Many tasks such as resizing, stretching, smoothening of images, etc., can be carried out using the library. Specifically, we have used it for resizing our images to desired size. Version used in our code is: 1.19.4.

Matplotlib.pyplot

Matplotlib is a library available for Python which provides plotting capability among other mathematical tools. We have used it to plot our results in easy-to-read fashion. Version used in our code is: 1.19.4.

Keras

Keras is open source available for Python, library that gives an interface for manipulating artificial neural networks. Keras also acts as an interface for the TensorFlow library which works with Tensors. Version used in our code is: 1.0.0.8.

B. *Importing Data*

We start by unzipping data from our local zip file. Size of infected images and uninfected images is (13,780, 13,780).

Dataset [9] is divided into two sets. Training and Validation Sets.

ImageDataGenerator is imported from Keras.

```
from keras.preprocessing.iamge.

import ImageDataGenerator.
```

Training dataset is created using ImageDataGenerator with a batch size of 256. Resizing is done to achieve target size of 64 $\times$ 64 pixels. Class mode is set to be 'binary' since there are only two classes, 'infected' and 'uninfected' (Figs. 10 and 11).

Validation set is created similar to the training set but with fewer images. Training set contains 23,426 images, whereas the validation set contains 4132 images (Figs. 12 and 13).

C. *Constructing Model*

Model is created using 'Model' submodule imported from Keras

```
from keras.models import sequential; model is created to be of Sequential type.

model = Sequential().
```

```
train_datagen = ImageDataGenerator(rescale = 1/255,
                                   horizontal_flip = True,
                                   width_shift_range = 0.2,
                                   height_shift_range = 0.2,
                                   fill_mode = 'nearest',
                                   validation_split = 0.15,
                                   zoom_range = 0.3,
                                   rotation_range = 30)


train_generator = train_datagen.flow_from_directory( directory =
    image_base_dir,
    class_mode = "binary",
```

Fig. 10 Code for creating training dataset

Fig. 11 Code for creating training set cntd

```
batch_size = 256,
target_size = (image_dims,image_dims),
subset = "training"
)
```

Then, the EfficientNet-B0 model is created (Fig. 14).

Weights are given to be 'imagenet' which corresponds to using pretrained CNN on ImageNet Dataset. Input shape is (64 × 64 × 3) pixels. Max pooling layer is used for it.

Then, we add our layer to the model we have created. model.add(efficient_net).

Subsequently, we add four layers (densely connected) with Eexponentially decreasing number of units. Each layer has LeakyReLU with alpha = 0.2 used (Fig. 15).

LeakyReLU is used so as to alleviate the 'vanishing gradient' problem. It also speeds up training (Fig. 16).

Finally, we add a densely connected layer with activation-type 'Sigmoid' for the final layer. This will have a single unit and will give output whether we have identified as 'infected' or 'uninfected' class.

model.add(Dense(units = 1, activation = 'sigmoid')).

```
val_datagen = ImageDataGenerator(rescale =
                                 1/255., validation_split = 0.15)

val_generator = val_datagen.flow_from_directory( directory =
    image_base_dir,
    class_mode = "binary",
    batch_size = 256,
    target_size =
    (image_dims,image_dims), subset =
    "validation" )
```

Fig. 12 Code for creating validation set

Finally, we print out a summary of our model:

```
model.summary().
```

We can see each layer with the corresponding number of parameters (Fig. 17).

Total parameters turn out as approximately 4.42M with 4.37M trainable and 42K non-trainable parameters.

D. *Training our Model*

From Keras, we first import our optimizer

```
from keras.optimizers import Adam
```

We use Adam optimizer with rate of learning of 1e-5. Loss function we have used as 'binary_crossentropy' which is suited for two class classification problems. Subsequently, we compile our model on our hyperparameters (Fig. 18).

Then, we set the number of epochs as 225, with 30 steps per epoch. Validation steps are as 15 (Fig. 19).

E. *Recording statistics and Saving our model*

We record statistics as follows (Fig. 20):

For saving our model, we use the dump function of Pickle library. Some examples of functionality are (Fig. 21):

F. *Plotting our results*

Using Matplotlib, we plot our results. An example is (Fig. 22):

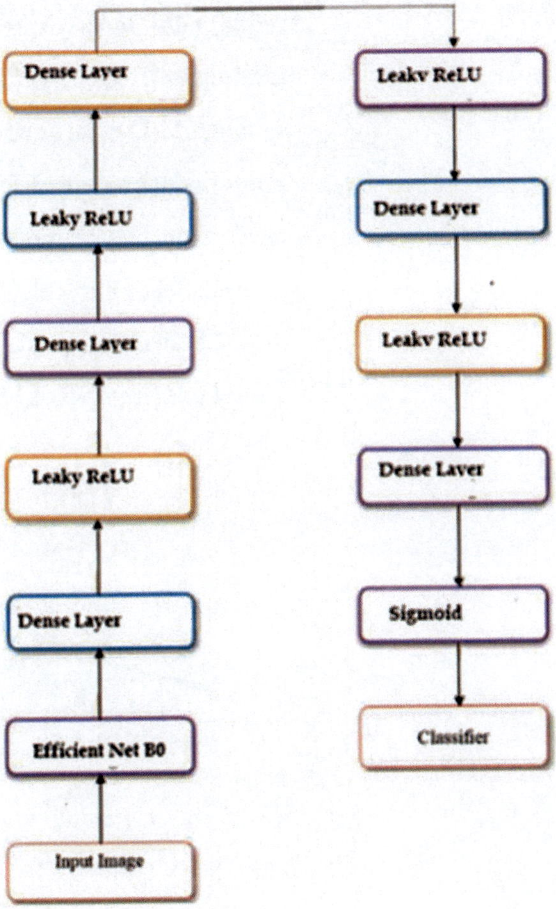

Fig. 13 Architecture of our model

```
efficient_net = EfficientNetB0(
        weights='imagenet',
        input_shape=(image_dims,image_dims,3),
        include_top=False,
        pooling='max'
    )
```

Fig. 14 Code for creating Efficient Net B0

Fig. 15 Code for adding layers to model

```
model.add(Dense(units = 256))
model.add(LeakyReLU(alpha=0.2))
model.add(Dense(units = 128))
model.add(LeakyReLU(alpha=0.2))
model.add(Dense(units = 64))
model.add(LeakyReLU(alpha=0.2))
model.add(Dense(units = 32))
model.add(LeakyReLU(alpha=0.2))
```

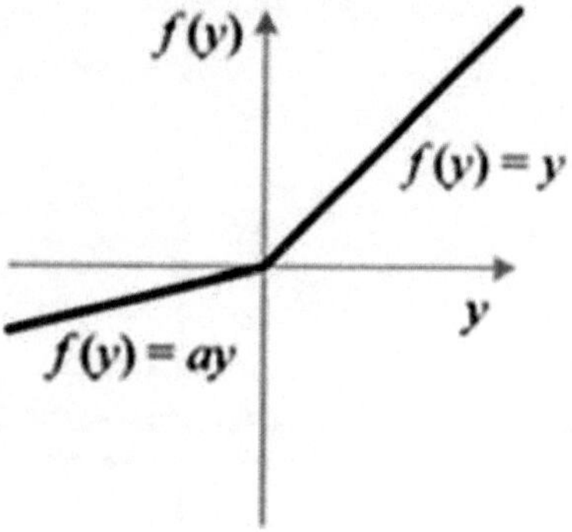

Fig. 16 Graph of LeakyReLU function

4 Task Achievement

The following are the results we got by running our model on training data:

1. Accuracy: 94.43%
2. Total number of parameters: 4,420,765
    a. Number of trainable parameters: 4,378,749
    b. Number of non-trainable parameters: 42,016
3. Area under the receiver operating characteristic (ROC) curve (AUC) 0.983
4. Precision: 92.648%
5. Recall: 96.398%
6. F-score: 94.486%
7. Matthews correlation coefficient (MCC): 0.889.

```
Model: "sequential"
_________________________________________________________________
Layer (type)                 Output Shape              Param #
=================================================================
efficientnet-b0 (Functional) (None, 1280)              4049564
_________________________________________________________________
dense (Dense)                (None, 256)               327936
_________________________________________________________________
leaky_re_lu (LeakyReLU)      (None, 256)               0
_________________________________________________________________
dense_1 (Dense)              (None, 128)               32896
_________________________________________________________________
leaky_re_lu_1 (LeakyReLU)    (None, 128)               0
_________________________________________________________________
dense_2 (Dense)              (None, 64)                8256
_________________________________________________________________
leaky_re_lu_2 (LeakyReLU)    (None, 64)                0
_________________________________________________________________
dense_3 (Dense)              (None, 32)                2080
_________________________________________________________________
leaky_re_lu_3 (LeakyReLU)    (None, 32)                0
_________________________________________________________________
dense_4 (Dense)              (None, 1)                 33
=================================================================
Total params: 4,420,765
Trainable params: 4,378,749
Non-trainable params: 42,016
_________________________________________________________________
```

Fig. 17 Summary of our model

Fig. 18 Code for adding metrics from Keras

```
model.compile(optimizer=Adam(lr=0.00001),
loss='binary_crossentropy',
metrics=['accuracy',
keras.metrics.Precision(),
keras.metrics.Recall(),
keras.metrics.AUC(),
keras.metrics.TruePositives(),
keras.metrics.TrueNegatives(),
keras.metrics.FalsePositives(),
keras.metrics.FalseNegatives()])
```

Fig. 19 Setting epochs and final settings for validation steps

```
history = model.fit_generator(
    train_generator,
    epochs = 225,
    steps_per_epoch = 30,
    validation_data = val_generator,
    validation_steps = 15
)
```

Fig. 20 Recording statistics through code

```
acc = history.history['accuracy']
val_acc = history.history['val_accuracy']
prec = history.history['precision']
val_prec = history.history['val_precision']
loss = history.history['loss']
val_loss = history.history['val_loss']
recall = history.history['recall']
val_recall = history.history['val_recall']
auc = history.history['auc']
val_auc = history.history['val_auc'] tp
= history.history['true_positives'] tn =
history.history['true_negatives'] fp =
history.history['false_positives'] fn =
history.history['false_negatives']
val_tp = history.history['val_true_positives']
val_tn = history.history['val_true_negatives']
val_fp = history.history['val_false_positives']
val_fn = history.history['val_false_negatives']
```

```
with open('/content/drive/MyDrive/results/acc.pkl','wb')
as f: pickle.dump(acc,f)
with
open('/content/drive/MyDrive/results/val_acc.pkl','wb')
as f: pickle.dump(val_acc,f)
with
open('/content/drive/MyDrive/results/prec.pkl','wb')as
f:pickle.dump(prec,f)
with
open('/content/drive/MyDrive/results/val_prec.pkl','wb')
as f: pickle.dump(val_prec,f)
with
open('/content/drive/MyDrive/results/loss.pkl','wb') as f:
pickle.dump(loss,f)
```

Fig. 21 Code for saving model through Pickle

Fig. 22 Code for plotting results through Matplotlib

```
plt.plot(epochs,acc,'bo',label =
'Training Accuracy',c='r')
plt.plot(epochs,val_acc,'b',label =
'Validation Accuracy',c='purple')
plt.title('Training and Validation Accuracy')
plt.legend()
plt.figure()
```

The following are the results we got by running our model on validation data:

1. Accuracy: 94.3755%
2. Total number of parameters: 4,420,765
    a. Number of trainable parameters: 4,378,749
    b. Number of non-trainable parameters: 42,016
3. Area under the receiver operating characteristic (ROC) curve (AUC) 0.984
4. Precision: 91.321%
5. Recall: 98.012%
6. F-score: 94.548%
7. Matthews correlation coefficient (MCC): 0.8899.

Table 1 Confusion matrix on training data

	Predicted: positive	Predicted: negative
Actual: Positive	3667	137
Actual: Negative	291	3585

Table 2 Confusion matrix on validation data

	Predicted: positive	Predicted: negative
Actual: Positive	1873	38
Actual: Negative	178	1751

See Tables 1 and 2.

Plotting graphs on our results (Fig. 23):

The above figure shows that our model is not overfitting as both training and validation accuracy are almost the same (Figs. 24, 25, 26 and 27; Table 3).

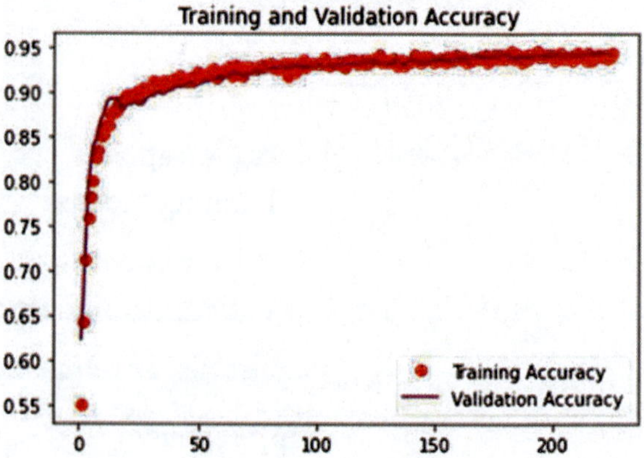

Fig. 23 Depiction and comparison of training and validation accuracy

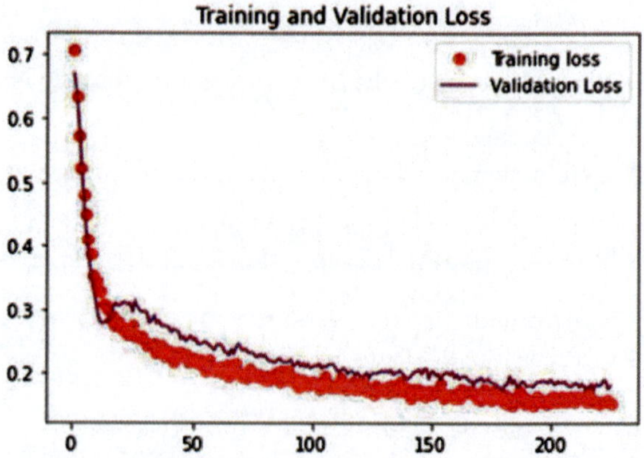

Fig. 24 Depiction and comparison of training and validation loss

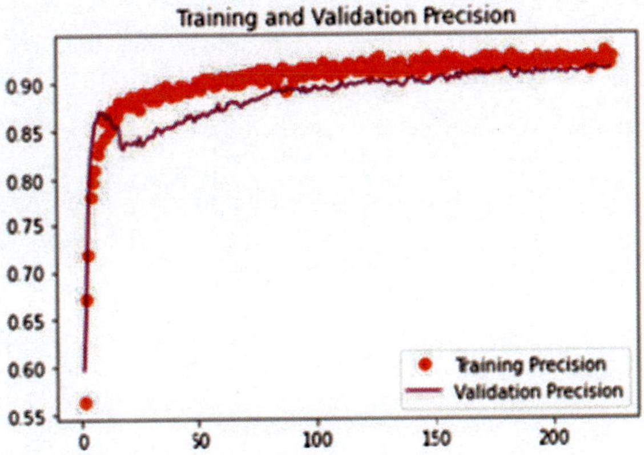

Fig. 25 Depiction and comparison of training and validation precision model

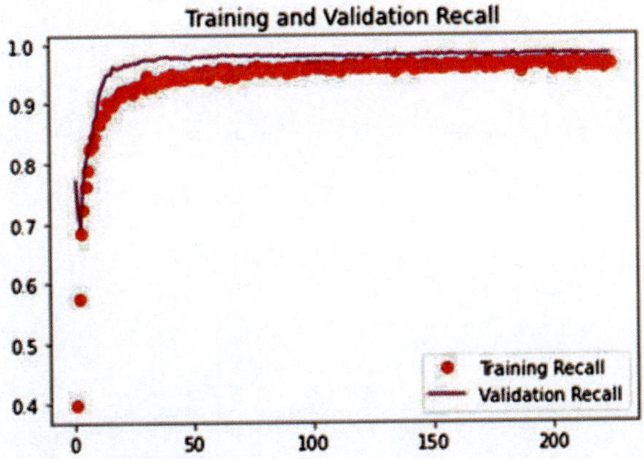

Fig. 26 Depiction and comparison of training and validation recall

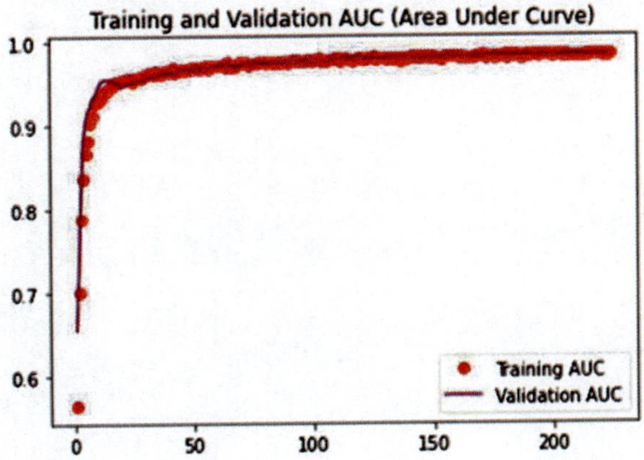

Fig. 27 Depiction and comparison of training and validation area under curve (AUC)

Table 3 Comparison of performance of prior models and our model

Models	Accuracy	No. of parameters	AUC	Sensitivity	Specificity	F1-score	MCC	Ratio-to-EfficientNet (#Parameters)
AlexNet	0.944	61M	0.983	0.940	0.933	0.937	0.872	13.9×
VGG-16	0.945	138M	0.981	0.939	0.951	0.945	0.887	31.4×
ResNet-50	0.957	23M	0.990	0.945	0.969	0.957	0.912	5.2×
Xception	0.890	24M	0.948	0.931	0.835	0.895	0.772	5.5×
EfficientNet-B0	0.945	4.4M	0.985	0.964	0.925	0.949	0.898	1×

From the above figure, it can be concluded that the use of EfficientNet-B0 model for this task achieved accuracy similar to existing models while drastically reducing the number of parameters. More concretely, it reduced the parameters from the existing model having the highest accuracy, i.e. Resnet-50 by 5.5×.

5 Conclusion and Future Work

A. *Conclusion*

We conclude that the EfficientNet-B0 model gives almost similar performance with less number of parameters. This is very important since in future hardware will require faster times and faster responses which are only available with either more computational power or efficient programmes.

B. *Future Work*

In future, we will look into working on the variants of EfficientNet model such as B1, B2, B3, B4, B5, B6 and B7. They offer similar and higher performance at cost of slightly more parameters. We will also try uploading our model on mobiles, hence creating a portable classifier.

References

1. World Health Organization Report for Malaria 2019. https://www.who.int/publications/i/item/9789241565721
2. Blood smear test information from https://www.uofmhealth.org/
3. Krizhevsky A, Sutskever I, Hinton GE (2012) Imagenet classification with deep convolutional neural networks. Adv Neural Inf Process Syst 25:1097–1105
4. Simonyan K, Zisserman A (2014) Very deep convolutional networks for large-scale image recognition. arXiv preprint arXiv:1409.1556
5. He K et al (2016) Deep residual learning for image recognition. In: Proceedings of the IEEE conference on computer vision and pattern recognition
6. Chollet F (2017) Xception: Deep learning with depthwise separable convolutions. In: Proceedings of the IEEE conference on computer vision and pattern recognition
7. Rajaraman S et al (2018) Pre-trained convolutional neural networks as feature extractors toward improved malaria parasite detection in thin blood smear images. PeerJ 6:e4568
8. Tan M, Le Q (2019) Efficientnet: rethinking model scaling for convolutional neural networks. In: International conference on machine learning. PMLR
9. Dataset used: https://lhncbc.nlm.nih.gov/LHC-publications/pubs/MalariaDatasets.html
10. Rajaraman S, Jaeger S, Antani SK (2019) Performance evaluation of deep neural ensembles toward malaria parasite detection in thin-blood smear images. PeerJ 7:e6977

Retinal Vessel Segmentation in Fundus Image Using Low-Cost Multiple U-Net Architecture

Supratim Ghosh, Mahantapas Kundu, and Mita Nasipuri

Abstract A low-cost solution to automatic segmentation of blood vessels in Fundus images is proposed here using deep learning (DL) methodology. Prior to vessel segmentation, each Fundus image is segmented into three zones, each of homogenous intensity, through a threshold-based algorithm designed under the present work. Thereafter, three U-Nets are separately trained for vessel segmentation in each of three zones. The models so trained are tested on two benchmark datasets, viz., DRIVE and STARE and have performed comparably with the present state-of-the-art methods.

Keywords Classification · U-Net · Neural networks · Fundus · Retina · Vessel segmentation · Retinopathy

1 Introduction

Automatic segmentation of blood vessels in fundus images is necessary for quick and effective detection of pathological cases. The most recent trend found in solving this problem is to apply convolutional neural networks (CNNs) under the framework of deep learning (DL) [1, 2]. Although such approach provides high segmentation accuracy, for its implementation, costlier high-end GPU-based machines are required. Under this situation, the work presented here aims at providing a low-cost solution of the problem within the same DL-based framework without hampering the segmentation accuracy of retinal blood vessels in Fundus images with both normal and pathological cases. A comprehensive study of the methods in the past can be found in the work by Fraz et al. [3]. In this section, we present a brief account of the methods [1, 2, 4–14] which have been considered as benchmark ones for comparison of performances of the recent methods.

S. Ghosh (✉) · M. Kundu · M. Nasipuri
Jadavpur University, Kolkata, West Bengal, India
e-mail: supratimghosh2772@gmail.com

M. Gupta et al. (eds.), *Artificial Intelligence on Medical Data*,
Lecture Notes in Computational Vision and Biomechanics 37,
https://doi.org/10.1007/978-981-19-0151-5_12

2 Literature Review

The vessel segmentation algorithms can be broadly categorized into unsupervised and supervised algorithms. The unsupervised algorithms comprise of techniques such as line detectors [4], matched filtering [15], morphological transformations [16–18], model-based methods [19–21] and multiscale segmentation methods [5, 22–24]. However, these algorithms are often computationally complex [25] and perform poorly on images with pathology. Supervised algorithms, in contrast, rely on the use of classifiers that learn to segment from labeled ground truth images. Algorithms such as the K-nearest neighbor [8], Gaussian mixture model (GMM) [10], support vector machine (SVM) [12], neural networks [13], decision trees [14], AdaBoost [26] and conditional random fields [6] have been proposed in the literature as classifiers for vessel and non-vessel pixels. In 2016, Liskowski et al. [1] published a paper highlighting the use of convolutional neural networks (CNNs) for automated self-learning of features and classification using ground truth images. DL-based approach for blood vessel segmentation in Fundus images is further pursued by Yan et al. [2]. In this work, they have introduced a joint loss function for the CNN by taking care of traditional pixel-wise losses treating all vessel pixels, thick or thin, with equal importance and the segment-level loss giving more importance on thin vessel pixels compared to thick ones.

3 Proposed Methodology

The work presented here has drawn inspiration from an observation made by Roychowdhury et al. [5]. In their observation, the authors have stressed on the need for a method with low computational complexity and high segmentation accuracy, especially for Fundus images with pathological cases. Although the DL-based method proposed by Liskowski et al. [1] achieves high segmentation accuracy, it requires costly, sophisticated and complex hardware systems for its realization. The proposed work deals with the problem of blood vessel segmentation in fundus images by partitioning it into three sub-problems, viz., by segmenting each image into three regions, optic disk region, high contrast region and low-contrast region. Shallow convolutional U-Nets [27] are employed individually on each of these three regions for segmentation of retinal blood vessels. As each region within the image comprises of a uniform intensity, it becomes easier for a shallower network to learn the features more aptly (Fig. 1).

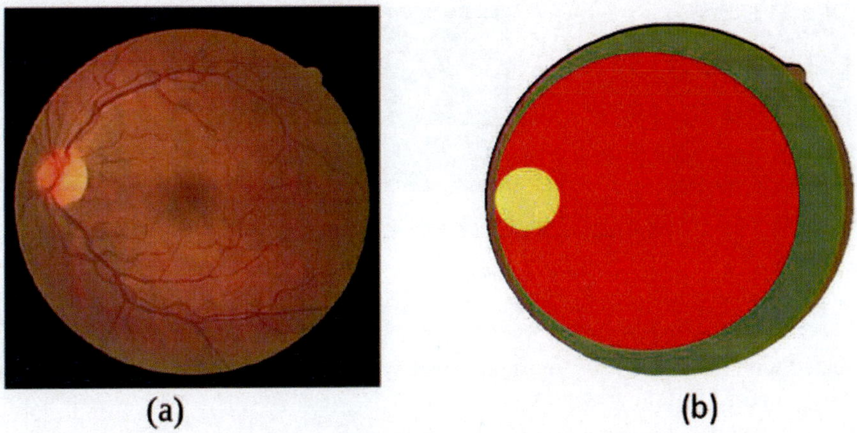

Fig. 1 **a** Original RGB Fundus image. **b** Highlighted homogeneous regions observable in the image (yellow—optic disk, red—high-contrast region, green—low-contrast region)

3.1 Region Segmentation

The first stage of the approach is the segmentation of the input Fundus image into three homogeneous regions. The contrast strength of the images degrades with increasing distances from the optic disk [1]. This property of the Fundus image has been utilized in this work for separation of the retinal image into three separate non-overlapping regions. For a normal Fundus image, high contrast between the background pixels and the vessel pixels is usually observed in the optic disk. For extracting the optic disk, the RGB Fundus image is first converted to Lab color space. In case of Fundus images, the optic disk region can be observed to comprise of the highest pixel intensities in the luminance channel of the Lab color space which makes the region distinct. Thus, the optic disk region can be separated from the rest of the Fundus image by designing a threshold-based approach. To calculate the optimal threshold for optic disk segmentation, the optic disk region in the luminous images of the training data is initially labeled manually to create a ground truth of the region. Let each luminance image of the Fundus image dataset be named I and the corresponding manually segmented optic disk image be named I_G. Let the threshold for isolating the optic disk region for image I be ε.

$$\varepsilon = \frac{\sum_{i=1}^{m} \sum_{j=1}^{n} I(i, j)}{mn} \quad (1)$$

where m and n represent the resolution of the image I and $I(i, j)$ represents the pixel intensity of the point (i, j) in the image I.

Image I is then thresholded to select all pixels above the threshold ε, and a new thresholded image I_ε is thereby formed. All pixels in I_ε with a pixel intensity greater than 0 are considered to be the optic disk pixels. After a thresholded image I_ε is

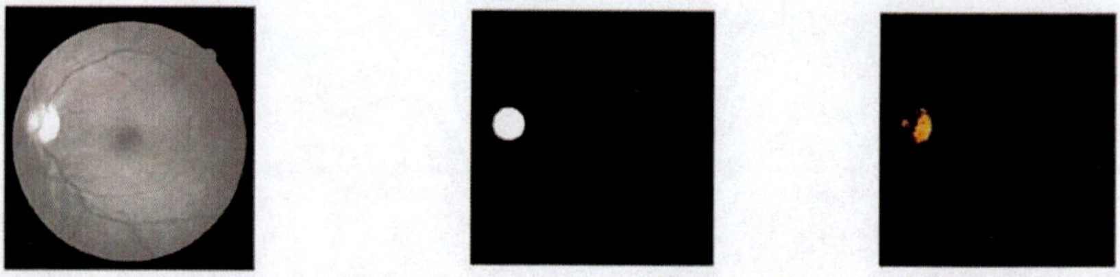

(a)Luminance Image (b)Manually Created ground truthI_G (c)Segmented ImageI_o

Fig. 2 **a–c** Illustration for optic disk segmentation

created with the threshold ε, an image overlap measure λ_ε between images I_G and I_ε is computed as follows:

$$\lambda_\varepsilon = \frac{\mathrm{TP}}{\mathrm{TP} + \mathrm{FN} + \mathrm{FP}} \tag{2}$$

where TP = true positives = the total number of disk pixels in I_ε correctly labeled as disk pixels, FP = false positives = the total number of non-disk pixels in I_ε falsely labeled as disk pixels, FN = false negatives = the total number of disk pixels in I_ε falsely labeled as non-disk pixels.

The value of ε is next time incremented by one to create a modified version of the thresholded optic disk image I_ε, and λ_ε is recalculated on I_ε. The process is repeated until ε is assigned its maximum value, i.e., 255 here. A (λ_ε vs. ε) curve is plotted to select the final value for ε at which λ_ε attains its maximum value. The process is repeated for all the images of the training set. Taking an average of all the maximum values of λ_ε calculated for all training images, the value of ε is finally selected. On the basis of the value of ε, optic disk pixels in all the training images are segregated (Fig. 2).

To isolate the low-contrast peripheral region in the non-optic disk region of the Fundus images, the remaining region of the fundus image is coarsely segmented into two more regions with the help of simple average thresholding. The following method for further segmentation of non-optic disk regions is calculated. Let I_N be the RGB image of the non-optic disk region. The equation for I_N may be expressed as:

$$I_N = I_F - I_0 \tag{3}$$

where I_F is the original RGB Fundus image and I_0 the RGB image of the optic disk region.

To isolate the low-contrast peripheral region of I_F, the red channel of the I_N image is extracted because the contrast between vessel and background pixels is minimal in red channel images. The red channel of the I_N image is used as it is easier to isolate the homogeneous regions without significant errors from the presence of vessel pixels. Let the red channel image of the non-optic disk region be named I_N^R. Because of the dark coloration of the peripheral region, the intensity of the pixels there would have lower values in the red channel compared to other channels. To

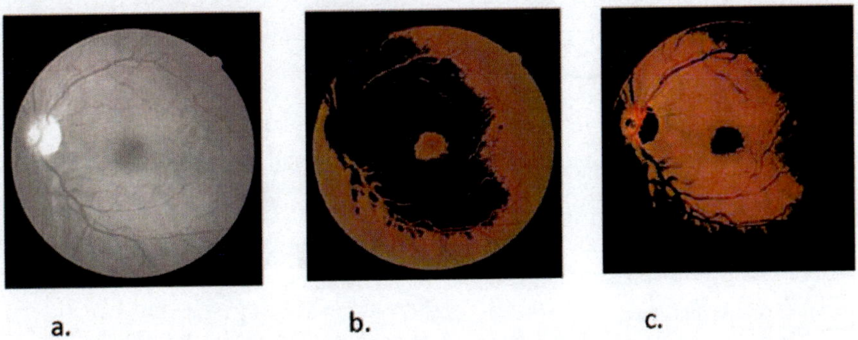

Fig. 3 **a** Red channel image. **b** Low-contrast region I_P. **c** High-contrast region I_H

separate this peripheral region, a thresholding operation is used with the limit set to the average pixel intensity of the image I_N^R. All pixels in I_N^R having intensity more than the threshold limit are removed. The remaining pixels constitute the low-contrast peripheral region of I_N^R. After isolating the optic disk region I_G and the low-contrast peripheral region I_P, the remaining region is taken to be the third homogeneous region in the input fundus image I. This region is denoted by I_H and expressed as:

$$I_H = I_N - I_P \tag{4}$$

In this manner, three homogeneous regions, viz., I_0, I_H and I_P, are obtained from the input Fundus image I. The outputs obtained in this stage are mainly used to facilitate training of three separate convolutional neural networks (CNNs) for vessel segmentation in the three homogenous regions (Fig. 3).

3.2 U-Net Architecture

A U-Net architecture acts similar to an encoder-decoder network and utilizes only convolutional layers for its realization which reduces the number of trainable parameters in the network. A detailed structure of our network is shown in Fig. 4.

Each layer of the network uses the ReLU activation to mitigate the vanishing gradient problem often associated with CNNs. In the final layer of the network, the Sigmoid activation function is used for classification. During testing, instead of considering all the outputs from the final layer, only the center pixel classification of the patch is considered. Thus, in our work, the classification of each pixel depends on its own respective patch.

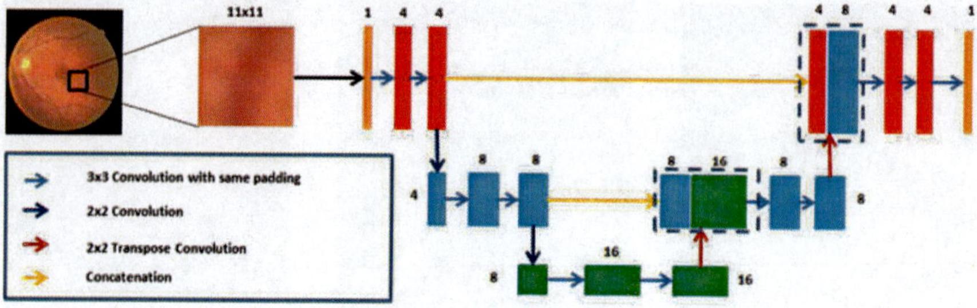

Fig. 4 U-Net Architecture

4 Experimental Observation

In this section, performances of the proposed method are presented tested on DRIVE [9] and STARE [15] datasets. The model proposed here works on overlapping image patches, and depending on the class of the central pixel in each image patch, the patch is labeled as positive or negative data. For obvious reasons, the number of negative patches is much higher than that of positive patches. In such situations, to pay equal attention to samples from both (positive and negative) classes during the training, the loss function is weighted by the class weight of each class. Performances of the proposed model are separately tested on DRIVE and STARE data with patch sizes 9×9, 11×11 and 13×13. The results so obtained are shown in Table 1. From Table 1, it can be observed that the proposed model works well with a patch size of 11×11 in respect to accuracy, sensitivity and specificity considered together. Hence, all the subsequent experiments are conducted with a patch size of 11×11. The value of specificity as obtained on DRIVE data by the proposed method is close to the second highest benchmark value obtained by Liskowski et al. [1]. Another important point to be noted here is that values of AUCs obtained by the proposed method trail behind some benchmark results obtained by Liskowski et al. [1] and Roy Chowdhury et al. [5] and Yan et al. [2]. Not withstanding that, in terms of sensitivity, the proposed method has performed more or less comparable with the work by Liskowski et al. [1] and outperformed the results obtained by Roy Chowdhury et al. [5] and Yan et al. [2].

As can be observed from the results shown in Table 1, the proposed model achieves comparable performances without necessitating high-end machines as required for the deep learning model proposed by Liskowski et al. [1] and Yan et al. [2]. This has been possible because three deep network models with far lesser parameters are separately trained here for vessel segmentation in optic disk, high-contrast and low-contrast regions of Fundus images.

Table 1 Performance comparison

Methods	AUC		Accuracy		Sensitivity		Specificity	
	DRIVE	STARE	DRIVE	STARE	DRIVE	STARE	DRIVE	STARE
Jiang et al. [7]	0.932	0.929	0.891	0.901	0.83	0.857	0.9	0.9
Niemeijer et al. [8]	0.93	–	0.942	–	0.689	–	0.969	–
Staal et al. [9]	0.952	0.961	0.944	0.952	0.719	0.697	0.977	0.981
Ricci et al. [12]	0.955	0.96	0.959	0.965	0.775	0.903	0.972	0.939
Roy Chowdhury et al. [5]	0.967	0.967	0.949	0.956	0.739	0.732	0.978	0.984
Liskowski et al. [1]	0.973	0.982	0.925	0.931	0.916	0.931	0.924	0.93
Yan et al. [2]	0.975	0.98	0.954	0.961	0.765	0.758	0.981	0.985
Proposed method	0.93	0.92	0.929	0.928	0.925	0.903	0.929	0.93

5 Conclusion

The proposed method provides a low-cost solution to automatic blood vessel segmentation in Fundus images using DL methodology without making any significant compromise in respect to performances. To achieve this, a threshold-based segmentation algorithm is designed to segment each Fundus image in three regions of homogenous intensity, and U-Nets are separately employed on these individual regions for vessel segmentation. As a result, training and testing of CNNs can be performed simply on a general purpose computer without looking for a costly GPU-based high-end machine. Hence, the method provides an easy achievable solution to blood vessel segmentation in Fundus images with performances comparable to benchmark results.

References

1. Liskowski P, Krawiec K (2016) Segmenting retinal blood vessels with deep neural networks. IEEE Trans Med Imaging 35(11):2369–2380
2. Yan Z, Yang X, Cheng KTT (2018) Joint segment-level and pixel-wise losses for deep learning based retinal vessel segmentation. IEEE Trans Biomed Eng 65(9):1912–1923
3. Fraz MM et al (2012) Blood vessel segmentation methodologies in retinal images–a survey. Comput Methods Programs Biomed 108(1):407–433
4. Nguyen UTV, Bhuiyan A, Park LAF, Ramamohanarao K (2013) An effective retinal blood vessel segmentation method using multi-scale line detection. Pattern Recognit 46(3):703–715
5. Roychowdhury S, Koozekanani DD, Parhi KK (2015) Iterative vessel segmentation of fundus images. IEEE Trans Biomed Eng 62(7):1738–1749

6. Orlando JI, Prokofyeva E, Blaschko MB (2017) A discriminatively trained fully connected conditional random field model for blood vessel segmentation in fundus images. IEEE Trans Biomed Eng 64(1):16–27
7. Jiang X, Mojon D (2003) Adaptive local thresholding by verification-based multithreshold probing with application to vessel detection in retinal images. IEEE Trans Pattern Anal Mach Intell 25(1):131–137
8. Niemeijer M, Staal J, van Ginneken B, Loog M, Abramoff MD (2004) Comparative study of retinal vessel segmentation methods on a new publicly available database, p 648
9. Staal J, Abràmoff MD, Niemeijer M, Viergever MA, van Ginneken B (2004) Ridge-based vessel segmentation in color images of the retina. IEEE Trans Med Imaging 23(4):501–509
10. Soares JVB, Leandro JJG, Cesar RM, Jelinek HF, Cree MJ (2006) Retinal vessel segmentation using the 2-D Gabor wavelet and supervised classification. IEEE Trans Med Imaging 25(9):1214–1222
11. Martinez-Perez ME, Hughes AD, Thom SA, Bharath AA, Parker KH (2007) Segmentation of blood vessels from red-free and fluorescein retinal images. Med Image Anal 11(1):47–61
12. Ricci E, Perfetti R (2007) Retinal blood vessel segmentation using line operators and support vector classification. IEEE Trans Med Imaging 26(10):1357–1365
13. Maràn D, Aquino A, Gegàndez-Arias ME, Bravo JM (2011) A new supervised method for blood vessel segmentation in retinal images by using gray-level and moment invariants-based features. IEEE Trans Med Imaging 30(1):146–158
14. Fraz MM et al (2012) An ensemble classification-based approach applied to retinal blood vessel segmentation. IEEE Trans Biomed Eng 59(9):2538–2548
15. Hoover A (2000) Locating blood vessels in retinal images by piecewise threshold probing of a matched filter response. IEEE Trans Med Imaging 19(3):203–210
16. Mendonca AM, Campilho A (2006) Segmentation of retinal blood vessels by combining the detection of centerlines and morphological reconstruction. IEEE Trans Med Imaging 25(9):1200–1213
17. Zana F, Klein J-C (2001) Segmentation of vessel-like patterns using mathematical morphology and curvature evaluation. IEEE Trans Image Process 10(7):1010–1019
18. Miri MS, Mahloojifar A (2011) Retinal image analysis using curvelet transform and multistructure elements morphology by reconstruction. IEEE Trans Biomed Eng 58(5):1183–1192
19. Vermeer KA, Vos FM, Lemij HG, Vossepoel AM (2004) A model based method for retinal blood vessel detection. Comput Biol Med 34(3):209–219
20. Lam BSY, Yan H (2008) A novel vessel segmentation algorithm for pathological retina images based on the divergence of vector fields. IEEE Trans Med Imaging 27(2):237–246
21. Lam BSY, Gao Y, Liew AW-C (2010) General retinal vessel segmentation using regularization-based multiconcavity modeling. IEEE Trans Med Imaging 29(7):1369–1381
22. Budai A, Michelson G, Hornegger J (2010) Multiscale blood vessel segmentation in retinal fundus images. Bild für die Medizin, pp 261–265
23. Budai A, Bock R, Maier A, Hornegger J, Michelson G (2013) Robust vessel segmentation in fundus images. Int J Biomed Imaging 2013:1–11
24. Palomera-Perez MA, Martinez-Perez ME, Benitez-Perez H, Ortega-Arjona JL (2010) Parallel multiscale feature extraction and region growing: application in retinal blood vessel detection. IEEE Trans Inf Technol Biomed 14(2):500–506
25. Roychowdhury S, Koozekanani DD, Parhi KK (2014) DREAM: diabetic retinopathy analysis using machine learning. IEEE J Biomed Heal Informatics 18(5):1717–1728
26. Lupascu CA, Tegolo D, Trucco E (2010) FABC: retinal vessel segmentation using AdaBoost. IEEE Trans Inf Technol Biomed 14(5):1267–1274
27. Ronneberger O, Fischer P, Brox T (2015) U-Net: convolutional networks for biomedical image segmentation

Detection of Coronavirus in Electron Microscope Imagery Using Convolutional Neural Networks

Apratim Shrivastav and Aaditya Narayan Subedy

Abstract Object recognition has always been one of the most crucial areas of research in modern times when it comes to industries like healthcare. Not only it is useful to automate certain kinds of jobs to improve efficiency, but also it also takes away the scope of human errors that are ever-present within such fields. A multitude of factors like differences in skills, experience, and more critically, basic human errors contributes to inadequacy and thereby inefficiency. These are not merely hypotheticals, but legitimate concerns raised in research papers and by professionals within their respective sectors as laid out in (Hopfer et al. in Histopathology 78:358–370, 2021 [1]). Yet, no significant work has been done so far in combining object detection with the identification of the coronavirus. This is due to some significant obstacles. Recognizing these viruses is a tedious task because substantial datasets of electron microscope images are not available. In this paper, we propose a method of detecting the virus through electron microscope imagery using the You Only Look Once (YOLO) convolutional neural network (CNN) model. We trained our model on an amalgamation of two datasets. The first is an open-access dataset named 'electron microscopy of SARS-CoV particles Dataset' by Laue, Michael; Kauter, Anne; Hoffmann, Tobias; Michel, Janine; Nitsche, Andreas. The second is 'The SARS-CoV-2 Dataset' by Northeastern University, made available to us on request by Prof. Jiawei Zhang. Overall, we achieved a mean average precision of 86.5%. Such a system is beneficial as it assists in a large amount of research work, both in terms of increasing efficiency and minimizing errors. However, we do believe there exists a lot of scope to build upon the work with a larger dataset in order to achieve higher accuracy.

Keywords Coronavirus · Object detection · YOLO

A. Shrivastav (✉) · A. N. Subedy
Delhi Technological University, New Delhi, India
e-mail: apratimshrivastav@gmail.com

M. Gupta et al. (eds.), *Artificial Intelligence on Medical Data*,
Lecture Notes in Computational Vision and Biomechanics 37,
https://doi.org/10.1007/978-981-19-0151-5_13

1 Introduction

Recognition of objects and understanding features that distinguish several objects are critical skills that human beings develop from a very young age. Humans are very good at recognizing objects in challenging environments, across a spectrum of lighting conditions. Object detection, therefore, plays a crucial role in automating that process and allowing computer programs with high accuracy to instead recognize such objects on humans' behalf. The use of object detection has increased in magnitude and frequency across industries and fields such as retail [2], marketing [3], and healthcare [4].

SARS-CoV-2 currently dominates all deadlines, being the cause of a global pandemic responsible for the deaths of millions of people across the world [5–8]. Consequently, there has been a concentrated effort globally to research and understand this virus. Virology allows us to deeply understand the biological basis of contagiousness and pathogenesis leading to the effects shown by the virus on the human body [1]. Transmission electron microscopy seems to be the perfect instrument for this purpose. There has been a lot of work done to identify the challenges in the current procedures; we employ to recognize this virus from the images gained through this process. This research predominantly highlights the lack of professionals in the fields of virology or electron microscopy dedicated to viral diseases [9–14]. In this paper, we propose a method for detecting the virus through electron microscope imagery using a convolutional neural network called You Only Look Once (YOLO).

2 Methodology

In this section, we explain the algorithm chosen to fit the data, the experimental setup, and processing computations performed before training and testing the model.

2.1 Data Preprocessing

The proposed work considered electron microscopic images of the coronavirus. Since collecting different electron microscopic images is difficult, we managed to gather 200 images with minimal reputation and augmented data through two distinct datasets [15, 16]. All images were converted to a common format. For greater analysis and understanding in this paper, all 200+ images were labeled by hand, and areas of interest were identified to ultimately train the weights that will be used in the final program script. All images were scaled down to a size of 416 × 416 pixels.

In order to make proper labeling in order to train the model accurately, we studied ways in which we could do it most accurately. We worked for weeks studying differences between these viruses including size, shape of crown, and multiple other features that distinguish coronavirus to other viruses [17].

2.2 YOLO Neural Network

YOLO neural networks are applied to the image, for the purpose of dividing it into regions and predicting the bounding boxes and probabilities of each given region within the image given. The model has a multitude of advantages over classifier-based systems.

The complete image is taken during test time, and hence, the predictions are informed by the general context present within the image itself. This makes the model incredibly fast. It is more than a 1000× faster than R-CNN [18] and 100× faster than Fast R-CNN [19].

The YOLO algorithm works on the principle of division of images into (N) number of grids and each one of those grids with side (S) having dimensions of $S \times S$. Each of these N grids contributes toward the detection and localization of the object it contains. This greatly lowers the computation as both detecting images and recognition of the image are handled by cells from the image.

Each grid cell works toward predicting the limits of B bounding boxes and assigns confidence scores to these bounds. The scores reflect how confident the model is about the boxes predicted. We define the confidence generated as the probability of the object multiplied by intersection over union (IOU); if no object exists, then the confidence would be zero; otherwise, we want the score to equal the IOU.

Each bound contains five values: x, y, w, h, and the confidence score. (x, y) refers to the coordinates of the center in relation to the bound of the box, whereas w and h represent width and height along with the box.

YOLO then predicts the object based on the boxes having several overlapping regions where it has identified that object. It then evaluates the highest probability to identify the object. It uses the detection mechanism both in the cell and the cells surrounding it. It can then generate a bounding box outside a given object with a label and an associated value for a degree of certainty.

2.3 Experimental Setup

We trained our model on an amalgamation of two datasets. First, being an open-access dataset named 'electron microscopy of SARS-CoV particles Dataset' by Laue, Michael; Kauter, Anne; Hoffmann, Tobias; Michel, Janine; Nitsche, Andreas. And, secondly, 'The SARS-CoV-2 Dataset' by Northeastern University, made available to us on request by Prof. Jiawei Zhang. The datasets contained close to 200 images of images from electron microscopes of SARS-CoV-2. These are high-quality images taken in laboratories by researchers to aid the study of this virus.

3 Result and Analysis

For the analysis and training, filters for every convolutional layer were set at 18, while a singular class was specified, using 2000 epochs with a batch size of 200. As the number of epochs increases, the training loss gradually decreases. Consequently, when the loss increases, the accuracy of the model increases, and effective results are obtained. The mean average precision (mAP) of the model amounted to 86.5%, while the loss stabilized after 1600 epochs.

The result was a highly accurate model capable of recognizing coronavirus particles from electron microscope imagery.

3.1 Future Scope

In this paper, we were successfully able to develop a YOLO v4 model to identify if a sample electron microscopic image of tissue contains the coronavirus or not. The final accuracy of the model (mAP) was 86.5%. A multitude of reasons currently prevents us from further optimizing this including but not limited to a severe lack of imagery available to analyze. We were grateful to have received an adequate number of images to build a basic model, but this model can be employed at a near-ideal accuracy level if the above-mentioned impediments with respect to the availability of data are eliminated.

Secondly, the errors can be minimized massively for more accurate prediction when the model is employed by virologists and professionals in identifying the virus and preparing the dataset, in order to better train and test it. Finally, we were successful in implementing a model and providing a framework on how automation of this specific task, i.e., object recognition can be done through machine learning.

References

1. Hopfer H, Herzig MC, Gosert R, Menter T, Hench J, Tzankov A, Hirsch HH, Miller SE (2021) Hunting coronavirus by transmission electron microscopy—a guide to SARS-CoV-2-associated ultra-structural pathology in COVID-19 tissues. Histopathology 78:358– 370. https://doi.org/10.1111/his.14264
2. Cai Y, Wen L, Zhang L, Du D, Wang W (2020) Rethinking object detection in retail stores
3. Shergill GH, Sarrafzadeh H, Diegel O, Shekar A (2008) Computerized sales assistants: the application of computer technology to measure consumer interest; a conceptual framework. J Electron Commer Res 9(2):176191
4. Esteva A, Robicquet A, Ramsundar B et al (2019) A guide to deep learning in healthcare. Nat Med 25:24–29
5. Li Q, Guan X, Wu P et al (2020) Early Transmission dynamics in Wuhan, China, of novel coronavirus—infected pneumonia. N Engl J Med 382:1199–1207
6. Huang C, Wang Y, Li X et al (2020) Clinical features of patients infected with 2019 novel coronavirus in Wuhan, China. Lancet 395:497–506

7. Battegay M, Kuehl R, Tschudin-Sutter S et al (2019) 2019-novel Coronavirus (2019-nCoV): estimating the case fatality rate—a word of caution. Swiss Med Wkly 150:w20203
8. Leuzinger K, Roloff T, Gosport R et al (2020) Epidemiology of SARS-CoV-2 emergence amidst community-acquired respiratory viruses. J Infect Dis 222:1270–1279
9. Kissling S, Rotman S, Gerber C et al (2020) Collapsing glomerulopathy in a COVID-19 patient. Kidney Int 98:228–231
10. Farkash EA, Wilson AM, Jentzen JM (2020) Ultrastructural evidence for direct renal infection with SARS-CoV-2. J Am Soc Nephrol 31:1683–1687
11. Su H, Yang M, Wan C et al (2020) Renal histopathological analysis of 26 postmortem findings of patients with COVID-19 in China. Kidney Int 98:219–227
12. Rousse C, Curtis E, Moran L et al (2020) Electron microscopic investigations in COVID-19: not all crowns are coronas. Kidney Int 98:505–506
13. Smith KD, Akilesh S, Alpers CE et al (2020) Am I a coronavirus? Kidney Int 98:506–507
14. Goldsmith CS, Miller SE (2020) Caution in Identifying coronaviruses by electron microscopy. J Am Soc Nephrol 31:2223–2224
15. Laue M, Kauter A, Hoffmann T et al (2021) Morphometry of SARS-CoV and SARS-CoV-2 particles in ultrathin plastic sections of infected Vero cell cultures. Sci Rep 11:3515
16. Li C, Zhang J, Kulwa F, Qi S, Qi Z (2021) A SARS-CoV-2 microscopic image dataset with ground truth images and visual features
17. Miller SE, Goldsmith CS (2020) Caution in identifying coronaviruses by electron microscopy. J Am Soc Nephrol 31(9):2223–2224. https://doi.org/10.1681/ASN.2020050755
18. R-CNN: regions with convolutional neural network features. https://github.com/rbgirshick/rcnn
19. Fast R-CNN: https://github.com/rbgirshick/fast-rcnn

A Survey on UWB Textile Antenna for Wireless Body Area Network (WBAN) Applications

Sapna Arora, Sharad Sharma, and Rohit Anand

Abstract The remote monitoring healthcare systems have tended to the advent of wearable antennas. To ensure that this monitoring should not make any effect on one's regular habits, the elements (i.e. antenna and sensor) should be embedded in clothes. Wearable antennas work in close proximity to human body. Since there is coupling of antenna radiations with human body tissues, the antenna performance parameters are affected. The performance of ultra-wideband (UWB) wearable antennas with various ground plane configurations such as partial ground plane, CPW-fed ground plane and full ground plane has been reviewed in this work. Moreover, this work also mentions the complexity and advantages related to antennas having full ground plane.

Keywords Wearable antenna · Ultra-wideband · Coplanar waveguide (CPW) · Wireless body area network (WBAN) · Ground plane

1 Introduction

In the era of information technology and communication, wearable devices are very important for on-body applications to establish Internet of Things [1] and wireless sensor network [2, 3]. These devices must be small, low-powered and capable of linking to a hub to access the Internet. The aim of such devices is to enhance the life quality by embedding electronics with fabrics. Due to the rapid growth of wearable devices, researchers are focusing on wireless body area network (WBAN) that connects sensors on body with Internet [4, 5]. However, the high-speed connectivity of wearable devices requires the WBAN system to possess ultra-wideband (UWB) antennas. Some of the applications of WBAN are given in Table 1 [6–9].

S. Arora · S. Sharma
MM(DU), Mullana, Ambala, India

R. Anand (✉)
DSEU, G.B.Pant Okhla-1 Campus, New Delhi, India
e-mail: roh_anand@rediffmail.com

M. Gupta et al. (eds.), *Artificial Intelligence on Medical Data*,
Lecture Notes in Computational Vision and Biomechanics 37,
https://doi.org/10.1007/978-981-19-0151-5_14

Table 1 WBAN applications [6–9]

Types of applications	Examples
Health care	Blood pressure, capsule endoscope, drugs delivery and electrocardiogram
Entertainment	Computer games, virtual reality, smart watches and LED dresses
Sports	Heartbeat, motion sensor, acetometer
Military	Fire detection, vital signs monitoring, smart suits and life jackets
Lifestyle	Posture detection and emotion detection

Wearable antenna is an important component of WBAN for wireless communication with different devices present on or off human body. Lots of challenges are involved in designing an antenna for wearable devices due to absorption of the radio signals by human body. Due to interaction of radio waves with body tissues, detuning of antenna resonant frequency occurs, which causes in the variation of radiation pattern of antenna. Also, a large amount of power is absorbed by body that causes reduction in gain [10–12]. Furthermore, uneven body surface and movement may deform the antenna shape if textile substrate is used. Bad weather conditions such as snow or rain can damage the textile antennas. However, special concern is specific absorption rate (SAR), which depicts the limit of safe exposure to radio waves energy absorbed by the body. FCC standard for SAR must be satisfied for safe communication on the body. Some more issues are related to wearable antenna such as comfortability, light weight and small size.

The wearable UWB antennas can be made with partial ground plane to fulfil the bandwidth requirement. In wearable UWB antennas with full ground plane, some optimization technique is to be applied to increase the bandwidth [13, 14]. In microstrip patch antennas, the bandwidth enhancement is a critical design issue. So, in this research paper, the related previous works are discussed on wearable UWB antennas with partial ground plane, coplanar waveguide and full ground plane in Sects. 3, 4 and 5 with a little description of the substrate selection and radiating element textile materials in Sect. 2. A brief conclusion is presented in Sect. 6.

2 Material Description for Wearable Antenna

The dielectric substrate provides the mechanical support to antenna, and radiating elements are required for patch, ground plane and feedline. Initially, it is important to select the suitable material for both substrate and radiating elements to provide flexibility, comfort and least amount of distortion with different weather conditions [15].

(a) Substrate for wearable antennas

The substrate is a crucial component because it affects the characteristics of antenna like bandwidth, reflection coefficient and efficiency. The important characteristics of substrate are loss tangent, thickness, permittivity and flexibility. The substrate with low loss tangent and permittivity tends to increase the efficiency. For the substrate, the choice of flexible materials gives a lot of comfort to the user. So, flexible substrate with low permittivity and loss tangent gives high efficiency, stable radiation pattern and acceptable gain at deforming conditions [10, 16]. Thus, flexible substrates due to their conformality are replacing the rigid substrates.

Textile materials are flexible and having low dielectric constant. This minimizes the surface losses and increases the bandwidth of antenna. Textile materials are having two categories—synthetic fibres and natural fibres. Common textiles are fleece, felt, leather, silk, jeans and cotton.

(b) **Radiating element for wearable antennas**

Conductive textiles are used to perform the function of radiators so that fully wearable antennas can be fabricated. The conductive material required for the antenna radiating element should be chosen to ensure the good radiation characteristics of antenna. Also, the flexibility should be maintained to ensure that the antenna is not distorted when worn. Some of the conductive textile flexible fabrics whose characteristics have been investigated under different environment conditions are Shield it Zelt, Flectron, Shield it Super and Shieldex Nora [17, 18].

3 Partial Ground Plane Wearable UWB Antennas

A compact elliptical shape planar UWB antenna with a size of 12 mm × 19 mm with bandwidth coverage from 2.9 to 12 GHz has been presented in [19] for WBAN applications. The narrow band interference from WiMAX, WLAN and X band satellite systems is suppressed with three notches created at 3.3–3.8 GHz, 5.1–5.825 GHz and 7.25–7.75 GHz, respectively, with two stubs of length 15.3 mm and 8.7 mm and complementary split ring resonator (CSRR) slot integration on the patch. The maximum gain obtained is 3.18 dB.

The size and shape of ground plane also affect the bandwidth of antenna. In [20], a compact novel antenna is presented with finite ground plane which provides wide bandwidth range of 2.96–11.6 GHz. This antenna's front and rear view is shown in Fig. 1. The radiating patch is fabricated on jeans substrate having $\epsilon_r = 1.7$ that is fabricated on by creating the slots and notches through the optimization process. This miniaturized UWB antenna is having dimensions 25 mm × 25 mm × 1 mm, and it provides gain of 5.47 dBi with omnidirectional radiation patterns. The SAR value measured is 1.68 W/Kg.

A novel technique for creating patch is flexible full embroidered topology that was used in [21] on felt textile substrate. The circular radiating patch made up of perfect electric conductor yarn with specific composition was embroidered by Singer Futura XL-550 embroidery machine with a T-shaped slot. A square slot was used in the partial ground plane to improve the bandwidth further. The measured UWB

Fig. 1 Textile UWB fabricated antenna **a** Front view **b** Rear view [20]

bandwidth is 2.6–11.3 GHz with dual-band characteristics at 3.36 and 8.78 GHz with the corresponding reflection coefficients −48.53 and −30.14 dB. Moreover, this antenna was tested under concave and convex bending conditions. The bending radius was taken according to human body. Slight variations in the bandwidth between experimental and simulated results are also justified.

An improved UWB antenna design with circular patch having square slot at centre and circular slot at upper part of antenna has been designed in [22]. The measured dimensions of design are $60 \times 60 \times 3$ mm^3 fabricated on flannel substrate with partial ground plane of size 50×27mm^2. Besides, the use of flannel substrate makes the antenna suitable for wearable applications and robust as it is consisting of 99% cotton. The antenna design was simulated on CST studio. The omnidirectional radiation pattern has been obtained at 9.7 GHz with gain 3.88 dB, return loss 28.99 dB and efficiency 89.06%. The directional radiation pattern has been achieved at 17.99 GHz with gain 6.365 dB, return loss 41.23 dB and efficiency 87.21%. So it can be used for wireless body area network sensor applications in military.

To enhance the directionality of WBAN antennas, different methods are used, i.e. reflecting layer, artificial magnetic conductor (AMC) structures and electromagnetic band gap (EBG) structures. These methods reduce the backward radiation towards body and hence improve the impedance matching, gain and SAR value.

In [23], a fork-slotted EBG reflector is printed at back side of substrate on the space above the partial ground plane. Due to this, the backward radiation is suppressed so that the antenna becomes most suitable for the wearable applications. This UWB antenna is fabricated on RO4003 substrate having dimensions $20 \times 25 \times 0.2$ mm^3. The bevelled Y-shaped patch with partial ground plane with size 20×5 mm^2 having inverted bell-shaped slot is designed. Further fork-slotted EBG cell is fabricated with size of 8×8 mm^2 and replicated in a 2×1 array on backside of the substrate. This structure having a peak gain of 6.25 dBi and SAR value of 0.695 W/Kg with directional radiation pattern can be used for wearable applications. Table 2 shows the comparison based on UWB wearable antennas with partial ground plane.

Table 2 Comparison based on UWB partial ground plane wearable antennas

References	Bandwidth	Size (mm^3)	SAR	Substrate	Gain (dBi)	Applications
[19]	2.9–12 GHz (with three notches)	12 × 19 × 1.6	–	FR4	3.18	WBAN
[20]	2.96–11.6 GHz	25 × 25 × 1	1.68 W/Kg,	Jeans substrate with $\epsilon_r = 1.7$	5.47	Telemedicine and mobile health systems
[21]	2.6–11.3 GHz	30 × 30 × 0.7	–	Felt substrate with $\epsilon_r = 1.45$	4	Wearable electronics applications
[22]	2–19.46 GHz	60 × 60 × 3	–	Flannel	3.88	Wireless body area network sensor (WBAN) applications in military
[23]	3.1–10.6 GHz	20 × 25 × 0.2	0.695 W/Kg	RO4003 with $\epsilon_r = 3.55$	6.25	Wearable applications

4 CPW-Fed UWB Antenna for WBAN Applications

A CPW-fed antenna [24] with a circular-shaped patch (shown in Fig. 2) was designed on polyethylene terephthalate substrate due to its flexibility, having dielectric constant of 3.2 with a loss tangent of 0.022 and thickness of 135 μm. In this design, UWB bandwidth 3.04–10.70 GHz was achieved by modifying the feedline after applying a cut-off length 11.44 mm. The average radiation efficiency achieved was 95.7% and peak gain at 10.6 GHz. This design is suitable for the various IoT applications due to omnidirectional radiation pattern. This design was also tested under bent

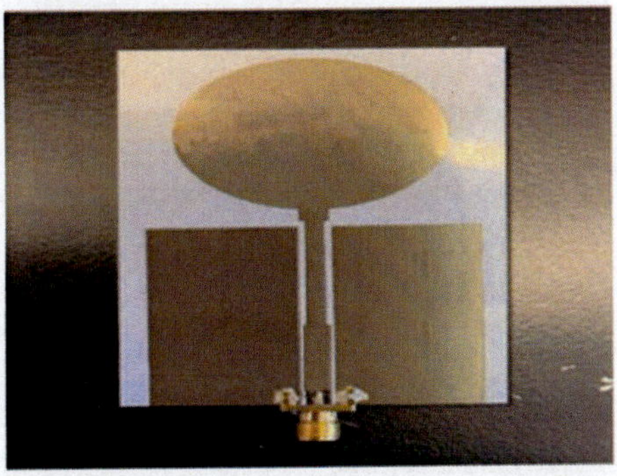

Fig. 2 Printed inkjet CPW-fed UWB antenna [24]

conditions to check for the real-time WBAN applications by placing it on a cylinder of radius 8 mm with permittivity = 2.7. The impedance matching was observed from 3.04–10.70 GHz making this design suitable for the practical WBAN application.

UWB applications demand that there should not be any interference from the coexisting narrowband. In [25], a CPW-fed semicircular radiator UWB textile antenna was designed with dual-band notched characteristics for the Bluetooth application band at 2.4 GHz and WiMAX band at 3.3–3.6 GHz with an impedance bandwidth of 1.8–10.7 GHz. The notch characteristics were implemented by etching two open ring slot structures in the patch. This antenna was designed on leather and denim textile materials, and the results were compared. To check the on-body performance, the antenna was tested on tissue of human body. The results were compared for on-body and off-body performance.

In [26], a UWB planar antenna with inverted cone was designed with CPW-fed transmission line having the dimensions 40 mm × 40 mm. To improve the radiation pattern of antennas, two holes having radius 2 mm each were placed on PICA. The ground plane was slotted due to which the bandwidth had raised up to 18 GHz. The results were calculated for the radiation efficiency by placing the antenna on the chest of human body modelled in HFSS. It was observed that at distance of 4 mm from human body, the efficiency reduces to 29% when compared to the antenna placed in free space. Also, the perfect radiation pattern was achieved due to the electromagnetic band gap (EBG) achieved due to the circular hole in the patch. Since antenna is placed on human body, there is change in radiation pattern due to the reflections from the human body. As frequency was increased from 3 to 9 GHz, the directional gain increased from 0.46 to 3.6 dB that shows the increase in reflections from human body.

A flexible wearable multiple-input multiple-output (MIMO) antenna was designed on liquid crystal polymer substrate for UWB application in [27]. Two orthogonal elements separated by F-shaped branch provided the port isolation greater than 23 dB. The radiation patches were fan shaped with C-shaped slots. The size of antenna is 30 mm × 56.5 mm fed with an asymmetric ground plane that is able to achieve bandwidth of 2.4–11.3 GHz with a notch of 3.1–4.3 GHz.

5 Full Ground Plane Wearable UWB Antennas

As discussed in above sections, antenna performance degrades in the presence of human body and radiation pattern changes which result into the unreliable performance of antenna. To enhance WBAN system performance, there is need to design a textile wearable antenna which counterparts the effects of detuning caused by the human body.

One of the most prominent solutions is to design a full ground plane between human body and radiating element. But narrow bandwidth is the main problem in antennas with full ground plane. To achieve UWB characteristics with full ground plane antennas, the combination of various bandwidth enhancement techniques is proposed. One such compact antenna with full ground plane has been built on textile substrate developed in [28] with dimensions of 39 mm × 42 mm × 3.34 mm. This antenna is shown in Fig. 3. The proposed antenna was printed on felt substrate with $\in_r = 1.45$ and 3 mm height. Broad bandwidth was achieved with multi-resonance technique and parasitic coupling technique between the two radiators, and radiators were fed with coaxial cable from its centre. Moreover, it works satisfactory in mandatory channels of low bands and high bands when evaluated on voxel model and simplified body model in bent condition at 30 degree and 60 degree. It shows bandwidth of 3.4 GHz collectively with a spatial radiation pattern having HPBW of 66.7 degree and realized gains of 4 dBi and 6 dBi. SAR evaluation was performed on detailed voxel model at 4 GHz and 8 GHz, and the values calculated are 0.335 W/ kg and 0.491 W/kg, respectively, which is lower than 1.6 W/kg over 1 g of tissue. The full ground planes ensure the high degree of resolution from human body and least body coupling that enables this antenna to work satisfactory in WBAN applications.

Full ground plane antennas [29] having dimension 88 mm × 97 mm are realized using textile felt substrate with thickness 3 mm, $\in_r = 1.45$, and Shield it Super as conductor with conductivity of 1.18×10^5 S/m and thickness of 0.17 mm. The design is formed by three primary radiators that are fed by transmission line of small length. Another patch that surrounds three primary patches is fed by capacitive coupling that leads to the increased bandwidth having range from 3.4 to 10 GHz. The gain obtained from all-textile antenna with full ground plane is having range between 3.15 dBi and 6.33 dBi when operated on back and between 5.09 dBi and 6.31 dBi when operated on chest of human body. Average radiation efficiency of 1.6% and front to back ratio of 35.7 dB are obtained through on-body measurements. The full ground plane between radiator and body provides isolation from electromagnetic radiation.

In [30], two stack patches composed of Shield it, where one is octagonal-shaped CPW-fed and another is a parasitic patch, are fabricated on two layers of felt substrate with 2 mm thickness. To further increase the bandwidth, a step is notched in both sides of ground plane near 50 Ω CPW feed line. Because of full ground plane on second layer of substrate, isolation is achieved between human body and radiators. The lowest value of SAR is 0.52 W/kg found at 7 GHz and maximum is 1.21 W/kg at

Fig. 3 Textile antenna with full ground plane [28]

3 GHz when evaluated on three-layer (skin–fat–muscle) human body model. When measured on the arm and chest of human body, the agreement is satisfactory in complete band from 3.18 to 11 GHz. The design and analysis have been done using CST Microwave studio. The fidelity factor is calculated by exciting the design with square root raised cosine pulses, and the achieved value is 96.9% to 99.6% in free space and on-body with unidirectional radiation pattern narrow band. The overall dimensions of designed antenna are $80 \times 61 \times 4.51$ mm^3.The bandwidth achieved is from 3.18 to 11 GHz with peak gain of 7.2 dBi at 9 GHz.

A flexible UWB antenna designed using conductive fabric on polydimethylsiloxane (PDMS) polymer as substrate was presented in [31]. The design contains two arc-shaped patches on PDMS substrate with $\in_r = 2.7$ and loss tangent from 0.02 to 0.07 in the bandwidth of 3.1 to 10.6 GHz. On other side of the substrate, the full ground plane is present. To enhance the bandwidth, two T-shaped slots having length less than $1 < \lambda_{\mathrm{eff}}/4$ were added in larger arc-shaped patch. The effect of full ground plane is that VSWR having value less than 2 is achieved in the range of 3.7–10.3 GHz in free space and on flat phantom that shows that the antenna's performance is not affected due to the presence of human body. This design also withstands the deformation when bent around the head and wrist of phantom and exhibits stable VSWR over UWB frequency range. Moreover, directional radiation pattern with average SAR of 0.132 W/kg and average efficiency of 27% is obtained when measured on the human muscle phantom.

Table 3 shows the comparison based on UWB wearable antennas with full ground plane.

6 Conclusion

This paper gives an overview of the UWB antennas used in the remote healthcare systems and their applications in WBAN systems. It was found that the wearable antennas fabricated on flexible substrates with low permittivity and loss tangent have very high efficiency. In addition, full ground plane antennas are not susceptible to on-body detuning due to minimal radiation in back side as compared to partial ground plane and CPW-fed antennas. The value of specific absorption rate of full ground plane antennas is very low as compared to the CPW-fed and partial ground plane antennas. Achieving UWB behaviour in full ground plane antenna is the main issue. This issue is solved at the expense of complex structure and large size with the various bandwidth enhancement techniques.

Table 3 Comparison based on UWB wearable antennas with full ground plane

References	Bandwidth (GHz)	Size (mm^3)	SAR	Substrate	Fidelity	Efficiency	Technique for bandwidth enhancement
[28]	3.1–10.6	39 × 42 × 3.34	0.491 W/kg	Felt substrate with $\in_r =$ 1.45	73.6%	–	Multi-resonance technique and active coupling
[29]	3.4–10	88 × 97 × 3	–	Felt substrate with $\in_r =$ 1.45	–	1.6%	Multi-resonance technique and capacitive coupling
[30]	3.18–11	80 × 61 × 4.51	1.21 W/kg	Felt substrate with $\in_r =$ 1.45	96.9% to 99.6%	–	Three stacking layers
[31]	3.1–10.6	80 × 67 × 3	0.132 W/kg	PDMS substrate, $\in_r =$ 2.7	–	27%	Arcs and T-shaped slots

References

1. Gupta A, Srivastava A, Anand R, Tomažič T (2020) Business application analytics and the internet of things: the connecting link. In: Shrivastava G, Peng SL, Bansal H, Sharma K, Sharma M (eds) New age analytics. Apple Academic Press, Canada, pp 249–273
2. Meelu R, Anand R (2010) Energy efficiency of cluster-based routing protocols used in wireless sensor networks. In: AIP conference proceedings, vol 1324, No 1. American Institute of Physics, pp 109–113
3. Meelu R, Anand R (2011) Performance evaluation of cluster-based routing protocols used in heterogeneous wireless sensor Networks. Int J Inform Technol Knowl Manage 4(1):227–231
4. Ashyap AY, Abidin ZZ, Dahlan SH, Majid HA, Kamarudin MR, Alomainy A, Abd-Alhameed RA, Kosha JS, Noras JM (2018) Highly efficient wearable CPW antenna enabled by EBG-FSS structure for medical body area network applications. IEEE Access 6:77529–77541
5. Vignesh NA, Kanithan S, Jana S, Konguvel E, Panigrahy AK (2021) Silicon wearable body area antenna for speech enhanced IOT and medical applications. SILICON. https://doi.org/10.21203/rs.3.rs-825439/v1
6. Salonen P, Hurme L (2003) A novel fabric WLAN antenna for wearable applications. In: IEEE antennas and propagation society international symposium digest held in conjunction with: USNC/CNC/URSI North American Radio Science Meeting (Cat. No. 03CH37450), pp 700–703. IEEE, Columbus, USA (2003). https://doi.org/10.1109/APS.2003.1219332
7. Li H, Sun S, Wang B, Wu F (2018) Design of compact single-layer textile MIMO antenna for wearable applications. IEEE Trans Antennas Propag 66(6):3136–3141
8. Ashyap AY, Abidin ZZ, Dahlan SH, Majid HA, Shah SM, Kamarudin MR, Alomainy A (2017) Compact and low-profile textile EBG-based antenna for wearable medical applications. IEEE Antennas Wirel Propag Lett 16:2550–2553
9. Chan M, Estève D, Fourniols JY, Escriba C, Campo E (2012) Smart wearable systems: current status and future challenges. Artif Intell Med 56(3):137–156
10. Salvado R, Loss C, Gonçalves R, Pinho P (2012) Textile materials for the design of wearable antennas: a survey. Sensors 12(11):15841–15857
11. Monti G, Corchia L, Tarricone L (2013) UHF wearable rectenna on textile materials. IEEE Trans Antennas Propag 61(7):3869–3873
12. Silva NL, Gonçalves LM, Carvalho H (2013) Deposition of conductive materials on textile and polymeric flexible substrates. J Mater Sci: Mater Electron 24(2):635–643
13. Anand R, Chawla P (2020) Optimization of inscribed hexagonal fractal slotted microstrip antenna using modified lightning attachment procedure optimization. Int J Microw Wirel Technol 12(6):519–530
14. Anand R, Chawla P (2016) A review on the optimization techniques for bio-inspired antenna design. In: 3rd International conference on computing for sustainable global development (INDIACom), IEEE Press, New Delhi, pp 2228–2233
15. Bashir S (2009) Design and synthesis of non uniform high impedance surface based wearable antennas. Doctoral dissertation, Loughborough University
16. Tsolis A, Whittow WG, Alexandridis AA, Vardaxoglou JC (2014) Embroidery and related manufacturing techniques for wearable antennas: challenges and opportunities. Electronics 3(2):314–338
17. Werner DH, Jiang ZH (eds) (2016) Electromagnetics of body area networks: antennas, propagation, and RF systems. Wiley
18. Bai Q, Rigelsford J, Langley R (2013) Crumpling of microstrip antenna array. IEEE Trans Antennas Propag 61(9):4567–4576
19. Doddipalli S, Kothari A (2018) Compact UWB antenna with integrated triple notch bands for WBAN applications. IEEE Access 7:183–190
20. Yadav A, Kumar Singh V, Kumar Bhoi A, Marques G, Garcia-Zapirain B, de la Torre Díez I (2020) Wireless body area networks: UWB wearable textile antenna for telemedicine and mobile health systems. Micromachines 11(6):558

21. El Gharbi M, Martinez-Estrada M, Fernández-García R, Ahyoud S, Gil I (2021) A novel ultra-wide band wearable antenna under different bending conditions for electronic-textile applications. J Textile Inst 112(3):437–443
22. Nikhil S, Singh AK, Kumar SV (2015) Design and performance of wearable ultra wide band textile antenna for medical application. Microw Opt Technol Lett 57(7):1553–1557
23. Sambandam P, Kanagasabai M, Ramadoss S, Natarajan R, Alsath MGN, Shanmuganathan S, Sindhadevi M, Palaniswamy SK (2019) Compact monopole antenna backed with fork-slotted EBG for wearable applications. IEEE Antennas Wirel Propag Lett 19(2):228–232
24. Kirtania SG, Younes BA, Hossain AR, Karacolak T, Sekhar PK (2021) CPW-fed flexible ultra-wideband antenna for IoT applications. Micromachines 12(4):453
25. Chilukuri S, Gogikar S (2019) A CPW fed denim based wearable antenna with dual band-notched characteristics for UWB applications. Progress Electromagnet Res C 94:233–245
26. Nadeem M, Khan AN, Ali Khan A, Azim T (2018) Low profile CPW fed slotted planar inverted cone ultra-wide band antenna for WBAN applications. Microw Opt Technol Lett 60(4):870–876
27. Du C, Jin G (2021) A compact CPW-fed band-notched UWB-MIMO flexible antenna for WBAN application. J Electromagnet Waves Appl 35(8):1046–1058
28. Samal PB, Soh PJ, Zakaria Z (2019) Compact microstrip-based textile antenna for 802.15. 6 WBAN-UWB with full ground plane. Int J Antennas Propagat. https://doi.org/10.1155/2019/8283236
29. Samal PB, Soh PJ, Vandenbosch GA (2013) UWB all-textile antenna with full ground plane for off-body WBAN communications. IEEE Trans Antennas Propag 62(1):102–108
30. Poffelie LAY, Soh PJ, Yan S, Vandenbosch GA (2015) A high-fidelity all-textile UWB antenna with low back radiation for off-body WBAN applications. IEEE Trans Antennas Propag 64(2):757–760
31. Simorangkir RB, Kiourti A, Esselle KP (2018) UWB wearable antenna with a full ground plane based on PDMS-embedded conductive fabric. IEEE Antennas Wirel Propag Lett 17(3):493–496

Effective Overview of Different ML Models Used for Prediction of COVID-19 Patients

Harinder Singh, Binay Kumar Pandey, Shaji George, Digvijay Pandey, Rohit Anand, Nidhi Sindhwani, and Pankaj Dadheech

Abstract The goal of this research is to see how well is a fast primary screening method for COVID-19 that relies only on cough sounds collected from 2200 clinically verified samples utilizing the laboratory molecular testing performs (1100 Covid-19 positive and 1100 Covid-19 negative). The clinical labels were applied to the results, and severity of the samples may be judged based on quantitative RT-PCR (qRT-PCR), cycle threshold, and patient lymphocyte counts. The fast spread of the COVID-19 virus poses a significant danger of serious pulmonary disease, and it also causes the most heinous harm to humanity. As a result, a quick and clear disease classification model to distinguish between normal and COVID-19 infected individuals is critical. In this article, we describe the various machine learning and other models that have been used to predict COVID-19 patients.

H. Singh
Department of CS and IT, Sant Baba Attar Singh Khalsa College, Sandaur, Punjab, India

B. K. Pandey
Department of Information Technology, College of Technology, Govind Ballabh Pant University of Agriculture and Technology Pantnagar, U S Nagar Uttarakhand, India

S. George
Department of Information and Communication Technology, Crown University, Int'l. Chartered Inc. (CUICI) Argentina Campus, Santa Cruz Province, USA

D. Pandey (✉)
IET, Dr. A.P.J. Abdul Kalam Technical University Uttar Pradesh, Lucknow 226021, India
e-mail: digit11011989@gmail.com

R. Anand
DSEU, G.B.Pant Okhla-1 Campus, New Delhi, India

N. Sindhwani
AIIT, Amity University, Noida, India

P. Dadheech
Computer Science and Engineering, Swami Keshvanand Institute of Technology, Management and Gramothan (SKIT), Jaipur, India

M. Gupta et al. (eds.), *Artificial Intelligence on Medical Data*,
Lecture Notes in Computational Vision and Biomechanics 37,
https://doi.org/10.1007/978-981-19-0151-5_15

Keywords SARS-CoV-2 · COVID-19 · Respiratory diseases · Cough · Convolutional neural network · Machine learning models · Classification

1 Introduction

COVID-19, an infectious illness caused by severe acute respiratory syndrome coronavirus 2 (SARS-COV-2), was first identified in China in December 2019. In March 2020, the World Health Organization (WHO) declared it a global pandemic. It affects the mucosa (mucous membrane) of the throat, resulting in moderate fever, difficulty in breathing, and exhaustion in approximately 59% of those affected [1]. It also causes the respiratory tract infections along the route to the lungs. When an infected individual coughs or sneezes, COVID-19 spreads via saliva droplets or nasal secretions. This cough is very much contagious in nature. The disease's consequences tend to impact millions of people in a short period of time, creating global economic upheavals. As a result, people's well-being is negatively impacted [2, 3]. As of May 26, 2021, according to World Meter data, there have been 168,597,013 COVID-19 positive cases and 3,501,691 fatality situations recorded [4], as depicted in Fig. 1.

As a result of the catastrophic effects of COVID-19 and the sad loss of life, it is critical to develop a solution for early detection of the illness. The illness spreads, but its impact may be reduced by improving the accuracy of prescreening and subsequent testing methods. This approach would assist in the creation of focused solutions to this problem. Since the start of the pandemic, artificial intelligence (AI) researchers all around the globe have been working to develop an AI-based testing and diagnostic tool for the speedier illness identification [1, 5].

The main objective of this article is to discuss briefly the various machine learning [6, 7] and other models that have been used to predict COVID-19 patients. Next section discusses the existing related work. Section 3 discusses some ML tools for

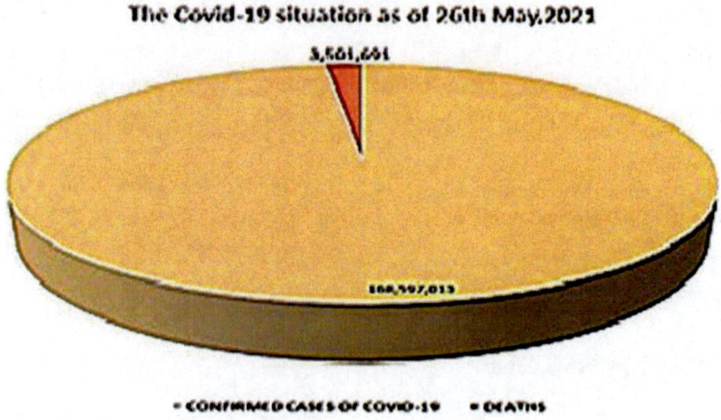

Fig. 1 COVID-19 cases as on 26th May 2021

prediction. The subsequent section discusses the analysis in terms of the performance metrics for some of the ML models. At last, a brief conclusion is drawn.

2 Literature Survey

Researchers and academics have recognized the use of sound as a possible characteristic in the prediction of behaviors and health. It all begins with digital stethoscopes along with external microphone recorders for assessing lungs and heartbeat sounds. Researchers have also been able to listen to and analyze the different techniques such as magnetic resonance imaging (MRI) and sonography using these devices so that things are simpler to study and understand. The current developments in detecting audio noises and their interpretations have an effect on altering the situation and offering a low-cost, easy-to-distribute alternative function for the sound. Several studies have been performed to see whether various coughs (such as wet and dry coughs) and audio analysis of the depth of the cough sound in patients can be classified. There are many kinds of illnesses in the pulmonary system, such as lung infection, pneumothorax, bronchitis, asthma, and so on that may be detected by signal processing of cough sounds.

Many studies on the identification of COVID utilizing cough sounds, CT scans, and X-rays have been published in the recent years. Several writers have tried to identify the virus using deep learning models, which are part of artificial intelligence (AI). In [8], the authors developed a technique to categorize the individuals infected with COVID-19. Individuals' breathing habits may be identified using this research by presenting the respiratory simulation (RS) model. By taking into account the features of correct respiratory signals, it is utilized to fill up the space in the large training dataset and the limited actual dataset from the real world. Six different respiratory trends have precision, accuracy, recall, and F1 of 94.5%, 94.4%, 95.1%, and 94.8%, respectively, according to the findings. The developed model has a huge potential for the use in large-scale screening. The findings of spirometry tests have been matched to the cough sounds collected in a clinical environment by the authors in. Spirometry is one of the mostly used tests for determining a patient's lung infection. It counts the amount of air breathed and exhaled by a person. The authors also used the time-domain waveform visualization to conduct the signal processing on the cough sounds, concluding that the cough sounds include some unique information from the spirometry tests. The author also showed that when the patients are given certain medications, the findings of spirometry may be altered, while the analysis of cough sounds remains unaffected.

There has been developed a model in [9] that incorporates signal and voice processing as well as image analysis techniques. The author proposed voice processing methods to help physicians anticipate TB in patients. The convolutional neural network is recommended in this model for making impartial judgments. The authors in [10] looked at the differences between pneumonia, asthmatic coughs, and bronchitis coughs. Using the characteristics such as time sequence statistics and

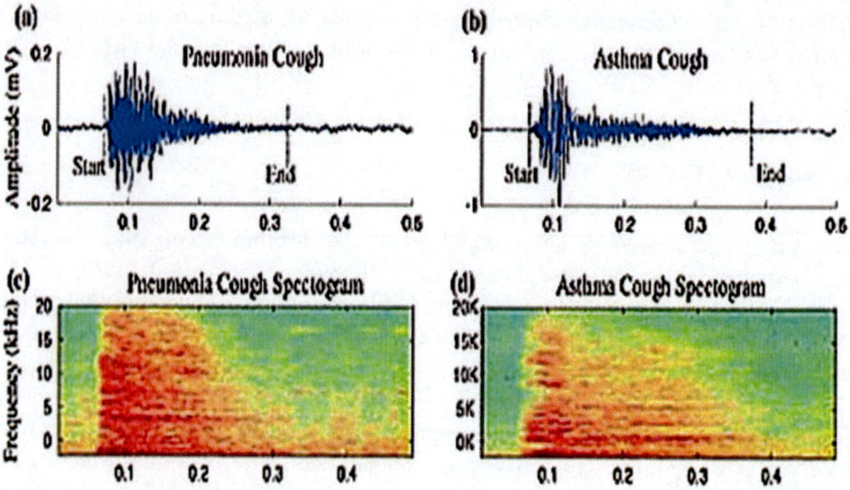

Fig. 2 COVID-19 cough samples

modeling formant frequency, they achieve a sensitivity of 93% and a specificity of 54%. They were able to accomplish the aforementioned characteristics by extracting the metrics from cough samples, as shown in Fig. 2.

The authors in analyzed the cough sounds using a variety of signal processing methods. It entails examining the characteristics such as energy spectrum and statistical waveform in a time-independent manner. By dividing the dataset into dry and wet coughs, the dataset of 536 samples may achieve 54% recall and 92% specificity. Further, the authors in [11, 12] utilized convolutional neural network (CNN) and recurrent neural network (RNN) models or architectures to forecast the asthma and general infections in the respiratory system based on breath as much as feasible.

3 General Machine Learning Model for Prediction

The general stages of the prediction of a COVID-19 patient utilizing an X-ray and a lung ultrasound classification technique are given in Fig. 3. Before being divided into training and testing datasets, the ultrasound videos were gathered and preprocessed. Then, from both the training and testing data, features were retrieved. Before the classification job, the retrieved characteristics are further processed. Finally, a supervised machine learning method was used to predict COVID-19 infected and healthy individuals. The extracted features are further processed before the classification task. Finally, the supervised machine learning approach was implemented to predict the COVID-19 infected people and normal people.

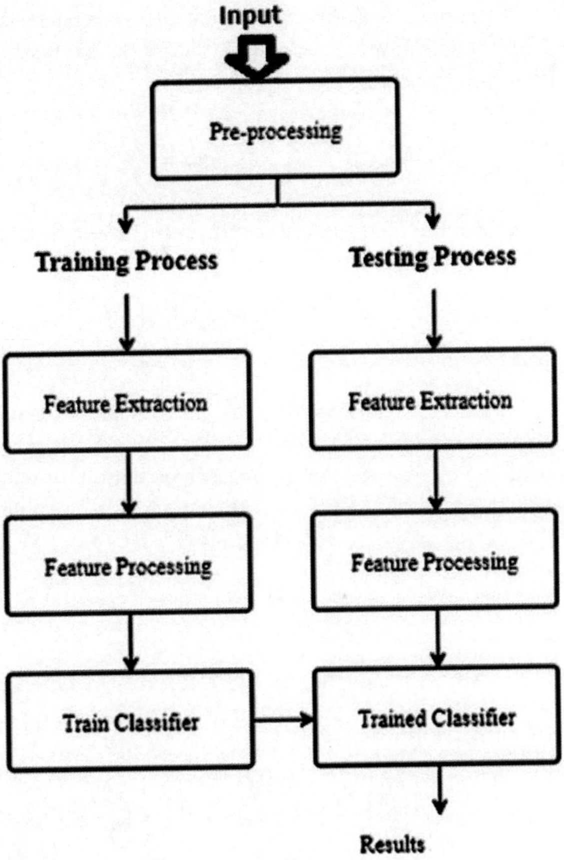

Fig. 3 General methodology for prediction of COVID-19

4 Analysis and Discussion

Precision, recall, accuracy, and F1-score [13] are some of the assessment measures used to evaluate the efficacy of this broad suggested approach for COVID-19 prediction. True positive (TP), true negative (TN), false positive (FP), and false negative (FN) are the four factors used to judge the performance metrics.

- TP determines whether the text parts are properly recognized as text.
- FN determines the text portion that is mistakenly classified as a non-text part.
- TN determines which non-text portion is properly recognized as non-text.
- FP determines the non-text portion that is mistakenly recognized as text.

Accuracy: Eq. (1) shows the formula that is used to determine the accuracy of this detection procedure.

$$\text{Accuracy} = \frac{\text{TP} + \text{TN}}{\text{TP} + \text{TN} + \text{FP} + \text{FN}} \tag{1}$$

Precision: Precision (p) is what determines the ratio of true positives (TP) to the total detections. Equation (2) shows the expression that is used to calculate this precision measure.

$$p = \frac{\text{TP}}{\text{TP} + \text{FP}} \tag{2}$$

Recall: Recall (r) determines the ratio between the detected true text and the whole TP. Equation (3) is used to calculate these recall metrics.

$$r = \frac{\text{TP}}{\text{TP} + \text{FN}} \tag{3}$$

F1-score: F1-score is the overall indication used to assess the performance of the suggested method. For F1-score assessment, precision and recall measures are used. Figure 4 shows the comparison depiction for this F1-score of the presented and previous methods. The F1-score may be calculated using Eq. (4).

$$F_1 = 2 \times \frac{p \times r}{p + r} \tag{4}$$

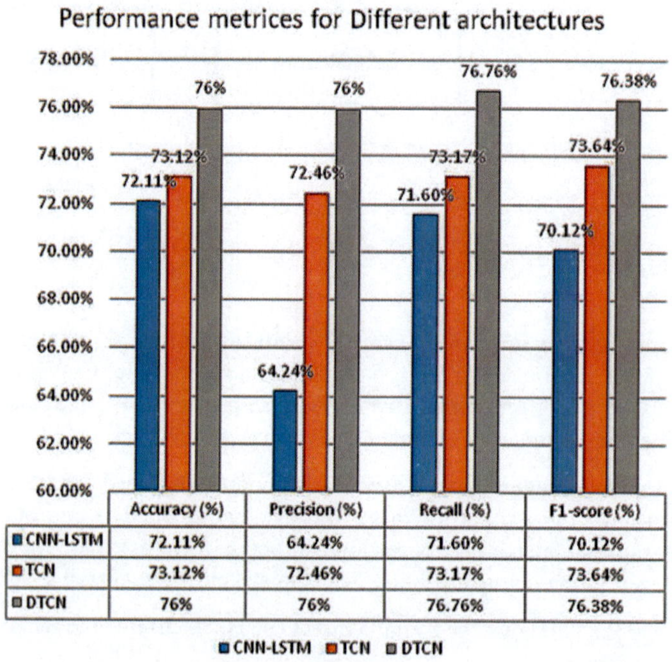

Fig. 4 Different models used for prediction of COVID-19 patients

Figure 4 shows the comparison between convolutional neural network [14, 15] with long short-term memory (CNN-LSTM), temporal convolutional networks (TCNs), and deep temporal context networks (DTCNs) in terms of the different performance metrics. It may be observed that DTCN model gives the best accuracy, precision, recall, and F1-score because of the temporal coherence across the various time-scales.

5 Conclusion

The cough sounds from COVID-19 patients and healthy people were collected from the COUGHVID dataset and utilized to train and evaluate the architectures mentioned. The features and characteristics produced by the convolutional layer were employed in the network's final layers. Maximum pooling of a one-dimensional layer is utilized in the model. With a learning rate of 0.2, accuracy of 76% in DTCN, 73.12% in TCN, and 72.11% in CNN-LSTM was achieved. We successfully evaluated three alternative algorithms and determined that DTCN was the best of the three. This study may be expanded to identify COVID-19 instances by including other data and characteristics such as patient breath, unique voice sounds, spontaneous speech signals, and more precise forecasts. By adding some additional data, performance may be improved even further.

References

1. Manne R, Kantheti SC (2021) Application of artificial intelligence in healthcare: chances and challenges. Current J Appl Sci Technol 40(6):78–89. https://doi.org/10.9734/cjast/2021/v40i631320
2. Khongsai L, Anal TSSC et al (2021) Combating the spread of COVID-19 through community participation. Glob Soc Welf 8:127–132. https://doi.org/10.1007/s40609-020-00174-4
3. Pandey D, Pandey BK, Wairya S (2021) Hybrid deep neural network with adaptive galactic swarm optimization for text extraction from scene images. Soft Comput 25(2):1563–1580
4. Kashif M, Javed MK, Pandey D (2020) A surge in cyber-crime during COVID-19. Indo J Social Environ Issues (IJSEI) 1(2):48–52
5. Pandey D, Ogunmola GA, Enbeyle W, Abdullahi M, Pandey BK, Pramanik S (2021) COVID-19: a framework for effective delivering of online classes during lockdown. Human Arenas 1–15
6. Singh H, Rehman TB, Gangadhar C, Anand R, Sindhwani N, Babu M (2021) Accuracy detection of coronary artery disease using machine learning algorithms. Appl Nanosci 1–7
7. Shukla R, Dubey G, Malik P, Sindhwani N, Anand R, Dahiya A, Yadav V (2021) Detecting crop health using machine learning techniques in smart agriculture system. J Sci Ind Res (JSIR) 80(08):699–706
8. Wang Y, Hu M, Li Q et al (2020) Abnormal respiratory patterns classifier may contribute to large-scale screening of people infected with COVID-19 in an accurate and unobtrusive manner. arXiv:2002.05534
9. Santosh KC (2019) Speech processing in healthcare: can we integrate? In: Intelligent speech signal processing, pp 1–4. Elsevier

10. Abeyratne UR, Swarnkar V, Setyati A, Triasih R (2013) Cough sound analysis can rapidly diagnose childhood pneumonia. Ann Biomed Eng 41(11):2448–2462
11. Manne R, Kantheti S (2020) Coronavirus impact on cardiovascular system of body-review. Int J Res Appl Sci Eng Technol (IJRASET) 8(11):276–280
12. Bokam Y, Guntupalli C, Gudhanti S, Kulandaivelu U, Alavala R, Alla N, Manne R (2021) Importance of pharmacists as a front line warrior in improving medication compliance in covid 19 patients. Indian J Pharm Sci 83(2):393–396
13. Sindhwani N, Verma S, Bajaj T, Anand R (2021) Comparative analysis of intelligent driving and safety assistance systems using YOLO and SSD model of deep learning. Int J Inf Syst Model Design (IJISMD) 12(1):131–146
14. Kamalraj R, Neelakandan S, Kumar MR, Rao VCS, Anand R, Singh H (2021) Interpretable filter based convolutional neural network (IF-CNN) for glucose prediction and classification using PD-SS algorithm. Measurement 183:109804
15. Kassaw C, Pandey D (2021) The current mental health crisis of COVID-19 pandemic among communities living in Gedeo Zone Dilla, SNNP, Ethiopia, April 2020. J Psychosoc Rehabil Mental Health 8(1):5–9
16. Pandey D, Islam T, Magray JA, Gulzar A, Zargar SA (2021) Use of statistical analysis to monitor novel coronavirus-19 cases in Jammu and Kashmir, India. Eur J Biol Res 11(3):274–282
17. Parthiban K, Pandey D, Pandey BK (2021) Impact of SARS-CoV-2 in online education, predicting and contrasting mental stress of young students: a machine learning approach. Augmented Hum Res 6(1):1–7
18. Ayenew B, Pandey D, Yitayew M, Etana D, Binay Kumar P, Verma N (2020) Risk for surge maternal mortality and morbidity during the ongoing corona virus pandemic. Med Life Clin 2(1):1012
19. Meivel S, Sindhwani N, Anand R, Pandey D, Alnuaim AA, Altheneyan AS, ... and Lelisho ME (2022) Mask detection and social distance identification using internet of things and faster R-CNN algorithm. Comput Intell Neurosci
20. Radwan E, Radwan A, Radwan W, Pandey D (2021) Prevalence of depression, anxiety and stress during the COVID-19 pandemic: a cross-sectional study among Palestinian students (10–18 years). BMC Psychol 9(1):1–12

Melanoma Cell Detection by Using K-means Clustering Segmentation and Abnormal Cell Detection Technique

Pratima Sarkar, Bijoyeta Roy, Mousumi Gupta, and Sourav De

Abstract Melanoma diagnosis in earlier stage is a challenging task, and it is ideally done by the pathologist to automate the detection of melanoma cells. Detection of melanocytes from histopathological image is difficult because melanocytes are very much similar to the keratinocytes. To detect melanoma cells at first, segmentation is required followed by calculation of perimeter of the cells. If it is greater than the threshold value, then it is identified as melanoma cells. K-means clustering unsupervised learning technique is used to segment the image followed by ellipse fitting algorithm to find out abnormally large cells. The centroid is then calculated and marked as nuclei. Experiment is performed on 30 images and is evaluated based on PPV, sensitivity, and NSR. The proposed work achieved 90% PPV, 85% sensitivity, and 13% SNR value.

Keywords Image segmentation · Melanoma detection · Skin cancer

1 Introduction

In last few years', computer-supported cancer detection has become one of the most popular area is skin lesion inspection [1]. Among all other cancer, increasing rate of melanoma is very high. Skin cancer increases 3–7% every year for fair skinned people. Earlier detection of melanoma is a crucial work because of less changes

P. Sarkar (✉) · B. Roy · M. Gupta
Sikkim Manipal Institute of Technology, SMU, Majitar, Sikkim, India
e-mail: psmoon2@gmail.com

B. Roy
e-mail: bijoyeta.r@smit.smu.edu.in

M. Gupta
e-mail: mousumi.g@smit.smu.edu.in

S. De
Department of Computer Science and Engineering, Cooch Behar Government Engineering College, Cooch Behar, India

M. Gupta et al. (eds.), *Artificial Intelligence on Medical Data*,
Lecture Notes in Computational Vision and Biomechanics 37,
https://doi.org/10.1007/978-981-19-0151-5_16

in intensity between melanocytes and keratinocytes of histopathological images. Skin lesions are present on skin in darker color; it is because of excessive melanin formation in skin. Initially, melanin present on epidermis of the melanoma is called "in situ" [2]and at this moment, no life risk is associated with the cancer patient. When malignant penetrates the dermis of skin and deposits melanin in the skin, life risk increases.

For earlier detection of melanoma cells, it is required to segment images to find out the melanoma cells. Segmentation of images can be done using different clustering algorithms. Here, we have used K-means algorithm to segment the images and then calculated perimeter of each of the cells. If it is greater than a calculated threshold value, then it is detected as melanoma cells. The contributory parts of this paper are given below:

1. Segmentation is performed based on unsupervised learning clustering technique, and so, the work is not dataset dependent.
2. Perimeter of the cells is used to detect abnormally large cells.
3. Threshold value is proposed to detect abnormal cells.

In this paper, in Sect. 2, previous related works are discussed followed by Sect. 3 where the proposed framework for detection of nuclei in skin histopathology images is discussed. Then, in Sect. 4, the experimental results obtained are shown followed by highlighting the performance evaluation in Sect. 5.

2 Background Study

According to a recent study, melanoma skin cancer affects around 70,000 individuals in the United States each year, with approximately 9000 people dying from it [3]. A deep learning-based approach for detecting malignant melanoma in whole-slide skin images has been developed. The nuclei in the image are first segmented using a convolutional neural network (CNN), and then melanoma area masks are created using the segmented nuclei. The dice score of the discovered melanoma area was over 85%, and the proposed approach could potentially segment the nuclei with an accuracy of 94%. The suggested technique also has a short execution time, making it suited for rapid clinical diagnosis. [4]. S. Alheejawi et al. reported the current state-of-the-art in computer-aided diagnosis and grading of skin cancer/melanoma. Melanoma identification and calculation of the proliferation index using deep learning networks like Unet and SegNet are described in detail and compared to classic feature-based techniques [5]. For image enhancement, the suggested method employs the CLAHE algorithm, followed by bilateral filtering. The initial segmentation is done using the fuzzy c-means algorithm, and the final segmentation results are generated using a local region recursive algorithm. To identify melanocytes from potential nucleus regions, an elliptical descriptor is utilized to calculate region ellipticity and local pattern features [6]. To separate the nuclei and identify the high nuclei density areas in the invasive breast carcinoma WSIs, Cheng Lu et.al. used grayscale conversion, adaptive thresholding, and morphological procedures. The tissue type is subsequently classified using several classifiers based on the characteristics collected from pre-segmented regions. On a database of 24 WSIs, a classification accuracy of 68% was reported [7]. Deep learning has emerged as a strong technique to nuclei segmentation; however, the performance of convolutional neural networks (CNNs) [8] is dependent on the quantity and quality of labeled histopathological data used for training. In 2019, Faisal Mahmood et.al. proposed a nucleus segmentation method that solves these difficulties by employing a conditional generative adversarial network (cGAN) trained with simulated and real data. This synthetic data along with real histopathology data from six different organs are used to train a conditional GAN with spectral normalization and gradient penalty for nuclei segmentation [9].

3 Proposed Framework

Segmentation of histopathological image is a challenging task as intensity values are very much similar throughout the image. In our proposed work, histopathological melanoma images are segmented using K-means clustering, and after segmentation, our aim is to find out perimeter of the segmented cells which is achieved using ellipse fitting algorithm. Our proposed work consists of following steps:

1. K-means clustering to identify cells on a histopathological image
2. Fit ellipse on small segments (assumed as cells)
3. Calculate perimeter of each ellipse
4. Identification of abnormal cells
5. Find out centroid of abnormal cells
6. Mark nucleus (Fig. 1).

3.1 *K-Means Clustering to Identify Cells on a Histopathological Image*

K-means is an unsupervised learning algorithm to make cluster-based partition on given information. K-means work computationally better than hierarchical clustering if value of K is considered to be small as instances are changing their cluster until it is getting optimal solution. In our experiment, it gives better segmentation than EM clustering techniques. So, we have selected K-means clustering to segment the image, and moreover, since it is unsupervised learning algorithm, so, it works even if single image is available. Here, we want to create cluster in a particular histopathological image that is basically segmented image. From segmented image, we can determine isolated cells.

Following algorithm is to identify cells:

Step 1: Read RGB image.

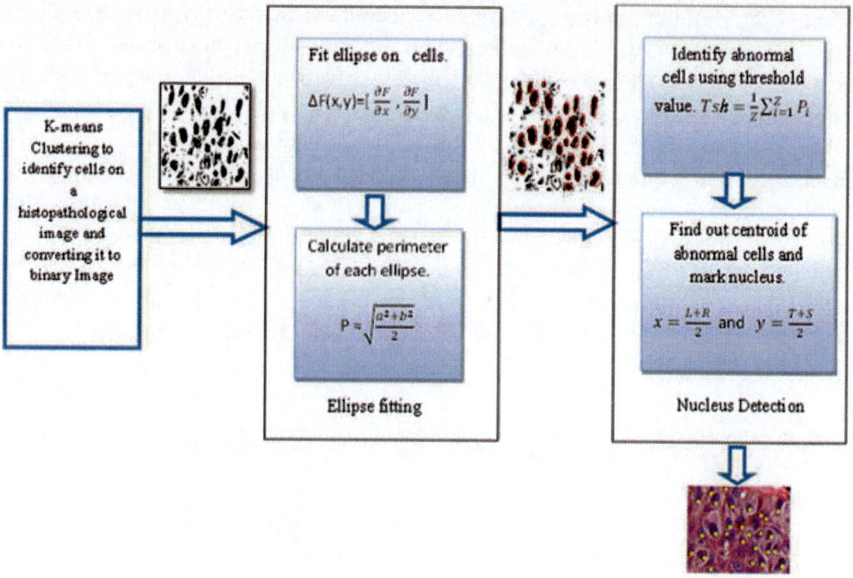

Fig. 1 Block diagram for proposed solution

Step 2: Segregate three different colors channels red, green, blue.

Step 3: To get better cluster, green channel is used for clustering the image.

Step 4: Initialize value of K = 3, where k is number of cluster.

Step 5: Select centroid randomly we have selected three centroids as number of cluster is 3.

Step 6: Assignment of Data points:

Calculate Euclidian distance from the centroid of all the clusters.

$$\mathrm{Dist(ci, x)} = \sqrt{(\mathrm{c_i} - \mathrm{x})^2} \tag{1}$$

where x is the intensity value of green component of the image and c_i is the centroid of ith cluster. If min (Dist(c_i, x)), then x is assigned to cluster $c_{i.}$

Step 7: Updating of centroid:

Let us assume S_j is number of data points assigned to a cluster $c_{i.}$ In this case, centroid is recalculated using following formula:

$$\mathrm{ci} = \frac{1}{S_i}\sum_{x=1}^{S_i} x_i \tag{2}$$

Step 8: Stopping criteria:

The algorithm iterates between steps six and seven until no data points change in particular clusters, and the sum of the distances with cluster centroid is minimized.

3.2 Fit Ellipse on Cell

Cluster resulted using k-means gives isolated cells where each of the cells is denoted as $\{X_i\}_{i=1....Z}$, where Z is number of cells on segmented image. Now, our aim is to identify nucleus of a particular cell. Based on previous literature, nuclei of cells have elliptical boundaries in case of histopathological image. To build the boundaries of the cells as ellipse, it is important to build elliptical model that must be fitted in the cell region. To generate elliptical model for each cell region $\{X_i\}_{i=1....Z}$, it is necessary to calculate gradient of the segmented image, and it is used to find out boundary points of the segmented region. To calculate the gradient it is needed to measure the change in image function, F(x, y) in x and y.

$$\Delta F(x, y) = \left[\frac{\partial F}{\partial x}, \frac{\partial F}{\partial y} \right] \tag{3}$$

3.3 Calculation of Perimeter of Ellipse

It is important to calculate perimeter of each of the ellipse to find out abnormal cells. Usually, abnormal cells have larger and abnormal shape with respect to normal shape. For calculating perimeter (P) of ellipse, the following formula is considered:

$$P \approx \sqrt{\frac{a^2 + b^2}{2}} \tag{4}$$

where a = R-L AND b = T-S.

L: is left most point of ellipse.

R: is left most point of ellipse.

T: is top most point of ellipse.

S: is lower most point of ellipse (Fig. 2).

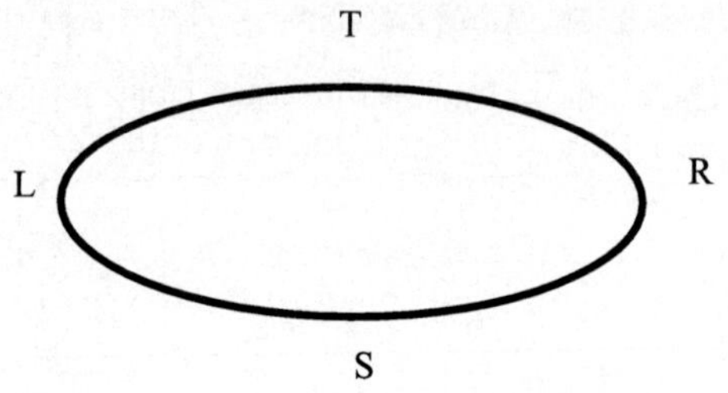

Fig. 2 Ellipse perimeter determination from L, R, T, and S

3.4 *Identification of Abnormal Cells*

After calculation of perimeter of each cell for identifying abnormal cells, we are considering abnormally large perimeter (ALP). An ellipse with ALP is selected depending upon a threshold value. Threshold (*Tsh*) is calculated by using following equation:

$$Tsh = \frac{1}{Z}\sum_{i=1}^{Z} P_i \tag{5}$$

where Z is total number of cells and P_i is the perimeter of ith ellipse. ALP ellipse is those which have perimeter >= *Tsh*, and these cells are termed as abnormal cells. Usually, nuclei of a cell are in the centroid of a cell. In next phase, we have calculated the centroid of the abnormal cells.

3.5 *Find Out Centroid of Abnormal Cells*

Let us assume centroid of a particular abnormal is C (x, y), where x and y are the coordinates.

$$x = \frac{L+R}{2} \text{and} y = \frac{T+S}{2} \tag{6}$$

The particular point is marked as nucleus of the abnormal cell.

4 Performance Evaluation

Here, our main objective of evaluation is to find out that our proposed algorithm is segmenting the image properly or not. We got final segmented result in binary image, and in that image, we are finding nucleus. The same nucleus is mapped in to original image. The proposed work is evaluated based on detection of nuclei with respect to manually counted nucleus.

We define N_M is total number of manually counted cells. N_D is total number of detected nucleus by proposed work. N_C is total number of nucleus correctly detected with respect to original number of nucleus. N_F is total number of falsely detected nucleus. N_u is total number of not identified cells.

The performance of the proposed technique is evaluated by using following metric [2]:

1. Positive predictive value (PPV)
2. Sensitivity (SEN)
3. Not segmented region (NSR)

$$\text{PPV} = \frac{N_C}{N_D}\text{x}100\% \tag{6}$$

$$\text{SEN} = \frac{N_C}{N_M}\text{x}100\% \tag{7}$$

$$\text{NSR} = \frac{N_U}{N_C}\text{x}100\% \tag{8}$$

5 Result and Discussion

The experimental results of the proposed segmentation approach are presented in this section. With the aid of Matlab (MATLAB 2015), the proposed segmentation approach is implemented. Figure 3 represents qualitative analysis of the work. First column consists of original images used in proposed technique. Column 2 of Fig. 3 shows intermediate result, i.e., segmented image. Blue-colored segmented regions are representing cells. The segmented image is the result of K-means clustering technique. It is observed from experiment that intensity value for cells is higher than the intensity value of the other region but for RGB image it happens only in green component. From manual identification of cells, we can conclude that blue color regions are representing same area. So, we have assumed each blue-colored region as cell.

Column three of Fig. 3 shows area-opening image for binary representation of the segmented image and each black-colored regions we assumed as area of a cell. These cells are might be normal or abnormal cells. As an observation of the image, from the image, we can identify all kinds of cells. After identifying cells, it is important to mark perimeter of a particular cell. To mark perimeter, we have used ellipse fitting algorithm.

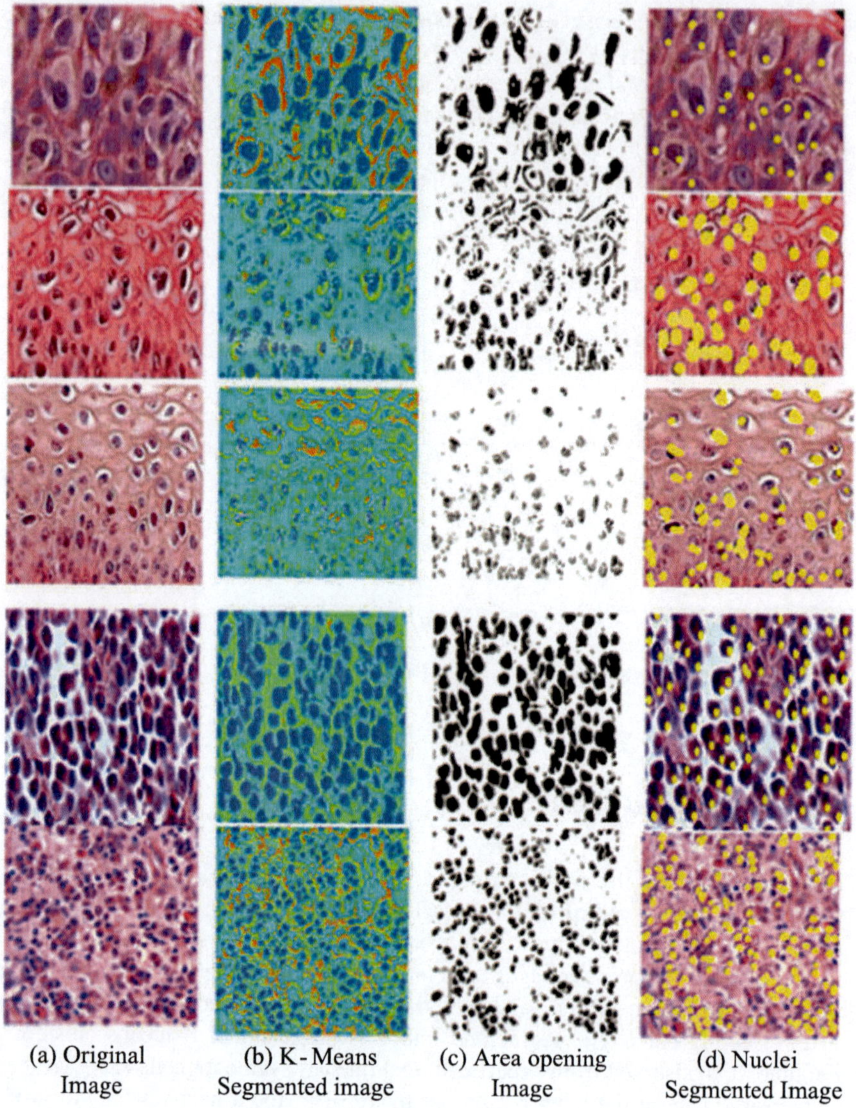

Fig. 3 Stages of nuclei segmentation

After fitting ellipse, we are able to calculate perimeter of cells. Depending on perimeter value, normal and abnormal cells are distinguished. If a cell having abnormally large perimeter (ALP), then those are marked as abnormal cells.

Final result of the proposed technique is shown in column four of Fig. 3. It is observed that nucleus of each of the cells is identified. The identified nucleus is marked yellow on original image. From final result, it is possible to conclude that

86% of the nucleus is marked properly with respect to manual identification of the cells.

We have used same performance metric as [2] to evaluate performance of our proposed work. This work was applied on 30 images. The following result we got from our experiment:

PPV = 90%

SEN = 85%

NSR = 13.11%.

6 Conclusion

In the proposed work, the problem related to melanoma detection is addressed. K-means clustering technique is applied on green channel of the RGB images. Then, gradient is used to fit ellipse on cells, and perimeter of the cells is calculated followed by marking the nuclei of abnormal cells which is our detected melanoma cells. We have used threshold value to detect abnormally large cells. PPV, SEN, and NSR are the considered performance parameter, and this approach achieved PPV = 90%, SEN = 85%, and NSR = 13.11%. The work can be extended to get exact size of each of the nucleus that might improve accuracy of the work.

References

1. Lu C, Manda M (2015) Automated analysis and diagnosis of skin melanoma on whole slide histopathological images. Pattern Recognition 48:2738–2750
2. Maglogiannis I, Doukas C (2009) Overview of advanced computer vision systems for skin lesions characterization. IEEE Trans Inf Technol Biomed 13(5):721–733
3. American Cancer Society (2008) What are the key statistics about melanoma? Technical report, American Cancer Society
4. Alheejawi S, Berendt R, Jha N, Maity SP, Mandal M (2021) Detection of malignant melanoma in H&E-stained images using deep learning techniques. Tissue Cell 101659
5. Alheejawi S, Mandal M, Xu H, Lu C, Berendt R, Jha N (2020) Deep learning-based histopathological image analysis for automated detection and staging of melanoma. In: Deep learning techniques for biomedical and health informatics. Academic Press, pp 237–265
6. Ray PJ, Priya S, Kumar TA (2015) Nuclear segmentation for skin cancer diagnosis from histopathological images. In: 2015 global conference on communication technologies (GCCT). IEEE, pp 397–401
7. Petushi S, Garcia FU, Haber MM, Katsinis C, Tozeren A (2006) Large-scale computations on histology images reveal grade-differentiating parameters for breast cancer. BMC Med Imaging 6:14

8. Dwivedi PD, Sandhu S, Zeeshan M, Sarkar P (2021) Data augmentation and CNN-based approach towards the classification of melanoma. In: Borah S, Pradhan R, Dey N, Gupta P (eds) Soft computing techniques and applications. Advances in intelligent systems and computing, vol 1248. Springer. https://doi.org/10.1007/978-981-15-7394-1_31
9. Mahmood F, Borders D, Chen RJ, McKay GN, Salimian KJ, Baras A, Durr NJ (2019) Deep adversarial training for multi-organ nuclei segmentation in histopathology images. IEEE Trans Med Imaging 39(11):3257–3267
10. Marks R Epidemiology of melanoma. Clin Exp Dermatol 25:459–463, 200

Intelligent Computing in Health Care

The Special Session Title 'Intelligent Computing in Healthcare' Dual Classification Framework for the Detection of Brain Tumor

Manu Singh and Vibhakar Shrimali

Abstract Cancer is a very common disease caused by mutation in DNA of cells and arises everywhere in the world. According to American Cancer Society, vast number of new cancer cases are predicted which are increasing every year. So, the early detection of brain tumor may decline the death percentage, and it would be helpful for the cancer specialist to cure the patients in primary stage. Image processing concepts are applied for the detection of normal and malignant brain scans using computer-aided tools. A new framework has been proposed to detect the brain tumor using magnetic resonance imaging (MRI) modality. The steps of the implementation of proposed framework are (a) image collection, (b) preprocessing, (c) segmentation, (d) statistical feature extraction, and (e) dual classification. Fuzzy K-mean segmentation is executed followed by statistical feature extraction in terms of texture. After that, dual classification is performed consequently one after another, i.e., autoencoder deep neural network followed by binary SVM. The primarily training has been completed by autoencoder, and the resultant of this is again trained by SVM. Finally, testing has been accomplished by SVM only. The performance of the proposed framework has been assessed with Multimodal Brain Tumor Segmentation Challenge dataset (BraTS 2013) in terms of sensitivity, specificity, AUC-ROC, and accuracy. The performance of the proposed framework is remarkable, and the accuracy of autoencoder + SVM achieved by this framework is 96.4% as compared to SVM alone.

Keywords Autoencoder network · Fuzzy K-mean · Image processing · MRI brain tumor · SVM

M. Singh (✉)
University School of Information and Communication Technology, Guru Gobind Singh Indraprastha University, Delhi 110078, India
e-mail: ersinghmanu06@gmail.com

V. Shrimali
Department of Electronic and Communication Engineering, G.B. Pant Government Engineering College, New Delhi 110020, India

M. Gupta et al. (eds.), *Artificial Intelligence on Medical Data*,
Lecture Notes in Computational Vision and Biomechanics 37,
https://doi.org/10.1007/978-981-19-0151-5_17

1 Introduction

In the recent years, introduction of well-established medical healthcare system in medical field has contributed the specialists or doctors to provide good medication to their patients. But still some diseases are most challenging task in day-to-day life because of unhealthy environment, junk food, bad habits, etc. [1]. Cancer is one of the diseases and is increasing among the adults and children both [2]. Early diagnosis of cancer may improve the treatment possibilities. Cancer can be described as uncontrolled and undesirable growth of abnormal cells called mutation inside the brain. This leads to the growth of cancer in the brain. Some cancer may be inherited from the parents or ancestors, and some are acquired from the present life. Gliomas are identified as primary tumor that mainly affects the spinal cord and brain. According to World Health Organization (WHO), gliomas are arranged as Grade I, Grade II, Grade III, and Grade IV where Grade I and Grade II are low-grade gliomas (LGG) [3]. LGG tumors are recoverable and cured if it could be treated on time, whereas Grade III and Grade IV are known to be high-grade gliomas (HGG) and they may be malignant tumors. These kinds of tumors may be non-curable and less survival rate. Brain tumor may be benign when the tissue cells in the tumor area are normal and malignant means cancerous cell tissues are abnormal, and uncontrolled growth of cells is present inside the tumor area [4].

There are number of medical imaging techniques like ultrasound, computed tomography (CT scan), magnetic resonance spectroscopy (MRS), positron emission tomography (PET), magnetic resonance imaging (MRI), etc., all are used to recognize the shape, volume, location, and many more parameters which may be helpful for the diagnosis if required. These modalities provide the valuable information about the brain tumor. MRI is evaluated as a standard technique to detect the normal and abnormal MRI brain scans. Paper representing different techniques in the image processing field includes [5, 6]. Initially, preprocessing of the MRI scans have been used to transform low-level image into high-level image scans by using different preprocessing techniques. After that, segmentation is widely used to segment the brain structure into different slices so that the detection and isolation of tumor can be easily identified. Thereafter, feature extraction methods are used to provide the useful information to the classifiers. At the end, classification methods are applied for the detection of brain tumor as normal or malignant.

The outline of the remaining paper is highlighted in the following sections: Section 2 has been focused on the methodology of proposed framework. Analysis and results have been discussed in Sect. 3. At last, Sect. 4 has been highlighted with conclusion.

2 Proposed Method

The main strategy of this framework is used to detect the normal and malignant MRI brain scans from the BraTS dataset. The proposed framework of the paper is shown in Fig. 1. The first step of the proposed framework is to acquire the medical images from different hardware sources and parameters. Thereafter, preprocessing is performed with MRI brain scans to diminish the problems of noise, contrast, visibility, and other outliers. After that, segmentation is applied on preprocessed MRI brain scans by fuzzy K-mean algorithm, and the outcome obtained from segmentation method undergoes feature extraction technique known as statistical feature parameter. Finally, dual classification has been employed which means autoencoder classifier is implemented to train the feature vectors obtained from the above method, and the outcome received from this has been trained by binary SVM again. At last, testing is executed by SVM only. The proposed structure has been discussed as follows.

2.1 *Image Collection*

It is the first and the basic step in the image processing system. The MRI brain scans have been retrieved from different hardware sources that are unprocessed. Hardware devices should be properly configured so that no noise or other artifacts can be produced. It will helpful for smooth processing which in return provides more accurate results. Here, MRI brain scans are collected from Multimodal Brain Tumor Segmentation repository [7].

2.2 *Preprocessing*

This section of the paper covers contrast adjustment, binarization, denoising, and morphological operations. Contrast adjustment has been used to provide new range of pixel intensity valued over the original image intensity using mapping function.

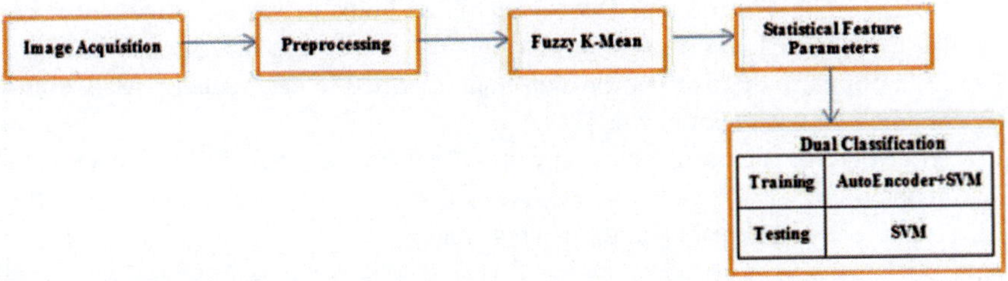

Fig. 1 Proposed framework

It actually distinguishes the darker and the lighter regions of MRI brain scans. Due to increasing contrast of the brain scans, dark region gets darker and light region gets lighter. After that binarization has been implemented which converts color or grayscale value into binary image (black–white image) based on threshold value of original brain scans [8]. After that, denoising has been used to get rid of noise generated by many factors like transformation, sensor heat, etc. Median filter has been executed to avoid the noise problem without any alteration of original brain scans. At last, morphological operations have been carried out which is helpful to preserve the structure of the MRI brain scans. Here dilation and erosion have been implemented one after another to separate the small regions of MRI scans and also smoothen the edges of the brain scans. These above methods have been performed in this proposed framework so that original or unprocessed MRI scans can be improved [9, 10]. Here the MRI scans are converted into standard dimensions for further processing.

2.3 Segmentation

This section plays an important role in image processing as it splits the digital MRI scans into various fragments. It would be helpful for further processing because segmentation provides you different angle or position of images. Here fuzzy K-mean algorithm has been implemented which is the extension of K-mean method [11, 12]. The K-mean algorithm converges at the faster rate for the smaller value of K clusters.

But when we increase the value of K, then the quality of this algorithm may vary. Hence, fuzzy K-mean has provided better results as compared to K-mean. It has been applied to unlabeled dataset, where particular pixel is present in more than one clusters [13, 14]. Figure 2 displays the evaluation of preprocessing and segmentation method using T2 and FLAIR MRI images. Fuzzy K-mean method uses weighted fuzzy average (WFA) instead of mean to get the new cluster center for maximum convergence. Steps of WFA algorithm are shown below:

1. Let the input Q be the set of data points $(x1, x2, \ldots, xq)$.
2. Initial mean should be taken by K-mean algorithm.
3. Gaussian is centered over the mean value.
4. Next, weight Wq is obtained for the set of data point's xq using formula of WFA: $\left(ur = \sum q = 1,\ q\ (Wq(r-1)xq)\right)$.
5. Data points in the distribution space are assigned to each cluster. Also, empty or small sets are eliminated.
6. After that, cluster center is changed with WFA method and feature vector is updated.
7. This process is repeated until convergence.

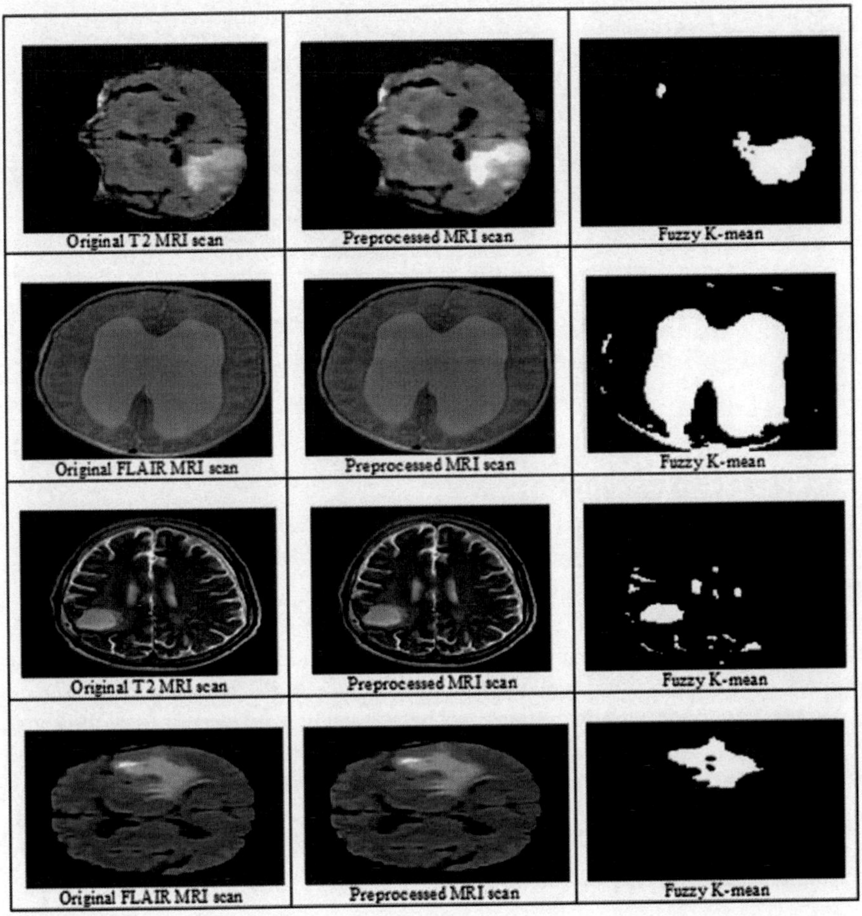

Fig. 2 Evaluation of preprocessing and segmentation method from MRI scan of T2 and FLAIR weighted images

2.4 *Statistical Feature Parameters*

Feature extraction is the process of reducing the unwanted or redundant features obtained from the above segmentation method. The main aim of the extraction is to limit the feature vectors and obtained only the useful vectors for classification. Here, statistical feature parameters have been used to extract the significant features [15]. The parameters are mean, maximum, variance, minimum, average frequency, energy, volume, kurtosis, skewness, and surface area. The value of the features obtained from statistical feature parameter of all the taken MRI scans is given in Table 1. The aim of using this method is used to check the quality estimation of the MRI images so that extraction of significant features is more precise.

Table 1 Feature extraction from statistical feature parameters

S. No.	Features	Value
1	Mean	0.64
2	Maximum	3.42
3	Minimum	4.43
4	Variance	0.95
5	Avg. frequency	0.01
6	Energy	6.08
7	Volume	0.42
8	Kurtosis	3.72
9	Skewness	1.49
10	Surface area	0.038

2.5 *Dual Classification*

This paper has presented the concept of dual classifier to get better performance. Here, autoencoder neural network and binary SVM have been used for classification purpose [16–18]. Deep autoencoder is unsupervised learning algorithm used to compress the MRI scans using encoder and decode it using decoder. The main purpose of using this autoencoder is to transform the high dimension data into low data representation. Initially, training has been done by autoencoder from the above obtained feature vectors, and the resultant outcome from this has been again trained by SVM. After that, testing has been performed using SVM only. The main purpose of using dual classification is to reduce the dimension space which minimizes the drawback of overlapping. Hence, this framework provides better performance compared to single classifier, i.e., SVM alone [19].

3 Analysis and Results

Here, the proposed system has been implemented on BraTS 2013 dataset, i.e., medical brain repository collected from Multimodal Brain Tumor Segmentation Challenge dataset [20]. From the above repository, 1800 2D MRI brain scans have been collected from BraTS 2013 dataset in which 1250 scans are used for malignant brain and 650 are used for normal brain scans. In this analysis, two categories of MRI scans have been taken, i.e., T2 and FLAIR weighted MRI scans. Training and testing have been implemented by tenfold cross-validation method during classification using 1450 MRI scans for training and 350 MRI scans for testing. The execution of complete framework has been completed by MATLAB R2018 software.

The analysis of dual classification has been assessed using parameters like sensitivity, specificity, AUC-ROC, and accuracy. Sensitivity defines the probability of exactly identifying the actual normal MRI scans. Specificity defines the probability

Table 2 Numerical analysis on T2 MRI scans using specificity, sensitivity, AUC-ROC, and accuracy

Parameters	T2 MRI scan			
	Sensitivity	Specificity	AUC-ROC	Accuracy
Proposed method (%)	97.02	95.8	96.66	96.4
SVM (%)	91.2	85	89.12	88.3

Table 3 Numerical analysis on FLAIR MRI scans using specificity, sensitivity, AUC-ROC, and accuracy

Parameters	FLAIR MRI scan			
	Sensitivity	Specificity	AUC-ROC	Accuracy
Proposed method (%)	96.20	94.03	95.60	95.30
SVM (%)	90.20	86.40	88.92	87.20

of exactly identifying the actual malignant MRI scans in the given dataset. True positive rate (TPR) enlightens the correct positive cases from all positive cases available in the dataset, whereas false positive rate (FPR) enlightens incorrect positive cases from all negative cases available in the dataset. AUC-ROC defines the graph between the different classification threshold value of two parameters, i.e., TPR and FPR. Accuracy defines the proposition of actual normal scans and actual malignant scans in the complete dataset [21–23]. The accuracy rate of dual classification, i.e., autoencoder + SVM achieved has been 96.4% for T2 MRI scans and 95.3% for FLAIR weighted MRI scans using BraTS 2013 dataset. The statistical examination in terms of sensitivity, specificity, AUC-ROC, and accuracy of proposed framework as compared to binary SVM is displayed in Tables 2 and 3 for T2 and FLAIR MRI scans, respectively. The graphical analysis for T2 and FLAIR MRI brain scans in terms of sensitivity, specificity, and accuracy is shown in Figs. 3 and 4, respectively. From the above analysis, it is apparent that through autoencoder + SVM, we obtain better accuracy as compared to binary SVM.

4 Conclusion

This paper has implemented the proposed framework to differentiate the normal and malignant brain scans using MRI modality. Image acquisition has been carried out from Multimodal Brain Tumor Segmentation Challenge, i.e., BraTS 2013 dataset using different sources. After that, preprocessing has been performed by using contrast adjustment, binarization, denoising, and morphological operations. Thereafter, segmentation has been executed by using fuzzy K-mean algorithm followed by statistical feature extraction method to get the essential and non-redundant feature vectors. Finally, dual classification has been implemented consequently one after

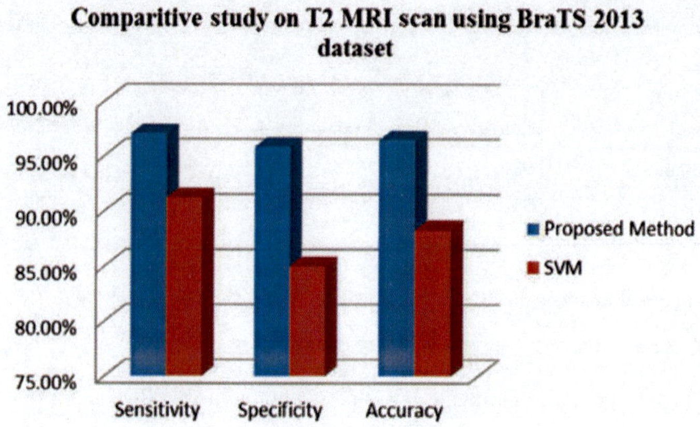

Fig. 3 Comparative study of proposed method and SVM on T2 MRI scans

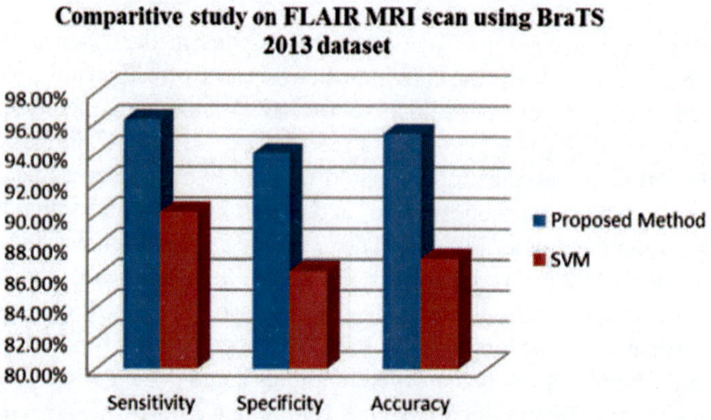

Fig. 4 Comparative study of proposed method and SVM on FLAIR MRI scans

another from the above resultant feature vectors. Here, training has been accomplished by autoencoder followed by SVM. At last, testing has been carried out by SVM only. We got 96.4% accuracy rate which is more preferable than binary SVM.

References

1. American Society of Clinical Oncology: The state of cancer care in America, 2014: a report by the American Society of Clinical Oncology. J Oncol Pract 10(2):119–142
2. Jenkins V, Fallowfield L, Saul J (2001) Information needs of patients with cancer: results from a large study in UK cancer centres. Br J Cancer 84(1):48–51

3. Louis DN, Perry A, Reifenberger G, Von Deimling A, Figarella-Branger D, Cavenee WK, Ohgaki H, Wiestker OD, Kleihues P, Ellison DW (2016) The 2016 World Health Organization classification of tumors of the central nervous system: a summary. Acta Neuropathologica 131(6):803–820
4. Tabatabai G, Stupp R, Van Den Bent MJ, Hegi ME, Tonn JC, Wick W, Weller M (2010) Molecular diagnostics of gliomas: the clinical perspective. Acta Neuropathol 120(5):585–592
5. Gonzalez RC, Woods RE (2002) Digital image processing, 3rd ed. Pearson Prentice Hall
6. Baxes GA (1994) Digital image processing: principles and applications. Wiley
7. Ghaffari M, Sowmya A, Oliver R (2019) Automated brain tumor segmentation using multimodal brain scans: a survey based on models submitted to the BraTS 2012–2018 challenges. IEEE Rev Biomed Eng 13:156–168
8. Natarajan P, Krishnan N, Kenkre NS, Nancy S, Singh BP (2012) Tumor detection using threshold operation in MRI brain images. In 2012 IEEE international conference on computational intelligence and computing research, pp 1–4
9. Borole VY, Nimbhore SS, Kawthekar DSS (2015) Image processing techniques for brain tumor detection: a review. Int J Emerg Trends Technol Comput Sci (IJETTCS) 4(5):2
10. Mzoughi H, Njeh I, Slima MB, Hamida AB, Mhiri C, Mahfoudh KB (2019) Denoising and contrast-enhancement approach of magnetic resonance imaging glioblastoma brain tumors. J Med Imaging 6(4):044002
11. Dhanachandra N, Manglem K, Chanu YJ (2015) Image segmentation using K-means clustering algorithm and subtractive clustering algorithm. Proc Comput Sci 54:764–771
12. Selvakumar J, Lakshmi A, Arivoli T (2012) Brain tumor segmentation and its area calculation in brain MR images using K-mean clustering and Fuzzy C-mean algorithm. In: IEEE-international conference on advances in engineering, science and management (ICAESM-2012), pp 186–190
13. Patel J, Doshi K (2014) A study of segmentation methods for detection of tumor in brain MRI. Adv Electron Electr Eng 4(3):279–284
14. Hossam MM, Hassanien AE, Shoman M (2010) 3D brain tumor segmentation scheme using K-mean clustering and connected component labeling algorithms. In: 2010 10th International conference on intelligent systems design and applications. IEEE, New York, pp 320–324
15. Padma A, Sukanesh R (2011) A wavelet based automatic segmentation of brain tumor in CT images using optimal statistical texture features. Int J Image Process 5(5):552–563
16. Raza K, Singh NK (2018) A tour of unsupervised deep learning for medical image analysis. arXiv preprint arXiv:1812.07715
17. Abdullah N, Ngah UK, Aziz SA (2011) Image classification of brain MRI using support vector machine. In: 2011 IEEE international conference on imaging systems and techniques, pp 242–247
18. Singh M, Shrimali V (2020) Two-level combined classification technique using Ranklet transformation for the detection of MRI brain tumor. In: 2020 IEEE 17th India council international conference (INDICON), pp 1–5
19. Othman MFB, Abdullah NB, Kamal NFB (2011) MRI brain classification using support vector machine. In: 2011 Fourth international conference on modeling, simulation and applied optimization, pp 1–4
20. Menze BH, Jakab A, Bauer S, Kalpathy-Cramer J, Farahani K, Kirby J, Burren Y, Porz N, Slotboom J, Wiest R, Lanczi L (2014) The multimodal brain tumor image segmentation benchmark (BRATS). IEEE Trans Med Imaging 34(10):1993–2024
21. Polly FP, Shil SK, Hossain MA, Ayman A, Jang YM (2018) Detection and classification of HGG and LGG brain tumor using machine learning. In: 2018 International conference on information networking (ICOIN). IEEE, New York, pp 813–817
22. Zhu W, Zeng N, Wang N (2010) Sensitivity, specificity, accuracy, associated confidence interval and ROC analysis with practical SAS implementations. In: NESUG proceedings: health care and life sciences, Baltimore, Maryland, 19, 67
23. Singh M, Shrimali V (2019) Exploring statistical parameters of machine learning techniques for detection and classification of brain tumor. Int J Innov Technol Explor Eng (IJITEE) 8(10)

Walking Assistant for Vision Impaired by Using Deep Learning

Amit Kumar, Gautam Nath, and P. Venkateshwari

Abstract In this paper, we address the people in the globe living with a visual impairment or living in isolation. Although it is already very difficult to live with sight loss, still we try to provide some support for them to access the environment freely. The lack of limited accessibility to activities and information, the societal stigma and the deficiency of joblessness are all factors frequently leading blind or low-vision individuals in isolation. We have used CNN technique and developed a method for processing an image and tells about the obstacle and depth of potholes to make the walking easy for visually impaired person. CNN has almost dominated computer vision and shows powerful performance on object classification. We create our own network on the basis of RCNN algorithm in which we consider 128 filters having size of (2×2). This network consists one input layer, nine middle layers, and five final layers and train the same network with CINIC-10 dataset having initial learn rate is 0.001, Learn rate drop factor is 0.1; learn rate drop period is 8, and maximum epochs are 40. After validation of our trained network, it gives the accuracy of 64.7%, and it wonderfully makes out results.

Keywords Visual impairment · Deep learning · CINIC database

1 Introduction

Blindness is considered the major sensory disability (it is estimated that 80% of the human sensorial information is provided by sight), which somehow cause our day-to-day life in very effective manner.

Sáez et al. [1] Visual impairment reduces a person's independent mobility and deteriorates the quality of life. According to the estimate from the World Health Organization, there are 1.3 billion people with visual impairment, of which 36 million is blind in 2018. For the reason, the leading cause of vision loss is age-related diseases,

A. Kumar (✉) · G. Nath · P. Venkateshwari
Department of Electronics and Communication Engineering, G.B. Pant Government Engineering College, New Delhi, India
e-mail: 1998amitkr@gmail.com

M. Gupta et al. (eds.), *Artificial Intelligence on Medical Data*,
Lecture Notes in Computational Vision and Biomechanics 37,
https://doi.org/10.1007/978-981-19-0151-5_18

and the world population is aging rapidly [2]. We will see more people experiencing visual impairment in the coming decades because more people going toward the digital world in various fields (e.g., online gaming, online learning, etc.). A sight-impaired person can remember things more than anyone else, but also, he/she cannot sense the current moving object as they can't see their surroundings or sense the moving object. Also, it is not possible to carry the cane, dog, or to take the help of another person everywhere, especially in sports which required the movement of the intact body [1].

Our project is basically divided into four parts, namely (1) Create deep understanding of neural networks and all the parameter used in code, (2) Create a neural network, (3) Training of neural network, and (4) Verify and validate our neural network on test data and some images and videos. After understanding all the concept deeply, we are ready to make our own neural network with some layers; here, we confirm our self about number of layers by the method of and creation of fully working network, lastly, we going to real images and video to satisfy our needs and provide a significant result. A convolutional neural network is a deep learning [3] algorithm, which take in an input image, assign importance (learnable weights and biases) to various features in the image, and be able to discriminate one from the other. The preprocessing required in a ConvNet is much lower than compared to other classification enough training, ConvNet can learn these filters/characteristics (Fig. 1).

We essentially include the toolbox of deep learning and compute vision offered by MATLAB; some necessary hardware includes camera and processor. The potential problem that occurs is just accuracy of output, but we have more advance algorithm based on RCNN with our own trained network using *CINIC-10* dataset [4], and it takes time in preprocessing but mostly depend upon system capability. After all this hindering and complex works, it will become the most valuable work for mankind. Since this research is auxiliary of human eyes that reflects continuous research in this field, so the camera and processor chip technologies have been increased rapidly in past few years.

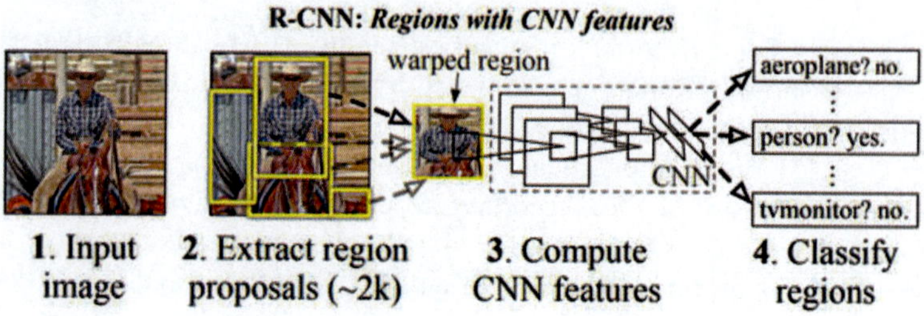

Fig. 1 Working example of RCNN [10]

2 Existing System

Currently, there are no fully functional system, especially for vision-impaired person around the globe. Although there are a lot of work present but in scattered manner. The most important work in this system is "object detection," i.e., observe the surrounding and provide the information in the audio form to the person. Previously, object detection is done by various computer vision techniques like HOG, cascade object detector, and stereo image processing [5–7]. The HOG will take the whole picture in processing and assigning weights to every pixel in the image [8], i.e., overall increase the processing time of the system, and the cascade object detector will only focused onto the region of interest (ROI) [4], i.e., only, target values are going into the processing that also cut the overall processing time, but these techniques didn't provide the pin point accuracy and always had a scattered output. These techniques are color based and only process the color saturation, eventually failed with similar surrounding as the target color. Overall, this produces some very good result if the color of target in picture is different than surrounding because the processing time reduces significantly. But, at the same time, this work is only effective in similar kind of surrounding, and if the surrounding chances, we have to do the whole work again from beginning.

3 Literature Review

In [8], the authors used CNN approach and HOG (histogram of oriented gradients) to increase a technique for processing a photograph and tell approximately the obstacle. The use of based light and grid pattern of the light they have been detecting the peak and form of impediment and depth of potholes to make the walking clean for a visually impaired person, so, they could stroll the tool keep in mind things or impediment for long-term or brief-time period as in keeping with the user need. The proposed technique can offer as a totally high-quality results with typical reputation and category accuracies of 98.37% for the binary case, and 95.97% and 92.62% for two one-of-a-kind multi-class situations. These effects represent the common quantity of sequences from four specific customers.

In [1], the authors purpose a unique approach with the help of 3D smartphones to detect objects that actually in the air, didn't touch the ground, like low tree branches to better assist visually impaired human beings. This kind of impediment is specifically hard due to the fact they cannot be detected by the on-foot stick or the manual canine. The algorithm captures the 3D information of the scene via stereo vision. The obtained 3D statistics are compressed after which linearized for detecting the potential barriers. Capability barriers are tracked to build up sufficient evidence to alert the consumer simplest, while a real impediment is determined.

The places like park and open-ground fields are the most difficult for VI because there they only depend upon the auditive response from surrounding human being and animals, etc. But, also, it shows difficultly while making system for those peoples because there are effectively low reference points for image or video processing.

In [4], the authors used deep convolution neural networks (DCNNs) as it offers many advantages over image classification, object detection, and semantic segmentation. In particular, presently, for object detection, proposal-based deep models are providing most effective performance and accurate results. There are many reasons of this outstanding performance, but the two reasons are most important, first, region proposal network (RPN) and, second, region of interest (ROI) pooling layer. The RPN is able to provide very long array of proposals, and to extract more features, they use ROI pooling layer. Apart from the results, there are some limitations, such as object deformation, which make hurdles in feature extract. Also, the objects, especially non-rigid, having different poses and environment result in different appearance.

4 Proposed System

This paper proposes a system for visually impaired person to observe surrounding in audio form with the use of convolution neural network [9] and MATLAB toolbox, i.e., deep learning toolbox.

4.1 Convolution Neural Network

First, we have specified the parameters like filter size and number of filters, and before going to create network, let us mention the whole network in three layers, namely input layer, middle layer, and final layer (Fig. 2).

1. The input layer constraints the images to the size of 32 × 32 × 1.
    - The middle layers of network are to be made up of repeated blocks of *convolutional, ReLU9 (rectified linear units),* and *pooling layers.* All the work of features extraction and weight assignment are accomplished in during execution of these three layers. The convolution layers define sets of filter weights, which are then fed onto the *ReLU9* layer, which cause non-linearity to the network; lastly, the pooling layers will down sample data as it flows throughout the network.
    - The final layers of a CNN are typically a combination of fully connected layers and SoftMax loss layer. The final layers use the output of the fully connected layer to compute the categorical probability distribution over the image classes. After all of these layers, finally, we will combine all of them to create fully connected neural network with 15 layers.

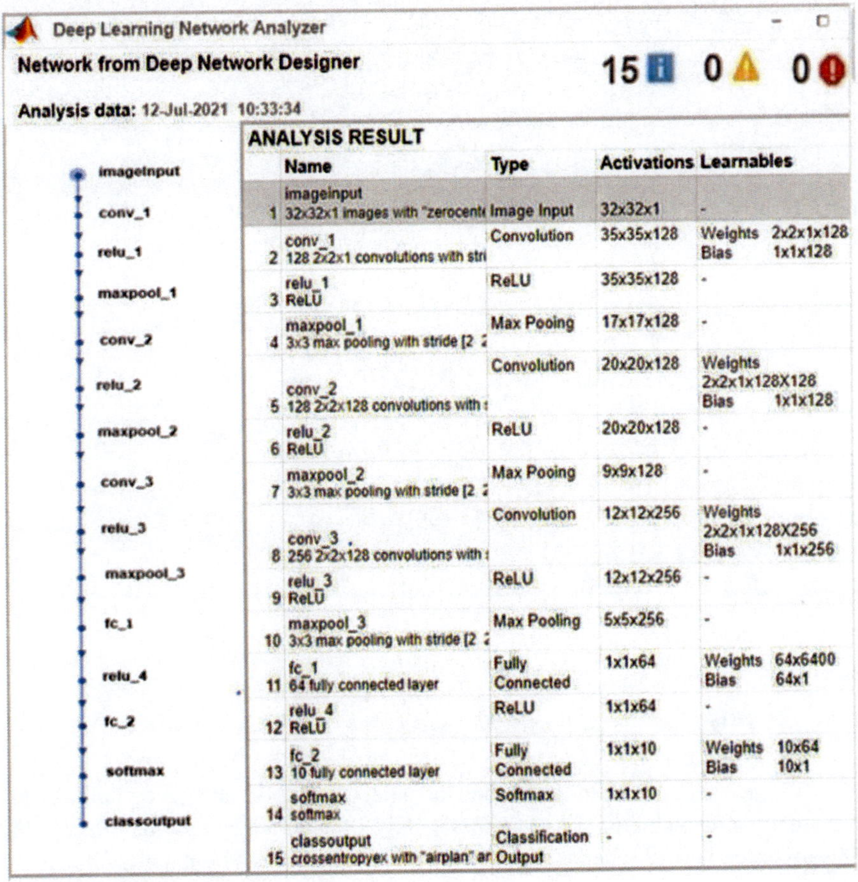

Fig. 2 Overview of CNN

4.2 *Training of the Network*

Now that the network architecture is defined; second, we are going to create an algorithm for network training using the *trainingOptions* function. We consider stochastic gradient descent with momentum (SGDM) having an initial learning rate of 0.001, and this learning rate is decrease after every eight epochs (algorithm consists total of 40 epochs). After defining these variables in the algorithm, we are going to train the network using *trainNetwork* function, while training the network will go through the entire training dataset of 90,000 images for 40 times called epochs (Fig. 3).

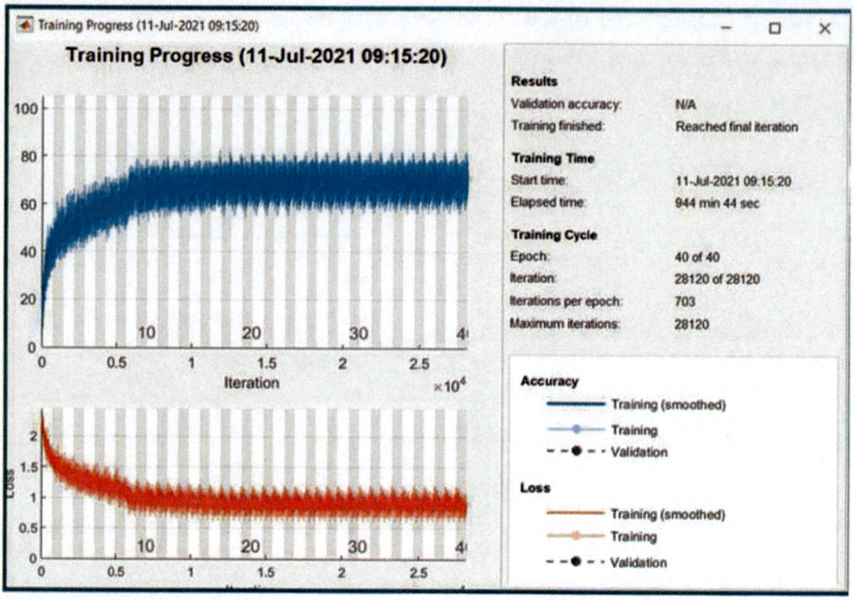

Fig. 3 Training process

4.3 *Results*

We all know about traditional method of object detection and recognition using computer vision techniques, and the percentage of accuracy is not more than 60%. This accuracy is totally depending upon the surrounding environment, and if there is change in environment, the whole algorithm is useless because all the image enhancement work is done on the basis of single environment. To overcome this issue, we take CNN into account, and thus, we approach 64.7% accuracy with 90,000 training image data and 180,000 validation image data (Fig. 4).

5 Conclusion and Future Work

In above images, we can see that it is easily classifying the input images correctly and giving the correct output as the requirement. The network is working properly and also giving accurate answers.

This project is a first step toward the fully developed device that use by the blind persons globally and reduce the blindness of any kind in near future. To complete this ambition, we may have work more than one deep learning to build neural network such as human brain so that sufficient amount of data can be processed and manage in real time, no such complex and time-consuming algorithm used again also the

Fig. 4 Verification

data storage capacity will increase. After all these ongoing efforts, we may surely see such device in coming years and more powerful techniques. One of the systems can be build having pictorial output directly fed to the brain of the person and thus overcome the requirement of vocal output, which effectively reduces the hearing ability. In future, this technology will divide for particular works like individual system for outside walking, in-house work sport, and different types of office work.

References

1. Sáez JM, Escolano F, Lozano MA (2014) Aerial obstacle detection with 3-D mobile devices. IEEE J Biomed Health Inform 19(1):74–80
2. Jarraya SK, Al-Shehri WS, Ali MS (2020) Deep multi-layer perceptron-based obstacle classification method from partial visual information: application to the assistance of visually impaired people. IEEE Access 8:26612–26622
3. Sindhwani N, Verma S, Bajaj T, Anand R (2021) Comparative analysis of intelligent driving and safety assistance systems using YOLO and SSD model of deep learning. Int J Inform Syst Model Des (IJISMD) 12(1):131–146
4. Wu S, Xu Y (2019) DSN: a new deformable subnetwork for object detection. IEEE Trans Circuits Syst Video Technol 30(7):2057–2066
5. Jain BD, Thakur SM, Suresh KV (2018, April) Visual assistance for blind using image processing. In: 2018 International conference on communication and signal processing (ICCSP). IEEE, New York, pp 0499–0503
6. Anand R, Shrivastava G, Gupta S, Peng SL, Sindhwani N (2018) Audio watermarking with reduced number of random samples. In: Handbook of research on network forensics and analysis techniques. IGI Global, New York, pp 372–394
7. Juneja S, Anand R (2018) Contrast enhancement of an image by DWT-SVD and DCT-SVD. In: Data engineering and intelligent computing. Springer, Singapore, pp 595–603
8. Cornacchia M, Kakillioglu B, Zheng Y, Velipasalar S (2018) Deep learning-based obstacle detection and classification with portable uncalibrated patterned light. IEEE Sens J 18(20):8416–8425
9. Kamalraj R, Neelakandan S, Kumar MR, Rao VCS, Anand R, Singh H (2021) Interpretable filter based convolutional neural network (IF-CNN) for glucose prediction and classification using PD-SS algorithm. Measurement 109804

10. Girshick R, Donahue J, Darrell T, Malik J (2014) Rich feature hierarchies for accurate object detection and semantic segmentation. In Proceedings of the IEEE conference on computer vision and pattern recognition (pp. 580–587)

Effective and Secure Transmission of Health Information Using Advanced Morphological Component Analysis and Image Hiding

Binay Kumar Pandey, Digvijay Pandey, Vinay Kumar Nassa, Shaji George, Bashiru Aremu, Pankaj Dadeech, and Ankur Gupta

Abstract The using morphological component analysis, deep learning, and steganography, this study examines the secure transmission, identification, and validation of textual pictures via this Internet of Things-based channel. To extract characteristics from text-based pictures, morphological component analysis is utilised. Each of these traits has a distinct morphological component. Without losing visual quality, the morphological component technique lowers duplicative but also uncorrelated characteristics, as well as the size in bytes of a text-based picture. The capacity to obtain full text-based picture texture characteristics, rather than relying on a single spatial texture feature, is considered to be a significant aspect of the proposed approach. Before being transferred over Internet of Things-based networks, the

B. K. Pandey (✉)
Department of Information Technology, College of Technology, Govind Ballabh Pant University of Agriculture and Technology Pantnagar, U S Nagar, Uttarakhand, India
e-mail: binaydece@gmail.com

D. Pandey
Department of Technical Education, IET, Dr. A.P.J. Abdul Kalam Technical University Uttar Pradesh, Lucknow 226021, India

V. K. Nassa
Department of Computer Science Engineering, South Point Group of Institutions, Sonepat 131001, India

S. George
Department of Information and Communication Technology, Crown University, Int'l. Chartered Inc. (CUICI) Argentina Campus, South America, Santa Cruz, Argentina

B. Aremu
Crown University, Int'l. Chartered Inc. (CUICI) Argentina Campus, South America, Santa Cruz, Argentina

P. Dadeech
Computer Science and Engineering, Swami Keshvanand Institute of Technology, Management and Gramothan (SKIT), Jaipur, India

A. Gupta
Department of Computer Science and Engineering, Vaish College of Engineering, Rohtak, Haryana, India

M. Gupta et al. (eds.), *Artificial Intelligence on Medical Data*,
Lecture Notes in Computational Vision and Biomechanics 37,
https://doi.org/10.1007/978-981-19-0151-5_19

morphological portions of a concealed text-based picture have been further divided and generally implanted interested in the least significant pieces of the cover pixels using spatial steganography. To retrieve the message from stored-images, the hidden key of an embedding algorithm can be exchanged with a text recovery approach on the recipient's side. To recognise an embedded text-based message, a hybrid convolution neural network technique was eventually employed. Aside from that an optimization approach is utilised to improve the hybrid convolution neural network's performance (HCNN's).

Keywords Effective secure transmission · Health information · Advanced morphological component · Image hiding · Text extraction

1 Introduction

The World Wide Web is one of the most well-known and simple means for people to transmit electronic information. Yet one of the most common threats to transmitting such information is that anybody may get such information, and the World Wide Web provides very little protection for such information on its own. Meanwhile, to prevent consumers from obtaining such digital material, a sender wants to execute a few more safety measures. Datagrams of various and defined sizes were used to transfer information amongst endpoints in a network connection. Most protected techniques were designed to apply data only at the application level, thus safe encrypted data was packetised and transmitted to the OSI architectural design's lower layers. When an intruder obtains all of the packets, he or she will receive encrypted data whilst appropriately arranging all of the data obtained from the packets. Following that attempts were made to break through the sender's secured technique. If it is difficult to erase information on the subject of the organisation of information inside the packet header during data transmission, an intruder will not identify the substance of information gathered or sent, or even the organising of important data from separate packets. As a consequence, the only job of safeguarding a transmitted signal from all sorts of threats whilst still safely delivering necessary data to a receiver has been accomplished. That is what a suggested technique, which would have centred on MCA steganography [1] and a hybrid convolution neural network, could have been able to complete. MCA is a method for distinguishing picture elements that appear to have diverse morphological core characteristics.

Morphological component analysis (MCA) should be used for picture in painting as well as scene segregation, and it is capable of dividing images into texture and smooth parts [2, 3]. When used with total variation regularisation, morphological component analysis looks to be a particularly effective approach for distinguishing a picture into its piecewise smooth content and texture. In morphological component analysis, the usage of the curvelet dictionary causes ringing artefacts. To eliminate ringing artefacts from piecewise smooth portions, a total variation regularisation approach was utilised. The Daubechies wavelet was used to forecast the TV regularisation method. The cartoon section of the picture has been transformed to the daubechies wavelet transform (DWT), which is helped by the coefficient's soft thresholding.

The researchers [4] presented a weighted Naive Bayes classifier (WNBC) deep learning model that can successfully recognise words and scripts in picture data. Real scene photos usually have a few tiny blemishes that are eliminated by utilising guided image filtering during the pre-processing step. The Gabor transformation (GT) and stroke width transformation (SWT) techniques were being used to retrieve meaningful data across the classifiers. Using these recovered features, textual identification, and character recognition are accomplished using WNBC and deep neural network-based adaptive galactic swarm optimization. To determine the competence of a suggested method, performance measures such as exactness, F1-score, precision, mean absolute error, mean square in accuracy, and memory assessments would be utilised. A novel technique for character identification [5] and automatic text extraction technique was presented. It required a linked component-based technique that made heavy use of a system for detecting the most stable external regions' characteristics. Contour-oriented and geometrical filters were used to identify non-text and text MSERs. The remaining textual areas were then separated into words and phrases. The surplus textual and non-textual regions that did not sufficiently correspond with predicted characteristics were then removed using creative filters. The words and phrases that remained throughout the final step were identified using OCR technology. Finally, a content discovery and distribution platform was used to accomplish this plan.

The goal of this study is to employ morphological component analysis, steganography, and text identification on the receiver side of a hybrid convolution neural network to improve the reliability of text-based picture bytes per second transfer. As a result, two efforts are proposed in this article. The first combines morphological component analysis with a steganographic technique, whilst the second mixes a hybrid convolution neural network with this steganographic algorithm. In reality, data protection is carried out twice. The proposed approach surpasses the existing technique in terms of performance parameters such as PSNR and accuracy, according to the results.

2 Methodology Used

At the sender's end, the algorithm for extracting characteristics [6] from a textual picture using morphological component analysis and then encoding these features using an eLSB-based steganography method is as follows:

The morphology component analysis technique was employed on the transmitter end to break the picture into individual parts, which were the real scene component and the pattern, as illustrated in Fig. 1. The curvelet transform improves the wavelet transform in experiments. In order to differentiate natural scene components, the wavelet transform is used. In fact, the vector variable will be set to the original

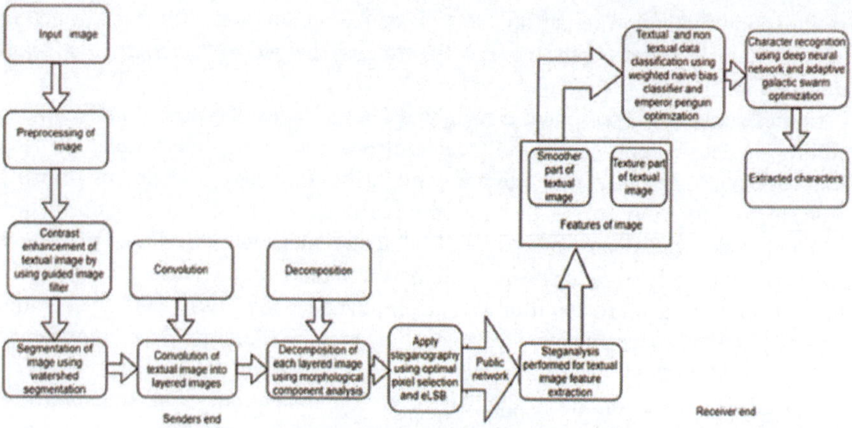

Fig. 1 Proposed methodology using advanced morphological component analysis and image hiding

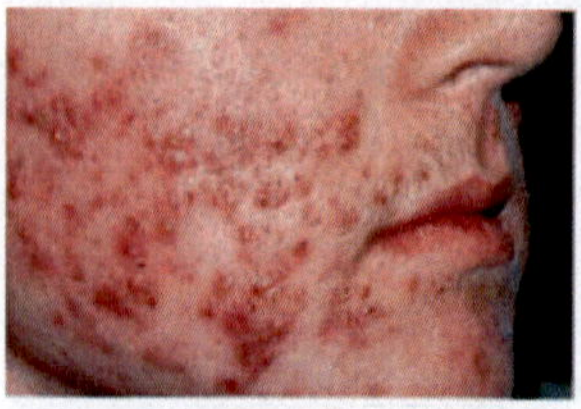

(a) Input Image of Skin Disease Infected Person

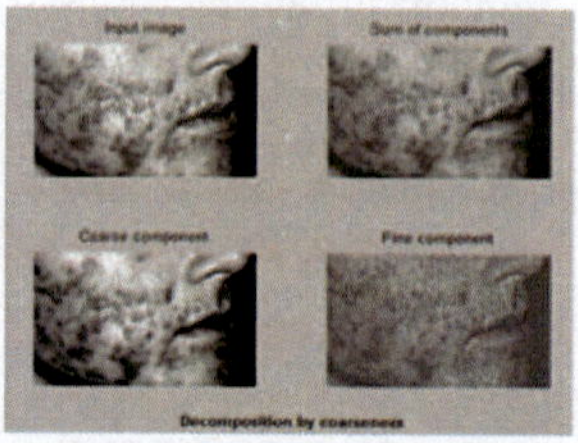

(b) Decomposition of input images by Coarseness

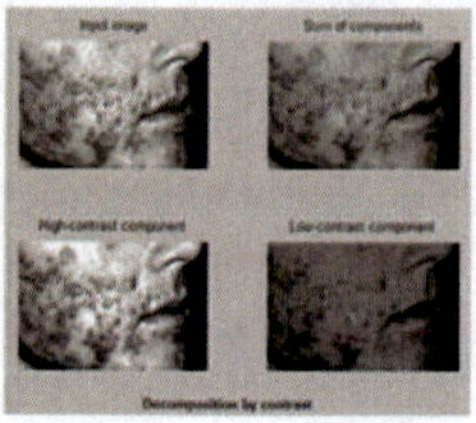

(c) Decomposition of input images by contrast

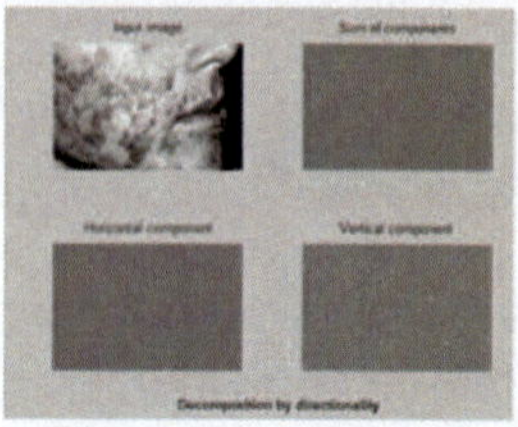

(d) Decomposition of input images by directionality

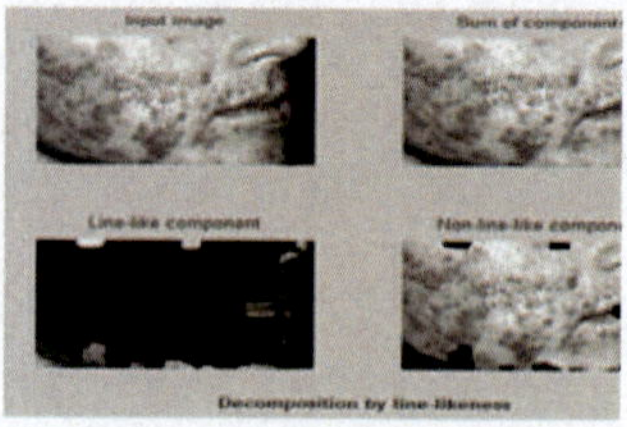

(e) Decomposition of input images by line-likeness

Fig. 2 a–e Feature extraction using morphological component analysis and encoding of textual image

value of the picture Curvelet transformation and discrete cosine transformation of a residual. Then it was calculated. The curvelet coefficient is obtained after getting a curvelet transformation of the residual. Hard edge detection was used. The curvelet transform returns the frequency component of the matrices.

(f) Component of input images to be embedded in cover image

(g) Cover image

(h) Stego-image obtained after embedding input image on cover image

Fig. 3 **f–h** Represents process of stenography

As illustrated in Fig. 2f–h, steganography is used to conceal a picture that has been contaminated with skin infection on the cover image (Fig. 3).

3 Results and Discussion

On the receiver side, the private key produced by the embedding method is sent to a strategy for extracting textual information. The secret key is utilised to acquire the text-based image, and the implanted text is extracted from the cover image using a hybrid convolutional neural network. An adaptive optimization technique is used to improve the effectiveness of machine learning [7] algorithms. The PSNR and accuracy are utilised to evaluate the effectiveness of the proposed secret text extraction procedure [8–10] (see Tables 1 and 2).

Figure 4 presents a comparative study of the accuracy of the Haar and Daubechies filters used in the MCA and steganography processes. Figure 5 shows the comparison of the PSNR for the two types of filters; one is the Haar Filtering Techniques and the other is the Daubechies Filtering Techniques used in the MCA and steganography processes.

Table 1 Effect on accuracy for cover image shown in Fig. 1

Threshold (γ)	Accuracy with Haar filtering techniques	Accuracy with Daubechies filtering techniques
5	95.645	96.343
9	97.923	98.126
12	96.976	97.997
16	98.695	99.182
18	98.456	99.385

Table 2 Effect on PSNR for cover image

Threshold (γ)	PSNR with Haar (existing methods)	PSNR with Daubechies (proposed methods)
5	40.12	40.75
9	40.50	40.25
12	41.25	42.00
16	42.70	45.70
18	43.80	47.20

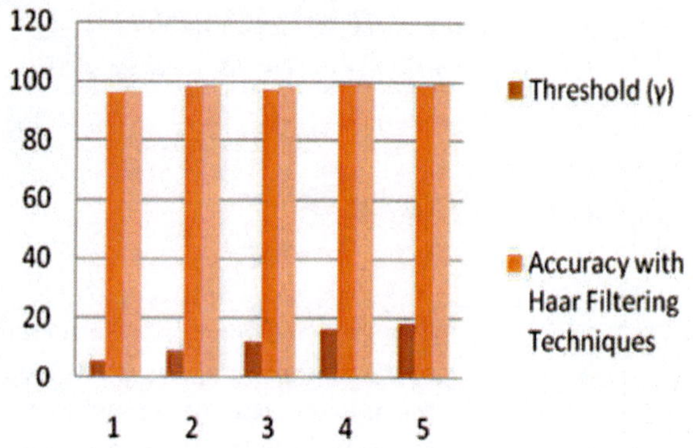

Fig. 4 Accuracy comparison for Haar and Daubechies filters used in MCA and steganography

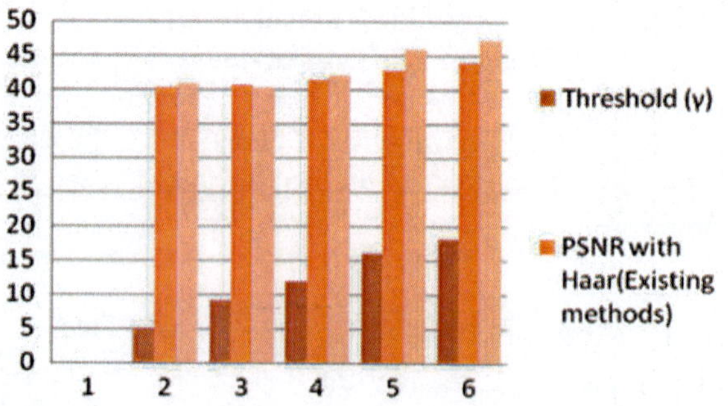

Fig. 5 Peak to signal ratio comparison for the Haar and Daubechies filters used in MCA and steganography

4 Conclusion

The textual image fetching algorithm on the receiving end was given the secret key produced via textual image implanting approach in order to get the textual image, and the integrated textual picture was eventually recognised using just a weighted Naive Bayes (WNB) classifier. A weighted Naive Bayes classifier's efficiency is also improved using an optimization [11] approach. The proposed technique outperforms the standard approach, which uses the HAAR filter to create text-contain scene elements by using a debauched filtering system. The proposed technique not only improves the PSNR of the method, but also outperforms it in terms of accuracy. The proposed approach correctly retrieves text at the receiving end, but the characters may be lost on occasion, or an equivalent character may be retrieved repeatedly, resulting in the recovery of an incorrect message. As a result, in the future, an effective approach for avoiding these sorts of text identification mistakes must be established. Various types of fuzzified systems are also addressed. Using an ant colony optimization approach, one can select to achieve better outcomes in the future.

References

1. Pramanik S, Ghosh R, Ghonge MM, Narayan V, Sinha M, Pandey D, Samanta D (2021) A novel approach using steganography and cryptography in business intelligence. In: Integration challenges for analytics, business intelligence, and data mining. IGI Global, pp 192–217
2. Anand R, Shrivastava G, Gupta S, Peng SL, Sindhwani N (2018) Audio watermarking with reduced number of random samples. In: Handbook of research on network forensics and analysis techniques. IGI Global, pp 372–394
3. Aujol JF, Aubert G, Blanc-Féraud L, Chambolle A (2003) Image decomposition application to SAR images. In: Griffin LD, Lillholm M (eds) Scale space methods in computer vision. Scale-Space 2003. Lecture notes in computer science, vol 2695. Springer, Berlin, Heidelberg. https://doi.org/10.1007/3-540-44935-3_21
4. Pandey D, Pandey BK, Wairya S (2021) Hybrid deep neural network with adaptive galactic swarm optimization for text extraction from scene images. Soft Comput 25:1563–1580. https://doi.org/10.1007/s00500-020-05245-4
5. Baran R, Partila P, Wilk R (2018) Automated text detection and character recognition in natural scenes based on local image features and contour processing techniques. In: International conference on intelligent human systems integration. Springer, Cham, pp 42–48
6. Bakshi G, Shukla R, Yadav V, Dahiya A, Anand R, Sindhwani N, Singh H (2021) An optimized approach for feature extraction in multi-relational statistical learning
7. Singh H, Rehman TB, Gangadhar C, Anand R, Sindhwani N, Babu M (2021) Accuracy detection of coronary artery disease using machine learning algorithms. Appl Nanosci, 1–7
8. Pandey D, Pandey BK, Wariya S (2020) An approach to text extraction from complex degraded scene. IJCBS 1(2):4–10
9. Pandey D, Pandey BK, Wariya S (2019) Study of various types noise and text extraction algorithms for degraded complex image. J Emerg Technol Innov Res 6(6):234–247
10. Pandey D, Pandey BK, Wariya S, Ventayen RJM, Khan B, Gupta M, Kumar T (2020) Analysis of text detection, extraction and recognition from complex degraded images and videos. J Crit Rev 7(18):427–433
11. Sindhwani N, Singh M (2017, March) Performance analysis of ant colony based optimization algorithm in MIMO systems. In: 2017 International conference on wireless communications, signal processing and networking (WiSPNET). IEEE, New York, pp 1587–1593
12. Pandey BK, Pandey D, Wariya S, Agarwal G (2021) A deep neural network-based approach for extracting textual images from deteriorate images. EAI Endorsed Trans Ind Netw Intell Syst 8(28):e3
13. Pramanik S, Samanta D, Dutta S, Ghosh R, Ghonge M, Pandey D (2020, December) Steganography using improved LSB approach and asymmetric cryptography. In: 2020 IEEE international conference on advent trends in multidisciplinary research and innovation (ICATMRI) (pp 1–5). IEEE

14. Pandey BK, Pandey D, Wairya S, Agarwal G (2021) An advanced morphological component analysis, steganography, and deep learning-based system to transmit secure textual data. Int J Distrib Artif Intell (IJDAI) 13(2):40–62
15. Meivel S, Sindhwani N, Anand R, Pandey D, Alnuaim AA, Altheneyan AS, ... and Lelisho ME (2022) Mask detection and social distance identification using internet of things and faster R-CNN algorithm. Comput Intell Neurosci

Bio-medical Image Encryption Using the Modified Chaotic Image Encryption Method

V. Vanitha and D. Akila

Abstract Protection of personal data from unwanted access while assuring speed and precision is essential in today's information and speed driven world. Recent research on picture encryption methods have increasingly been based on chaotic systems, although in one-dimensional chaotic cryptosystems there are apparent disadvantages of a tiny key space and low security. The chaotic approach has been used to suggest a novel picture encryption algorithm to avoid the aforesaid limitation. A chaotic dynamic mechanism is employed in the synchronization of two chaotic systems and controllers. The picture is cramped by adding the chaotic sequence to the disturbance term associated with the plaintext. Then the dynamic index's cipher text feedback device is adopted at the diffusion stage. This means that the cipher text cipher index utilized to provide feedback is locative indexed. The proposed stenography with the chaotic map technology, as regards the time required to mark a certain collection of images and with greater space, smaller iteration times and high security analysis such as key space analysis, statistical analysis and sensitivity analysis, is simpler and more effective. Image encryption is accompanied with a confused logic to create a secret key. In addition, experimental findings like key space analysis, analysis of key sensitivity, differential analysis, histogram, entropy information and correlation coefficients reveal a safe and reliable image encryption technique with significant potential for application.

Keywords Steganography · Chaotic logistic map · Image encryption · Chaos shift keying · Chaotic masking · Secret key

---

V. Vanitha
Department of Computer Science, Vels Institute of Science, Technology & Advanced Studies (VISTAS), Chennai, India

D. Akila (✉)
Department of Information Technology, Vels Institute of Science, Technology & Advanced Studies (VISTAS), Chennai, India
e-mail: akiindia@yahoo.com

M. Gupta et al. (eds.), *Artificial Intelligence on Medical Data*,
Lecture Notes in Computational Vision and Biomechanics 37,
https://doi.org/10.1007/978-981-19-0151-5_20

1 Introduction

With the advancement of information science, it is now possible to gain fast and convenient access to the majority of communication networks. The art and science of hiding a message behind a cover medium are known as steganography. This approach is generally used for secure and confidential communications and is required in security applications [1]. The least important nuance (LSB) steganography is a basic spatial domain data hiding strategy in which the least important bits of some cover image hidden pixels hold a series of secret binary information [2]. As a cover picture, we may utilize the image, sound and text to communicate the message (host image). The graphic arts as the wrapper for this post. The image pixel LSB is one of the most prominent ways for data steganography in photographs (least significant bits) [3].

In today's modern world, the protection of digital images is becoming increasingly relevant as digital product messages over networks become more common [4]. To satisfy the demand for encrypted image exchange in real-time via the Internet and cellular networks, camera encryption technologies are continually being explored [5]. Traditional image encryption techniques, such as data encryption standards (DES), perform poorly when the image is large. A fresh and successful method to the thorny problem of simple and highly reliable picture encryption has been offered using chaos-based encryption. Since Matthews suggested the chaotic encryption algorithm in 1989, further picture encryption application research has been focused on chaotic systems. Many articles on chaotic encryption schemes have recently been published [8, 9].

In the last decade, chaos has emerged as one of the most critical research areas. Almost all researchers are attempting to implement it in the field of reliable correspondence and cryptography using disorderly regulation or synchronization [10]. The value of medical image security is growing as the focus on medical privacy and medical identity safety grows. Medical images provide the greatest proportion of medical records. At the same time, the medical picture, as an essential transporter of medical knowledge delivery, necessitates high protection and confidentiality. Although the standard encryption algorithm can be used to encrypt images, it is inefficient and easily broken. Furthermore, the medical picture has several unique properties. Orthodox image encryption methods are not appropriate for medical images. As a result, it is critical to developing a stable and effective picture encryption algorithm [11, 12].

We employed chaos in this paper to present a chaotic stenography method in the spatial domain that was not only secure against multiple assaults but also performed well statistically. Messages are broken down into LSBs in binary data and positioned in various areas of a carrier, with arbitrary locations within these areas identified by disorderly sequences [13]. Using chaotic map logic for image encryption. Logistic map generation, as well as random hidden column and secret row index generation. It is possible to construct a logical operation of an input image with disorderly images. Stenography using the disorderly map approach is primarily useful for effective and precise data analysis in real-time applications [14]. In terms of time taken to label a specific collection of Bio-images, the proposed stenography with chaotic map technique is simpler and more effective. Bio-image encryption is accompanied by the generation of a secret key using disorderly logic [15, 16].

2 Related Work

Xie et al. [17] Fridrich's approach is vulnerable to a chosen-cipher text attack that uses an influence network between cipher-pixels and their plain-pixel counterparts. Based on their inventive work, this study explored specific characteristics of Fridrich's design using compact mathematical jargon. Then, several minor shortcomings in Solak's attack method's real performance were highlighted. The thesis sets the framework for additional in-depth examinations of Fridrich's scheme and its variations.

Istanbul et al. [18] suggested a method for revealing the invisible permutation used to shuffle the pixels of a circular input. We use examples and simulation outcomes to illustrate the efficacy of our assault. We also demonstrate that our proposed attack applies to other well-known chaotic image encryption algorithms.

Ye et al. [19] the generalized Arnold map is used to create an effective image encryption algorithm. Rather than the conventional periodic location permutation, a complete circular mechanism is used in the permutation stage. It has the potential to significantly minimize the association between neighboring pixels. Then, during the diffusion phase, double diffusion functions, i.e., positive and opposing modules, are combined with a new keystream generation. Because the keystream is reliant on the processed picture, the suggested technique can survive known-plaintext and chosen-plaintext assaults. The utility of our approach is shown by experimental findings and theoretical studies. The suggested algorithm's expansion to other chaotic systems is also explored.

Wang et al. [20] the proposal includes a novel technique for encryption based on DNA sequence operations (Deoxyribonucleic acid) and a messy approach. First, the Pixel Scale is carried out with the pseudorandom sequences of the space-timed chaos scheme, i.e., CML, which are bitwise exclusive OR operations (coupled map lattice). Secondly, by applying a DNA rule for encoding the jumbled picture, a DNA matrix is generated. This DNA matrix plus the prior starting conditions will then be used to construct fresh CML initial circumstances, which can lead to the encryption result being very dependent on every pixel on the single image.

Ullah et al. [21] for picture encryption, a chaotic method with a replacement box is used. The group operation of the projective general linear group over a finite field creates a substitution box. The main feature of this scheme is that on repeated attempts with the same encryption key, the host image returns a new encrypted image. As seen in simulation and security research, this concept of chaos-based image encryption combined with replacement box yields comparatively better results for safe communication [22, 23]. The scheme shows resistance to image recognition threats.

3 Proposed Method

In this research, we present a novel steganography-based secret communication technique based on the chaotic logic map encryption approach. Application of a chaotic dynamical system with a connection to the synchronization and control of two chaotic systems. Random secret column, secret row index generation and Logistic map generation. Logical operation of the input image with chaotic data created. For real-time applications, stenography utilizing the chaotic map approach is most effective in terms of efficient and precise data processing. In terms of the time necessary to mark a certain collection of photos, the proposed stenography with chaotic map methodology is faster and more efficient. Secret key generation using chaotic logic and followed by image encryption.

A. *Steganography of images*

The incorporation of data into an image carrier is called image steganography. The four forms of steganography pictures are spatial domain, transforming the domain, scope spread and model-oriented image steganography [25–26].

i. *New Modified Chaotic behavior of the Logistic Map*:

A modified chaotic function known as Logistic Map is used in this study:

$$f(x) = \lambda x(1 - x), 3.57 < \lambda < 4 \quad (1)$$

The value of λ is governed by the chaotic behavior of the function. It is restricted to the range (3.99, 4) to operate in the messy field of the feature. The logistic map is presented in Fig. 1 the bifurcation chart. The contribution of the disorderly function, as can be seen in Fig. 1, when the value of approaches λ and 4, takes on increasingly distinctive values between 0 and 1. Equal (2) and (3), give the disordered function in the discrete form (3).

ii. *Encryption/decryption algorithm*

For encryption and decryption operations, the recommended crypto-system utilizes two distinct logistic maps. Equations (2) and (3) create logistical maps in where x_{i+1} and y_{i+1} are a state value with $I = 0, 1, 2, \ldots$, λ_1 and λ_2 are parameters which are used as a crypto-system key, which depict the chaotic behavior of maps.

$$x_{i+1} = \lambda 1\ xi(1 - xi), \quad i \geq 0 \quad (2)$$

$$y_{i+1} = \lambda 2\ xi(1 - yi), \quad i \geq 0 \quad (3)$$

*x*0 and *y*0 are the initial seeds, which take on actual values between 0 and 1. The different values of *xi*+1 and *yi*+1 between 0 and 1 are supplied by iterating Eqs. (2)

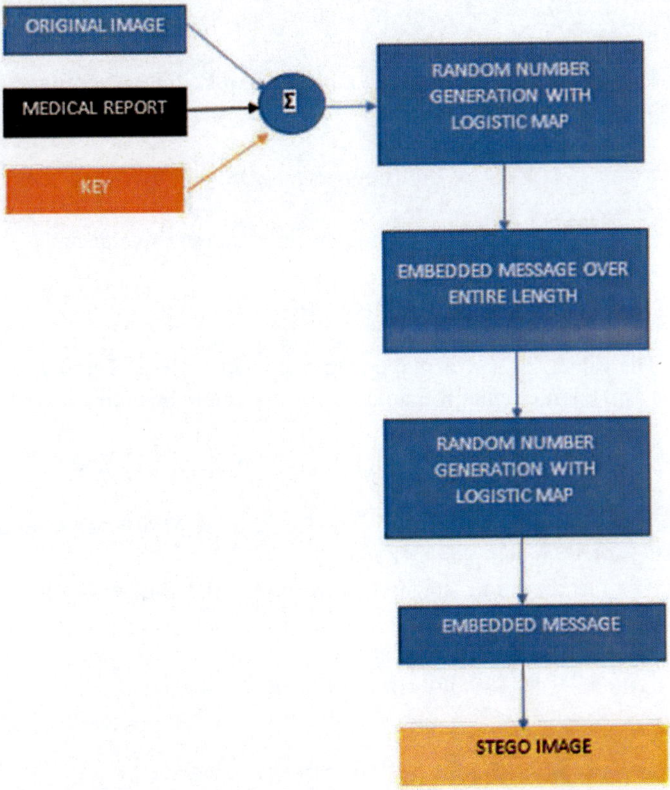

Fig. 1 Proposed secure steganography model

and (3). Pixel coordinates are permuted using Eq. (2), while pixel strength values are modified using Eq. (3). The steps of the encryption algorithm are outlined below.

Step 1: The original image I of the pixel intensity of size $M\,N$ are combined into a one-dimension array P of the length of $L = MN$.

Step 2: The recommended crypto-keys system consists of two initial seed sequences of the employed chaotic functions, $x10$, ..., $xR0$, $y10$, ..., $yR0$, selected to be (0, 1), where $R$is is the number of encryption/decryption rounds.

Step 3: To eliminate the chaotic functions' temporary influence [10], the first r iterations of Eqs. (2) and (3) are eliminated in each round. The round loop vector (r) has the values 1, 2, 3, ..., R. The r value, which is also a secret key, is advised to be in the range (300, 600). The highest limit of this spectrum increases the encryption/decryption time by expanding the keyspace.

Step 4: Array P elements are done using a loop vector I from 1 to L 2 consecutively. Pixels created from Pis and Xored consecutively with the pixel value indicated by Eq's second term (4).

$$V1 = \text{Pr}[i] \oplus \text{Pr}\left[i + \text{floor}(\alpha x r_{i+1}) \bmod S\right] \tag{4}$$

Step 5: The range 0–255 is mapped to the result of Eq. (3). After that the mapped value is XORed with $V1$, yielding $V2$.

$$V2 = V1 \oplus \left[\text{floor}(\alpha y r_{i+1}) \bmod 256\right] \tag{5}$$

Step 6: $V2$ is XORed with the previously ciphered byte's value as

$$V3 = V2 \oplus \text{Pr}[i - 1] \tag{6}$$

Step 7: The value $Yr\ i + 1$ is mapped to a number between 1 and 8 to decide as many times as the left circular rotations of the $V3$ should be duplicated via itself.

$$V4 = \text{Le f tCircularRotate}\left(V3, \left(1 + \text{floor}\left[\alpha y r_{i+1}\right] \bmod 8\right)\right) \tag{7}$$

The combinations of steps 6 and 7 enhance plaintext resistance substantially to prevent preferred or known-plaintext attacks.

Step 8: For the preceding stages to take occur, the following procedures are required.

$$\text{Pr}\left[\left(i + \text{floor}\left(\alpha x_{i+1}^{r}\right)\right) \bmod S\right] = \text{Pr}[i] \tag{8}$$

$$\text{Pr}[i] = V4 \tag{9}$$

Step 9: Repeat steps 3–8 until the penultimate byte of the byte array Pr is reached. To now, the actions reflected just one cycle of encryption.

Step 10: The current encrypted byte array Pr is used as an input byte array in the subsequent encryption cycle if R is higher than one. As a result, the following task is done following increase r.

$$\text{Pr} = \text{Pr} - 1 \tag{10}$$

Finally, a keystream made up of R sets of quintuple values is used to construct an encrypted byte array PR ($x0$, $y0$, 1, 2,). To help understand how the algorithm works, a flowchart of the encryption method is included. The decryption phase reverses the R sequence of quintuple ($x0$, $y0$, 1, 2,) (Table 1).

iii. *Secret key*

The fact that the message bits are arbitrarily inserted in the picture is critical for protection. As a result, the logistic map technique, the most often used approach for the production of random numbers, was selected. This approach generates different random numbers one by one till the pixel number of the picture is reached. Nonlinear dynamic Eq. 11 expresses the unstable.

Table 1 Encryption and decryption algorithms

Encryption Algorithm	Decryption Algorithm

INPUT: IE ; L = M ∗ N; R(roundnumber) x1,...,R 0 , y1,...,R 0 (initial seeds); δ1,...,R ;λ1,...,R 1 ; λ1,...,R; 2
INITIAL ASSIGNMENTS: r = R;α = 1014; Pr = IE ;
while(r > 0)
S = L − 1;i = 0; x = xr 0; y = yr 0; λ1 = λr 1; λr 2; δ = δr;
for h=1 to δ do
x = λ1 ∗ x ∗ (1 − x); y = λ2 ∗ y ∗ (1 − y);
end for
while(i < L − 2)
x = λ1 ∗ x ∗ (1 − x); y = λ2 ∗ y ∗ (1 − y);
y_norm = f loor(α ∗ y)mod256; y_seq[i] = y_norm;
j = i + (f loor(α ∗ x)modS); Z[i] = j; i = i + 1; S = S − 1;
end while
19: j = L − 3;
while(j >= 0)
i = Z[j]; V1 = Pr[j];
if(j > 0) then
prev_chiphered_byte = Pr[j];
else
prev_chiphered_byte = 0;
end if
y_norm = y_seq[i]; V2 = Le f tCircularRotate(V1, 7 − (y_normmod8));
V3 = V2 ⊕ Pr[i] ⊕ y_norm; V4 = V3 ⊕ prev_chiphered_byte;
	INPUT: IE ; L = M ∗ N; R(roundnumber) x1,...,R 0 , y1,...,R 0 (initial seeds); δ1,...,R ;λ1,...,R 1 ; λ1,...,R; 2
	INITIAL ASSIGNMENTS: r = R;α = 1014; Pr = IE ;
	while(r > 0)
	S = L − 1;i = 0; x = xr 0; y = yr 0; λ1 = λr 1; λr 2; δ = δr;
	for h=1 to δ do
	x = λ1 ∗ x ∗ (1 − x); y = λ2 ∗ y ∗ (1 − y);
	end for
	while(i < L − 2)
	x = λ1 ∗ x ∗ (1 − x); y = λ2 ∗ y ∗ (1 − y);
	y_norm = f loor(α ∗ y)mod256; y_seq[i] = y_norm;
	j = i + (f loor(α ∗ x)modS); Z[i] = j; i = i + 1; S = S − 1;
	end while
	19: j = L − 3;
	while(j >= 0)
	i = Z[j]; V1 = Pr[j];
	if(j > 0) then
	prev_chiphered_byte = Pr[j];
	else
	prev_chiphered_byte = 0;
	end if
	y_norm =y_seq[i]; V2=LeftCircularRotate(V1,7−(y_normmod8));
	V3 = V2 ⊕ Pr[i] ⊕ y_norm; V4 = V3 ⊕ prev_chiphered_byte;

$$xn + 1 = \mu.xn.(1 - xn) \tag{11}$$

In Eq. 11, is the chaotic character of the equation's parameter, and *xn o* is the start value in the [0-1] range. Usually, the value is picked to represent the chaos of the equation between [3.5699356 and 4].

The suggested model divides the encryption method into four stages. These steps are listed in detail below.

Step 1: In the proposed model, different random numbers from [0–1] until the number of pixels in the image were created using Eq. 11 with $S1$ initial value that is to insert message length into the band B of a colored picture, which is the Square root of the Ascii values the secret key.

Step 2: Stage 2: After receiving the values from the logistic diagram, the following method was done to insert message length $M1$ into the color graphic B band.

- The random number created has been sorted by the margin sort algorithm based on the sequence of their arrival.
- The rows and columns of the embedded pixels were decided to utilize the arrivals. The division and other lines and columns values were derived by dividing the pixel obtained at the time of arrival by the number of columns. As a result, the row and column values were computed of the pixel into which the message bit is inserted.

The color image's B band contains the message length $M1$ bits.

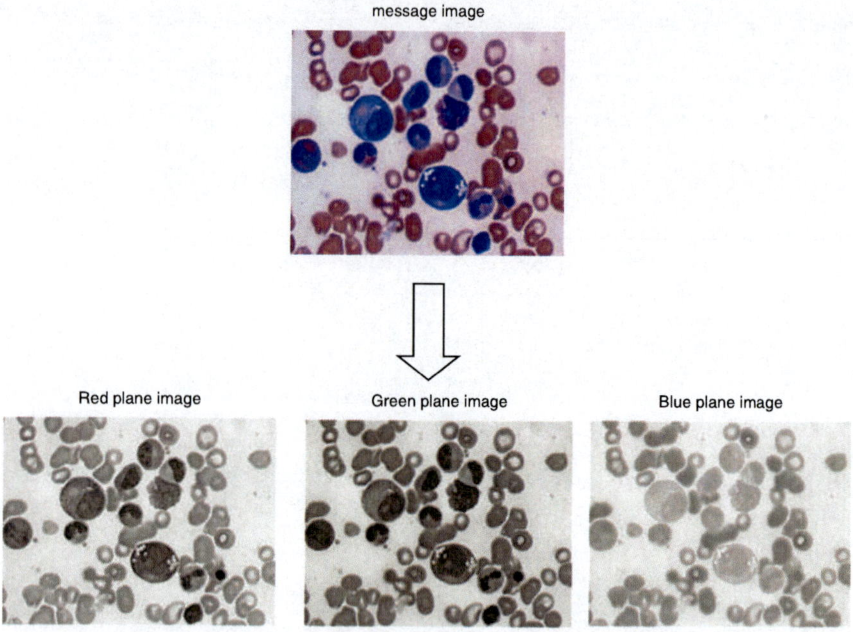

Fig. 2 Input image of blood smear and *R*, *G* and *B* channels of the input image

Step 3: With Eq. 1, fresh chaotic values from different, alterations of picture number were established, which included the sum of the value of the ascii, hidden keys and the message length, to insert the message into the *G* band of the color image with *S*2 start value as shown in Fig. 2.

Step 4: The bits of message *M*2 has been placed in the *G* band of the color picture as specified in stage 2 after receiving fresh random numbers with the logistic map as shown in Fig. 3 (Table 2).

Figure 4 depicts the *RGB* histogram of plain image, cipher image and extracted image from Table 1 entropy of our proposed scheme is very much close to "8' compares to others. The absolute values of proposed scheme's correlation coefficient of plain text is close to "1" whereas for cipher it is close to "0". Rate of change of cipher and NCPR (%) is very high showing high sensitivity toward change in key. showing better results compared to existing chaotic systems.

4 Conclusion

In this research, we provide a novel approach for covert communication in bio-images that employs steganography and a modified chaotic logic map encryption approach. Disorderly masking, shift keying and modulation using inverse systems all need additional computation time and power when applied to bio-images. Because of the

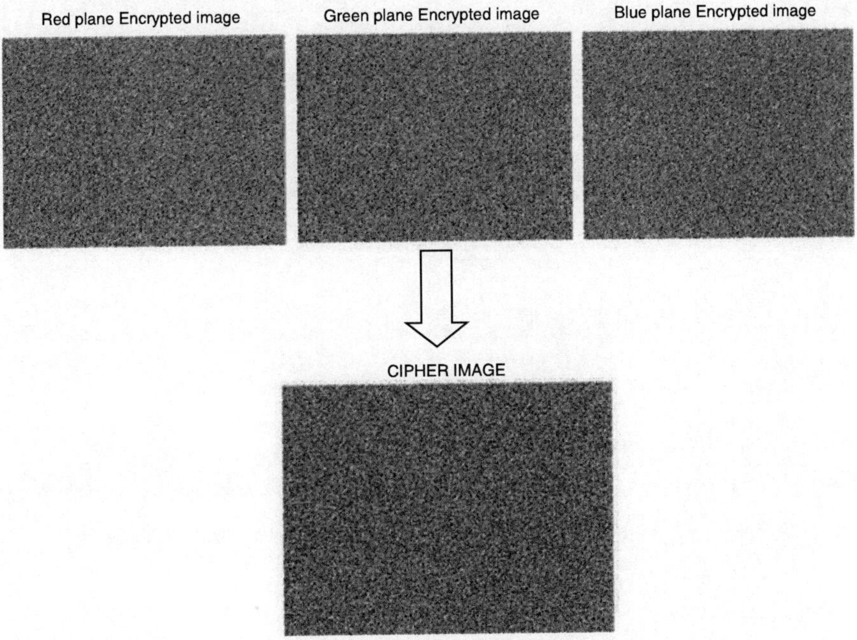

Fig. 3 *R*, *G* and *B* channels encrypted with chaos method and final encrypted image with combined *RGB* channels

Table 2 Performance comparison of proposed and existing methods

Method	Entropy	Rate of change of cipher		Correlation coefficient		NCPR (%)
		Key1	Key2	Plain	Cipher	
Proposed	7.9998	0.9964	0.9963	0.9315	−0.0031	99.9576
Cyclic shift Chaos	7.9864	0.9674	0.9612	0.7645	0.2467	99.8812
Baptista-type chaotic method	7.9643	0.9433	0.9498	0.8132	0.32501	99.6623

inclusion of huge and unnecessary blocks of image files, it is unsuitable for image applications. Using modified chaotic map logic for image encryption we made the security higher than related models.

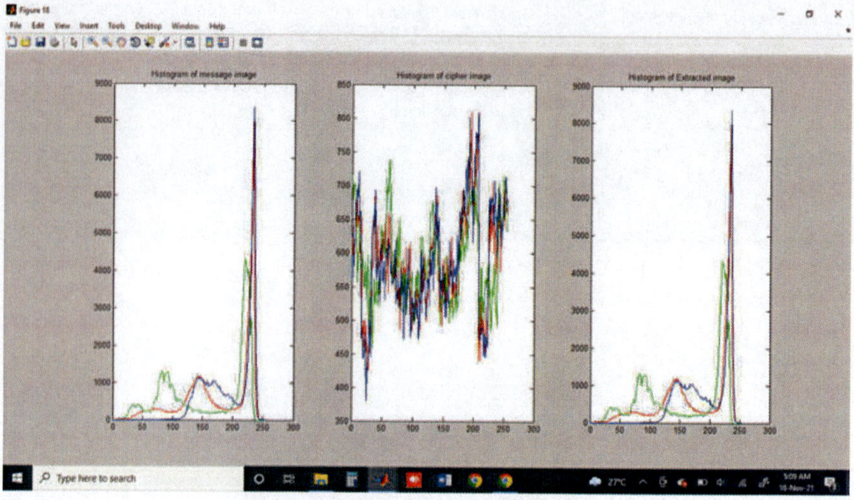

Fig. 4 Comparison of *RGB* histogram of plain image, cipher image and extracted image

References

1. Valandar MY, Ayubi P, Barani MJ (2017) A new transform domain steganography based on modified logistic chaotic map for color images. J Inform Secur Appl 34:142–151
2. Hussain MA, Bora P (2018, September) A highly secure digital image steganography technique using chaotic logistic map and support image. In: 2018 IEEE international conference on information communication and signal processing (ICICSP). IEEE, New York, pp 69–73
3. Enayatifar R, Mahmoudi F, Mirzaei K (2009, April) Using the chaotic map in image steganography. In: 2009 International conference on information management and engineering. IEEE, New York, pp 491–495
4. Krishnagopal S, Pratap S, Prakash B (2015) Image encryption and steganography using chaotic maps with a double key protection. In: Proceedings of fourth international conference on soft computing for problem solving. Springer, New Delhi, pp 67–78
5. Ganesan K, Venkatalakshmi B, Moothy RK (2004, December) Steganography using enhanced chaotic encryption technique. In: International conference of communication systems
6. Kumar S, Kumari S, Patro S, Shandilya T, Acharya AK (2015) Image steganography using index based chaotic mapping. Int J Comput Appl 975:8887
7. Alvarez G, Montoya P, Pastor G, Romera M (1999, October) Chaotic cryptosystems. In: Proceedings IEEE 33rd Annual 1999 International Carnahan conference on security technology (Cat. No. 99CH36303). IEEE, New York, pp 332–338
8. Li S, Zheng X (2002, May) Cryptanalysis of a chaotic image encryption method. In: 2002 IEEE international symposium on circuits and systems. proceedings (Cat. No. 02CH37353), vol 2. IEEE, New York, pp II–II
9. Gao H, Zhang Y, Liang S, Li D (2006) A new chaotic algorithm for image encryption. Chaos, Solitons Fractals 29(2):393–399
10. Wang XY, Yang L, Liu R, Kadir A (2010) A chaotic image encryption algorithm based on perceptron model. Nonlinear Dyn 62(3):615–621
11. Han B, Jia Y, Huang G, Cai L (2020, June) A medical image encryption algorithm based on hermite chaotic neural network. In: 2020 IEEE 4th information technology, networking, electronic and automation control conference (ITNEC), vol 1. IEEE, New York, pp 2644–2648

12. Prabhavathi K, Sathisha CP, Ravikumar KM (2017, December) Region of interest based selective medical image encryption using multi Chaotic system. In: 2017 International conference on electrical, electronics, communication, computer, and optimization techniques (ICEECCOT). IEEE, New York, pp 1–5
13. Jeong HS, Park KC, Cho SJ, Kim ST (2018, July) Color medical image encryption using two-dimensional chaotic map and C-MLCA. In: 2018 Tenth international conference on ubiquitous and future networks (ICUFN). IEEE, New York, pp 501–504
14. Choi US, Cho SJ, Kang SW (2020, July) Color image encryption algorithm for medical image by mixing chaotic maps. In: 2020 12th International symposium on communication systems, networks and digital signal processing (CSNDSP). IEEE, New York, pp 1–5
15. Ye G, Pan C, Huang X, Mei Q (2018) An efficient pixel-level chaotic image encryption algorithm. Nonlinear Dyn 94(1):745–756
16. Pareek NK, Patidar V, Sud KK (2006) Image encryption using chaotic logistic map. Image Vis Comput 24(9):926–934
17. Xie EY, Li C, Yu S, Lü J (2017) On the cryptanalysis of Fridrich's chaotic image encryption scheme. Signal Process 132:150–154
18. Solak E, Çokal C, Yildiz OT, Biyikoğlu T (2010) Cryptanalysis of Fridrich's chaotic image encryption. Int J Bifurcation Chaos 20(05):1405–1413
19. Ye G, Wong KW (2012) An efficient chaotic image encryption algorithm based on a generalized Arnold map. Nonlinear Dyn 69(4):2079–2087
20. Wang XY, Zhang YQ, Bao XM (2015) A novel chaotic image encryption scheme using DNA sequence operations. Opt Lasers Eng 73:53–61
21. Ullah A, Jamal SS, Shah T (2018) A novel scheme for image encryption using substitution box and chaotic system. Nonlinear Dyn 91(1):359–370
22. Vanitha V, Akila D (2021) Detection and diagnosis of hepatitis virus infection based on human blood smear data in machine learning segmentation technique. In: 2021 9th International conference on reliability, Infocom technologies and optimization (Trends and Future Directions) (ICRITO), 2021, pp 1–5. https://doi.org/10.1109/ICRITO51393.2021.9596482. ISBN No: 978-1-6654-1703-7
23. Vanitha Carmel V, Akila D (2020) A survey on biometric authentication systems in cloud to combat identity theft. J Crit Rev 7(3):540–547. ISSN: 2394-5125

Analysation of Big Data in Smart Healthcare

Sritha Zith Dey Babu, Digvijay Pandey, G. Taviti Naidu, S. Sumathi, Ankur Gupta, Malik Bader Alazzam, and Binay Kumar Pandey

Abstract Large information analytics can improve affected person outcomes, enhance and customise care, enhance company collaborations with treatment holders and deduction the huge cost of clinical processes. This study defines the proper smart system of clinical records, huge information about clinical and diagnostics structures, the terminology of health data with structure. We applied the linear regression policy for a better understanding of data. Hence, we have provided the theoretical analysis regarding the challenge of data, its prediction, and the reduction of overfitted model. The difficulties of large dataset for analysing have also discussed in this paper using the features of Internet of thinking.

Keywords Smart system · Electronic healthcare · Prediction · Analysis · Naïve Bayes

---

S. Z. D. Babu
Chandigarh University, Chandigarh, India

D. Pandey (✉)
IET, Dr. A.P.J. Abdul, Kalam Technical University Uttar Pradesh, Lucknow 226021, India
e-mail: digit11011989@gmail.com

G. T. Naidu
Department of Management, Dr. L. Bullayya College, Visakhapatnam, Andhra Pradesh, India

S. Sumathi
Department of Information and Technology, St. Joseph's College of Engineering, Chennai, Tamil Nadu, India

A. Gupta
Department of Computer Science and Engineering, Vaish College of Engineering, Rohtak, Haryana, India

M. Bader Alazzam
Faculty of Computer Science and Informatics, Amman Arab University, Amman, Jordan

B. K. Pandey
College of Technology, Govind Ballabh Pant University of Agriculture and Technology Pantnagar, Udham Singh Nagar, Pantnagar, Uttarakhand, India

M. Gupta et al. (eds.), *Artificial Intelligence on Medical Data*,
Lecture Notes in Computational Vision and Biomechanics 37,
https://doi.org/10.1007/978-981-19-0151-5_21

1 Introduction

Exactly patterned patients and clinical patients' data remedies about to precise the illnesses may lessen unimportant side consequences, enhance serving quality of practitioners will keep away from mistaken service and inclusions of scientific offerings. Therefore, this could deliver latest clinical services via handler of latest drugs and the usage of present pills for revolutionary or more focused uses [1]. Biotechnology is the unique and précised method that can intercept more than one records source and studies of biologic tactics. A great deal looks at offers the terminal fashions and the usage of apex database features can reduce the fractional mistakes of examiner called biofield test makers. Many forms of scientific system, in particular wearable devices, vast data size exclusions and the excessive acceleration of dynamic source of information mostly demands a speedy progress of diagnosis using Hadoop techniques [2] The basic methodology of cloud model fitted with healthcare system can be an updating of smart healthcare projects because of its unique computational method. Now a days the smart technology has developed in the basis of computational method using different machine learning [3–6] approaches and artificial intelligence systems. We can now visualise the patients mind with the help of patient's data learning model including a huge number of database storage, statistics, mind computational methods and regular fit neural network methods [7]. All axons are intercepting with the primary layer of patient's information which is the main base of read the mind of the treatment holders and also the clinical stage makers [8]. Every local and national distribution of cloud can be localised now only for the blessings of smart and green management of the technology. The technology gives us the simple and use friendly platform for both users and admins to control the visualisation and customise this by own self [9, 10]. We can analyse every data with our own method which with which we are very comfortable [11]. The total package of the healthcare system can be defined in both technology and business field in this era of biotechnology also. Now, we can predict the symptoms, cases and issues regarding recent COVID-19 pandemics also using machine learning and smart healthcare system [12, 13]. Because we have the new existing resources for analysis any situation with its records [14] (Figs. 1 and 2).

So that we can easily reach to any emergency decisions or can take immediate proper steps for the evaluation of disease control [15]. Disease control section of many countries is also getting the advantages of this smart healthcare approach [16]. They developed many healthcare application and web pages by which people can get the real time information about kind of serious or pandemic situation [17]. They can get the information of affected peoples around them [18]. So that they can help or inform the health section or ministry of health office.

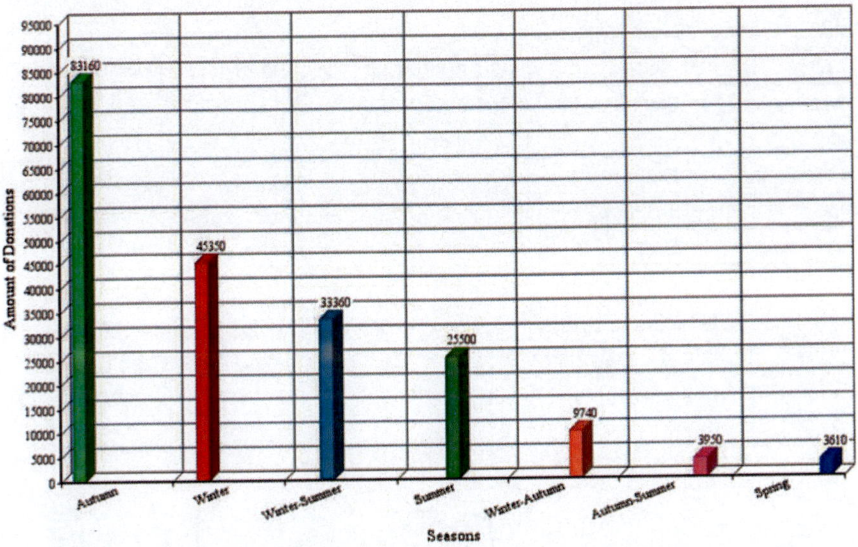

Fig. 1 Amount of donations for healthcare by seasons

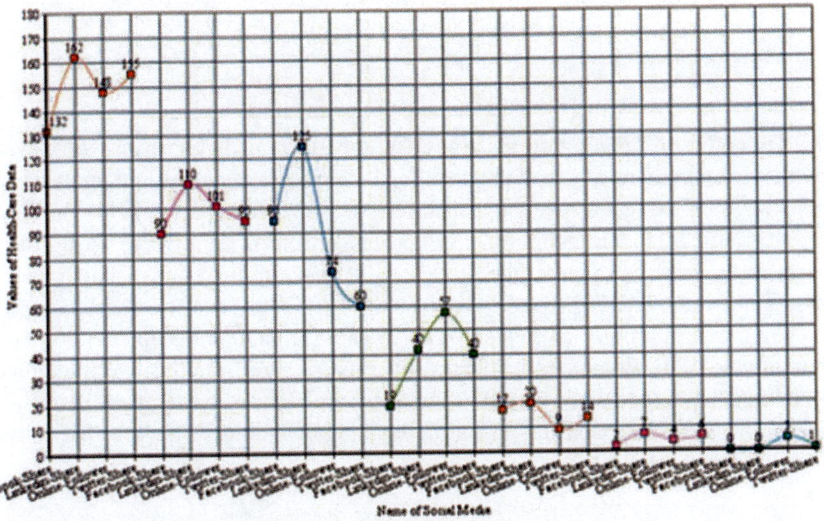

Fig. 2 Social media sharing comparison of healthcare data

2 Healthcare System

Here in Table 1, we have shown the science of disease diagnosis in this platform of smart healthcare. After that we can understand the improved places by smart technology. Hypertension, Obesity, Heart Disease, Stress, Respiration problem these

Table 1 Scene of disease diagnosis

Disease	Method	Health measure via IoT	Layers	Description
Hypertension	Frequency	Blood pressure	1	HER
Obesity	Scale	ECG pattern	2	IoT
Heart disease	Frequency	EPOC sensor	3	Purchase
Stress	Frequency	Body pressure	4	Clinical
Respiration	Scale	Body mass index	5	HER

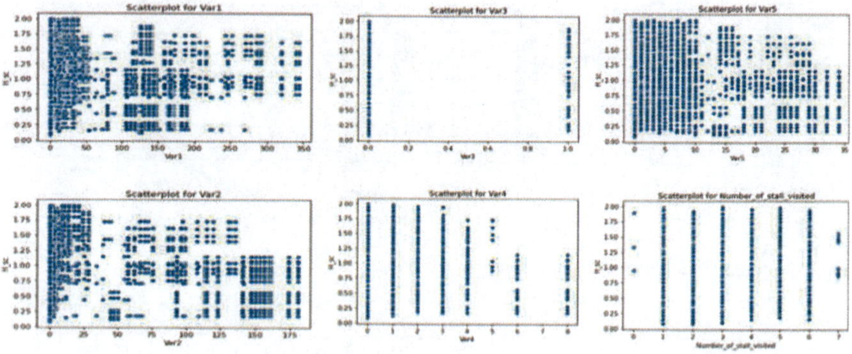

Fig. 3 Electronic health shop visitors number per day

are, Ost common dangerous disease now. We measure and diagnose the whole system by the measurement of blood pressure [19], electromagnetic pattern analysis [20], body mass indexing, etc. We use multilayers for the regression also [21]. Multilayers are for getting hidden layers for accurate prediction of the disease. Within a few seconds now we can predict the disease as we have trained our machine learning model like that way [22] (Fig. 3).

The maximum noise of data and trained garbage value intercepts with the overfitted model [23]. But smart healthcare gives us the cure from this overfitted model. In this case developers often need to connect more hidden layers rather than requirements that's why we are now using the web application also for user friendly and location approach [24].

3 Big Data Analytics

As we use blood pressure measurement techniques, systolic measurement, etc. we use the basic model which are existing such as NN models and KNN model. But now we want to aware the patients and doctors about the situation or the condition of patients so that we use the techniques of emotional classifiers for generating the maximum rendering optimization [25, 26]. We can predict and analyse the situation and the

```
# corr = musk.corr()
cmap = cmap=sns.diverging_palette(5, 250, as_cmap=True)

def magnify():
    return [dict(selector="th",
                 props=[("font-size", "7pt")]),
            dict(selector="td",
                 props=[('padding', "0em 0em")]),
            dict(selector="th:hover",
                 props=[("font-size", "12pt")]),
            dict(selector="tr:hover td:hover",
                 props=[('max-width', '200px'),
                        ('font-size', '12pt')])
]

corr.style.background_gradient(cmap, axis=1)\
    .set_properties(**{'max-width': '80px', 'font-size': '10pt'})\
    .set_caption("Hover to magify")\
    .set_precision(2)\
    .set_table_styles(magnify())
```

	Patient_ID	Health_Camp_ID_x	Donation	Health_Camp_ID_y	score
Patient_ID	1.00	0.02	0.02	0.02	0.02
Health_Camp_ID_x	0.02	1.00	0.88	1.00	0.96
Donation	0.02	0.88	1.00	0.88	0.88
Health_Camp_ID_y	0.02	1.00	0.88	1.00	0.96
score	0.02	0.96	0.88	0.96	1.00

Fig. 4 Electronic score of patients score system

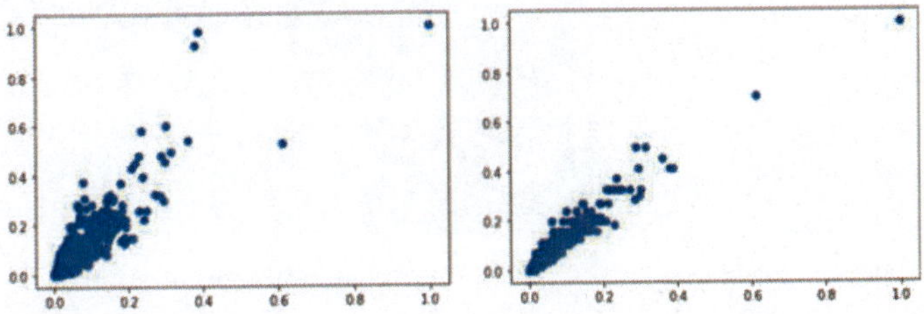

Fig. 5 Data scatter value of health monitoring system

proper steps of cure there. We have added one important system in healthcare platform that is make appointment and get cure from the doctors electronically (Fig. 4).

Sometimes, we face such situation that we are unable to get the treatment physically. So, on that time we use the electronic treatment service, telemedicine service, online medicine shop service, etc. [27]. The main focus of disease detection and making health protected from any kind of disease is to build our health system monitoring-based. Many organisations are now using many customised machines learning model for making the thing possible in early based [28] (Fig. 5).

4 Challenges

There is a huge challenge of smart healthcare system as we face problems such as data string, data cleaning and data searching. [29]. To recover these problems, we have created our machine learning techniques [30]. Here Rae steps given below:

Step1: Gather the fraud cases in electronic health service provider platform.
Step 2: Make a list of unprovoked billing cases.
Step 3: Duplicate documents and reports.
Step 4: Misrepresentation of the service provider.
Step 5: Renal disease indicator frequency.
Step 6: Accuracy measurement.

Step 7: Determine the value of prediction and.
Step 8: Print the graph (Figs. 6 and 7).

The raw package of this smart healthcare specially creed for the blessings of the patients as they can response in time and can get their reports electronically [31]. Different association can also use the patient's data with ethical permission for any research purpose. By electronically the dataset is going to be processed and researched by many data science organisations [32–40] which are the main cure of any uneven disease or pandemics.

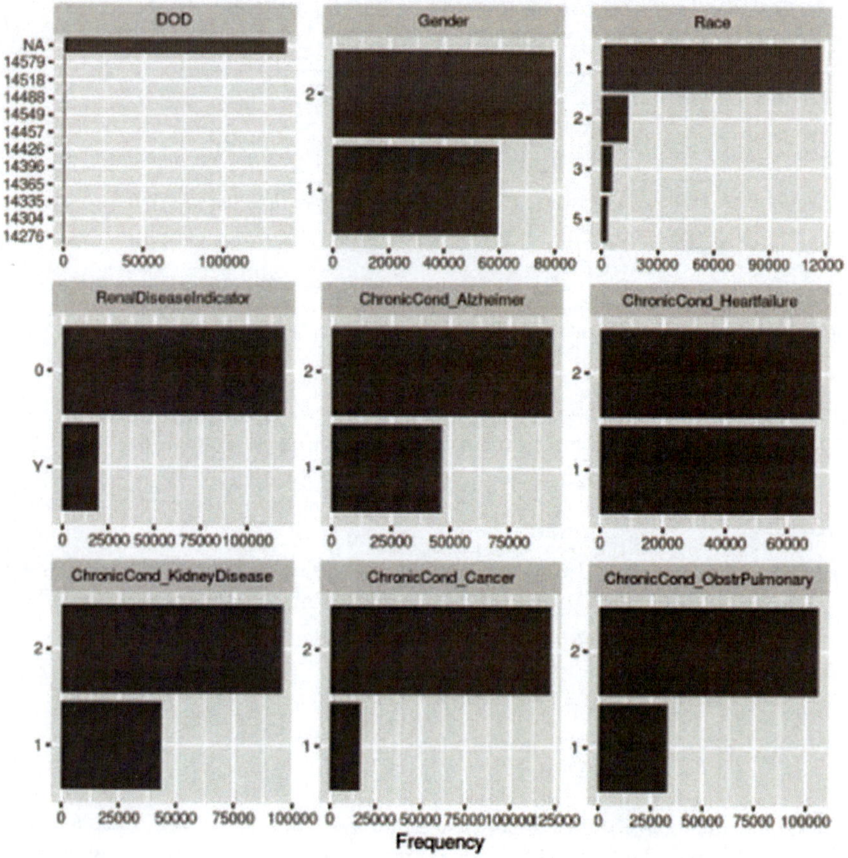

Fig. 6 Disease's indicator frequency

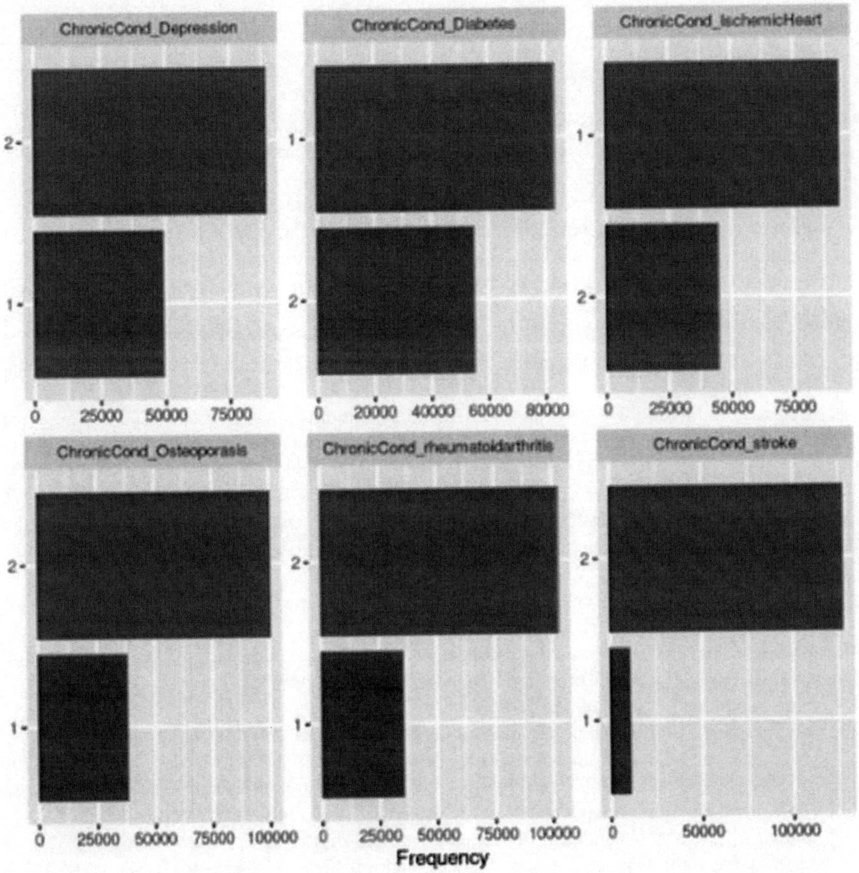

Fig. 7 Chronic type disease detection frequency

5 Conclusion

Healthcare is the most vital field for making ourselves healthy in this pandemic. That's why for we are always giving the full concentration into the physical condition of the patients and we are not much thinking about the mental condition. Smart healthcare helps us to go closer to the patients for getting the recent mental condition of patients by which we can predict the psychological terms of patients. We can make the decision by using the smart decision makers algorithms indifferent classifiers such as KNN, Naïve Bayes and FP Growth.

References

1. Andreu-Perez J, Poon CC, Merrifield RD, Wong ST, Yang GZ (2015) Big data for health. IEEE J Biomed Health Inform 19(4):1193–1208
2. Capobianco E (2017) Systems and precision medicine approaches to diabetes heterogeneity: a big data perspective. Clin Transl Med 6(1):23
3. De Silva D, Burstein F, Jelinek HF, Stranieri A (2015) Addressing the complexities of big data analytics in healthcare: the diabetes screening case. Australas J Inf Syst 19:S99–S115
4. Farid DM, Nowe A, Manderick B (Dec 2016) A feature grouping method for ensemble clustering of high-dimensional genomic big data. In: Future technologies conference (FTC). IEEE, pp 260–268
5. Shukla R, Dubey G, Malik P, Sindhwani N, Anand R, Dahiya A, Yadav V (2021) Detecting crop health using machine learning techniques in smart agriculture system. J Sci Ind Res (JSIR) 80(08):699–706
6. Singh H, Rehman TB, Gangadhar C, Anand R, Sindhwani N, Babu M (2021) Accuracy detection of coronary artery disease using machine learning algorithms. Appl Nanosci 1–7
7. Hernandez I, Zhang Y (2017) Using predictive analytics and big data to optimize pharmaceutical outcomes. Am J Health Syst Pharm 74(18):1494–1500
8. Istephan S, Siadat MR (Nov 2015) Extensible query framework for unstructured medical data-a big data approach. In: Data mining workshop (ICDMW), 2015 IEEE international conference on. IEEE, pp 455–462
9. Lo'ai AT, Mehmood R, Benkhlifa E, Song H (2016) Mobile cloud computing model and big data analysis for healthcare applications. IEEE Access 4:6171–6180
10. Mathew PS, Pillai AS (March 2015) Big data solutions in healthcare: problems and perspectives. In: Innovations in information, embedded and communication systems (ICIIECS), 2015 international conference on. IEEE, pp. 1–6
11. Mendelson D (2017) Legal protections for personal health information in the age of Big Data–a proposal for regulatory framework. Ethics, Med Public Health 3(1):37–55
12. Murphy SN, Avillach P, Bellazzi R, Phillips L, Gabetta M, Eran A, Kohane IS (2017) Combining clinical and genomics queries using i2b2–Three methods. PloS one 12(4)
13. Juneja S, Juneja A, Anand R (2020) Healthcare 4.0-digitizing healthcare using big data for performance improvisation. J Comput Theor Nanosci 17(9–10):4408–4410
14. Ni J, Chen Y, Sha J, Zhang M (Nov 2015) Hadoop-based distributed computing algorithms for healthcare and clinic data processing. In: Internet computing for science and engineering (ICICSE), 2015 eighth international conference on. IEEE, pp 188–193
15. Olaronke I, Oluwaseun O (Dec 2016) Big data in healthcare: prospects, challenges and resolutions. In: Future technologies conference (FTC). IEEE, pp 1152–1157
16. Özdemir V, Hekim N (2018) Birth of industry 5.0: making sense of big data with artificial intelligence, "the internet of things" and next-generation technology policy. Omics: J Integr Biol 22(1):65–76
17. Anand R, Sindhwani N, Saini A (2021) Emerging technologies for COVID-19. Enabling healthcare 4.0 for pandemics: a roadmap using AI, machine learning, IoT and cognitive technologies. pp 163–188
18. Pramanik MI, Lau RY, Demirkan H, Azad MAK (2017) Smart health: big data enabled health paradigm within smart cities. Expert Syst Appl 87:370–383
19. Reddy AR, Kumar PS (Feb 2016) Predictive big data analytics in healthcare. In: Computational intelligence and communication technology (CICT), 2016 Second International Conference on. IEEE, pp 623–626
20. Ren G, Krawetz R (2015) Applying computation biology and "big data" to develop multiplex diagnostics for complex chronic diseases such as osteoarthritis. Biomarkers 20(8):533–539
21. Sabharwal S, Gupta S, Thirunavukkarasu K (April 2016) Insight of big data analytics in healthcare industry. In: Computing, communication and automation (ICCCA), 2016 International Conference on. IEEE, pp 95–100

22. Salavati H, Gandomani TJ, Sadeghi R (Sept 2017) A robust software architecture based on distributed systems in big data healthcare. In: Advances in computing, communications and informatics (ICACCI), 2017 international conference on. IEEE, pp 1701–1705
23. Simonsen L, Gog JR, Olson D, Viboud C (2016) Infectious disease surveillance in the big data era: towards faster and locally relevant systems. J Infect Dis 214(suppl_4):S380-S385
24. Ta VD, Liu CM, Nkabinde GW (July 2016) Big data stream computing in healthcare real-time analytics. In: Cloud computing and big data analysis (ICCCBDA), 2016 IEEE international conference on. IEEE, pp 37–42
25. Vanathi R, Khadir ASA (Feb 2017) A robust architectural framework for big data stream computing in personal healthcare real time analytics. In: Computing and communication technologies (WCCCT), 2017 world congress on. IEEE, pp 97–104; Verma P, Sood SK (2018) Cloud-centric IoT based disease diagnosis healthcare framework. J Parallel Distrib Comput 116:27–38
26. Anand R, Chawla P (March 2016) A review on the optimization techniques for bio-inspired antenna design. In: 2016 3rd international conference on computing for sustainable global development (INDIACom). IEEE, pp 2228–2233
27. Wang C, Li F, Wang L, Zhou W, Zhu B, Zhang X, Jin C (2017) The impact of population aging on medical expenses: a big data study based on the life table. Biosci Trends 11(6):619–631
28. Wooden B, Goossens N, Hoshida Y, Friedman SL (2017) Using big data to discover diagnostics and therapeutics for gastrointestinal and liver diseases. Gastroenterology 152(1):53–67
29. Wu PY, Cheng CW, Kaddi CD, Venugopalan J, Hoffman R, Wang MD (2017) Omic and electronic health record big data analytics for precision medicine. IEEE Trans Biomed Eng 64(2):263–273
30. Pandey BK, Pandey SK, Pandey D (2011) A survey of bioinformatics applications on parallel architectures. Int J Comput Appl 23(4):21–25
31. Pandey D, Pandey BK, Wairya S (2021) Hybrid deep neural network with adaptive galactic swarm optimization for text extraction from scene images. Soft Comput 25(2):1563–1580
32. Pandey D, Ogunmola GA, Enbeyle W, Abdullahi M, Pandey BK, Pramanik S (2021) COVID-19: a framework for effective delivering of online classes during lockdown. Hum Arenas 1–15
33. Khongsai L, Anal TS, AS R, Shah M, Pandey D (2021) Combating the spread of COVID-19 through community participation. Glob Soc Welfare 8(2):127–132
34. Kassaw C, Pandey D (2021) The current mental health crisis of COVID-19 pandemic among communities living in Gedeo Zone Dilla, SNNP, Ethiopia, April 2020. J Psychosoc Rehab Mental Health 8(1):5–9
35. Kassaw C, Pandey D (2021) COVID-19 pandemic related to anxiety disorder among communities using public transport at Addis Ababa, Ethiopia, March 2020: cross-sectional study design. Hum Arenas 1–10
36. Ayenew B, Pandey D, Yitayew M, Etana D, Binay Kumar P, Verma N (2020) Risk for surge maternal mortality and morbidity during the ongoing corona virus pandemic. Med Life Clin 2(1):1012
37. Parthiban K, Pandey D, Pandey BK (2021) Impact of SARS-CoV-2 in online education, predicting and contrasting mental stress of young students: a machine learning approach. Augmented Hum Res 6(1):1–7
38. Radwan E, Radwan A, Radwan W, Pandey D (2021) Prevalence of depression, anxiety and stress during the COVID-19 pandemic: a cross-sectional study among Palestinian students (10–18 years). BMC Psychol 9(1), 1–12
39. Pandey D, Babu SZD, Noibi TO, Pandey BK, Dey PTD, Kassaw C, Canete JJO et al (2020) A study on novel covid-19 epidemic threat for universal physical situation in 2020. Asian J Adv Res 27–33
40. Pandey D, Islam T, Magray JA, Gulzar A, Zargar SA (2021) Use of statistical analysis to monitor novel coronavirus-19 cases in Jammu and Kashmir, India. Eur J Biol Res 11(3):274–282

IoT and RFID-Based Smart Card System Integrated with Health Care, Electricity, QR and Banking Sectors

Ankita Gupta, Afshan Asad, Laxmi Meena, and Rohit Anand

Abstract With the vast growing population and the immense and rapid transformation to the digital world after COVID-19 pandemic, everyone from all age groups has now turned to using technology as never before. In many terms, 2020–2021 has become the era of digital boom in all sectors as well in all age groups of the world. Many business ideas and technologies have surfaced to help make users deal with small daily things like payment, health care, electricity bills and others readily at hand without the actual need of presenting the data in the form of hard copy. This work is based on IoT and RFID technology to help users reduce the manual burden to carry so many cards for every unique purpose. The proposed research aims to automate and integrate as many industries as possible into a single card system. So, with just a tap of our RFID card on the system, we are good to go. Integrating all the cards of daily use into a single RFID card will help people free from the all the different bank cards into one single card. The proposed smart card system works on an electronic device that is specifically loaded with the required software and collects only the necessary data from our card, eliminating the need for us to carry the documents and also the different bank cards, user identification cards and medical cards.

Keywords Smart card system · Internet of things (IoT) · One card system · RFID smart card · Banking cards

1 Introduction

Internet of things (IoT) [1–3] in a broad scenario is a technological paradigm to help in the communication and transmission of data between electronic devices extending the capabilities to interaction between humans and devices [4]. IoT not only establishes the connection between man and machine but also machine to machine. These technological advancements when used to other technologies provide numerous applications that are exceptionally valuable for individuals, industries and the world as

A. Gupta (✉) · A. Asad · L. Meena · R. Anand
G.B. Pant Engineering College, New Delhi, India
e-mail: ankitagupta1331@gmail.com

M. Gupta et al. (eds.), *Artificial Intelligence on Medical Data*,
Lecture Notes in Computational Vision and Biomechanics 37,
https://doi.org/10.1007/978-981-19-0151-5_22

a whole [5]. IoT has opened various evolutionary concepts in every sector [6–8], helping in establishing the connections between large industries, agricultural sector, transportation, business, education, banking sector, etc. [9].

IoT along with RFID has opened new doors to smart cities connecting everything together to work in synchronization. Likewise in the proposed smart card system, the use of IoT and RFID has been predominant. The fundamental concepts of IoT, i.e., giving a digital address to everything, are thus making it feasible for connection. The remote monitoring feature, smart metering and process automation have led to IoT being a super successful technology [10].

IoT and RFID applications in smart card systems are mainly based on an automated interaction between the various sectors and real-time data transfer for ease and transparency in the system. The aim of using IoT is to establish the connections between the databases of different platforms and integrate them into a single card. It may be used in the healthcare sector. Also, carrying different banking cards for every different purpose can be a cumbersome task. Further, with the concept of IoT, prepaid electricity meter is employed and customers can recharge through the Internet. If energy usage reaches the set point value of the unit, then automatically it disconnects the power. It continuously analyzes the meter and sends data through the Internet [11].

The focus of this paper is on the design, analysis and evaluation of IoT and RFID-based frameworks in the context of "smart card system". It aims to implement the ultimate innovative approach of IoT and RFID technology in achieving a system known as smart card system that integrates the four major card systems present today. The overall paper aims to achieve digitalization and automation to the maximum possible extent (as of now) in the following areas:

1. Healthcare department cards like CGHS card, medical cards and insurance cards
2. Electricity management through a tap of our card
3. Various bank cards into a single card
4. QR code.

The proposed work presents research on various sectors that the work targets to integrate. The smart card system covers entirely new domains of making the world more digital and taking the user experience to the whole new level. This system would further establish the transparency among the various sectors and reduce the chances of flaws.

2 Related Work

The literature survey may be studied under four three different subsections.

2.1 Healthcare Sector

The authors in [12] have proposed a system that intends to store patient's personal information and transfer health-related data. In such a system, VLSI technology was used to integrate the fingerprint verification into the smart card-based healthcare information system. This system was based on a personal identification number (PIN) and a user password. In [13], a smart card has been proposed that can store the healthcare data as well as biometric information. Hence, the system has been observed to be safer and more secure.

2.2 Banking and Online Transactions

In [14], the authors proposed a single card system consisting of all the banks. Later in 2015, the authors in [15] proposed the idea of interbank transfer. The authors in [16] explained how the online transactions are increasing, but they are lacking the required security. Further in [17], an algorithm known as secured online transaction (SONT) has been deployed for the protection against fake electronic cards. Unauthorized access to the network and to keep a check on IoT data transfer to maintain security has been presented in [18], where management of technologies using secure sockets layer (SSL) protocol was discussed.

2.3 QR Code

The concept of quick response (QR) code was first designed by an automotive company in Japan. It has been further explained in detail in [19]. A review on QR code analysis has been proposed in [20] in which quick response code is usually authenticated with the help of the embedded camera of a mobile phone. The encrypted string is converted to a QR image using an algorithm. The QR code algorithm is made up of two different stages. The first one is similarity transformation, where the novel matrix gets transformed into a limited number of steps to real tri-diagonal or Hessenberg form. The first stage of the QR algorithm prepares for the next stage which refers to the actual iterations of tri-diagonal or Hessenberg matrix form [21]. The simulated output in [22] demonstrates that the picture of the QR code can be masked well and can be efficiently reconditioned.

2.4 Electricity

The prepaid energy meter described in [23] is a single-phase 230 V/40A energy meter which consists of a metering device designed according to the IEC 1036 (1996–09) standard and a prepaid module that uses GSM/GPRS technology to communicate with the utility server. The system suggested in [24] employs the ATMega328p and ESP8266 to operate a dual-core microprocessor unit with one core dedicated to energy sensing and measurements, while the other handles the network connectivity, storage, computations and overall system performance. An automated billing system [25] keeps tracks of the real-time consumption and leaves very little scope for the disagreement on consumption and billing.

3 Proposed Work

The smart card system proposed in this paper mainly consists of two parts: the software for the user and the admin and the hardware, i.e., the device on which the software will be installed. Installation of the software part would totally depend on the sector. Each sector will have only the required part of the software installed for the users to use, so that they will not be able to access the other details that are not required by them. The system will adapt the algorithm based on the application type. For example, for the traffic police device, when the user taps on his device, the options of Aadhaar, driving license and the transport documents can only be selected and seen. For the bank device, only the user identification details can be seen.

4 Software Implementation

4.1 Healthcare Sector

The proposed smart card system can be used in hospitals to avail discounts and other benefits as a medical card of a particular hospital would provide. Along with it, the proposed card aims to integrate the cards like CGHS and medical insurance and life insurance cards so that the user would not be required to keep and carry all these different cards in times of need. This module will work on a simple tap of the card on the machine, and it will display options to move ahead with. User can select among the hospital medical cards and even have access to the medical insurance and the life insurance data through this smart card. The entire flow of the smart card for medical sector integration is shown in Fig. 1.

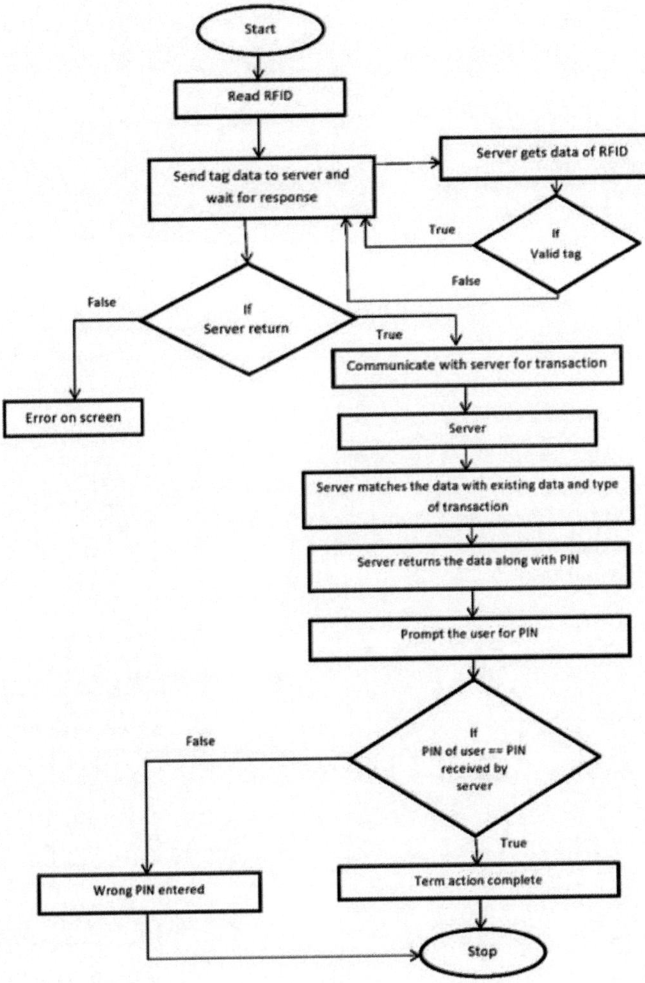

Fig. 1 Flowchart for healthcare sector

4.2 Bank Cards/ATM Transaction Integration

The proposed smart card can be used even at ATMs for transactions. This module contains all our registered bank cards. Through this portal, we can easily manage all our bank cards and transactions. The interface would show the screen with the various registered bank options to choose from and proceed with our transactions or withdrawal. We can authenticate as a user and keep a track of all our transactions and payments of all the banks. We can see the transaction history of a particular bank separately after authenticating with a six-digit user PIN. The admin in this case would be the ATM machine that will show the screens for proceeding with our transactions. The flowchart explaining the entire process is shown in Fig. 2.

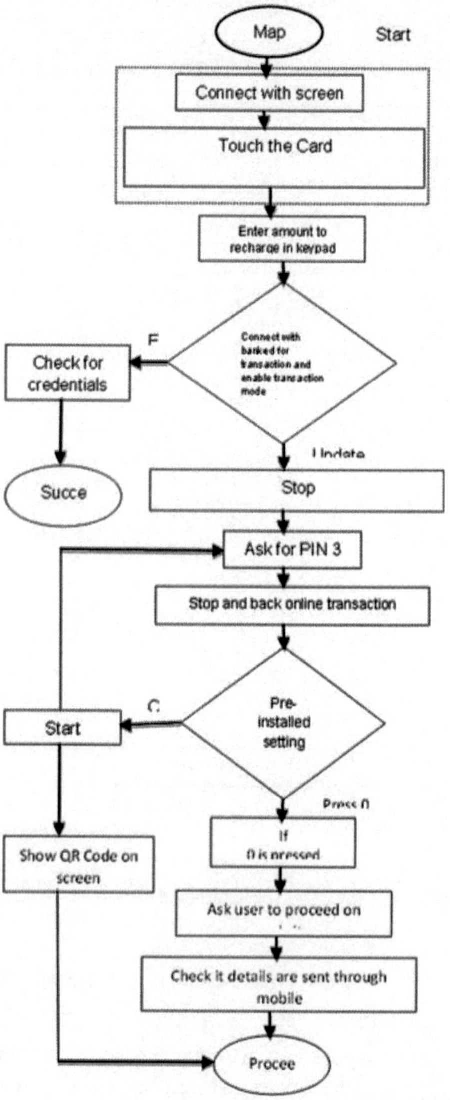

Fig. 2 Flowchart for bank cards and transactions

4.3 Electricity Meter Bill Integration

With electricity meters going prepaid, this paper aims to integrate that feature as well into the smart card. Now, the user can easily view its electricity details on the same website and can easily recharge it using the same card by tapping it on the electric meter device. The device would then start acting as an interface, and we can type the amount using the keypad and pay. This feature would be helpful for the old people and the kids who are digitally not so well equipped with. Further to keep the security [26, 27] at check, the user would be required to authenticate with the six-digit PIN to proceed with the payment. The flow of this complete process is shown in Fig. 3.

Fig. 3 Flowchart for electricity

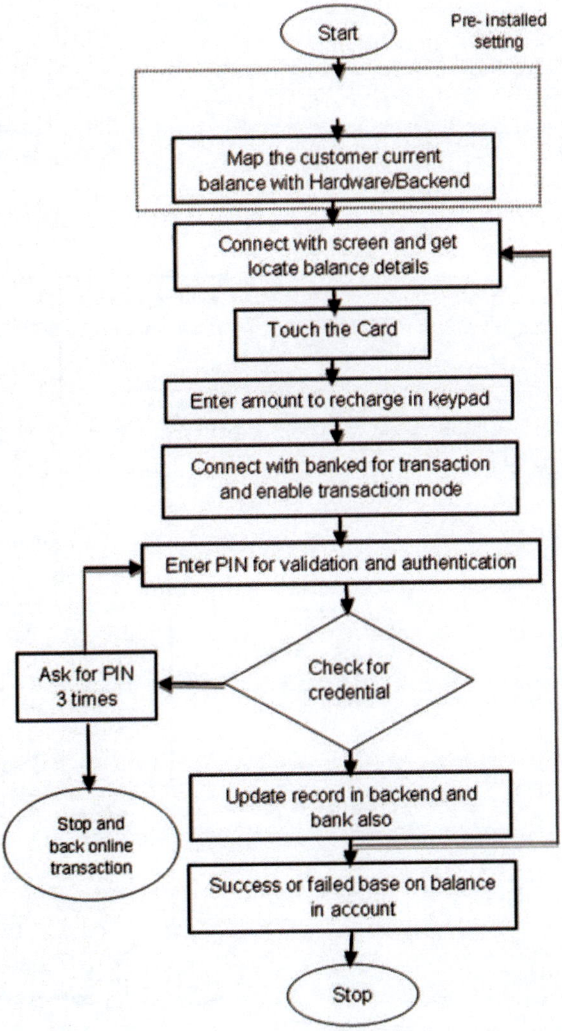

4.4 QR Code

The step-by-step procedure regarding the proposed QR code sector is explained in Fig. 4.

The main aim of this feature is to consider the situation when the user forgets his card at home or he loses it and it is taking some time for the new card to be issued. In that case, we can easily show our QR code and proceed with our transactions and everything else. This feature is implemented using the camera, tag ID, Node.js and Vue.js. The user can select the QR code option from the portal screen. On opening, it will show the button to generate the QR code. By clicking on this feature, a new unique QR code will be generated that can be used while not carrying the smart card.

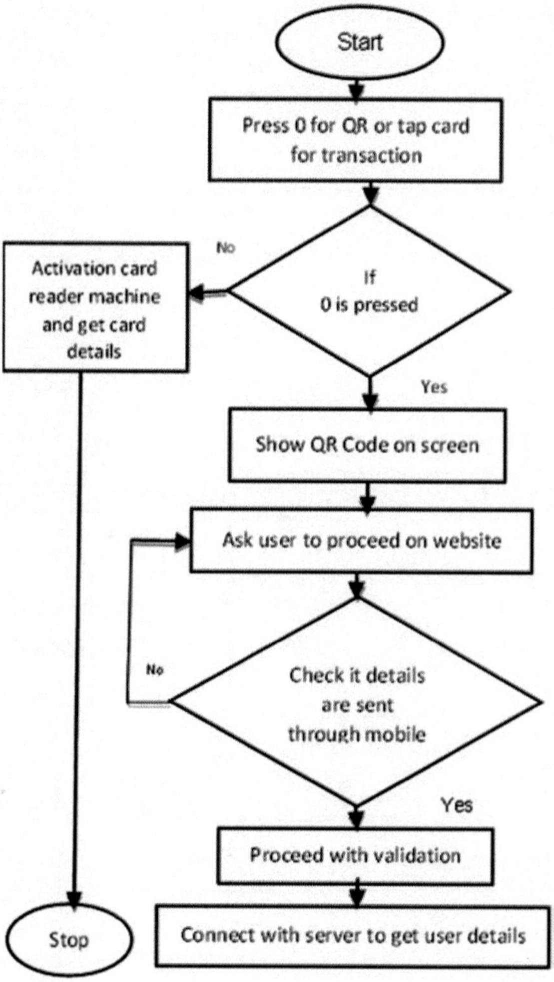

Fig. 4 Flowchart for QR code

5 Results

The proposed smart card system can be utilized to club all the banks into a single card, thus making it easy for the user to manage his resources and keep a track of the transactions from a single portal. The portal can also be used to pay the electricity bill or recharge the prepaid electricity meter in advance with just a single tap of the RFID card and a six-digit secure user PIN. In order to increase accessibility, QR code option is being provided to be used in case the user is not carrying his smart card.

5.1 *Experimental Results*

Following are some of the major screens that indicate that the user will interact to proceed with the transactions and to recharge his electricity meter. Figures 5, 6, 7, 8, 9, 10 and 11 show the entire process.

(1) Secure PIN of user to access all the features.
(2) Health insurance after proceeding with the selected option.
(3) RFID-based smart card scanning option for transaction.
(4) Bank options to choose from to proceed.
(5) Bank transactions.

Fig. 5 Enter PIN screen

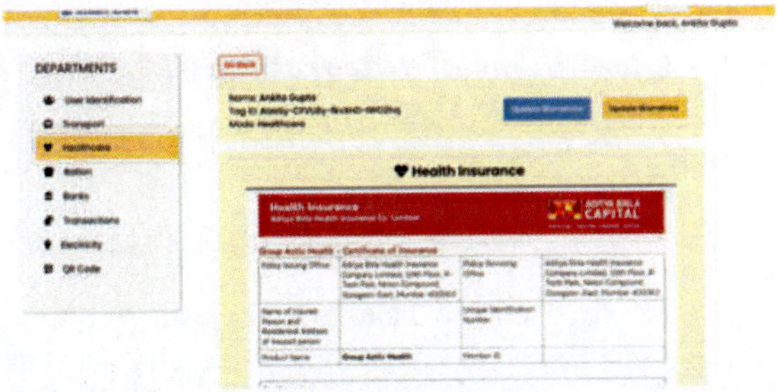

Fig. 6 Health Insurance under health care

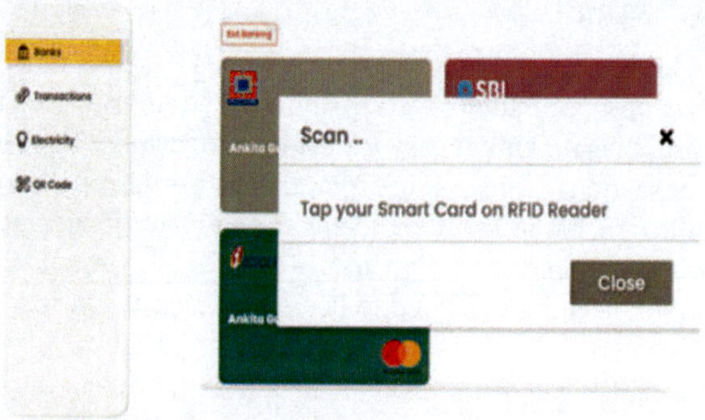

Fig. 7 Scan RFID smart card screen

Fig. 8 Choose bank option screen

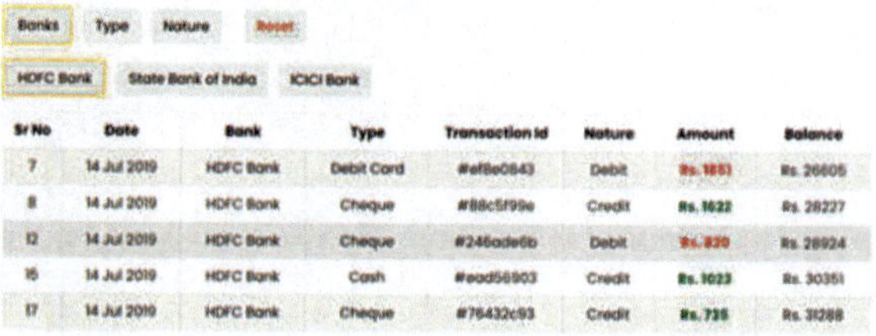

Fig. 9 View bank transaction screen

(6) Electricity portal.
(7) QR code option.

5.2 Advantages of Integrating Electricity Sector and All the Bank Cards into One

Table 1 shows the comparison and advantages of the integrated smart card over the existing cards [28].

So, the proposed RFID and IoT-based smart card model increases the transparency of the user's financial records and avoids frauds. The model further adds extensive feasibility as integration of all cards ensures less confusion and proper management.

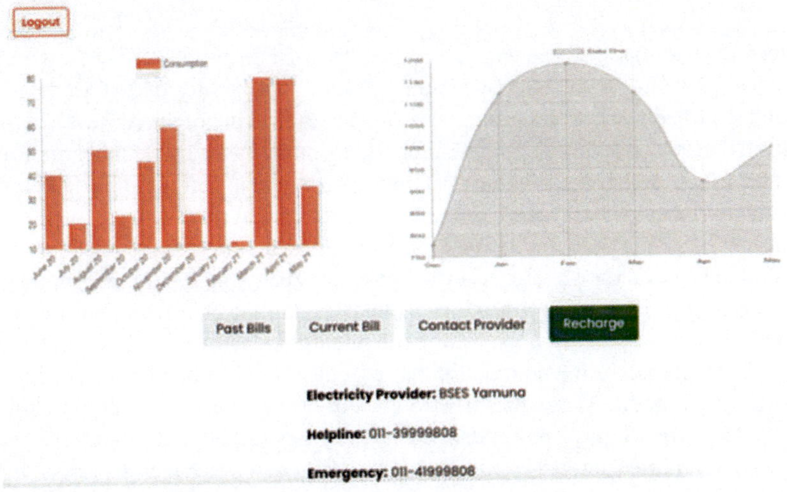

Fig. 10 Electricity portal screen

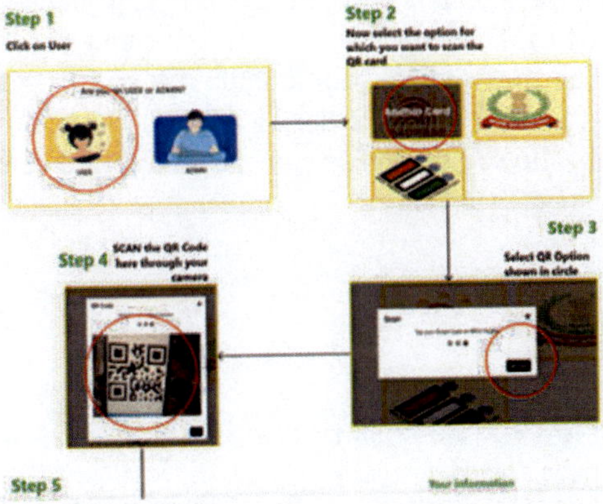

Fig. 11 QR code scanning instructions

Table 1 Advantages of ALL-IN-ONE smart card

Feature	Existing	Smart card
Transparency of credit records	Low	High
Positive identification	Medium	High
Feasibility	Negligible	High
Mismanagement	High	Low

6 Conclusion

The Internet of things has changed the entire era of computing technology and converted it into a universal global neural network. The ability of IoT to establish a connection between materials and humans lies at the core of the success of this technology. The proposed model based on RFID technology including other add-on technologies such as GSM, GPS and Arduino is economical and optimized [29] in all the aspects of implementation. The proposed smart card system will help us to not bother about forgetting different medical cards in times of emergencies. Also, it will help both the user and the government on a broader level to maintain transparency. It will also help in preventing any kind of delay in the medical treatment due to late submission of these very important medical cards. It will not only reduce the manual load and increase the automation but will also provide a better user experience, keeping people free from the daily troubles of electricity and the need to carry different bank cards for payments. Also, even if one has forgotten the card at home or misplaced it and registered a complaint to get a new one, we can easily make use of the QR code facility.

The scope of the proposed smart card system in this paper can be further extended in the QR code sector. We can integrate the feature of online payments and UPI setup using the same. Integrating all the banking cards into a single smart card requires effective PIN security than the existing. Multicomponent security that is using the biometric for enhancing the security can be used [30]. This can be done in a manner that only the user and his allowed nominees can access the card details or proceed with the transaction at the next step.

References

1. Anand R, Sinha A, Bhardwaj A, Sreeraj A (2018) Flawed security of social network of things. In: Handbook of research on network forensics and analysis techniques. IGI Global pp 65–86
2. Gupta A, Srivastava A, Anand R, Tomažič T (2020) Business application analytics and the internet of things: the connecting link. In: New age analytics. Apple Academic Press, pp 249–273
3. Gupta R, Shrivastava G, Anand R, Tomažič T (2018) IoT-based privacy control system through android. In: Handbook of e-business security. Auerbach Publications, pp 341–363
4. Elijah O, Rahman TA, Orikumhi I, Leow CY, Hindia MN (2018) An overview of internet of things (IoT) and data analytics in agriculture: benefits and challenges. IEEE Internet Things J 5(5):3758–3773
5. Gupta A, Srivastava A, Anand R, Chawla P (2019) Smart vehicle parking monitoring system using RFID. Int J Innovative Technol Exploring Eng 8(9S):225–229
6. Kumar AS, Suseendran G, Akila D, Priya C (2018) Computerized agricultural storage manipulation system using IOT technique. Network 4(12)
7. Suseendran G, Chandrasekaran E, Akila D, Kumar AS (2020) Banking and FinTech (Financial Technology) embraced with IoT device. In: Data management, analytics and innovation. Springer, Singapore, pp 197–211
8. Shukla R, Dubey G, Malik P, Sindhwani N, Anand R, Dahiya A, Yadav V (2021) Detecting crop health using machine learning techniques in smart agriculture system. J Sci Ind Res 80:699–706

9. Xiao K, Li W, Kadoch M, Li C (2018) On the secrecy capacity of 5G mmWave small cell networks. IEEE Wirel Commun 25(4):47–51
10. Maple C (2017) Security and privacy in the internet of things. J Cyber Policy 2(2):155–184
11. Ramalingam H, Venkatesan VP (Oct 2019) Conceptual analysis of internet of things use cases in banking domain. In: TENCON 2019–2019 IEEE region 10 conference (TENCON). IEEE, pp 2034–2039
12. Kardas G, Tunali ET (2006) Design and implementation of a smart card based healthcare information system. Comput Methods Programs Biomed 81(1):66–78
13. Yang W, Wang S, Hu J, Zheng G, Chaudhry J, Adi E, Valli C (2018) Securing mobile healthcare data: a smart card based cancelable finger-vein bio-cryptosystem. IEEE Access 6:36939–36947
14. Ranjit PK, Kulkarni RV, Ghodke BS (2014) ARM7 based smart ATM access and security system using fingerprint recognition and GSM technology. Int J Emerg Technol Adv Eng 4(2)
15. Syed AKR, Vijay RW, Theft identification and remote information of ATM card by a unique system. Int J Sci Res (IJSR) ISSN (Online) 2319–7064
16. Uchechi I, Obinna O, Modesta E, Chidinma AI, Chimezie UB (2020) Secured online transaction (SONT): analysis and solution of secure electronic transaction. IOSR J Electron Commun Eng 15(2):32–39
17. Alghazo JM, Kazmi Z, Latif G (2017) Cyber security analysis of internet banking in emerging countries: user and bank perspectives. In: 2017 4th IEEE international conference on engineering technologies and applied sciences (ICETAS). IEEE, pp 1–6
18. Ken Nwogbo (11 Jan 2019) Worrying security situation of critical infrastructure. https://guardian.ng/businessservices/communications/worrying-security-situation-of-critical-Infrastructure. Assessed on 13 Dec 2019
19. Shamal S, Monika K, Neha N (2014) Secure authenticating for online banking using QR code. Int J Emerg Technol Adv Eng 4(3):778–781
20. Jathar C, Gurav S, Krantee J, A review on QR code analysis. Int J Appl Innovation Eng Manag (IJAIEM) 8
21. Shah K, Shah S, Shah T, Shinde R, Gharat S (2016) Application of fuzzy inference system to evaluate elapsed time for generation, insertion and retrieval of QR code from database. Int J Sci Eng Res 7(10):1570–1574
22. Cao X, Feng L, Cao P, Hu J (Nov 2016) Secure QR code scheme based on visual cryptography. In: Sehiemy RE, Reaz MBI, Lee CJ (eds) 2016 2nd international conference on artificial intelligence and industrial engineering (AIIE). Beijing, China, pp 20–21
23. Rodrigo WDAS, Lakmal HKIS, Roshani N, Samarakoon SWMGS, Samararatne S (2016) A prepaid energy meter using GPRS/GSM technology for improved metering and billing. Int J Comput Sci Electron Eng 4:65–69
24. Surajudeen-Bakinde NT, Ayodele SO, Oloruntoba TD, Otuoze AO, Faruk N (Sep 2017) Development of an internet based prepaid energy meter. In: 2017 IEEE AFRICON. IEEE, pp 1370–1373
25. Dange KM, Patil SS, Patil SP (2017) Prepaid energy meter using GSM module. Int J Eng Sci Invention 6(2):80–85
26. Anand R, Shrivastava G, Gupta S, Peng SL, Sindhwani N (2018) Audio watermarking with reduced number of random samples. In: Handbook of research on network forensics and analysis techniques. IGI Global, pp 372–394
27. Sindhwani N, Verma S, Bajaj T, Anand R (2021) Comparative analysis of intelligent driving and safety assistance systems using YOLO and SSD model of deep learning. Int J Inf Syst Model Design (IJISMD) 12(1):131–146
28. Jain Y, NamrataTiwari S, Jain S (2019) A comparative analysis of various credit card fraud detection techniques. Int J Recent Technol Eng 7(5S2):402–407
29. Sindhwani N (2017) Performance analysis of optimal scheduling based firefly algorithm in MIMO system. Optimization 2(12):19–26
30. Sanchez-Reillo R, Sanchez-Avila C (2002) Fingerprint verification using smart cards for access control systems. IEEE Aerosp Electron Syst Mag 17(9):12–15

A Database Application of Monitoring COVID-19 in India

Vishali Pathania, Sritha Zith Dey Babu, Shahanawaj Ahamad, P. Thilakavathy, Ankur Gupta, Malik Bader Alazzam, and Digvijay Pandey

Abstract A quick user-friendly application of any pandemic situation can reduce the huge value of mortality with producing the graph of cases. Simple database application can make sense to the people about the pandemic and transmissions. This research aimed to develop a simple application which shows the real-time cases of COVID-19 and analyzes different states condition of India and a proper graphical prediction of cases. This application notifies people to get alert about the transmission and precautions to get rid of this pandemic. This application also helps clinical doctors, ministry and decision makers to improve the gap of any unfilled section. We have used the platform of APEX Oracle to develop this application and analyzed the dataset. The accuracy of the data is 78% rather than any other existing techniques. Combining the application and advance techniques, this study can create a vital framework for the prediction of any pandemics.

Keywords Oracle · Database application · COVID-19 · Apex · SQL

V. Pathania · S. Z. D. Babu
Chandigarh University, Chandigarh, India

S. Ahamad
College of Computer Science and Engineering, University of Hail, Hail, Saudi Arabia

P. Thilakavathy
Department of Information and Technology, St. Joseph's College of Engineering, Chennai, Tamil Nadu, India

A. Gupta
Department of Computer Science and Engineering, Vaish College of Engineering, Rohtak, Haryana, India

M. B. Alazzam
Faculty of Computer Science and Informatics, Amman Arab University, Amman, Jordan

D. Pandey (✉)
Department of Technical Education, IET, Dr. A.P.J. Abdul Kalam Technical University, Lucknow, Uttar Pradesh 226021, India
e-mail: digit11011989@gmail.com

M. Gupta et al. (eds.), *Artificial Intelligence on Medical Data*,
Lecture Notes in Computational Vision and Biomechanics 37,
https://doi.org/10.1007/978-981-19-0151-5_23

1 Introduction

When we consider that in the start of 2020, investigators have posted many vital researches and dataset about coronavirus disease 2019 (COVID-19) pandemic [1, 2], expanding the developing body of literature on the disorder and its causative agent, excessive acute respiration and spatial syndrome of this virus (SARS-CoV-2). The studies establishments have invested giant sources into closing key understanding gaps about this initial epidemic and pandemic as the scope of contemporary research stays unclear [3]. In this paper, we present the goal to become aware of which subjects COVID-19 studies have aimed at the Indian states and which areas have seemed to require additional attention. This task is one of the coronaviruses-associated theme tasks. It's miles a system studying-based application for an information dashboard. A statistics dashboard is a statistics control device that visually tracks, analyzes and shows key performance signs (KPI) [4], metrics and key records factors to display the particular procedure. The dashboard consists of fronts: back and front. The back stop includes statistics collecting, data coaching, facts evaluation, cases of COVID and device gaining knowledge of, all of which implemented the use of APEX [5]. The front quit includes making the application, changing the processed facts at back end to a consumable form and deploying most of these functions online. On the lower back, give up facts for prediction and showcasing data for extraordinary purposes turned into accumulated from the reliable repository of Kaggle dataset online platform [6]. For application and the information changed into taken from the Kaggle database of COVID-19. At the front end, the documents were processed into consumable cloth for Internet site building functions on the usage of APEX-based totally open-source framework Oracle [7]. The application turned into made presentable and interactive the use of SQL and Its App Builder platform. These kinds of capabilities blended formed a live information dashboard as an application updating itself daily, displaying overall wide variety of cases for every Indian state at a time and in shape of a global map for better relative knowledge of the state of affairs. It additionally snaps shots the improving and infected instances for each state in a graphical shape for distinctive view [8]. The dashboard offers you an option to install your queries and get the solutions to them in the form of an application together with giving prediction of overall quantity of instances for each Indian states in near future. The application additionally gives you the output of COVID-19 outbreaks in Indian states using the techniques of data mining and SQL (Figs. 1 and 2).

2 Methods and Materials

The dataset that has been applied in prediction is fetched from facts archive for highest accurate COVID-19 prediction dataset analysis, or structures technology and engineering help analyzers and predictors for getting the accurate output after preprocessing the data. The facts fetched may be seemed as parameterized dataset

Fig. 1 Architecture of data application

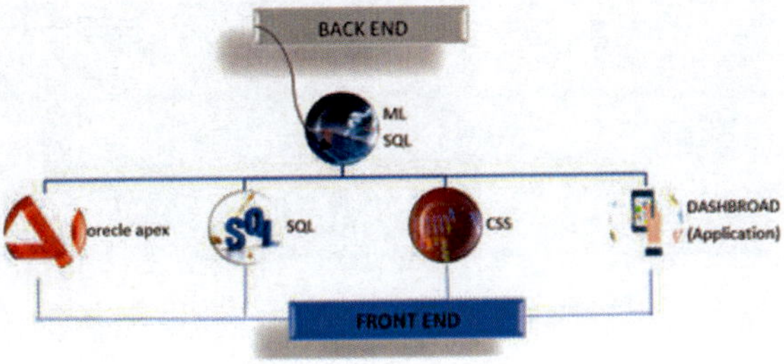

Fig. 2 Highbred model of application

having pertinent parameters which include nation, country, latitude rely, longitude depend and dates. To create autility, we use APEX. Oracle application allows you to construct scalable, at ease, application enterprise software, with global-magnificence features [9, 10]. Oracle APEX minimizes the complexity involved with multifaceted programs and gives developers with the capabilities to remedy troubles without needing to become experts in substantial Internet technologies. Oracle APEX allows to build higher programs via taking care of those six sides of software development:

- Statistics
- User interface
- Protection
- Accessibility
- Tracking
- Globalization.

We have created our login page for authenticating the users. We have created this page for blocking third-party entries. The login credentials include user name or email and the unique password. As we are using the database application, the password remember option is thereby default (Figs. 3 and 4).

After that, we have created a simple user interface for getting the user friendly with this simple application. Here, users can see the options in front view of the application as top list bar and left navigation bar. Front view of this application shows

Fig. 3 Code of login page

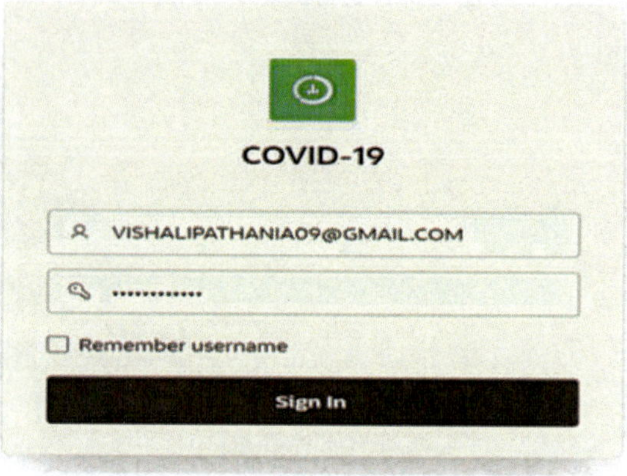

Fig. 4 Login page view of COVID-19 application

parameters such as COVID-19 CASES, COVID-19 REPORT and PREDICTION, and left navigation bar shows parameters such as HOME APPLICATION, COVID-19 CASES, COVID-19 REPORT, PREDICTION and ADMINISTRATION (Figs. 5 and 6).

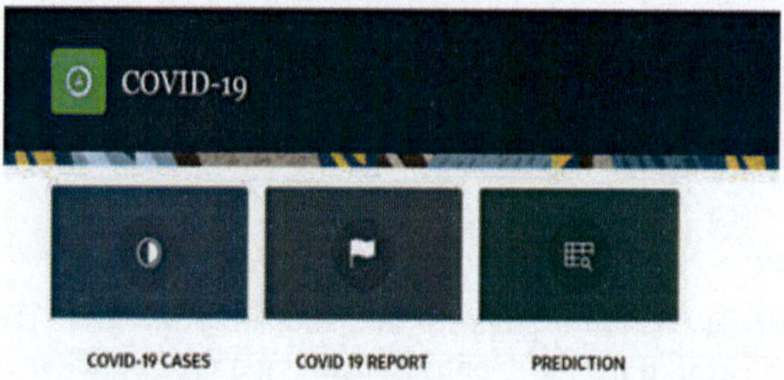

Fig. 5 User interface of home application

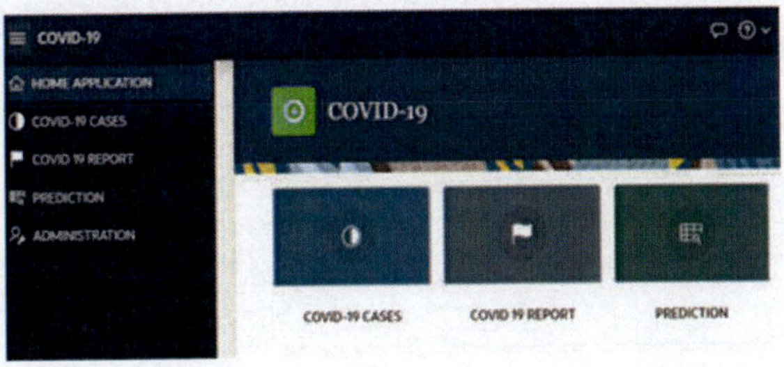

Fig. 6 Left navigation menu bar view

Hence, we have featured the dashboard of this application with COVID-19 cases divided into four categories such as state cases, active cases, average ratio and discharge. Here, state cases represent the number of real-time cases over all the Indian states, and active cases define the active number of non-recovered patients. Average ratio shows the graph of the representation of average on daily basis transmission, and discharges evaluate the recovered patients from hospitals with the view pie chart. We have given the output of this feature below (Fig. 7).

At the third section of this application, we added the option of COVID-19 reports. Here, we have made the report by calculating different parameters such as total cases, active cases, active ratio, discharges, discharge ratio, total death and death ratio. We have shown the report of 36 Indian states. Here is the output of this section given below (Fig. 8).

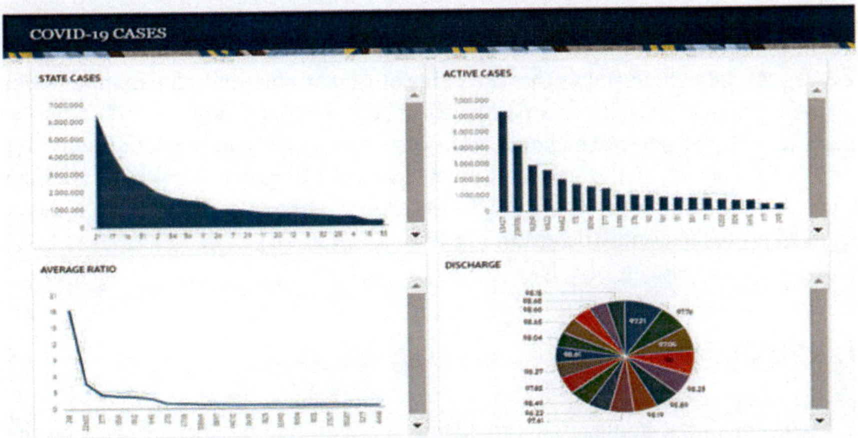

Fig. 7 Output view of COVID-19 cases

COVID 19 REPORT

Id	State UTs	Total Cases	Active	Discharged	Deaths	Active Ratio	Discharge Ratio	Death Ratio
1	Andaman and Nicobar	7,584	8	7,447	129	0.11	98.19	1.7
2	Andhra Pradesh	2,030,849	14,652	2,002,187	14,010	0.72	98.59	0.7
3	Arunachal Pradesh	53,807	533	53,004	270	0.99	98.51	0.5
4	Assam	595,669	5,615	584,296	5,758	0.94	98.09	1.0
5	Bihar	725,833	77	716,098	9,658	0.01	98.66	1.3
6	Chandigarh	65,160	32	64,311	817	0.05	98.70	1.3
7	Chhattisgarh	1,004,902	376	990,968	13,558	0.04	98.61	1.4
8	Dadra and Nagar Haveli and Daman and Diu	10,670	5	10,661	4	0.05	99.92	0.0
9	Delhi	1,438,250	377	1,412,790	25,083	0.03	98.23	1.7
10	Goa	174,891	702	170,972	3,217	0.40	97.76	1.8
11	Gujarat	825,629	161	815,386	10,082	0.02	98.76	1.2
12	Haryana	770,659	351	760,501	9,807	0.05	98.68	1.3
13	Himachal Pradesh	215,893	1,521	210,733	3,639	0.70	97.61	1.7
14	Jammu and Kashmir	326,990	1,247	321,329	4,414	0.38	98.27	1.4
15	Jharkhand	348,079	117	342,829	5,133	0.03	98.49	1.5
16	Karnataka	2,962,408	16,269	2,908,622	37,517	0.55	98.18	1.3
17	Kerala	4,390,489	209,335	4,158,504	22,650	4.77	94.72	0.5
18	Ladakh	20,608	30	20,371	207	0.15	98.85	1.0
19	Lakshadweep	10,353	5	10,297	51	0.05	99.46	0.5
20	Madhya Pradesh	792,353	131	781,705	10,517	0.02	98.66	1.3
21	Maharashtra	6,500,617	53,427	6,309,021	138,169	0.82	97.05	2.1
22	Manipur	117,230	2,617	112,801	1,812	2.23	96.22	1.6
23	Meghalaya	78,359	1,859	75,141	1,359	2.37	95.89	1.7
24	Mizoram	72,883	13,369	59,273	241	18.34	81.33	0.3

Fig. 8 COVID-19 report section output view

3 Result

COVID-19 reports show the report with the basis of population and cases. This graph, which we have shown below in Fig. 9, shows the real-time remission cases of peoples. As this graph shows that the cases of Maharashtra in India are in the pick value of this data, this is not a permanent report. But, somehow COVID-19 cases increase with the number of populations also. Our graph and prediction will help the clinical doctors, analyzers, decision makers and Ministry of Health for getting a smooth information regarding COVID-19. Here we have also shown the predicted death and actual causality deaths in Fig. 10.

4 Conclusion

A simple predictor application for getting the real-time number of cases in this COVID-19 pandemic is an essential idea that can reduce the number of deaths with the aid of giving information regarding cases of COVID-19 with the analyzation of different states. The actual casual cause prediction capacity of laboratory biomarkers is wonderful from those of clinical and demographic information. The percentages and the graphs are crucially needed for a quick analysis and help us to make a proper step. A simple application can always remind people regarding his next step or what

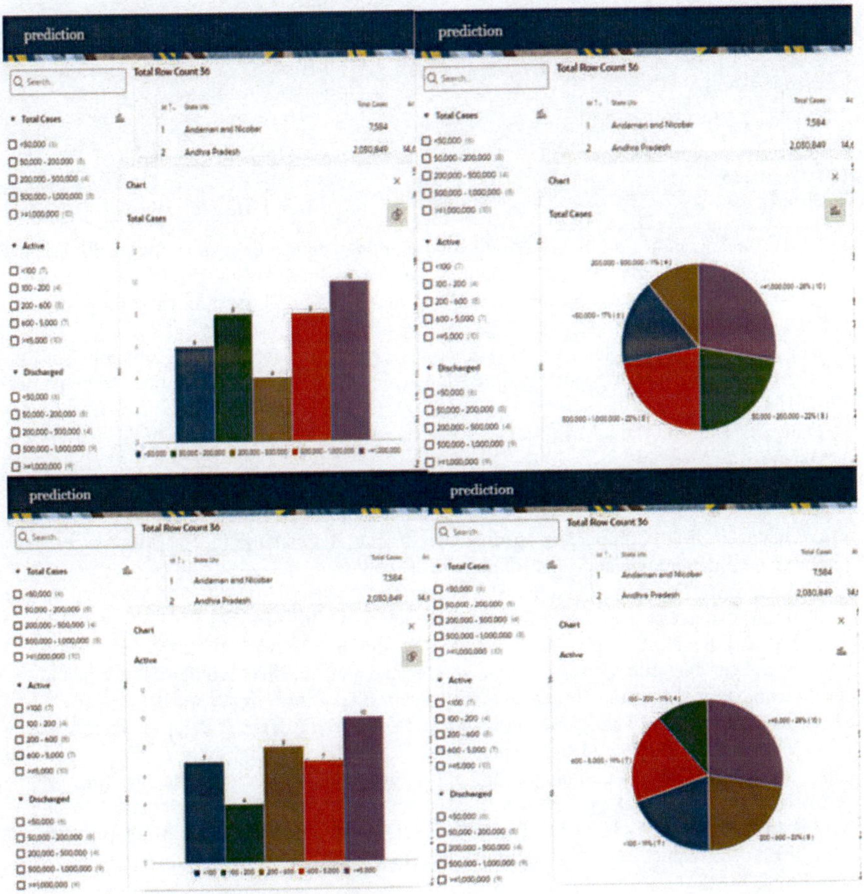

Fig. 9 Predicted graphical output of total and active cases

Fig. 10 Predicted death and actual casual value output

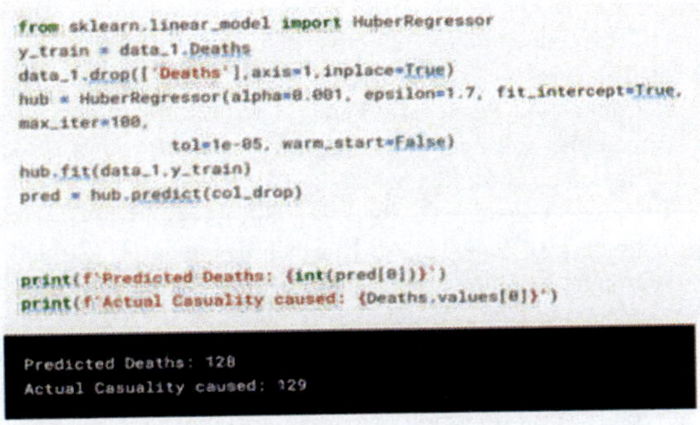

```
from sklearn.linear_model import HuberRegressor
y_train = data_1.Deaths
data_1.drop(['Deaths'],axis=1,inplace=True)
hub = HuberRegressor(alpha=0.001, epsilon=1.7, fit_intercept=True,
max_iter=100,
             tol=1e-05, warm_start=False)
hub.fit(data_1,y_train)
pred = hub.predict(col_drop)


print(f'Predicted Deaths: {int(pred[0])}')
print(f'Actual Casuality caused: {Deaths.values[0]}')
```

```
Predicted Deaths: 128
Actual Casuality caused: 129
```

kind of things he actually needs to do in this pandemic. To control this pandemic situation, a simple database application plays a vital role.

References

1. W. H. Organization (2020) WHO director-generals opening remarks at the media briefing on COVID-19—11 March 2020. [Online]. Available: https://www.who.int/dg/speeches/detail/who-director-general-s-opening-remarks-at-the-media-briefing-on-/covid-19---11-march-2020
2. Anand R, Sindhwani N, Saini A (2021) Emerging technologies for COVID-19. Enabling healthcare 4.0 for pandemics: a roadmap using AI, machine learning, IoT and cognitive technologies, pp 163–188
3. Pandey D, Islam T, Malik MA (2021) Novel coronavirus disease (SARS-COV-2): an overview. Asian J Adv Med Sci 8–12
4. Pandey D, Islam T, Magray JA, Gulzar A, Zargar SA (2021) Use of statistical analysis to monitor novel coronavirus-19 cases in Jammu and Kashmir, India. Eur J Biol Res 11(3):274–282
5. de Dombal F (1984) Computer-aided diagnosis of acute abdominal pain. The British experience. Revue d'epidemiologie et de santepublique 32(1):5056
6. Miller RA, McNeil MA, Challinor SM, Masarie FE, Myers JD (1986) The internist-1/quick medical reference project—status report. West J Med 145(6):816–822
7. Kumar MS, Sankar S, Nassa VK, Pandey D, Pandey BK, Enbeyle W (2021) Innovation and creativity for data mining using computational statistics. In: Methodologies and applications of computational statistics for machine intelligence. IGI Global, pp 223–240
8. Meslie Y, Enbeyle W, Pandey BK, Pramanik S, Pandey D, Dadeech P, Belay A, Saini A (2021) Machine intelligence-based trend analysis of COVID-19 for total daily confirmed cases in Asia and Africa. In: Methodologies and applications of computational statistics for machine intelligence. IGI Global, pp 164–185
9. Bakshi G, Shukla R, Yadav V, Dahiya A, Anand R, Sindhwani N, Singh H (2021) An optimized approach for feature extraction in multi-relational statistical learning. J Sci Ind Res (JSIR) 80(6):537–542
10. Shukla R, Dubey G, Malik P, Sindhwani N, Anand R, Dahiya A, Yadav V (2021) Detecting crop health using machine learning techniques in smart agriculture system. J Sci Ind Res (JSIR) 80(08):699–706
11. Pandey D, Ogunmola GA, Enbeyle W, Abdullahi M, Pandey BK, Pramanik S (2021) COVID-19: a framework for effective delivering of online classes during lockdown. Hum Arenas 1–15
12. Parthiban K, Pandey D, Pandey BK (2021) Impact of SARS-CoV-2 in online education, predicting and contrasting mental stress of young students: a machine learning approach. Augmented Hum Res 6(1):1–7
13. Babu SZD, Pandey D, Sheik I (2020) Acts Of COVID19. Int J Adv Eng 3:2457-0397
14. Khongsai L, Anal TS, AS R, Shah M, Pandey D (2021) Combating the spread of COVID-19 through community participation. Glob Soc Welfare 8(2):127–132
15. Zargar SA, Islam T, Rehman IU, Pandey D (2021) Use of cluster analysis to monitor novel corona virus (Covid-19) infections In India. Asian J Adv Med Sci 1–7
16. Meivel S, Sindhwani N, Anand R, Pandey D, Alnuaim AA, Altheneyan AS, Lelisho ME et al (2022) Mask detection and social distance identification using Internet of Things and faster R-CNN algorithm. Comput Intell Neurosci 2022

Health Monitoring

Review on Lung Nodule Segmentation-Based Lung Cancer Classification Using Machine Learning Approaches

Shazia Shamas, S. N. Panda, and Ishu Sharma

Abstract Lung cancer or lung carcinoma is one of the major reasons for non-accidental death in the world with a high fatality rate in both men and women. The major cause of lung cancer is wrong lifestyle choices such as consumption of beedi, cigarette, and hukka. Lung cancer is broadly categorized as small cell lung carcinoma (SCLC) and non-small cell lung carcinoma (NSCLC). It is very difficult to detect lung cancer well in advance. However, technological advances in medical imaging have resulted in the diagnosis and prediction of various stages of lung cancer by analyzing CT scans. In the present paper, a constructive review of the existing approaches for lung nodule detection and classification using machine learning approaches is presented. Authors have analyzed the articles published in the last decade to access the current status of the research in the field of lung cancer classification. The survey study concluded that the involvement of optimization approaches to improve the feature extraction and segmentation stage has been involved in recent years. Further, it is observed that the integration of the neural network architecture has become the first choice of numerous researchers for lung cancer classification.

Keywords Lung cancer · SCLC · NSCLC · Machine learning · Classification · Introduction

1 Introduction

The disease lung cancer is happening due to the growth of uncontrolled cells inside the lung tissues. This growing criterion leads to metastasis that termed as the adjacent tissue invasion and infiltration ahead of the lungs. The prognosis and treatment are specifically relying on the histological kinds of cancer, and the degree of spread or, stage and the status of the patient's performance. The treatment that is possible involves chemotherapy, surgery, and radiotherapy. In medical science, data mining plays an important role because it is very useful to detect different types of diseases

S. Shamas (✉) · S. N. Panda · I. Sharma
Chitkara Institute of Engineering and Technology, Chitkara University, Chandigarh, Punjab, India
e-mail: shaziabhat6115@gmail.com

M. Gupta et al. (eds.), *Artificial Intelligence on Medical Data*,
Lecture Notes in Computational Vision and Biomechanics 37,
https://doi.org/10.1007/978-981-19-0151-5_24

with the help of algorithms. It is very successful in cancer disease. Lung cancer is a major issue all over the world which increases the daily death rate of men and women.

According to the WHO report, in the year 2020, 1.80 million death were registered due to lung cancer. World Wellbeing Association is predicted that approximately 17 million cases of lung cancer overall the world in 2030 [1]. A sample CT scan of human lung is given in Fig. 1.

Tobacco is the major issue of increases in the death rate due to lung cancer. There are a large number of symptoms that arise in the human body when he/she is suffering from lung cancer. Lung cancer is divided into two categories:

1. Small cell lung carcinoma SCLC
2. Non-small cell lung carcinoma NSCLC (Fig. 2).

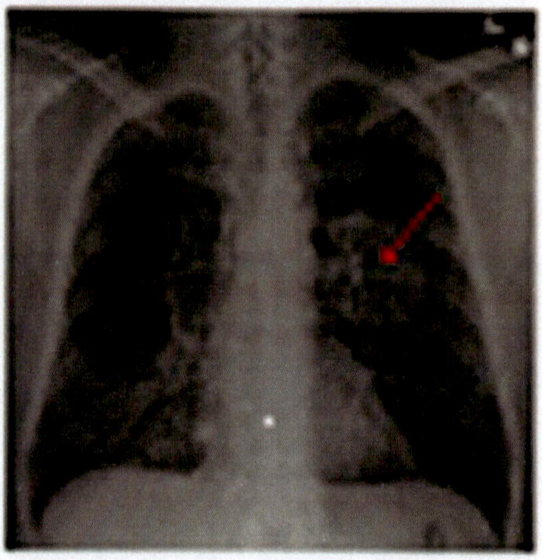

Fig. 1 CT scan of lung

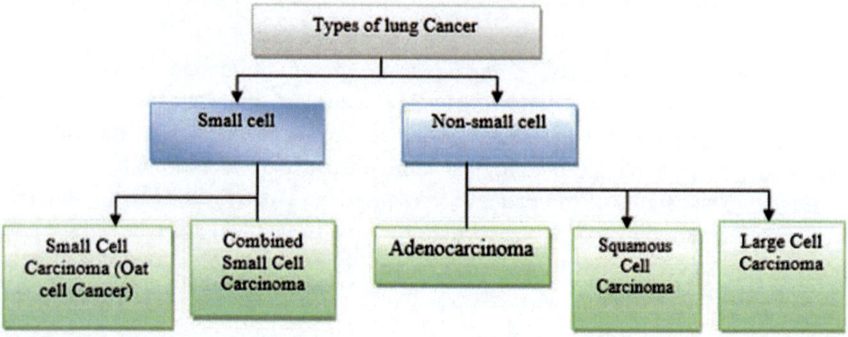

Fig. 2 Classification of lung cancer

Fig. 3 Lung cancer prediction data mining process

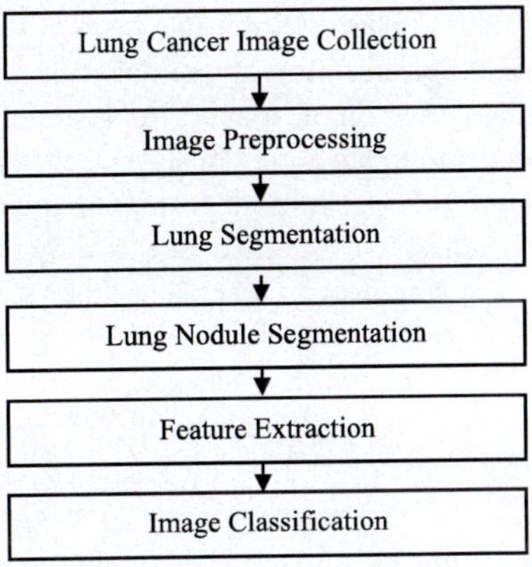

Small cell lung cancer is that type of cancer which is always seen in the cancer patient. It spreads very fastly in the human body. It is very faster than small cell lung carcinoma. This type of cancer is also seen in those patients who are smoking. The treatment of those patients who are suffering from SCLC is done with the help of chemotherapy and therapy. Surgery of SCLC patients is not done because it spreads very quickly in the human body. It is a very harmful cancer. If it is not detected within the time, then the patient will die soon [2]. Non-small cell lung cancer is that type of cancer which are mostly seen in the lung cancer patient. It spreads very slowly in the human body. It is slower than small cell lung carcinoma. This type of cancer is also seen in those patients who have never smoked. The treatment of those patients who are suffering from NSCLC is done with the help of surgery or radiation therapy [3].

The data mining process used for automatic lung cancer detection is shown in the following figure. It is broadly divided into five steps that are used to predict lung cancer (Fig. 3).

Data mining also involves the novel advancement in network security techniques which utilizes a multi-phase heuristic for the detection of damage from the network [4].

In this work focused on disease detection apart from this to determine the deficiencies of present technologies, software-defined networking (SDN) has been taken into account [5–7].

To get a better learning outcome, the animation-based teaching style has been adopted, for example, teaching through two-dimension and three-dimension animation or animated pictures which tends to improved learning outcomes [8].

2 Literature Survey

Chaudhary and Singh, 2012, have presented a lung cancer detection system using image processing for the classification of lung cancer in computed tomography (CT) images. The research has been exploring processes such as segmentation, preprocessing, and extraction of data. The authors have divided the work into three sections: segmentation of images, image enhancement, and extraction of features. The cancer is restricted to a lung in stage 1, the cancer is constrained to the chest with more invasive and large tumors in stage II and stage III, although the cancer is distributed from the chest to other parts of the body in stage IV [9]. Hadavi et al. (2014) have studied automatic lung cancer identification using cellular learning automatons. The images have some form of unwanted data with features that are important for preprocessing and image improvisation by separating the distortion and enhancing the important features. For enhancing CT images, the Gabor filter with the region-widening method is used. The lung cancer nodule was extracted according to the characteristics after the preprocessing stage. The patterns were obtained with the cellular learning automata (CLA) method and the CAD model. CLA includes several multi-agent process communication automates [10]. Kuruvilla and Gunavathi (2014) have presented the computer-aided classification method in computer tomography (CT) images of lungs formed through the artificial neural network (ANN). The whole lung has been segmented from the CT images, and some parameters have been computed from these obtained segmented images. The parameters utilized for classification purpose are named as mean, standard deviation, skewness, kurtosis, fifth central moment, and sixth central moment. For classification purpose, the neural network used feed-forward and back propagation neural network. The Traingdx method gives the highest classification accuracy of 91.1 percent of the thirteen training functions of the back propagation neural network already available and has provided two new training features. The results show that the proposed learning function 1 offers 93.3% precision, 100% specificity, 91.4% sensitivity, and 0.998 mean square error [11]. Kurkure and Thakare (2016) have emphasized developing and designing the initial diagnosis and detection method for CT, PET, and X-ray lung cancer. A genetic algorithm is used to improve the results, allowing the detection of the nodules in the lungs at the early stages of lung cancer. The use of Naïve Bayes and genetic algorithms to solve the challenge of selecting the best features is used to accurately and quickly identify the different stages of the cancer images. After the overall analyzed reliability of classification has obtained 80% [12]. Chauhan and Jaiswal, 2016, have proposed an approach for studying how to spot lung cancer. Automatically generated classification of diseases through machine learning is often based on features derived from the segmentation of individual objects that can be complicated and time consuming. The author worked on predicting and preventing various medical diseases using PCA; preprocessing and post-processing Canny edge detection techniques were used. Firstly, edge detection is performed, then extraction of the feature is done to obtain the optimal number of features for identification between infected and non-infected diseases. The developed method has been introduced in MATLAB and evaluated with actual CT scan images. The data set comprises five classes, and each class has about 120 cancer and 200 kb non-cancerous images. A connection has been done with the SURF system, and it was found that the PCA approach was better than the SURF method [13]. Razmjooy et al. (2018) have proposed a novel efficient methodology to detect malignancy in melanoma from images. Firstly, the scales that are not needed are removed through the edge detection and smoothing technique. After getting extracted and optimized outcomes, this method becomes able to segment the cancer images. In the end, information of objects that are not required in this task has to be filtered out through morphological operations and utilized to focus on the area, where melanoma boundaries are present. To perform this operation, world cup optimization algorithm has been used to optimize an multi-layer perceptron (MLP) and artificial neural network (ANN). This proposed scheme WCO is a derivative-free, meta-heuristics and global search algorithm; however, gradient-based back propagation method is local search. The MLP employs the problems constraint, and WCO algorithms have utilized to reduce the root mean square error [14]. Arulmurugan and Anandakumar (2018) have presented a system based on using the descriptor of the wavelet function and combined with the classification of an artificial neural network. After applying wavelet transformation, measured statistical attributes named as autocorrelation, entropy, contrast, and energy are obtained and used as input parameters for neural network classification. The NN classifier is configured to use feed-forward neural network and feedback propagation network to recognize training operations, i.e., Traingd, Traingda, Traingdm, and Traingdx. Neural network feedback propagation provides better findings than feed-forward classification. The potential classification system provided 92.6% precision, 100% specificity, 91.2% flexibility, and 0.978 mean square error [15]. Shakeel et al. (2020) have proposed an optimized image processing and DNN-based machine learning technique that is used to detect lung cancer. The data set is collected from the NSCLC CT scan images. The simulation results computed using the MATLAB simulator achieved an average accuracy of 96.2% with a sensitivity of 97.4%. The paper is mainly focused to increase the brightness of the images while reducing the noise levels [16]. Priyadharshini et al. (2021) introduced a bat optimization technique to optimize the feature extraction step involved in the detection of lung cancer. The classification was performed using fuzzy C-means (FCM) against the LIDC-IDRI data set demonstrating an average accuracy of 97.43% with 95.5% specificity [17].

The above literature illustrated different classification approaches that were applied for classification. Segmentation is also a major part of this research work. Several segmentation approaches were preferred by the different authors. But, some common segmentation approaches are discussed in Table 1.

Table 1 Various segmentation approaches

Year, citation	Authors	Segmentation approach
2012, [18]	Aarthy and Ragupathy	Multiscale wavelet-based edge detection
2018, [19]	Silva et al.	Threshold-based clustering
2018, [20]	Gong et al.	OTSU threshold algorithm
2018, [21]	Khosravan et al.	3D segmentation using dice similarity coefficient
2021, [17]	Priyadharshini	Fuzzy c-means clustering algorithm
2021, [22]	Surendar and Ponni	MasiEMT-SSA approach

3 Classification Approaches

For lung cancer classification, several approaches are available now that are implemented by different researchers. Some provide improvement in classification but some fail. But, most of the researchers applied SVM and neural networks for classification because these techniques provide better accuracy as compared to other techniques.

3.1 *Support Vector Machine (SVM)*

In machine learning algorithms, SVM performs an important part; therefore, it is applied in several applications to solve complex problems, especially applied classification problems. SVM comes under the supervised learning algorithms that analyze various data for classification. The major objective of the SVM algorithm is to find a hyperplane that depends on the N-dimensional space. Here N defines the different number of features that are used to classify data points. The main target of the SVM is to find a hyperplane that contains maximum margin means the maximum distance that lies between the two classes. For lung cancer classification, several researchers proposed different classification approaches in which SVM was applied [18, 23–27]. Some researchers applied hybrid approaches by integrating SVM with other approaches like GA, clustering, and KNN for classification [28–30].

3.2 *Neural Networks (NN)*

For lung cancer classification, neural network has been the second choice of the researchers. NN is considered as an electronic network-based that works like a human brain. The number of input records is processed at once after that the network learns by comparing the classification based on records. NN works in the form of layered architecture in which three layers are present that are, namely defined as input layer, hidden layer, and output layer. The weights are assigned to the network input layer that is further passed to the next layers, and at the end, the output layer generates the output. There are several variants of neural networks but commonly used DNN and CNN. Some researchers used NN variants for the classification of data [16, 19, 21, 22, 31, 32].

4 Survey Outcomes

The critical analysis of the literature survey is given in Table 2. It shows that SVM and neural networks have been the first choice of most of the highly accurate lung cancer diagnostic techniques.

Table 2 Comparative analysis of performance of various classifiers

Year, citation	Authors	Data set used	Classification approach	Performance parameters
2010, [23]	da Silva Sousa et al.	33 CT scans comprising benign (23) and malignant (10)	RBF kernel—SVM	ACC = 95.21%, SPE = 96.15% and SEN = 84.84%
2012, [18]	Aarthy and Ragupathy	LIDC for 28 CT scan images comprising cancerous (13) and non-cancerous (15) nodules	RBF kernel—SVM	SPE = 85.71% and SEN = 92.86%
2013, [24]	Choi et al.	58 CT scans representing 151 nodules	RBF kernel—SVM	ACC = 97.61%, SEN = 95.28%
2013, [25]	Sivakumar and Chandrasekar	54 CT scans comprising 36 cancerous and 18 non-cancerous	RBF kernel—SVM	ACC = 80.36%, SEN = 82.05% and SPE = 76.4%
2016, [28]	Abduh et al.	112 CT scans from LIDC	KNN and SVM	SPE = 0.946 and ACC = 97.3% using KNN, SPE = 0.982 and ACC = 99.1% using SVM
2016, [33]	Kuppusamy	302 CT scan images	K-means with GA	ACC = 64%
2017, [29]	de CarvalhoFilho et al.	LIDC for 833 CT scan images comprising benign (1011) and malignant (394)	RBF kernel—SVM with GA	ACC = 93.19%, SEN = 92.75% and SPE = 93.33%
2017, [30]	Hussein et al.	LIDC-IDRI	SVM and clustering	ACC = 78.06% and SEN = 77.85%
2018, [19]	Silva et al.	LIDC-IDRI comprising 219,522 CT slices	Particle swarm optimization and CNN	ACC = 97.62%, SEN = 92.20% and SPE = 98.64%
2018, [20]	Gong et al.	LUNA16 ANODE09	Random forest	SEN 79.3%

(continued)

Table 2 (continued)

Year, citation	Authors	Data set used	Classification approach	Performance parameters
2018, [21]	Khosravan et al.	LUNA16 ANODE09	3D deep CNN	SEN = 98%
2018, [26]	Naqi et al.	LIDC-IDRI	Support vector machine-based ensemble	SPE = 98.2% and SEN = 98.6%
2018, [27]	Zhang et al.	LIDC—IDRI	Support vector machine	SEN = 84.62% with 2.1 FP per scan
2019, [31]	Monkam et al.	LIDC-IDRI comprising 13,179 micro-nodules	3D—CNN and ELM classifier	ACC = 97.35%, SEN = 96.57% and *F*-score = 96.42%
2020, [32]	Masood et al.	LIDC-IDRI comprising 892 CT scan images	Enhanced faster RCNN	ACC = 97.35%, SEN = 98.1%
2020, [16]	Shakeel, et al.	NSCLC CT scan images	DNN	ACC = 96.2%, SEN = 97.4% SPE = 98.4% and *F*-score = 98.4%
2021, [22]	Surendar and Ponni	LIDC-IDRI	Hybrid DNN	ACC = 98.2%, SEN = 97.3%
2021, [17]	Priyadharshini	LIDC-IDRI	Fuzzy C-means	ACC = 97.43%, SPE = 95.5%

The number of parameters, such as accuracy, precision, recall, sensitivity, specificity, f-score, true positive rate, and area under curve, has been used for the performance analysis of lung cancer classification works. The observed popularity based on the literature surveyed in the presented work is shown in Fig. 4.

5 Conclusion

Lung cancer is one of the commonly affecting cancer types that has a high death rate among various cancer types. The presented survey illustrates that despite many advances in the field of biomedical imaging and image processing, still precise detection of lung cancer is a challenging task and requires skillful interpretation of lung nodules. In the past few years, the development of computer-aided techniques had significantly improved lung cancer classification due to better and precise lung and lung nodule segmentation. This improves the overall lung cancer classification with earlier detection and diagnosis for the timely planning of medical treatment. The presented paper provides an overview and solid background of the till date research about lung cancer segmentation to the researchers to design future research.

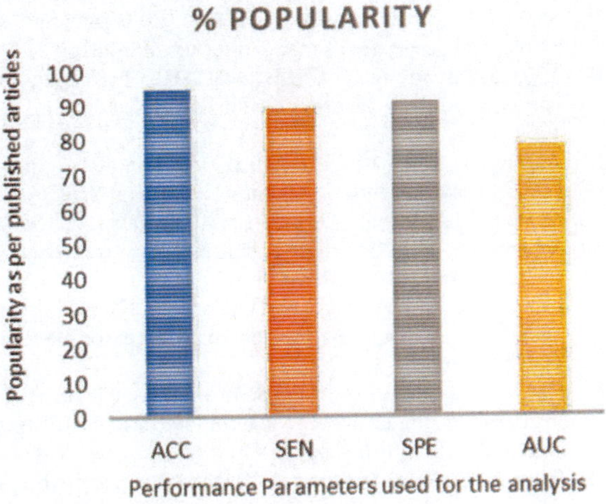

Fig. 4 Popular performance parameters as depicted by the surveyed articles

References

1. WHO (2021). Accessed online at https://www.who.int/news-room/fact-sheets/detail/cancer
2. Chandra EYV, Teja KR, Prasad MHCS (2019) Lung cancer prediction using data mining techniques. Int J Recent Technol Eng 8
3. Krishnaiah V, Narsimha G, Chandra NS (2013) Diagnosis of lung cancer prediction system using data mining classification techniques. Int J Comput Sci Inf Technol 4(1):39–45
4. Panda SN, Mangla V (2010) Protecting data from the cyber theft—a virulent disease. J Emerg Technol Web Intell 2(2)
5. Bhardwaj S, Panda SN (2019) SDWSN: software-defined wireless sensor network. Int J Innov Technol Explor Eng 2278–3075
6. Badotra S, Panda SN (2019) A review on software-defined networking enabled iot cloud computing. IIUM Eng J 20(2):105–126
7. Baggan V, Panda SN (2019) Enhancing network path restoration with software defined networking 14:1910–1916
8. Kaushal RK, Panda SN (2019) A meta analysis on effective conditions to offer animation based teaching style. Malays J Learn Instr 16(1):129–153
9. Chaudhary A, Singh SS (2012) Lung cancer detection on CT images by using image processing. In: 2012 international conference on computing sciences. IEEE, pp 142–146
10. Hadavi N, Nordin MJ, Shojaeipour A (2014) Lung cancer diagnosis using CT-scan images based on cellular learning automata. In: 2014 international conference on computer and information sciences (ICCOINS). IEEE, pp 1–5
11. Kuruvilla J, Gunavathi K (2014) Lung cancer classification using neural networksfor CT images. Comput Methods Programs Biomed 113:202–209
12. Kurkure M, Thakare A (2016) Lung cancer detection using genetic approach. In: International conference on computing communication control and automation (ICCUBEA). IEEE, pp 1–5
13. Chauhan S, Jaiswal M (2016) Determinants of acceptance of ERP software training in business schools: empirical investigation using UTAUT model. Int J Manage Educ 14(3):248–262
14. Razmjooy N, Sheykhahmad FR, Ghadimi N (2018) A hybrid neural network–world cup optimization algorithm for melanoma detection. Open Med 13(1):9–16
15. Arulmurugan R, Anandakumar H (2018) Early detection of lung cancer using wavelet feature descriptor and feed forward back propagation neural networks classifier. In Computational vision and bio inspired computing. Springer, Cham, pp 103–110

16. Shakeel A, Kirichek A, Chassagne C (2020) Effect of pre-shearing on the steady and dynamic rheological properties of mud sediments. Mar Petrol Geol 116:104338
17. Priyadharshini P, Zoraida BSE (2021) Hybrid semantic feature descriptor and fuzzy C-means clustering for lung cancer detection and classification. J Comput Theor Nanosci 18(4):1263–1269
18. Aarthy KP, Ragupathy US (2012) Detection of lung nodule using multiscale wavelets and support vector machine. Int J Soft Comput Eng (IJSCE) 2(3)
19. da Silva GLF, Valente TLA, Silva AC, de Paiva AC, Gattass M (2018) Convolutional neural network-based PSO for lung nodule false positive reduction on CT images. Comput Methods Programs Biomed 162:109–118
20. Gong J, Liu JY, Wang LJ, Sun XW, Zheng B, Nie SD (2018) Automatic detection of pulmonary nodules in CT images by incorporating 3D tensor filtering with local image feature analysis. Phys Med 46:124–133
21. Khosravan N, Bagci U (2018) Semi-supervised multi-task learning for lung cancer diagnosis. In 2018 40th Annual international conference of the IEEE engineering in medicine and biology society (EMBC). IEEE, pp 710–713
22. Surendar P (2021) Diagnosis of lung cancer using hybrid deep neural network with adaptive sine cosine crow search algorithm. J Comput Sci 53:101374
23. da Silva Sousa JRF, Silva AC, de Paiva AC, Nunes RA (2010) Methodology for automatic detection of lung nodules in computerized tomography images. Comput Methods Programs Biomed 98(1):1–14
24. Choi WJ, Choi TS (2013) Automated pulmonary nodule detection system in computed tomography images: a hierarchical block classification approach. Entropy 15(2):507–523
25. Sivakumar S, Chandrasekar C (2013) Lung nodule detection using fuzzy clustering and support vector machines. Int J Eng Technol 5(1):179–185
26. Naqi SM, Sharif M, Yasmin M (2018) Multistage segmentation model and SVM-ensemble for precise lung nodule detection. Int J Comput Assist Radiol Surg 13(7):1083–1095
27. Zhang W, Wang X, Li X, Chen J (2018) 3D skeletonization feature based computer-aided detection system for pulmonary nodules in CT datasets. Comput Biol Med 92:64–72
28. Abduh Z, Wahed MA, Kadah YM (2016) Robust computer-aided detection of pulmonary nodules from chest computed tomography. J Med Imag Health Inf 6(3):693–699
29. de Carvalho Filho AO, Silva AC, de Paiva AC, Nunes RA, Gattass M (2017) Computer-aided diagnosis system for lung nodules based on computed tomography using shape analysis, a genetic algorithm, and SVM. Med Biol Eng Comput 55(8):1129–1146
30. Hussein S, Gillies R, Cao K, Song Q, Bagci U (2017) Tumornet: Lung nodule characterization using multi-view convolutional neural network with gaussian process. In 2017 IEEE 14th international symposium on biomedical imaging (ISBI 2017). IEEE, pp 1007–1010
31. Monkam P, Qi S, Xu M, Li H, Han F, Teng Y, Qian W (2018) Ensemble learning of multiple-view 3D-CNNs model for micro-nodules identification in CT images. IEEE Access 7:5564–5576
32. Masood A, Sheng B, Yang P, Li P, Li H, Kim J, Feng DD (2020) Automated decision support system for lung cancer detection and classification via enhanced RFCN with multilayer fusion RPN. IEEE Trans Industr Inf 16(12):7791–7801
33. Kuppusamy V, Gopalakrishnan RC (2016) Feature extraction based lung nodule detection in CT images. Int J Appl Eng Res 11(4):2697–2700

Hyper Parameters Tuning ResNet-50 for Alzheimer's Disease Classification on Neuroimaging Data

Monika Sethi and Sachin Ahuja

Abstract One of the most prominent types of dementia is Alzheimer's disease (AD). It's among the main causes of death in elderly individuals in all developed nations. Numerous deep learning (DL) models for image classifications and object identification have been developed. However, DL techniques prepare the network model from scratch, which has some downfalls including demanding a massive amount of labeled training data source, which might be a trouble in the medical world, one in which practitioners annotate the data, being very expensive, and usually requires high computational resources. Transfer learning methods are currently being utilized to address these difficulties. In this research, a transfer learning-based ResNet 50 model that was pre-trained on the ImageNet dataset was adjusted on different hyper parameters using the ADNI dataset to provide the best possible results. There are four distinct optimizers utilized, SGD, Adagrad, rmsProp, and Adam, as well as two different batch sizes. The results show that the optimizers' rmsProp and Adamax performed well with batch sizes of 16 and 32 when compared to the SGD and Adam optimizers. For batch size 32, the classification accuracy for AD vs. NC using rmsProp is 74.22%. However, using batch size 16 resulted in a 1.01% relative improvement with 75.65% classification accuracy.

Keywords Alzheimer's disease · Neuroimaging data · ResNet · Hyper parameters tuning · Classification

M. Sethi (✉) · S. Ahuja
Chitkara Institute of Engineering and Technology, Chitkara University, Rajpura, Punjab, India
e-mail: monika.sethi@chitkara.edu.in

S. Ahuja
e-mail: sachin.ahuja@chitkara.edu.in

M. Gupta et al. (eds.), *Artificial Intelligence on Medical Data*,
Lecture Notes in Computational Vision and Biomechanics 37,
https://doi.org/10.1007/978-981-19-0151-5_25

1 Introduction

Medical concerns seem to be the foremost discussed topics worldwide, and both healthcare professionals and academics continue working actively to advance clinical diagnosis, cures, and assessments in order to maintain individual lives and improve healthier lifestyles. Vascular dementia and Alzheimer disease (AD) are both prevalent neuro disorder diseases [1, 2]. AD has become one of the world's deadliest in developed nations, primarily in Australia. It is an uncontrolled, progressive neurodegenerative disease that harms neurons in the brain. Neurological disease develops progressively over time, creating difficulties with speech, feeding, driving, and, eventually, death. Numerous therapies have been studied to delay the course of AD, although with limited effectiveness. Nonetheless, experts believe that AD-related brain alterations begin around 20 years before symptoms actually occur [3]. According to the facts and figures mentioned in [4]; by the mid-century, the population of citizens with AD is estimated to increase to 13.8 million. AD was the 6th leading cause of mortality in the World and the fifth cause of morbidity and mortality among Citizens of US countries. It is predicted to be 10 million people living with AD by 2025 [5]. Research for definitive diagnosis of AD is being carried out to prevent premature brain deterioration, decrease medical care costs of AD sufferers and assure better treatment [6]. However a lot of study has been done, there is just no clear diagnosis because of certain conflicting indicators in the existing medical practice. Only after death through some kind of post-mortem reports investigation on the brain's tissue could report the existence of AD clearly. Consequently, precise early identification of AD has great significance in patient and social care [7].

AD diagnosis has several effective approaches, including computer-aided systems. These techniques proposed different machine learning methods and model recognition using neuroimaging data such as CT scans, functional and structural MRI (fMRI, sMRI), positron emission tomography (PET) etc. [8]. AD and Normal Control (NC) may be detected by numerous approaches, most of which can be classified as computer vision and deep learning (DL) based approaches. Numerous characteristics may be utilized in neuroimaging for diagnosis and discrimination between AD and NC. The percentage of brain atrophy is one of the characteristics that in AD sufferers are considerably higher than the one without AD. Some other markers mostly used by specialists include ventricular enlargement and hippocampal atrophy [9].

The tremendous development in computational power has made it possible to develop breakthrough strategies for DL. DL can handle complex strategic decision issues as a subclass of machine learning and these approaches have been adopted for cognitive computing in several domains, notably for the analysis of medical imaging [10]. Furthermore, CNNs have already shown remarkable outcomes for the segmentation of tissue and disease diagnosis. CNNs integrate feature extraction, selection, and classification—as opposed to traditional machine learning methods—into three major steps of the classifier [11].

In [12], Islam and colleagues had been using a pre-trained Inception-V4 model to define the hyper parameters of its model instead of just randomized initialization to extract attributes from such a limited OASIS sample. Researchers changed the raw MRI images to 299 * 299 * 1 in a way to align it to the Inception-V4 model. With their other study [13], three DenseNet systems having various depths which were pretrained on ImageNet dataset were fine-tuned on OASIS MRI scan. Finally the outcome from three different DenseNet models was combined and an overall decision was made by calculating votes by some voting technique.

In this paper, we used CNN architecture (ResNet 50) to categories normal healthy individuals from AD patients using MRI neuro-scans which are obtained from ADNI dataset. More precisely, we employed the ResNet-50 framework, since this enables the detection of hidden patterns of the brain, relationships between various parts and recognition of patterns that are associated with AD in MRI brain images. The significance of this paper is the implementation of ResNet-50 the paradigm of transfer learning. The ResNet-50 transfer learning model has also been used to find optimum hyper parameters such as batch size and optimizers during training to identify the efficacy of each parameter in AD classification.

2 Background

The standard CNN is composed of a series of layers involving convolution, batch standardization, activation, max/avg. pooling and finally fully connected layers with dropout and softmax function. CNN learning incorporates chain principles to estimate the loss from the projected outcome to the actual outcome repeatedly, and then to implement an error propagation task to learning parameters. The accuracy of CNN networks is heavily dependent on the layer configuration. Therefore, numerous researches focused on developing various designs to attain the required results.

CNN's convolutional layer models merge the input image with a kernel smaller in size in comparison to the actual input image to be convolved. Initial convolutional layers are responsible for features extraction from the local areas whereas, classification task is performed by the last convolution layers of the model. Batch normalization layers follow convolutional layers. After this, to learn various complex representations, a non-linear ReLU function is introduced. Pooling layer reduces the number of parameters while preserving the most important features. The next step is to vector the extracted features into a single dimensional feature vector, where a drop-out layer is added. Dropout, a strategy often employed to avoid a network from being over fitted by dropping out the neurons randomly and also resulting in decreased computational problems. To identify spatial complex relationships among the local extracted features by the convolutional layers, finally a fully connected layer is added in the model. Undoubtedly, CNN model allows for automated low- to high-level complex feature extraction; trained from scratch with certain drawbacks. As it involves a huge volume of labeled training datasets, which might be an issue in the medical profession since clinicians annotate the data, and it could also be highly expensive in terms of processing power and time. One promising option to these challenges is to use transfer learning to fine-tune a deep network (especially CNN) instead of training from scratch. This approach reduces the requirement for a lot of data, which prompts researchers to use it to overcome the scarcity training data. In radiology, transfer learning is also being addressed, which gives a detailed explanation and comparison of training from beginning vs. fine-tuning on certain clinical applications. Researchers demonstrate that fine tuning improves training from beginning in the majority of situations. CNNs that have been fine-tuned used to recognize patterns in CT scans, MRI, PET images. All of these techniques demonstrate that using transfer learning throughout the medical imaging area offers significant utility and the chance to attain maximum efficiency in AD diagnosis using smaller training samples than training from scratch [14, 15].

ResNet, abbreviation as Residual Networks, a transfer learning model has become a traditional neural network that serves as the foundation for many visual recognition tasks. In 2015, this model won the ImageNet challenge. The main innovation with ResNet was that it enabled us to effectively train incredibly deep convolutional neural networks over 150 layers. Creating extremely deep learning models became challenging prior to ResNet owing to the difficulty of vanishing gradients.

3 Methods and Materials

3.1 Dataset

In this paper, we acquired the MRI data from Alzheimer's disease Neuroimaging Initiative (ADNI). The description of ADNI investigations is available online at http://adni.loni.usc.edu/wp-content/uploads/howtoapply/ADNIAcknowledgementList.pdf ADNI is a multi-center project presently undertaken in which several types of biomarkers are discovered for early AD recognition. ADNI is now supported through 2021 [16] and has been active since 2004. In each of the two classes: NC and AD, a total of 18 subjects are included in the dataset applied in this research as shown in Table 1. The downloaded dataset was initially in NiFti format (3 dimensional). Each subject had around 3–4 scans in NiFti. We employed an Algorithm 1 to translate a NIfTI to 2D images as it takes a long time and are operationally expensive

Table 1 Dataset splitting into training and testing

Subject type	Total subject count	Training subject count	Number of MRI scans	Extracted images	Testing subject count	Number of scans	Extracted images
NC	18	14	51	3366	4	12	792
AD			33	2178		8	528

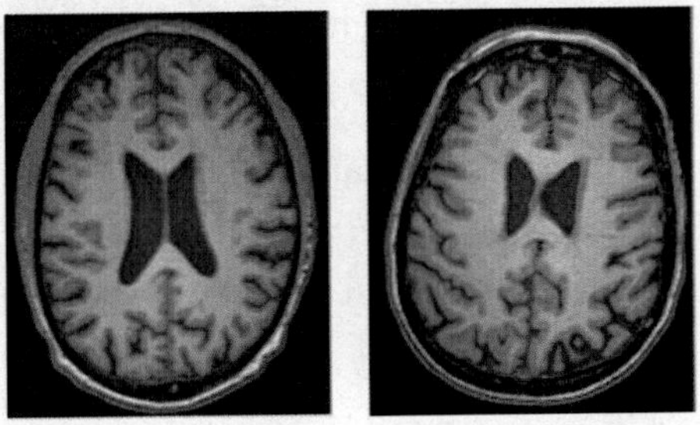

Fig. 1 Processed images after NiFti to PNG conversion (left: NC, right: AD)

to train a 3D CNN on 3D MRI images (NIfTI) [17]. Corresponding to each scan, the number of extracted images were 256. Out of 256 only middle 66 slices were taken into account, rest were discarded keeping in view that they do not contain any meaningful features. In Fig. 1, the processed NC and AD class slices are shown.

Algorithm 1: NIfTI to png Conversion

```
Pseudocode: Nifti to png conversion
Input: 3D MRI scan (NiFti extension)
import nibabel//To read NiFti
import numpy //To modify numpy array
for each 3D MRI scan
        read image ← nibabel.load(scan)
        imageShape ← image.shape()
        x ← imageShape[0] //height
        y ← imageShape[1] //width
        z ← imageShape[2] //depth or no of slices
        for n in range 0 to z
                save cohort_n ← image.save(x, y) //where cohort is either AD or NC
        end for loop
end for loop
Output: 2D images in png extension
```

3.2 *Proposed Model*

DL techniques support solving the issue from beginning to end rather than breaking it down into segments. Because of the unclear activities undertaken by the group of

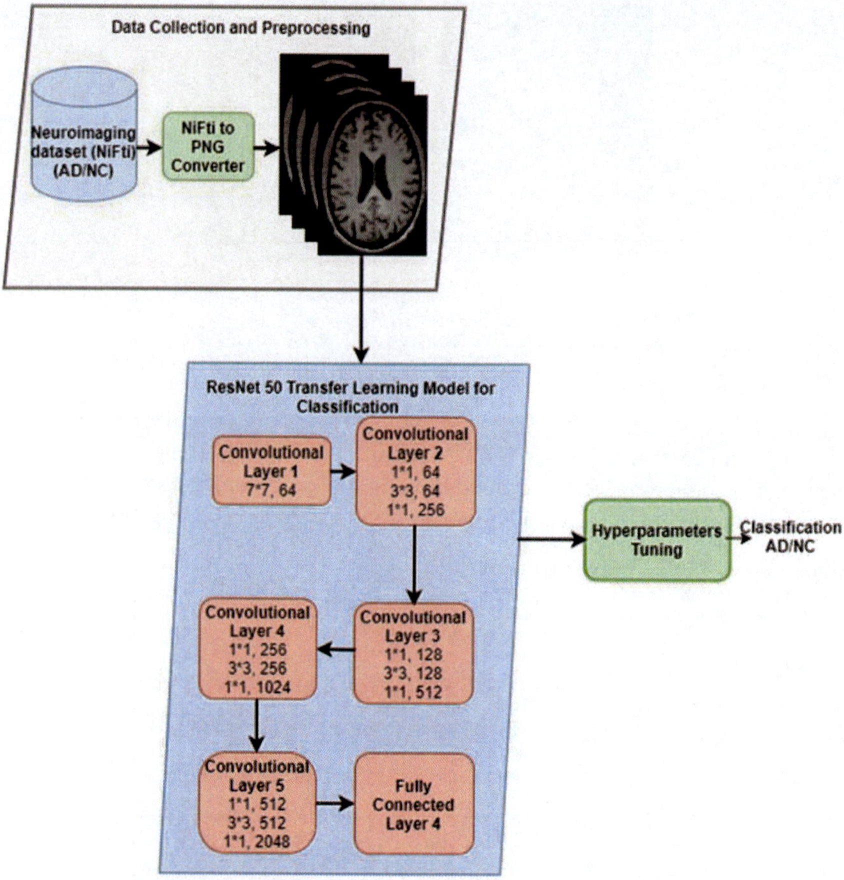

Fig. 2 Proposed model architecture

neurons while behind complex structure of the model, it fails to comprehend the logic behind the outcome achieved. As a result, to increase the model's performance, the following hyper parameters are accessible: a series of optimizers, batch size, number of layers, dropout rate, number of epochs, data augmentation. The Loss Function is an important factor of DL models. Loss is simply a Neural Net prediction failure. And the approach for calculating the loss is known as the Loss Function which is used to compute the gradients. Gradients are often employed to adjust the Neural Network's weights. Optimizers are programs that are used to modify the parameters of a neural network, such as weights and learning rate, and aim to minimize these losses. The batch size defines the total number of samples that must be processed prior to the model's weights and biases being calculated. In this research, various optimizers and batch size are taken into consideration.

Figure 2 depicts the workflow of the proposed method based on the ResNet 50 transfer learning model, which occurs in three phases: data collection and preprocessing, features extraction and classification using ResNet 50, and then hyper parameter tuning to improve the performance for AD classifiers.

4 Results and Discussion

Researchers need to somehow change the hyper parameters for certain system layers to make the model optimally run throughout classification processes. First, by analyzing estimated output to the actual outcome, this is essential to be aware of how well model performance is. Cross-entropy is one of the key performance indicators. Cross-entropy is always a positive continuous function, which is zero when the predicted result of the model coincides with the desired output. The aim of optimization is therefore to decrease cross-entropy as near as feasible to nil. There are several optimizers which may be employed during models such as Stochastic Descent Gradient (SGD), Adamax, Adam, RmsProp. Another hyper-parameter batch size can also substantially influence the model's learning performance. The algorithms used for various optimizers are varied in terms of their learning rates [18]. The batch size dictates the volume of samples disseminated all across the networks.

The effect of optimizer algorithms with batch size 16 with the remaining parameters such as learning rate, epochs fixed on ResNet-50 is shown in Fig. 3a–d. All the optimizers corresponding to batch size 16 and 32 are analyzed based on the performance of the network such that the training process is guided by a number of hyper-parameters. Because of the exorbitant computing expense of exploring the parameter, there has also been surprisingly limited study on design options for hyper-parameter quantities and loss-functions. The results of the experiments indicate that hyper-parameter choices have a big impact on training. We selected significant hyper-parameters to investigate further using multi-objective analysis. Since the various metrics have been used as the performance measurements, we set the parameter 'metrics' for evaluation as name = 'accuracy', name = 'precision', name = 'recall' name = 'auc' and fi_score in the compilation step. Table 2 shows the results generated from several optimizers (SGD, Adam, rmsProp, Adamax) while preserving the batch size fixed at 16. The bold text represents the best results collected inside the same table using different optimizers with fixed batch size (either 16 or 32). The text in bold italic represents the best results achieved from the two distinct tables Tables 2 and 3 with different optimizers varied with batch size as 16 and 32. The ResNet 50 model performance for batch size 32 is depicted in Fig. 4a–d.

Overall, the collected data showed that rmsProp with batch size = 16 was the most accurate with highest accuracy as 75.49%, whereas Adamax optimizers with the batch size 32 achieved maximum auc (82.69%) and precision (77.65%). In contrast with the SGD and Adam Optimizers, rmsProp and Adamax optimizers showed higher results for all optimizers in batch sizes 16 and 32.

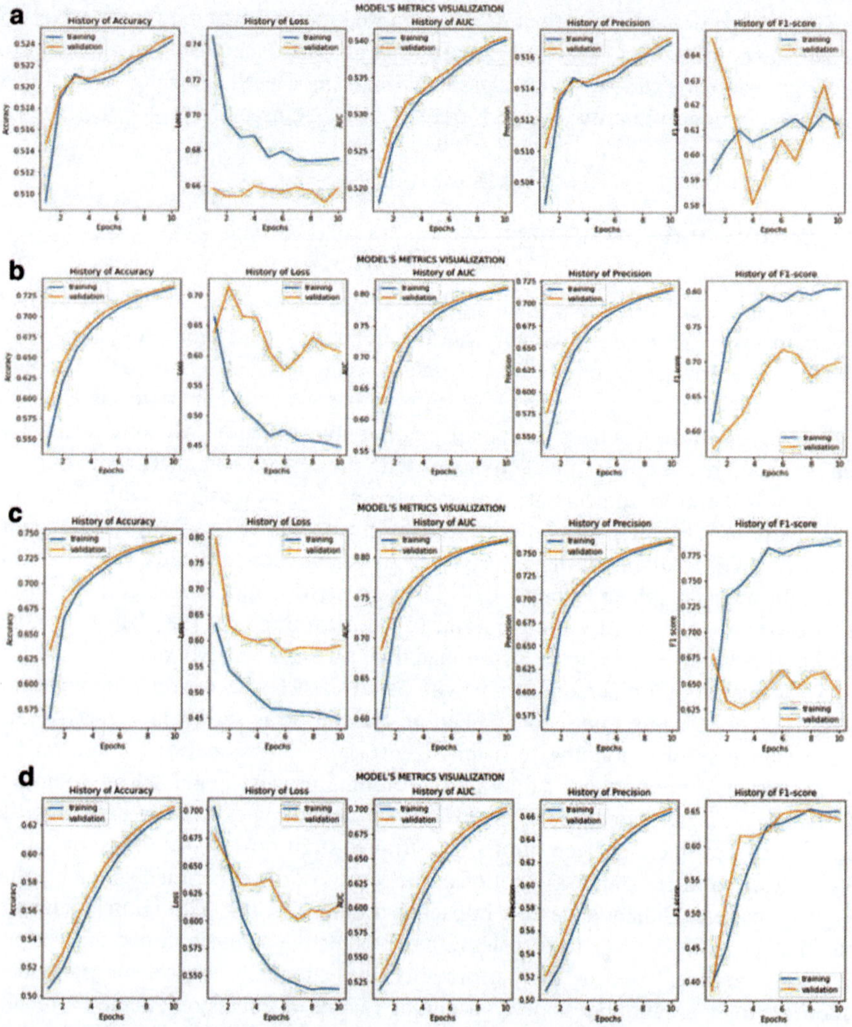

Fig. 3 **a** Graphical visualization performance results ResNet-50 with batch size 16 (optimizer: SGD), **b** graphical visualization performance results ResNet-50 (optimizer: Adam), **c** graphical visualization performance results ResNet-50 (optimizer: rmsProp), **d** graphical visualization performance results ResNet-50 (optimizer: Adamax)

Table 2 Alzheimer's disease classification evaluation results using batch size 16

Optimizers→		SGD	Adam	rmsProp	Adamax
Performance metrics	Accuracy	52.47	71.38	75.49	63.26
	Precision	51.71	69.67	75.52	66.66
	Recall	74.30	75.72	75.43	53.04
	AUC	54.05	78.61	82.38	70.12
	*F*1-score	60.67	71.70	71.98	63.79

Table 3 Alzheimer's disease classification evaluation results using batch size 32

Optimizers→		SGD	Adam	rmsProp	Adamax
Performance metrics	Accuracy	50.63	72.77	74.22	74.80
	Precision	51.11	75.89	75.47	77.65
	Recall	28.74	66.73	71.78	69.64
	AUC	51.24	80.17	81.21	82.69
	*F*1-score	42.04	66.81	67.80	67.04

5 Conclusion

AD is the most severe neurological disease that affects several organs of the body, especially the brain, and causes cognitive decline. The neuroimaging data, mostly MRI scans, was thus utilized as the basis for detecting Alzheimer's disease using various deep learning algorithms. In this work, we developed a hyper-tuned ResNet 50 model to perform binary classification between normal and AD individuals. Our suggested architecture was tested on several optimizers with varying batch sizes in order to obtain a substantial improvement in classification accuracy. In this study, the learning from the ImageNet dataset successfully transferred into the given ADNI MRI data set to classify AD. ImageNet dataset comprises millions of natural images, whereas the ADNI collection includes a handful of AD patient MRI scans. ResNet 50 on rmsProp optimizers with batch size 16 performed better than batch size 32, with a substantial relative improvement of 1.01 percent. It is possible to conclude that reducing the batch size increases model accuracy. Furthermore, the generated outcomes from the various models show the potential of future study utilization to test the influence of optimizers and batch size instead of random selection of these hyper parameters. However, this technique requires more development, as other prior studies have promised great accuracy in comparison to our proposed model. Other hyper parameters, such as learning rate and number of epochs, can also be adjusted.

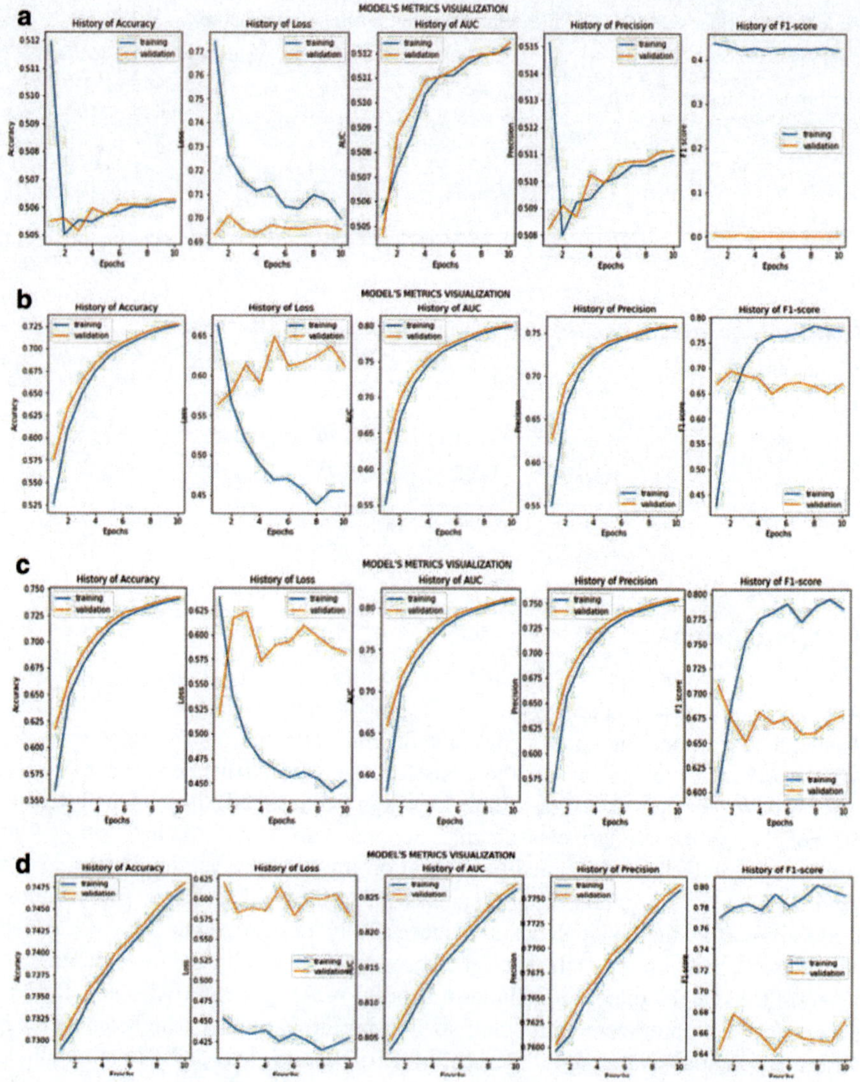

Fig. 4 **a** Graphical visualization performance results ResNet-50 with batch size 32 (optimizer: SGD), **b** graphical visualization performance results ResNet-50 (optimizer: Adam), **c** graphical visualization performance results ResNet-50 (optimizer: rmsProp), **d** graphical visualization performance results ResNet-50 (optimizer: Adamax)

References

1. Ajagbe SA, Amuda KA, Oladipupo MA, AFE OF, Okesola KI (2021) Multi-classification of Alzheimer disease on magnetic resonance images (MRI) using deep convolutional neural network (DCNN) approaches. Int J Adv Comput Res 11–53

2. Choi BK, Madusanka N, Choi HK, So JH, Kim CH, Park HG, Bhattacharjee S, Prakash D (2020) Convolutional neural network-based MR image analysis for Alzheimer's disease classification. Curr Med Imag 16(1):27–35
3. Ebrahimi A, Luo S, Chiong R (2020) Introducing transfer learning to 3D ResNet-18 for Alzheimer's disease detection on MRI images. In: 2020 35th international conference on image and vision computing New Zealand (IVCNZ). IEEE, pp 1–6
4. Association A (2018) 2018 Alzheimer's disease facts and figures. Alzheimers Dement 14(3):367–429
5. Punjabi A, Martersteck A, Wang Y, Parrish TB, Katsaggelos AK, Alzheimer's Disease Neuroimaging Initiative (2019) Neuroimaging modality fusion in Alzheimer's classification using convolutional neural networks. PloS One 14(12):e0225759
6. Ahmed S, Choi KY, Lee JJ, Kim BC, Kwon GR, Lee KH, Jung HY (2019) Ensembles of patch-based classifiers for diagnosis of Alzheimer diseases. IEEE Access 7:73373–73383
7. Jabason E, Ahmad MO, Swamy MNS (2019) Classification of Alzheimer's disease from MRI data using an ensemble of hybrid deep convolutional neural networks. In: 2019 IEEE 62nd international midwest symposium on circuits and systems (MWSCAS). IEEE, pp 481–484
8. Amin-Naji M, Mahdavinataj H, Aghagolzadeh A (2019) Alzheimer's disease diagnosis from structural MRI using Siamese convolutional neural network. In: 2019 4th international conference on pattern recognition and image analysis (IPRIA). IEEE, p. 75–79
9. Taheri Gorji H, Kaabouch N (2019) A deep learning approach for diagnosis of mild cognitive impairment based on MRI images. Brain Sci 9(9):217
10. Basaia S, Agosta F, Wagner L, Canu E, Magnani G, Santangelo R, Filippi M, Alzheimer's disease neuroimaging initiative (2019) Automated classification of Alzheimer's disease and mild cognitive impairment using a single MRI and deep neural networks. NeuroImage Clin 21:101645
11. Ebrahimi A, Luo S, Chiong R (2020) Introducing transfer leaming to 3D ResNet-18 for Alzheimer's disease detection on MRI images. In 2020 35th international conference on image and vision computing New Zealand (IVCNZ). IEEE, pp 1–6. https://doi.org/10.1109/IVCNZ51579.2020.9290616
12. Islam J, Zhang Y (2017) A novel deep learning based multi-class classification method for Alzheimer's disease detection using brain MRI data. In: International conference on brain informatics. Springer, Cham, pp 213–222
13. Wang S, Shen Y, Chen W, Xiao T, Hu J (2017) Automatic recognition of mild cognitive impairment from MRI images using expedited convolutional neural networks. In: International conference on artificial neural networks. Springer, Cham, pp 373–380
14. Hon M, Khan NM (2017) Towards Alzheimer's disease classification through transfer learning. In 2017 IEEE international conference on bioinformatics and biomedicine (BIBM). IEEE, pp 1166–1169
15. Amini M, Pedram M, Moradi A, Ouchani M (2021) Diagnosis of Alzheimer's disease severity with FMRI images using robust multitask feature extraction method and convolutional neural network (CNN). Comput Math Methods Med 2021
16. Weiner MW, Veitch DP, Aisen PS, Beckett LA, Cairns NJ, Green RC, Harvey D, Jack CR Jr, Jagust W, Morris JC, Petersen RC (2017) The Alzheimer's Disease neuroimaging initiative 3: continued innovation for clinical trial improvement. Alzheimers Dement 13(5):561–571
17. Xing X, Liang G, Blanton H, Rafique MU, Wang C, Lin AL, Jacobs N (2020) Dynamic image for 3d mri image alzheimer's disease classification. In: European conference on computer vision. Springer, Cham, pp 355–364
18. Vani S, Rao TM (2019) An experimental approach towards the performance assessment of various optimizers on convolutional neural network. In: 2019 3rd international conference on trends in electronics and informatics (ICOEI). IEEE, pp 331–336

Stability of Feature Selection Algorithms

Dipti Theng and K. K. Bhoyar

Abstract Feature selection is in great demand as it is a crucial phase in classification, clustering, and prediction tasks all around the different sectors. The stability of the feature selection process determines the sensitivity of the selection to dataset variation. In bioinformatics, a task of biomarker identification requires selection of disease associated genes (relevant feature subset), to be insensitive to the variations in the training set, which can give confidence to the domain experts and can prevent result manipulation in the real-world applications. Traditional feature selection strategies are ineffective to capture this high sensitivity to the numerous perturbations. This motivates the need of robust feature selection strategy which produces stable feature set across datasets from multiple domains.

Keywords Machine learning · Feature selection · Stability · Stability measures

1 Introduction

A feature is an attribute or variable used to characterize some property of individual data objects in data mining, machine learning, and data analytics [1]. In high-dimensional datasets, each individual pattern vector can contain tens to hundreds, or even thousands, of features. Recognition tasks in biology, diagnostics tasks, and biomarkers identification in health care and genetic engineering, character, text, and face recognition from digital images, spam email identification, astronomy, and economics are all examples of such data. Feature selection (FS) is a challenging topic in pattern recognition with many potential applications. The number of samples required for every dataset grows exponentially in proportion to the dimensionality of the feature space [1]. Thus, huge data storage and computational resources are required for such datasets. The curse of dimensionality is the term coined to this

D. Theng (✉) · K. K. Bhoyar
Department of Computer Technology, YCCE, Nagpur, India
e-mail: deepti.theng@gmail.com

M. Gupta et al. (eds.), *Artificial Intelligence on Medical Data*,
Lecture Notes in Computational Vision and Biomechanics 37,
https://doi.org/10.1007/978-981-19-0151-5_26

phenomenon [2]. Reducing the dimensionality of the input feature space is a prominent solution to this problem and to employ only those features that are most relevant to the given problem. There are two types of feature reduction:

Feature Selection: It is the process of selecting a minimal number of features from a vast range of features. Because of the smaller feature set, certain techniques are computationally possible to implement.

Feature Extraction: The goal of feature extraction is to create new features that are not always the consequence of feature transformations. For example, the features defined from patterns are generated features.

One successful method for identifying important features for dimensionality reduction is feature selection. It is a key requirement of every data analytics or machine learning task. Informative features are the useful features for describing the underlying objectives. They are vital for producing accurate and easy to explain predictive models and yield good results in various data analytics part [1]. Feature selection is a method for selecting the best subset of features based on a set of criteria. Feature selection is needed:

- To reduce the size of feature set.
- To evaluate the usefulness of features and feature set.
- To improve model performance in terms of predictive power, speed, and model simplicity.
- To assist in model selection by visualizing data.

The stability of the feature selection process determines the sensitivity of the selection to dataset variation. The stability of the results can be determined by comparing them pair by pair. The stability is higher when the similarity is larger. Stability of feature selection methods is an important criterion indicating how sensitive the solution (i.e., selected features) is to agitate in the input training data. The output of the feature selection algorithms can be represented in three ways: indexing, ranking, and weighting. The stability of the feature selection is a relatively ignored concern in work on high-dimensional problems and in general for tasks requiring feature selection. Moreover, the manuscript will concentrate on the stability and general design of feature selection method and on the evaluation methodology applied in feature selection.

1.1 Feature Selection from High-Dimensional Data

Major emerging areas like IoT, AI, and Big Data have tremendously increased data size and dimensionality of the data. As pointed by the curse of dimensionality, for any dataset with the size of the feature space, the number of samples required grows exponentially. It has fetched many challenges for traditional approaches in the area of machine learning and data analytics. This has fetched an attention of researchers to need to work on dimensionality reduction for the high-dimensional data. Another motive behind high-dimensional data future set reduction is to reduce the computational overhead. Processing massive amount of data for high dimensionality (large number of observed features) consumes more resources and time as reported by many literatures [3–8].

1.2 *Need of Stable Feature Selection*

Stability of feature is defined as the selection of most robust feature set which does not change with data or function perturbations. Recent research on feature selection has gained an attention on stability of features a major claim by the bioinformatics and biomedical fields. Feature selection is in great demand as it is a crucial phase in classification, clustering, and prediction tasks all around the different sectors including finance, health care, biomedical, industry, and social media. Such diversified application area for feature selection requires an approach which is independent of the classifier type, feature selection approach, and unbiased to the dataset. A rational of 'No Free Lunch Theorem' no algorithm that can outperform stochastic search on all problems can be applied to feature selection. Goodness of an algorithm for feature selection is defined by the way it address the optimal trade-off between accuracy and stability [9]. Past work presented in [10] has analyzed that the existing algorithms are deficient to evaluate joint performance to stability and accuracy on the dataset at hand. Stable feature selection has been emerged as a strong evidence for disease associated biomarker selection (stable feature selection) to bioinformatics problems of microarray and mass spectrometry datasets [11].

In bioinformatics, a task of biomarker identification requires selection of disease associated genes (relevant feature subset), to be insensitive to the variations in the training set, which can give confidence to the domain experts and can prevent result manipulation in the real-world applications [12]. Traditional feature selection strategies are ineffective to capture this high sensitivity to the numerous perturbations. This motivates the need of robust feature selection strategy which produces stable feature set across datasets from multiple domains.

1.3 *Factors for Instability: Why Instability May Occur?*

Major cause of instability in many feature selection algorithm is that, they are classifier dependent (targeting to achieve highest accuracy) and do not consider stability into account, thus, causing limited reproducibility of the result. Presence of high noise, variance, and outlier samples in the dataset may cause instability in the feature selection. Another major cause of instability in bioinformatics applications is due to high dimensional with low sample size dataset which is more susceptible to the irrelevant and redundant feature inclusion in the feature subset generation [13]. Apart from the insensitivity to the change in training dataset (data perturbation), the stability is also highly dependent on the choice of feature selection approach (i.e., function perturbation) considered for a task in hand. Beside of small sample size, other causes of instability such as type of feature selection algorithm and different partitioning of the training–testing sets have also been proved in [14]. An algorithm may hold one or more causes of instability.

In ensemble selection methods, the choice of ensemble parameters like, aggregation function, ensemble size, method selected in ensemble, etc. may critically affect the feature selection performance and stability of the feature selection [15]. Also, feature selection algorithm stability is significantly affected by the dataset characteristics. Authors in [2] have presented comprehensive analysis on dependency of feature selection stability to dataset characteristics, which take account of feature set dimensionality, sample size, and variation of data distribution across folds. Many research studies claimed that, the feature selection outcome (feature subset) stability has a strong dependency on the size of sample. In [16], authors have discussed that the dataset with large number of redundant features is more prone to the unstable feature selection. Removal of redundant features is necessary in order to obtain stable feature set. Several techniques that gained popularity in improving feature selection stability are: prior feature relevance (domain or expert knowledge), ensemble feature selection, group feature selection, and sample injection.

2 Literature Review

2.1 Stability Measures

In the similarity-based method, Dunne et al. looked at the idea in the context of wrapper-based feature selection, defining stability as the average pairwise similarity between the $M(M-1)$ possible pairs of feature sets. The choice of a similarity measure has a significant impact on the Dunne definition for stability measure and its properties [17]. Regardless of whether feature selection approach was employed, Kalousis et al. were among the first to discuss stability in depth. Authors have not addressed how best to quantify stability [3]. Yu et al. introduced a stability metric based on the Dice similarity measure, which shows that stability grows predictably with the number of features picked, favoring bigger feature sets [4].

Zucknick et al. create a correlation-accounting form of the Jaccard index. It is not possible to compare feature sets of various sizes [5]. Although Shi et al. used the Percentage of Overlapping Genes (POG) index, the measured percentage of overlapping genes (POG) of DEG lists from two research is most likely to be low. It is used only in the case of no redundancy [6]. Kuncheva's stability index does not make use of correlation coefficients of sampling experiments. Instead, all subsampling experiments are combined using a simple accumulation of feature subsets [7].

Lustgarten et al. suggested a modified version of Kuncheva's index that allows feature sets of various sizes to be compared. It is ineffective to satisfy the property of correction-for-chance [8]. Wald et al. projected a stability measure based on similarity-based measure. Constants are not a constraint on Wald's stability measure, as it has been observed [9]. Zhang et al. devise a variant of the POG as normalized POG (nPOG). The stability measurements based on POG and nPOG similarity measures have been found to be unbounded [15]. Goh and Wong formulated a frequency-based stability measure. Goh's measure violates the backward implication. It employs variable stability values that are determined by the number of features selected (k) [18]. Davis et al. penalize frequency in order to account the increase in stability by artificial which tends to occur with gene signatures get longer. Monotonicity is a quality that this metric lacks [19].

Each conceivable feature set of k features is treated as a random variable by Krizek et al., who estimate its Shannon entropy. This measure records lower values corresponding to higher stability [20]. Guzman et al. defined feature sets stability measure of fixed cardinality k. Guzman's stability measure produces weaker stability as well as a smaller average variance, violating the monotonicity requirement [13]. Somol et al. suggested CWrel metric which mandates a stability measure that is constrained by constants. This measure violates the forward implication and thus fails the property of correction-for-chance [21]. Lausser et al. proposed a metric for feature sets of k dimensional. Lausser's measure does not satisfy the property 'fully defined' because they are only defined when the quantity of features chosen is predetermined [22].

The researchers have proposed new metrics for measuring stability, experimental designs for testing the stability of feature selection, and new strategies to increase feature selection stability, as described above. This section has briefly analyzed the gap in present work on the feature selection stability.

Many researchers have studied an effect and applications of the stability on feature selection. Table 2 summarized the important finding by these authors for the stability of feature selection. It has listed important finding by the researchers for the effect of stability on feature selection. From these findings, it is observed that stable feature selection improves the performance and confidence of the feature subset selection for various applications. Major application of the stable feature selection is in the area of bioinformatics where stable feature subset represents the true biomarkers.

3 Stable Feature Selection Algorithms

Most machine learning and data analytics techniques found not effective for high-dimensional data. It has been observed that some approaches perform well for some datasets but not for others. This is mostly due to the dataset characteristics. As a result, studying the features of the dataset is critical for recommending a suitable feature selection algorithm. Review of the work done on stability index is designed considering the characteristics of specific dataset and its applications. No single

Table 1 Stability measures

S. No.	Type of measure	Name	First used in	Measure	[min; max]	Fully defined	Monotonicity	Bounds	Maximum	Correction
1	Similarity-based approach	Hamming	Dunne et al. (2002)	$1 - \frac{\lvert s_i \setminus s_j \rvert + \lvert s_i \setminus s_j \rvert}{d}$	[0; 1]	√	√	√	√	
2		Jaccard	Kalousis et al. (2005)	$\frac{r_{i,j}}{\lvert s_i \cup s_j \rvert}$	[0; 1]	√	√	√	√	
3		Dice	Yu et al. (2008)	$\frac{2r_{i,j}}{\lvert k_i + k_j \rvert}$	[0; 1]	√	√	√	√	
4		Ochiai	Zucknick et al. (2008)	$\frac{r_{i,j}}{\sqrt{k_i k_j}}$	[0; 1]	√	√	√	√	
5		POG	Shi et al. (2006)	$\frac{r_{i,j}}{k_i}$	[0; 1]	√	√	√	√	
6		Consistency	Kuncheva (2007)	$\frac{r_{i,j} - \frac{k^2}{d}}{k - \frac{k^2}{d}}$	[-1; 1]		√	√	√	√
7		Lustgarten	Lustgarten et al. (2009)	$\frac{r_{i,j} - \frac{k_i k_j}{d}}{\min(k_i, k_j) - max(0, k_i + k_j - d)}$	[−1; 1]	√	√	√		√
8		Wald	Wald et al. (2013)	$\frac{r_{i,j} - \frac{k_i k_j}{d}}{\min(k_i, k_j) - \frac{k_i k_j}{d}}$	[1 − d; 1]	√	√			√
9		nPOG	Zhang et al. (2009)	$\frac{r_{i,j} - \frac{k_i k_j}{d}}{k_i - \frac{k_i k_j}{d}}$	[1 − d; 1]	√	√		√	√

(continued)

Table 1 (continued)

S. No.	Type of measure	Name	First used in	Measure	[min; max]	Fully defined	Monotonicity	Bounds	Maximum	Correction
10	Frequency-based approach	Goh	Goh and Wong (2016)	$\hat{\Phi}_{\text{Goh}}(\mathcal{Z}) = \frac{1}{d}\sum_{f=1}^{d} p_f$	[0; 1]	✓		✓		
11		Davis	Davis et al. (2006)	$\hat{\Phi}_{\text{Davis}}(\mathcal{Z}) = \max\left(0, \frac{1}{f}\sum_{f=1}^{d} \hat{p}_f - \alpha \frac{\text{median}(k_1, k_2, \ldots, k_M)}{d}\right)$	[0; 1]	✓		✓		
12		Krizek	Krizek et al. (2007)	$\hat{\Phi}_{\text{Krizek}}(\mathcal{Z}) = -\sum_{s_i \in \mathcal{Z}} \hat{p}(s_i) \log_2 \hat{p}(s_i)$	$\left[0, \log\left(min\left(M, \binom{d}{k}\right)\right)\right]$				✓	
13		Guzman	Guzman et al. (2011)	$\hat{\Phi}(\mathcal{Z}) = 1 - \frac{D_{\text{JS}}(q_1, \ldots, q_M)}{D_{\text{JS}}^{*}(q_1, \ldots, q_M)}$	[0; 1]			✓	✓	✓
14		CWrel	Somol et al. (2010)	$\hat{\Phi}(\mathcal{Z}) = \frac{d\left(M\bar{k} - D + \sum_{f=1}^{d} M\hat{p}_f(M\hat{p}_f - 1)\right) - (M\bar{k})^2 + D^2}{d(H^2 + M(M\bar{k} - H) - D) - (M\bar{k})^2 + D^2}$	[0; 1]	✓	✓	✓		
15		Lausser	Lausser et al. (2013)	$\hat{\Phi}(\mathcal{Z}) = \frac{1}{M^2 k}\sum_{i=1}^{M} i^2 a^{(i)}$	[0; 1]		✓	✓	✓	

measure exists claiming to work on multiple datasets of varying characteristics. It is required to study trade-off between dataset and feature selection algorithm for suitable stability measure selection. Designing such a measure is a challenging task. Furthermore, it has been observed that feature selection methods are not wisely selected for stability analysis. The feature selection methods need to be studied for classifier-independent and classifier-dependent categories.

Table 2 Review of past stability research

S. No.	References	Major area/keywords	Important findings
1	[23]	Feature subset selection, filter, wrapper, hybrid methods, multi-objective optimization, particle swarm optimization	According to research, using a heuristic, evolutionary FSS technique repeatedly fails to yield consistent outcomes. Proposed a new hybrid (wrapper-filter) FSS technique, COMB-PSO-LS that selects the least dependent and most relevant feature subsets by combining particle swarm optimization (PSO) with a local search strategy. COMB-PSO-LS outperforms typical PSO approaches in terms of providing gene subsets that are stable and non-redundant relevant to the classification procedure, according to research
2	[24]	Feature selection, stability selection, sample-specific analysis, random lasso, L1-type regularization	A new technique termed robust sample-specific stability selection is proposed. A sample-specific random lasso based on kernel-based L1-type regularization and weighted random sampling is also proposed. According to major study, the proposed methodology performs sample-specific analysis efficiently and precisely, yielding biologically meaningful gene selection results
3	[25]	Feature selection, ensemble methods, classification, stability, high dimensionality	The reliability of feature selectors is being used to formulate an ensemble feature selection strategy. Its goal is to provide a distinctive and reliable feature selection while also considering forecasting accuracy. According to a major research finding, homogeneous ensembles constructed with unstable base learners are superior to heterogeneous ensembles for enhancing stability because they produce optimal stability results

(continued)

Table 2 (continued)

S. No.	References	Major area/keywords	Important findings
4	[26]	Dimensionality reduction, feature selection, filter method, wrapper method, gene expression data	In the research study, five feature selection techniques: relief, chi-square, information gain, random forest, and recursive feature elimination for SVM (RFE-SVM) are compared on the basis of Kuncheva index stability. It is noted that the performance of any feature selection approach on high-dimensional microarray data is dependent not only on the accuracy of the classifier, but also on the method's stability. In terms of stability, the two filter methods information gain and chi-square method beat other methods, and their stability grows with subset size
5	[27]	Stability, data selection bias, interpretability	Developed a framework for assessing the influence of deliberate stability, as well as trials that were parametric to a variety of pre-processing procedures and classification algorithms. Stability, overfitting, and accuracy are contrasting objectives, which then require a trade-off analysis. According to the findings of the study, a stability impact evaluation should be considered alongside predictive performance analysis
6	[28]	Machine learning, feature selection, stability, fMRI	A unique feature selection approach based on a single-layer neural network that integrates cross-validation in feature selection as well as stability selection via iterative sub-sampling is proposed. A comparative analysis of the proposed approach demonstrated improved classifier accuracy, lower processing costs, and more consistency in the selection of important features. Future study can focus on figuring out how to best optimize neural networks for stable detection of relevant features in a variety of datasets

(continued)

Table 2 (continued)

S. No.	References	Major area/keywords	Important findings
7	[29]	Stacking, stacked, generalization, bagging, bootstrap, algorithmic, stability, generalization	Dag-stacking and bag-stacking, the hypothesis stability of stacking is investigated by the authors. Demonstrated a relationship between weighted bagging and bag-stacking, showing that bag-stacking equals weighted bagging. Research study showed that stacking improves the stability of the stacked learning algorithms by a factor of $1/m$
8	[30]	Feature selection, evolutionary algorithm, multi-objective problem, stability ensemble	The authors presented feature selection wrapper technique based on a multi-objective evolutionary algorithm. In order to assess the stability of the suggested wrapper technique, a novel feature ranking procedure is proposed
9	[31]	Algorithmic stability, hypothesis stability, pointwise hypothesis stability, decision tree, logistic regression, stability measures	For decision trees and logistic regression, derive two ideas of stability: hypothesis and pointwise hypothesis stability. Also, presented a stability measuring framework to measure these notions of stability. The stability of decision trees has been shown to be dependent on the tree's depth, and that it is determined by the cross-entropy loss's smallest eigenvalue of the Hessian matrix. Finding of the study: The learning algorithm logistic regression is not stable because it is dependent on an uncontrollable parameter for its stability
10	[32]	Stability, generalization, stochastic gradient descent, kernel activation functions	The generalization capabilities of non-parametric activation functions, such as kernel activation functions, were investigated (KAFs). KAFs add further parameters to the learning process, allowing nonlinearities to be adjusted per neuron independently. When trained with SGD for a finite number of steps, a NN endowed with KAFs is stable and generalizes well, according to a major research conclusion

(continued)

Table 2 (continued)

S. No.	References	Major area/keywords	Important findings
11	[33]	Feature selection, stability, perturbation, knowledge discovery, robustness instability	An overview of feature selection strategies is given, as well as discussed the issue of instability of the feature selection algorithm. Few solutions for dealing with various sources of instability are presented. In the domains of health care, bioinformatics, image analysis, and business analytics, feature selection strategies and stability analysis were discussed
12	[34]	Feature selection, stability, feature correlations	Presented estimation of feature selection stability, on given correlations in the data or domain knowledge. Proposed a new measure that enables information regarding correlated and/or semantically relevant variables to be incorporated. Defined a new stability measure which satisfies these properties
13	[35]	Feature selection, wrapper method, stability, classification accuracy	Proposed a stability improved feature selection algorithm. The sequential forward selection approach is utilized in research to define the upper bound of the confidence interval for classification accuracy. The results of the experiments revealed that this approach can meet classification accuracy and stability requirements simultaneously

The most often widely used methods are group feature selection and ensemble feature selection (EFS). For relevant feature selection, these methods necessitate a thorough examination of the ideal number of components in ensembles. In the case of ensemble feature selection, a priori ensemble size determination has yet to be addressed; hence, statistical tests are currently utilized to determine the appropriate number of components. Present EFS is computationally expensive for implementation. There is a potential for optimization in the future work.

Existing literatures have addressed stability in presence of feature correlation. A similarity metric is required to analyze the overlap of two feature subsets in order to assess the stability of feature selection strategies. It requires the creation of new ways to help eliminate the redundancy that can occur when aggregating the partial feature subsets obtained from individual feature selectors. It has been noted that, feature subset is not stable with respect to the properties of stability measures. It is necessary to investigate function perturbation on each bootstrap dataset, which will result in M ranked lists for each EFS. Stability of unsupervised feature selection algorithms has not been well studied yet.

The primary key for the improved stability and performance of the solution (i.e., selected features) is a good estimator of the search criterion functional values. A better estimate means that a feature selection algorithm is more likely to converge to its (sub)optimal solution on the given problem with respect to the criterion function and the search strategy [11, 16, 23, 36]. A proper use of the available data can help to improve the reliability and performance of feature selection results to some extent.

3.1 Major Application Areas Demanding Stable Feature Selection

Different application domains obsessive of stable feature selection are: study of biomarkers in genomics, proteomics, metabolomics, and mass spectrometry data; biomedical signal processing, medical imaging; data analysis for text mining, social media, and industrial applications; audio and video analysis, etc. Each one of it demands stable attribute selection (feature selection), which helps in generating consistent results irrespective to the background noise in data. Feature dimensionality reduction in multivariate time series is explored extensively in [23].

High-dimensional disease data to identify microbial biomarkers for disease prediction, e.g., mental illness (e.g., Alzheimer, depression), migraine, inflammatory bowel disease, and type-2 diabetes [22]. Omics data such as genomics, proteomics or metabolomics is available, which provides ample evidence to study and build models for healthcare customization [37]. Study and application of machine learning in genomic microarray dataset have largely been used to determine the risk of getting cancer. Affected genes from the gene expression profile show prominent patterns linked to disease and help in effective personalized diagnosis [38]. Objective of stable feature selection to the cancer prediction is to identify highly stable and most informative features (genes) which are responsible for causing cancer [39, 40]. On cancerous RNA-Seq data, stable feature selection allows for the discovery of strand-specific information, which is a crucial element of gene regulation. The analysis of somatic clonal growth in normal tissues to determine the occurrence and extent of clonal expansion across human tissues is another application of stable feature selection. Application of stable feature selection to build a correlation between the microbiome and disease syndromes can help with personalized medication.

Feature selection has been in great demand for predicting cloud computing resource utilization and intrusion detection. Task of predicting future cloud resource usage is considered a time series analysis task. In this task, resource metrics are termed as feature set which usually includes running job count, memory usage, and cache, time of disk input/output, disk space, and CPU usage. These features are observed as independent but correlated to each other and hold cause-effect relationship [17]. Synthetic datasets are important and widely used in feature selection algorithm evaluation due to their flexibility as they provide precise information of relevant features with respect to the label/output variable [41].

3.2 *Implementation of the Stability Measures*

Implementation of the stability measures for the dataset and algorithm used in experimentation is quite easy as many libraries are defined for popular stability measures. Thanks to the research community making these stability measures packages available for the researchers to use and study them. Most preferred programming languages for implementation of the feature selection stability measures are; Python, MATLAB, and R-programming. These programming languages have rich packages and libraries for machine learning and stability measures, thus, making it very easy for the researchers to use and apply them to their datasets focusing more on the research objectives without worrying about the implementation part.

Some of the most popular packages for stability measures are listed in Table 3. These are the most cited packages and programming languages for the feature selection stability measures (Table 4).

4 Conclusions

An appropriate use of the available data can help to improve the reliability and performance of feature selection results to some extent. From the study presented in this manuscript, it is inferred that the existing algorithms are deficient to evaluate joint performance to stability and accuracy on the dataset at hand. Thus, there is a good future scope to apply stability for feature selection and study its effect on the accuracy. There are very few researches that presented the joint study of stability and accuracy of the feature selection. Major demanding areas for stable feature selection are bioinformatics gene expression (RNA-Seq) data analysis, biomedical image analysis, industrial optimization problems, network traffic classification, and personalized medicine, etc [42–46].

Table 3 Classes for stable feature selection approach

Class	Approach	Subtype	Description
Incorporate stability into algorithm	Ensemble feature selection	Function perturbation	It combines the results of various feature selection techniques to find a stable feature subset
		Data perturbation	Feature selectors are used in the first approach using distinct sample subsets selected from the same dataset applying random sampling, such as bagging and boosting
			The second method employs adverse feature subspaces, such as random subspace
	Prior feature relevance	Prior knowledge of the dataset is being used for the feature selection process in this approach, implying that some features are more significant than others. For instance, transfer learning	
Handling data with highly correlated features	Group feature selection	Knowledge-driven group formation	Methods that use pathway information can be used to identify groupings of related features
		Data-driven group formation	The feature selection process in this approach groups the features into clusters of similar density
Address curse of dimensionality	Sample injection	Transductive learning	In transductive learning, test data is combined with training data for feature selection
		Artificial training sample	On the basis of the available training data distribution, an artificial training sample is generated

Table 4 Implementation setup and programming languages used

S. No.	Refs.	Stability/performance measures	Experimental setup environment
1	3	Executed pairwise Spearman's Rho to calculate stability	Implemented in Python
2	6	Jaccard similarity is used to measure stability	Python, developed Asaph: an open-source toolkit for variant analysis
3	10	Implemented Kuncheva index for stability measure	Used Weka machine learning library
4	15	Implemented Jaccard index, Hamming distance, Kuncheva index	MATLAB, Weka, R, KEEL, RapidMiner, Scikit-learn, Apache Spark MLlib
5	24	Proposed a robust sample-specific stability selection method	Monte Carlo simulations
6	25	Proposed and implemented proper consistency index with pairwise comparisons	Weka machine learning workbench
7	26	Measure stability by Kuncheva index stability measure	R 3.5.1
8	27	Jaccard coefficient, Pearson's correlation coefficient	Python
9	28	SNR of the relevant features, Jaccard index, mean accuracy, standard deviation (STD)	Experiments are carried using Python version 2.7.6
10	43	The 20 stability measures were implemented in an open-source manner: Davis, Dice, Hamming, IntersectionCount, IntersectionGreedy, IntersectionMBM, IntersectionMean Jaccard Kappa, Lustgarten, Nogueira, Novovicova, Ochiai, Phi, Sechidis, Somol, Unadjusted, Wald, Zucknick, and Yu stability index	Implemented 'stabm' package, as an R (R Core Team, 2020) package that allows you to quantify the similarity of two or more feature sets
11	44	Similarity-based measure using the Jaccard index, the cosine similarity, and Lin's method	Defined a package GSimPy in Python. https://github.com/curlyae1995/GSimPy
12	45	Kuncheva index (KI), weighted consistency index (WCI), weighted aggregation: ensemble-weighted mea, Ensemble-weighted stability (top s) selection, ensemble-weighted-exponential, ensemble mean, ensemble stability (top s) selection	Used R: h2o library for LASSO, ElastNet. DNN and XGBoost library for GBM
13	46	Proposed a length adjusted stability stab	Implemented in R-programming, source code: https://github.com/zhxiaokang/EFSIS

(continued)

Table 4 (continued)

S. No.	Refs.	Stability/performance measures	Experimental setup environment
14	47	Proposed new stability measure satisfying the properties defined by the authors for stability measure	Implemented it in R, Python, and MATLAB. Implementation can be accessed at https://github.com/nogueirs/JMLR2018/tree/master/python/

References

1. Dong G, Liu H (2018) Feature engineering for machine learning and data analytics. CRC Press
2. Goswami S, Chakrabarti A, Chakraborty B (2016) A proposal for recommendation of feature selection algorithm based on data set characteristics. J UCS 22(6):760–781
3. Nogueira S, Sechidis K, Brown G (2017) On the use of Spearman's rho to measure the stability of feature rankings. In: Iberian conference on pattern recognition and image analysis. Springer, Cham
4. Chelvan P, Perumal K (2017) A comparative analysis of feature selection stability measures. In: 2017 international conference on trends in electronics and informatics (ICEI). IEEE
5. Chelvan PM, Perumal K (2017) The effects of change in statistical properties of datasets on feature selection stability. In: 2017 international conference on information communication and embedded systems (ICICES). IEEE
6. Nowling RJ, Emrich SJ (2017) Stable feature ranking with logistic regression ensembles. In: 2017 IEEE international conference on bioinformatics and biomedicine (BIBM). IEEE
7. Liu T et al (2017) Algorithmic stability and hypothesis complexity. In: Proceedings of the 34th international conference on machine learning, vol 70. JMLR. org
8. Zomaya AY (2013) Stability of feature selection algorithms and ensemble feature selection methods in bioinformatics. In: Biological knowledge discovery handbook: preprocessing, mining and postprocessing of biological data, vol 23, p 333
9. Brown G et al (2012) Conditional likelihood maximisation: a unifying framework for information theoretic feature selection. J Mach Learn Res 13:27–66
10. Pes B (2017) Feature selection for high-dimensional data: the issue of stability. In: 2017 IEEE 26th international conference on enabling technologies: infrastructure for collaborative enterprises (WETICE). IEEE
11. Ahmed S et al (2014) Multiple feature construction for effective biomarker identification and classification using genetic programming. In: Proceedings of the 2014 annual conference on genetic and evolutionary computation. ACM
12. Khoshgoftaar TM et al (2013) A survey of stability analysis of feature subset selection techniques. In: 2013 IEEE 14th international conference on information reuse & integration (IRI). IEEE
13. Yang P et al (2013) Stability of feature selection algorithms and ensemble feature selection methods in bioinformatics. In: Biological knowledge discovery handbook: preprocessing, mining and postprocessing of biological data. Wiley, Hoboken, New Jersey, pp 333–52
14. Kalousis A, Prados J, Hilario M (2007) Stability of feature selection algorithms: a study on high-dimensional spaces. Knowl Inf Syst 12(1):95{116}
15. Bolón-Canedo V, Alonso-Betanzos A (2019) Ensembles for feature selection: a review and future trends. Inf Fusion 52:1–12
16. Suriyamurthi D (2017) Stability of indexed microarray and text data. Int J Comput Algorithm 06(02):64–68
17. Xu Z et al (2017) An empirical study on the equivalence and stability of feature selection for noisy software defect data. SEKE
18. Alelyani S (2021) Stable bagging feature selection on medical data. J Big Data 8(1):1–18

19. Zhang L (2007) A Method for improving the stability of feature selection algorithm. In: Third international conference on natural computation (ICNC 2007), vol 1. IEEE
20. Kalousis A, Prados J, Hilario M (2005) Stability of feature selection algorithms. In IEEE international conference on data mining (ICDM'05)
21. Kamkar I, Gupta SK, Phung D, Venkatesh S (2015) Stable feature selection with support vector machines. In: Australasian joint conference on artificial intelligence (AI 2015), volume 9457 of LNCS, pages 298{308}
22. Ludwig L, Christoph M, Markus M, Kestler HA (2013) Measuring and visualizing the stability of biomarker selection techniques. Comput Stat 28(1):51{65}
23. Dhrif H et al (2019) A stable hybrid method for feature subset selection using particle swarm optimization with local search. In: Proceedings of the genetic and evolutionary computation conference. ACM
24. Park H et al (2019) Robust sample-specific stability selection with effective error control. J Comput Biol
25. Pes B (2019) Ensemble feature selection for high-dimensional data: a stability analysis across multiple domains. Neural Comput Appl 1–23
26. Tatwani S, Kumar E (2019) Effect of subset size on the stability of feature selection for gene expression data. World Congress on Engineering
27. Guidotti R, Ruggieri S (2018) Assessing the stability of interpretable models. arXiv preprint arXiv:1810.09352
28. Deraeve et al (2018) Fast, accurate, and stable feature selection using neural networks. Neuroinformatics 16(2):253–268
29. Arsov N, Pavlovski M, Kocarev L (2019) Stacking and stability. arXiv preprint arXiv:1901.09134
30. González J et al (2019) A new multi-objective wrapper method for feature selection—accuracy and stability analysis for BCI. Neurocomputing 333:407–418
31. Arsov et al (2019) Stability of decision trees and logistic regression. arXiv preprint arXiv:1903.00816
32. Cirillo M et al (2019) On the stability and generalization of learning with kernel activation functions. IEEE Trans Neural Netw Learn Syst arXiv preprint arXiv:1903.11990
33. Khaire et al (2019) Stability of feature selection algorithm: a review. J King Saud Univ Comput Inf Sci
34. Sechidis K et al (2019) On the stability of feature selection in the presence of feature correlations. In: European conference on "machine learning and principles and practice of knowledge discovery in databases" (ECML/PKDD)
35. Zhang et al (2018) A stability improved feature selection method for classification of ship radiated noise. OCEANS 2018 MTS/IEEE Charleston. IEEE
36. Suriyamurthi D, Velmurugan T (2018) Benchmarking attribute selection techniques for microarray data. ARPN J Eng Appl Sci 13:3740–3748
37. Goh WWB, Wong L (2016) Evaluating feature-selection stability in next generation proteomics. J Bioinform Comput Biol 14(05):1650029
38. Gaudelet T et al (2019) Unveiling new disease, pathway, and gene associations via multi-scale neural networks. arXiv preprint arXiv:1901.10005
39. Haury A-C, Gestraud P, Vert J-P (2011) The influence of feature selection methods on accuracy, stability and interpretability of molecular signatures. PloS One 6(12):e28210
40. Chavez A et al (2019) Identify statistical similarities and differences between the deadliest cancer types through gene expression. arXiv preprint arXiv:1903.07847
41. Loscalzo S, Yu L, Ding C (2009) Consensus group stable feature selection. In: Proceedings of the 15th ACM SIGKDD international conference on knowledge discovery and data mining. ACM
42. Bommert A, Lang M (2021) stabm: stability measures for feature selection. J Open Source Softw 6(59):3010
43. Zhang Y, Cao J (2020) GSimPy: a python package for measuring group similarity. SoftwareX 12:100526

44. Song X et al (2019) Robust clinical marker identification for diabetic kidney disease with ensemble feature selection. J Am Med Inform Assoc 26(3):242–253
45. Zhang X, Jonassen, I (2018) EFSIS: ensemble feature selection integrating stability. arXiv preprint arXiv:1811.07939
46. Nogueira S (2018) Quantifying the stability of feature selection. The University of Manchester (United Kingdom), Diss

Mobile Healthcare Application for Virtual Patient Monitoring System

Suneeta Raykar and Vinayak Shet

Abstract Health care is one of the most significant areas to be focused on and which can be developed through innovative technology. Digitization of the health sector has been on for a long time. The use of smartphones, tablets, and other mobile devices is now used extensively. Communication between patients and their doctors has been made much easier and flexible through available mobile applications (app). So, the barrier of distance between patient and doctor has been eliminated. Mobile apps are convenient in clinical diagnosis and evaluation, timely advice, medicine prescription, and the distant monitoring of patients from various areas. The patient can able to share the medical information about the health condition securely through the developed mobile application. By this means, traveling a long distance and visiting the doctor for a regular checkup may be evaded. This chapter stretches a thorough description of a mobile app developed for the virtual patient monitoring system. A novel solution is proposed to monitor the most important health parameters of a patient using wearable sensors and images obtained using medical equipment. The mobile app supports receiving vital parameters such as body temperature, pulse rate, and oxygen saturation level from the wearable sensor node placed in the vicinity of a patient. Hence, developed application assists the doctor in the early prediction of the disease. It can able to analyze a patient's health status by comparing it with their previous history. Further, the application helps to send alert notifications or short messaging services (SMS) to the patient for emergency physical examination. An elderly patient finds difficulty in visiting clinics for a regular checkup and also finds difficulty in staying in hospital for a longer duration. Thus, proposed and developed mobile application makes it convenient to monitor health status for any patient who are staying with limited access to the medical facility.

Keywords Health care · Mobile applications · Smart health · Virtual patient · Wearable sensor

---

S. Raykar (✉) · V. Shet
Department of E&E Goa college of Engineering, Goa University, Ponda, Goa, India
e-mail: suneeta@gec.ac.in

V. Shet
e-mail: vns@gec.ac.in

M. Gupta et al. (eds.), *Artificial Intelligence on Medical Data*,
Lecture Notes in Computational Vision and Biomechanics 37,
https://doi.org/10.1007/978-981-19-0151-5_27

1 Introduction

Mobile application is gaining popularity among people due to its usefulness and simplicity in their work. The designed application helps to connect the patient to the physician when a patient cannot physically visit the clinic or does not have a good facility to take expert opinion from the nearest doctor. The timely communication with the doctor helps to tackle the problem faced by people who are not able to discuss or share their health conditions with the local doctor. In developed countries, the proportion of elderly people is increasing. Asia as a whole, the proportion of the elderly people is expected to increase from 10.5 to 22.4% during 2012–2050 [1]. One of the biggest problems for elderly patients is forgetting to take their prescription drugs, skipping a regular checkup, or staying in the hospital for a longer duration. It is estimated that only 50% of patients take their medication as prescribed [2]. As per the statistics gathered, 71% of the elderly population reside in rural areas, while 29% is in urban areas of India [3]. People staying in places where accessibility of medical facilities (services) are not available and traveling a long distance for a medical checkup will find this mobile application very convenient. The use of the mobile application in health care assists in terms of early symptom detection and to avail remote assistance from expert doctors. With a mobile application, secure communication between a patient and doctor can be established and obtain an opinion from the doctor about the patient's health report. The parameters from the sensors are shared through the designed application, or the patient can manually enter these parameters using a mobile application. It is a boon to the aging people staying in rural parts of the country. Mobile application in health care serves two major advantages, one is early diagnosis of any disease by sharing vital body parameters, thereby obtaining advice from medical professionals to avoid complications. Secondly, it saves time and cost in traveling to the clinic. The Internet of Things (IoT) can be functional to continuous monitoring of healthcare data which can be analyzed to improve the quality of care. Figure 1 depicts the various benefits of smart healthcare applications. Mobile application helps to tackle problems faced by people, in terms of nonavailability of medical services, afford to pay high cost and time.

Main features incorporated with the mobile app will help to educate, guide, and assist people from expert doctors. Figure 1 represents the various functionality delivered to the patient using a smart healthcare application.

2 Existing Mobile Applications in Healthcare Service

Mobile applications provide a unique opportunity in terms of their ability to allow distant connectivity with the flexibility of function, design, and accessibility [4]. Various mobile applications that are utilized during COVID-19 pandemic situation in healthcare sectors were reviewed in terms of their functionality, strength, and weakness.

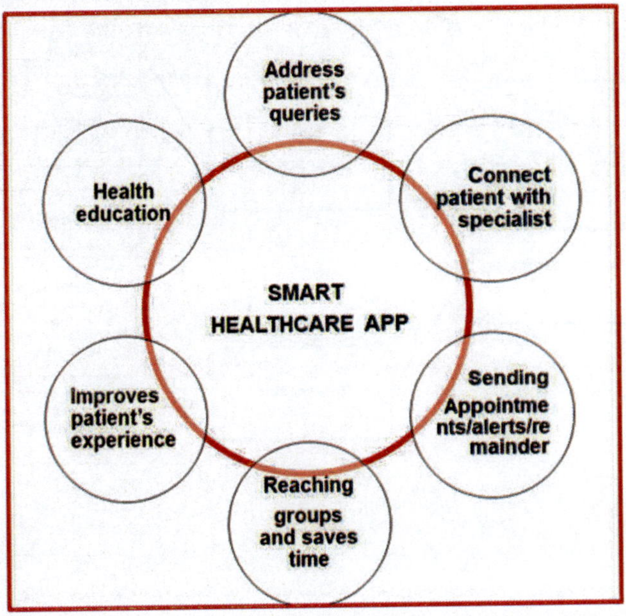

Fig. 1 Applications of smart healthcare applications

Many research papers were published between January 2020 and January 2021 in the area of contribution and remedial measures engaged in the healthcare industry. Digital contact tracing via smartphone apps was established as a new public health intervention in many countries during this timeframe [5]. Such applications are categorized into five broad categories. The mobile applications used to combat and help during the pandemic situation are listed as

1. Contact tracing application.
2. Health monitoring application.
3. Educate application.
4. Regular monitoring application.
5. Alert messengers.

- Contact tracing applications: This application is designed to know if any patient's test reports are diagnosed as corona positive [6]. With this information, nearby people or people who come in contact with the COVID-19 patient can take sufficient care in isolating from the patient or get tested at a very early stage.
- Health monitoring applications Regular health monitoring applications used in health care are classified into two categories, regular checkups and prescription management. Figure 2 represents mobile applications used for regular monitoring and guiding the health status of the patients. Some of the regular checkup applications are Teledoc, Doctor on demand, K-health, ZocDoc, and medical apps for prescription management are Good Rx, Medisafe [7].

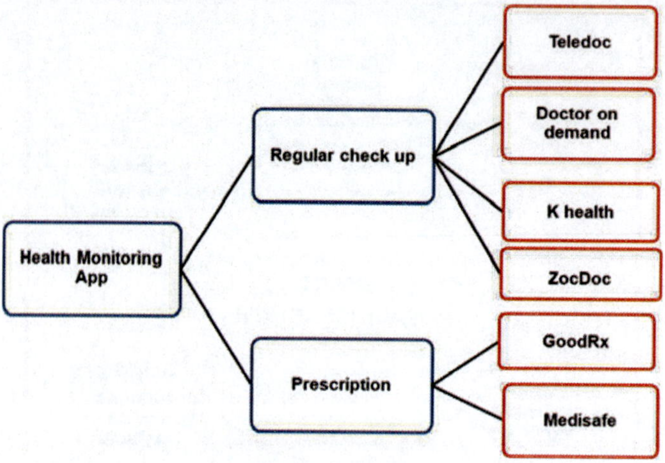

Fig. 2 Classification of mobile applications in health care

- Teledoc is designed for a virtual care system that helps to monitor wellness, acute care to complex health problems. It helps the people and practitioners using questions and answers (chat) facilities.
- Doctors on demand application helps to connect the doctor with the registration. It helps to provide video consultation with the doctor by taking appointments. The patient can get a medical prescription of the medicine online duly signed by the doctor.
- K-health helps to provide medical assistance free of cost. People can chat and find a solution to their health issues by connecting professionals using a virtual mode of communication.
- ZocDoc is used for scheduling appointments with the doctor. This type of app helps in reducing waiting time and avoids crowding in the clinics, thereby allowing proper social distancing.
- Good Rx and Medisafe apps help in prescription management for the patient.

2.1 *Mobile Applications Used in a Pandemic Situation*

The most important mobile app used during the COVID-19 pandemic situation was summarized in Table 1 [8].

- Educate Application: Mobile app is developed toward educating people in terms of health care, general awareness, vaccination, camp details, blood test facility, and other important information. Medicine Plus, MyFitnessPal, Argus, the Centers for Disease Control and Prevention (CDC), and HIV/AIDS drug Database [9].

Table 1 Mobile applications in COVID-19 health care

Name of the app	Country	Salient features
Aarogya Setu	India	App uses Bluetooth and GPS technologies to alert a user when the COVID-19 patient is in closer proximity
Trace together	Singapore	Bluetooth mechanism to trace the presence of COVID-19 patients in their close vicinity by exchanging the normalized ID between two mobiles. It uses the protocol Blue trace
Private Kit Safe Paths	USA	Uses overlapped GPS and Bluetooth trails that allow an individual to check if they have crossed paths with someone who was later diagnosed with positive for the virus
Corona 100m	South Korea	Helps to alert people who are within a distance of the 100 m radius of an infected person
Corona Warn	Germany	Bluetooth-enabled application helps to detect person infected with COVID-19. QR code received by the test report of the laboratory helps to detect the patient status to others
Immuni	Italy	Medical staff sends an alert message to people who came in contact with the patient tested positive
TousAntiCovid	France	Extended version of Stop Covid app uses Bluetooth low energy when a patient tested positive, they will receive a QR code or set of numbers from the test center. A patient can share this with the list of ephemeral IDs of people who interacted with the patient
NHS Covid	England	People to self-isolate if their phone detects they were near someone later determined to have the virus. Bluetooth handshake signals exchange between the two users when came across
Ranking C-19	Iceland	Route tracking is a GPS logger app for Android and iOS

- Regular monitoring Application: VitalHubChart is a regular checkup application and is also integrated with the Apple health kit.
  It helps to seek information about the patient using Electronic Health Records (EHRs), and it can place laboratory (lab) tests and orders and reporting of laboratory results, prescription management, and notifications about critical patient health results. The cost of using the application is high, and it is not affordable for all patients.
- Alert messengers: This type of mobile application enables the patient to receive alert messages during a critical time, or at a specific time to send alert remainders.

2.2 General Healthcare Using the Mobile Application

Some of the existing mobile healthcare applications were reviewed for proposing a virtual patient monitoring.

Most of the app uses two layouts one for healthcare professionals and one for patients. Patients can share images of their laboratory tests, reports on X-rays, MRI scans, CT scan report with doctors and get an opinion about their health status.

Doctors can also recall the patient's health details and prescriptions used earlier for a faster and accurate diagnosis of the illness. This kind of application will not help to collect real-time data of a patient on the mobile terminal. The application is not providing secured video and chat applications for instant messaging.

Smartphone-based Fundus cameras for the diagnosis of retinal diseases are used in diabetes detection. But, they are having a drawback in the quality of video captured. Early detection is not providing satisfactory results due to its low resolution [10].

A Smart Healthcare Monitoring System Using Smartphone Interface illustrates that smartphone in-built sensors are used in tracking the health of a patient. The accuracy and reliability of the smartphone built-in sensor used for tracking health parameters were found not satisfactory [11].

COVID-19 and applications of smartphone technology in the current pandemic help and play a key role in the future of health medicine, patient referral, consultation, ergonomics, and many other extended applications of health care. But, they are facing privacy and data ownership issues [12]. Gamification effect in using health care increases the use of mobile health care by attracting gamification factor. The effects show a slower response for elderly patients [13].

By taking into consideration the factors such as security, user-friendly approach, the resolution, the accuracy of the sensor used, online real-time capture of data a design, development of health care in virtual patient monitoring is proposed.

3 Proposed Layout of Developed Mobile Application

The virtual patient monitoring application is proposed for regularly monitoring the patient without physical interaction with doctors. The most important utilities included in this application are recording parameters through wearable sensors and saving them in the cloud server using Wi-Fi communication, transferring laboratory reports, images, sharing the health parameters recorded using available medical gadgets, sharing the symptoms of the sickness using audio and video format features that are incorporated. The medical professional can view and analyze the reports for obtaining the decision about early detection of disease, notification to the patient for an emergency condition, sharing prescription details, any other specific guidelines to be followed by the patient through a short messaging service and also have online chat option with registered patients.

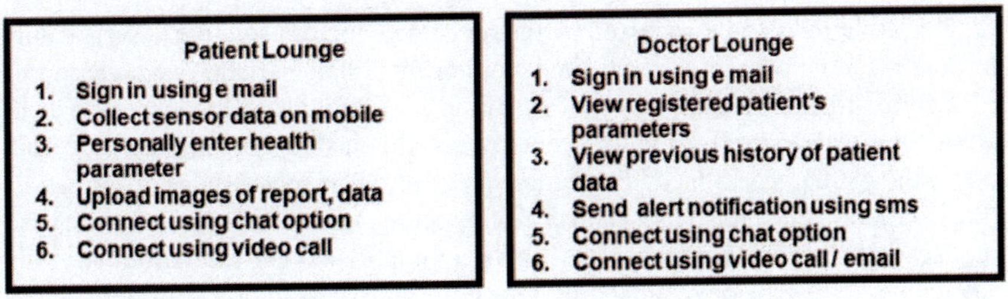

Fig. 3 Proposed layout of mobile application in virtual patient monitoring

The proposed and developed virtual patient monitoring mobile application is represented in Fig. 3.

The layout for virtual patient monitoring application is having two distinct features, the doctor's lounge and patient lounge are the two sections with diverse features.

Algorithm: Virtual patient monitoring application

Result: Virtual patient monitoring features

Step 1: Start the registration process

Step 2: Classify the user as a patient or doctor

Step 3: If the user is a patient

- Upload the wearable sensor information to cloud storage.
- View the information uploaded.
- Upload laboratory reports and readings using the image upload option.
- Add the health parameters manually.
- Connect the registered available doctor using email or mobile number.
- Go to step 8, else if a user is a doctor go to step 5.

Step 5: Find patient information from cloud storage or view real-time data from the wearable sensor unit

Step 6: If present data > Reference value

generate an ALERT notification to the patient using SMS, chat, or email

Step 7: Communicate to patient using chat room facility to advice or alert.

Step 8 End.

The algorithm explains the procedure used in developing the mobile application. The threshold values used in monitoring and alerting the patients are listed as below.

1. Body temperature: 97–98.6 °F is referred to as normal.
2. Heart rate beats per second: 70–80 BPM as normal range.
3. Oxygen saturation level: 95–100% normal value.

Features developed under the patient's lounge will accept the data from patients using sensor nodes or through images captured using an imaging device. The patient can also manually enter all the health details using the option add health parameter. The collected information from the patient is saved in the cloud using a unique patient ID. The cloud storage helps to retrieve the last saved information of the patient. The stored information can be accessed by the registered doctor using the designed app. The doctor can view online or real-time monitoring of patient data. Wearable sensor nodes communicate to the gateway for storing real-time data. Previous records of the patient are also saved for reference purposes. It helps the doctor to compare and analyze the patient's health condition more precisely. This mobile application assists in the interaction between patient and doctor remotely. Wearable sensors that are used to collect real-time data are heart rate (beats per minute), oxygen saturation level, and body temperature. The developed prototype unit is having a microcontroller ESP-32 as a gateway node which is having a built-in Wi-Fi module.

3.1 Features of the Developed Mobile Application

The proposed Android app was developed using the Thunkable app. Firebase was used as cloud storage in the backend for storing users' health information. The mobile application's main feature includes Login, View data, Save data, Sending short messaging service (SMS), Communicate using Chat room/email, and Viewing the history of the patient.

1. Login screen: Mobile applications were authorized to use only by the registered user. All the users (patient/doctor) should sign using a valid email ID in the application. Acknowledgment to registered users will be communicated using email. Figure 4 shows the Login screen used to register the application.
   It illustrates the sign-in option and option for the user to confirm as a patient or doctor. Confirm option allows the doctor to use the available services designed for this application. The cancel option directs toward the facility created for the patient to use the application efficiently.
2. Patient screen: The various features incorporated for the patient under this application are shown in Fig. 5. A patient can acquire the data using a hardware unit attached to the person. Wearable sensors can collect the current information and upload them to the cloud. This information can be accessed using the mobile application.
   A patient can also enter the parameters manually with the option available in the application. Laboratory reports, images from commercial instruments (blood pressure meter, blood sugar meter) can be captured using the designed app and stored in the cloud.
   Figure 6 shows sensor information gathered from the hardware unit and displayed on the design app.
   The display option on the mobile screen helps the users to obtain the sensor values by clicking the "Display value" button.
   The hardware unit captures the data from the patient in online mode. The proposed system was designed to collect three parameters from the hardware unit.
   Figure 6 displays temperature as 98.6 F, heart rate (HR) as 72 beats per minute, and oxygen saturation as 99%, respectively.

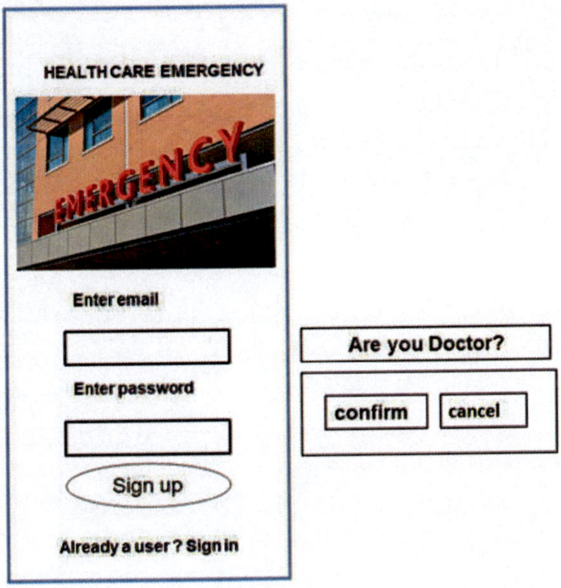

Fig. 4 Login screen to register

3. Doctor/Physician Screen: The designed application facilitates the doctor's lounge to view patient's current information, history, email, send messages to the patient, and connect with the chat room facility. A doctor can communicate to the patient using online or offline mode. Figure 7 represents the doctor lounge indicating the features included for the doctor in assessing patient data monitoring.
   Patient/doctor can chat or connect using chat/email/short messaging services. The application also consists of various other features for the patient and the doctor to interact using short messaging service, chat, and email. The patient and doctor communication details and personal information are maintained secretly by considering security and safety. This ensures to send alert messages and advice to the patient at an early stage thereby reducing complications.

3.2 Various Software Services Used in the Design of Mobile Applications

The preliminary requirement used in the design of healthcare mobile applications is security, authorization, and confidentiality. The following software services are used while designing the application. Thunkable application, Firebase cloud, sheetDB, and Cloudinary services.

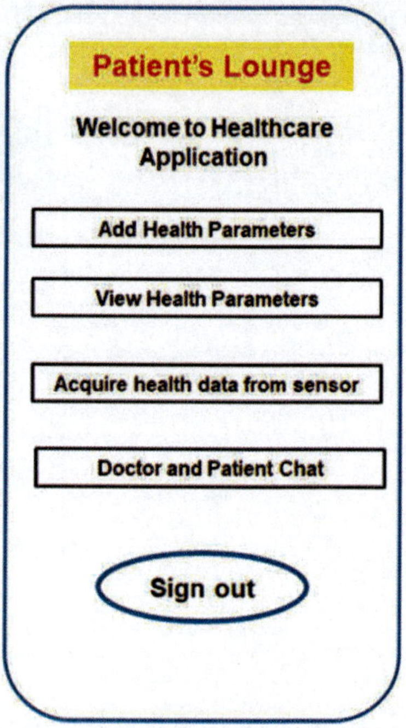

Fig. 5 Patient's lounge

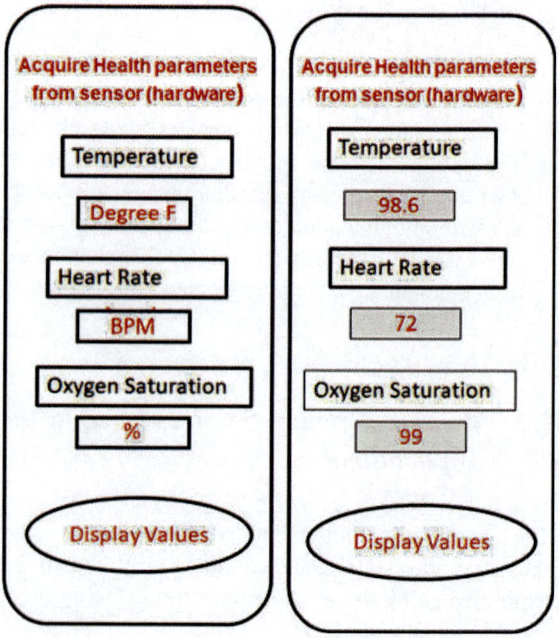

Fig. 6 Mobile app for collecting sensor data

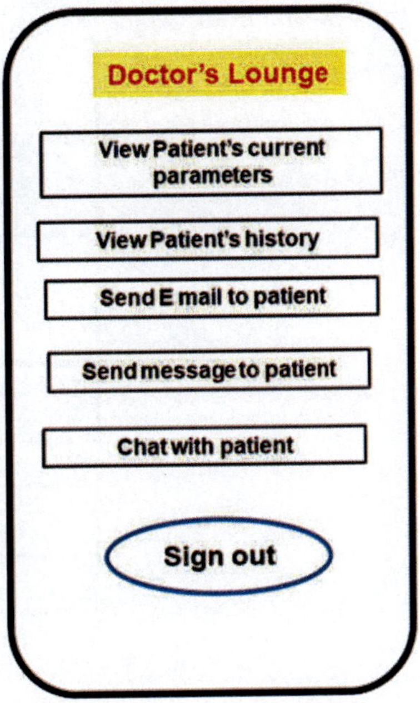

Fig. 7 Doctor's lounge

1. Thunkable Application: Thunkable gives a wide range of ways to create its cross platform apps. Several online available app inventors such as MIT App inventor, Thunkable, AppyBuilder, AppsGeyser, and Infinity Monkey were studied. Thunkable App [14] inventors were selected due to their support to the Android and iOS platform.
2. Firebase: This is a real-time database, and this service was used to create an account and safeguard user's identities. The registered user information was stored under the Firebase cloud. The cloud database provides a security key and APP ID to get access to the contents of the database. Firebase authentication services are used for creating a mobile application using Thunkable applications.
3. Cloudinary: It is used for saving images of patient records for remote access. Services provided by Cloudinary enable to link of the patient information stored in the Firebase cloud. This feature helped the user to upload the laboratory reports and readings from instruments. Certain parameters like blood pressure, blood sugar, laboratory report, X-ray, and images of the other report are uploaded using the designed application. Figure 8 below represents the pictures captured using available instruments before uploading the image into Cloudinary API. Thunkable application code could access the image uploaded in the Cloudinary API.
4. sheetDB: Service provided by Google was used to store patient health parameter history securely for further reference purposes. This helps to create a log file that contains the details of patient email and health parameters. The sample log file generated is represented in Fig. 9.

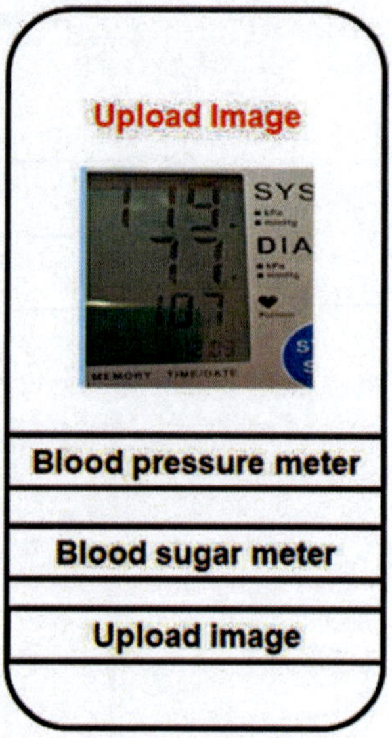

Fig. 8 Uploading patient data using an image

Email ID	Temperature	Heartrate	Spo2
abc@gmail.com	100	73	99
xyz@yahoo.com	99	80	99
ppp@gmail.com	99	72	99
a123@gmail.com	98.7	77	99

Fig. 9 Log entry into sheet DB database

This log file is generated as a backend service to the mobile application. The mobile application utilizes the log file to retrieve the past data stored in the sheet DB. The doctor will be able to co-relate the current parameters with earlier patient history. Therefore, mobile app can display patient history to analyze patient health details in an appropriate way.

4 Results and Discussion

The performance and benefits of the proposed mobile application are compared with the presently available mobile applications. The existing mobile application Health Assistant keeps a record of health parameters like weight, blood pressure, body temperature, and other physical activities. This information is privately accessible by the user not shared with the doctors to obtain an expert opinion. The proposed mobile application assists in sharing the information with healthcare specialists and seeking the right advice.

Healthy Children mobile app is used for scheduling appointments with the pediatrician in a nearby location and obtaining a quick response. The application is having the facility to share current information. It will not able to compare the previous history or statistics collected from the user, whereas the proposed application helps to analyze the data more scientifically by comparing it with the patient history stored in the database.

Freeletics, Sworkit, Nike Training Club, Daily Burn application using sensors are available for fitness tracking; they automatically track walking, running, and cycling activities. These are mainly designed for workout exercises. They help only to educate people.

Noom Walk is a health and fitness app that assists in counting the walking distance of the user and calorie burned, and this application has limited usability as compared to the developed mobile application.

Heart Rate Monitor is an app that allows a user to measure and monitor the heart rate at the actual time and keep a record for later analysis.

The proposed mobile applications are very useful in virtual patient monitoring to record and analyze the health parameters of a registered patient. The scope is incorporated to interface the type of sensor based on the patient's requirement. Patient information is securely shared through the registered application instead of sharing through the WhatsApp application. Chat room facility, SMS, email messages in the designed application help in alerting the patient at an early stage through this application.

5 Conclusion

The proposed application helps to take precautionary measures as per the requirement. The doctor can advise the patient to take sufficient care. These applications find it helpful for rural people to take the assistance of an expert doctor. The application provides help for the early detection of disease. Virtual patient health monitoring application fulfills the requirements of patient and doctor in healthcare monitoring. People could utilize this application for a regular checkup, during the pandemic as well as during normal situations.

The application was uniquely designed to obtain wearable sensor data parameters directly, or patients can manually enter and send the data. It also enables the user to upload the laboratory reports, X-ray images, images captured from the instruments. The communication between the doctor and patient was established using secure communication. More features can be upgraded to the current application according to the end-user/physician demands.

References

1. Rajan SI (2014) Social security for the elderly: experiences from South Asia. Routledge
2. Sabate E, Sabate E (eds) (2003) Adherence to long-term therapies: evidence for action. World Health Organization
3. Nayak S, Blumenfeld NR, Laksanasopin T, Sia SK (2017) Point-of-care diagnostics: recent developments in a connected age. Anal Chem 89(1):102–123
4. Davalbhakta S, Advani S, Kumar S, Agarwal V, Bhoyar S, Fedirko E, Misra DP, Goel A, Gupta L, Agarwal V (2020) A systematic review of smartphone applications available for corona virus disease 2019 (COVID19) and the assessment of their quality using the mobile application rating scale (MARS). J Med Syst 44(9):1–15
5. Budd J, Miller BS, Manning EM, Lampos V, Zhuang M, Edelstein M, Rees G et al (2020) Digital technologies in the public-health response to COVID-19. Nat Med 26(8):1183–1192
6. Yasaka TM, Lehrich BM, Sahyouni R (2020) Peer-to-peer contact tracing: development of a privacy-preserving smartphone app. JMIR mHealth and uHealth 8(4):e18936
7. Banskota S, Healy M, Goldberg EM (2020) 15 smartphone apps for older adults to use while in isolation during the COVID-19 pandemic. Western J Emerg Med 21(3):514
8. Akinbi A, Forshaw M, Blinkhorn V (2021) Contact tracing apps for the COVID-19 pandemic: a systematic literature review of challenges and future directions for neo-liberal societies. Health Inf Sci Syst 9(1):1–15
9. Hicks JL, Althoff T, Kuhar P, Bostjancic B, King AC, Leskovec J, Delp SL (2019) Best practices for analyzing large-scale health data from wearables and smartphone apps. NPJ Digit Med 2(1):1–12
10. Monjur M, Hoque IT, Hashem T, Rakib MA, Kim JE, Ahamed SI (2021) Smartphone based fundus camera for the diagnosis of retinal diseases. Smart Health 19:100177
11. Marques B, McIntosh J, Valera A, Gaddam A (2020) Innovative and assistive ehealth technologies for smart therapeutic and rehabilitation outdoor spaces for the elderly demographic. Multimodal Technol Interact 4(4):76
12. Iyengar K, Upadhyaya GK, Vaishya R, Jain V (2020) COVID-19 and applications of smartphone technology in the current pandemic. Diab Metab Syndr Clin Res Rev 14(5):733–737
13. Lee C, Lee K, Lee D (2017) Mobile healthcare applications and gamification for sustained health maintenance. Sustainability 9(5):772
14. ThunkableX Cross Platform, [online]. Available https://community.thunkable.com/thunkable-cross

Channel-Based Similarity Learning Using 2D Channel-Based Convolutional Neural Network

Ravi Shekhar Tiwari

Abstract Object identification is one of the major aspects of computer vision. In recent years, the development the computing as well as the storage capacity has increased drastically. These breakthroughs in the technology have blessed us with various data storage technologies and computational engines. Because of the breakthrough in recent years, we are generating humongous amounts of data of which 80% of data is unstructured data and only 20% of data is structured. Unstructured data are mainly composed of images, video and as well as the natural language, i.e. text. These 80% unstructured data consist of the enormous information, but it is difficult to unravel the information contained in these data. Convolution neural network (CNN) is backbone of computer vision and deals with extracting information from the image and video, before the invention of recurrent neural network (RNN), CNN was also employed for natural language processing (NLP) task such as classification and text generation, but the specialty of CNN lies where the dataset consists of sound signals, images or sequence of frames. On Internet, we can find 60% of the unstructured dataset consists of images or sequence of image or text. Basically, image consists of the features which is the orientation of the pixels in a well-defined pattern which can be extracted by using kernel's known as the feature maps and Maxpooling layers to extract the underlying feature present in the image to train the neural network. CNN is one of the parts of supervised learning techniques which uses labelled data, but it is difficult to label huge number of images. The similarity-based learning enables us to control the similarity percentage as well as it has minimum labelling procedure, i.e. labelling of the dataset is to be labelled 0 or 1. Similarity learning is used to compute the percentage of the features which are similar in the target image with respect to the input image. Image consists of three channels, i.e. red, green and blue channels, which is basically a 2D vector with pixel values in range of 0–255. These individual channels contribute to the features present in the images, and if we can calculate the similarity between input image and the query image, then we can be able to present

R. S. Tiwari (✉)
AI Engineer, Chadura Tech, Bangalore, Karnataka, India
e-mail: tiwari11.rst@gmail.com
URL: https://www.rstiwari.com

M. Gupta et al. (eds.), *Artificial Intelligence on Medical Data*,
Lecture Notes in Computational Vision and Biomechanics 37,
https://doi.org/10.1007/978-981-19-0151-5_28

the unstructured images in relation to the similarity with respect to the input image by using the channels in channel-based CNN tower.

Keywords Modified softmax · Distance learning · Similarity learning · Channel-based similarity

1 Introduction

Artificial intelligence (AI) is being implemented everywhere industries sectors such as health care, network security, network maintenance, finance and many others [1]. We are surrounded by the gadgets in which AI is being implemented or is implemented from laptops, tablets, smart watch and ATM card transactions; these devices help us to perform our activity seamlessly. Machine learning (ML) is a subset of AI which deals with the limited data and can predict as well as perform supervised as well as unsupervised tasks [2]. Deep learning (DL) is a subset of ML, which is composed artificial neural network (ANN), convolution neural network (CNN), recurrent neural network (RNN) and long short-term memory (LSTM) as a backbone of neural network and is capable of estimating any complex function [2]. The idea behind AI is to mimic the human brain and work accordingly [3]. We have several epitomes of AI, which can outperform humans in certain fields such as Alexa by Amazon, Cortana by Microsoft, Siri by Apple and the pioneer in self-driving cars Tesla and Uber.

Last year, we saw pandemic, i.e. COVID-19 and still it is with us, it had adverse effects on our life, but numerous technological advancements took place, and some of them are online classes, robots delivery system/robot assistant integration of chatbots in almost all applications. But the generation of data increased exponentially, and as a result, we generated 59 zettabytes of data, out of which 80% was unstructured data and 20% of the was structured data [4]. Structured data has a specific schema like forms we fill to open a bank account or for admission in school, college or at any other place in a well-defined way, whereas unstructured data do not have any schema; hence, it is very difficult to unravel the underlying information [5]. This is acting as a major roadblock in artificial intelligence because most of the applications which are currently deployed are supervised learning, i.e. dataset on which the model is trained is labelled. Labelling is a tedious time-consuming task; it requires manpower to label the dataset, and if anything goes wrong, the objective of the model will not be achieved which can lead to a catastrophe [6].

Researchers have invented various techniques to deal with these unstructured image datasets such as clustering techniques K-means clustering, K-++ means clustering, DB scan and several other techniques which are state-of-the-art methods/algorithms for unsupervised data [7]. But these algorithms/methods have defects or are based on some assumption like in K-means we need to specify number of clusters beforehand, and we use elbow method to determine optimal cluster for the given dataset; similarly, K-++ means and DB scan have their own demerits [8].

Recently, network architecture known as Siamese network was invented which is based on the distance metric between the input image and the target image using triple loss, and it was trained on the MNIST dataset. It is being implemented in many multinational organizations and requires comparatively less dataset to train [9, 10].

Unstructured data mostly consist of image, video and a text [11]. Image and videos are a collection of pixel which ranges from 0 to 255. Generally, images are composed of one channel image, i.e. grey or black and white image, where pixel whose numeric value is 0 is considered as zero, whereas pixel with numeric value 255 is considered as white or three channels, i.e. red, green and blue (RGB), where each channel has its own pixel value which has a range of 0 to 255 [12]. Pixels are the smallest part of the image, and these pixels combine to form a feature which is extracted by the filters present in the CNN layer [13]. This features the combination of the pixels when passed through the convolutional layer, i.e. a series of filters that give output to a set of the matrix which is composed of features such as horizontal lines, vertical lines and many other features which is the building block of the image [14].

The contributions of the paper are as follows:

1. The image consists of channel, i.e. RGB and feed into CNN, where the feature maps are extracted through channel-based CNN tower.
2. The distance is computed from each channel-based CNN tower, and the similarity is aggregated which denotes the similarity percentage of the feature map between the images.

The rest of this paper is organized as follows: in Sect. 2, we have discussed the related works. Channel-based similarity learning using 2D convolution neural network based on feature map is discussed in detail in Sect. 3. Section 4 depicts the results obtained and the discussions that are followed. Section 5 concludes the work.

2 Related Works

In the field of AI, the problem related to the unstructured data is existing from very beginning [15]. But in recent times with the increase in the computational capacity as well as the storage capacity, the condition became worse because humans were generating huge amount of unstructured data by employing machines [4]. There are several algorithms which were invented to counter this problem and were not able to eradicate the problems related to the unstructured but paved the path for the other scientists/researchers to innovate these algorithms [16].

Over the most recent couple of many years, the view of likeness got a developing consideration from mental specialists and, more as of late, how to get familiar with a likeness measure has pulled in additionally the AI people group. A few approaches are proposed to figure iteratively the comparability measure, tackling an arched advancement issue, utilizing a little arrangement of sets to characterize the difficult limitations different procedures such as EM-like calculations, hidden Markov random fields and compelled portion mappings [17]. The existing methodologies are by and large rigorously identified with semi-directed bunching. Furthermore, the presence of some class names or pairwise limitations on a subset of information is misused to improve the bunching cycle [18].

Clustering is the process of associating the data points in their respective groups based on their similarity with other data points. The similarity is calculated using distance matrix. Distance and similitude are the reason for developing bunching calculations [19]. Concerning quantitative information highlights, distance is liked to perceive the relationship among information [20].

There are various types of the clustering algorithm which associate data with group of data; some of them are K-means, K-medoid's, K++ means, DB scan, agglomerative clustering and many more, but they all have various disadvantages [21] such as we have to predefine number of clusters, time complexity is high, some are sensitives to outliers and easily drawn to local minima, clustering is sensitive to the initial parameter values, and if the density of the data is not even, then low quality of the cluster is obtained [22–24].

A Siamese neural organization (also called a twin neural organization) is a counterfeit neural organization that utilizes similar loads while working pair on two diverse info vectors to figure equivalent yields vectors [25]. Often one of the yield vectors is precomputed, accordingly shaping a benchmark against which the other yield vector is looked at. This is like contrasting fingerprints however can be depicted more in fact as a distance work for area delicate hashing. It is feasible to fabricate an engineering that is practically like a Siamese organization yet executes a somewhat unique capacity. This is normally utilized for contrasting comparable occurrences in various sort sets. By and large, most of the profound learning techniques are designed in a Siamese style [26–29]. Li et al. plan a novella blending neural organization (FRNN) to coordinate with patches in different horizontal stripes [30] and afterwards accomplished astonishing advancement by improving the Siamese model [26, 28, 29]. Ahmed et al. use deduction to compute cross-input neighbourhoods' differences and sum up the differences by a fix rundown including layer. In view of a novel inexact coordinating with strategy, a more extensive area search technique is employed by Subramaniam et al. [28]. Varior et al. [29] propose a gating capacity to specifically accentuate new neighbourhoods' designs. All the three techniques have a conspicuous hindrance: during test time, the question needs to combine with every exhibition picture, which is time incident particularly in huge datasets. Another sort of the Siamese model is removing worldwide CNN including vectors (e.g. from the last completely associated layer) as highlight portrayals and optimize pairwise or trio misfortune [30, 31]. This kind of model is not important to take picture sets as contribution during testing, but it infrequently considers closeness examination between picture sets at low level.

3 Similarity Learning Using 2D CNN-Based on Feature Map

3.1 Datasets

To train this network, we have used two different datasets which consist of ten classes, respectively. Cifar-10 is an open-source dataset consisting of 60,000 images with ten classes, where each class has 6000 images of aeroplane, automobile, bird, cat, deer, dog, frog, horse, ship and truck [32]. The Cifar-10 dataset is divided into two sets, where the training set consists of 50,000 images and the test set consists of 10,000 images. Further, this dataset is divided into five batches—each batch composed of 10,000 images randomly selected from the dataset. Animal-10 is an open-source dataset consisting of 28,000 images with ten classes, where each class has images between 2000 and 5000 of dog, horse, elephant, butterfly, chicken, cat, cow, sheep, squirrel, and spider. Animal-10 dataset is collected using Google images under human supervision [33]. Every image of this dataset is in an RGB channel with shape of (32, 32, 3), where 32 represents rows and columns of the image and 3 represents RGB channel of the image, respectively. Pixel which is the building block of these channels has intensity between 0 and 255 (inclusively). The combination of these pixels gives rise to the feature in the respective channel which results in a specific feature map [32, 33] (Figs. 1 and 2).

Fig. 1 Cifar-10 dataset

Fig. 2 Animal-10 dataset

3.2 *Pre-processing of Images and Labelling*

In image channels typically, image consists of three channels—red, green and blue channel stacked together—where each pixel has intensity between 0 and 255(inclusively) or greyscale image with the pixel intensity between 0 and 255 (inclusively) but consists of only one channel, where each pixel has intensity between 0 and 255(inclusively) [34, 35]. If we consider greyscale images, then there are only 255 ways we can create the pixel intensity, i.e. 255 shade of grey that is why it is called greyscale image. If we analyse RGB image, it consists of three channels, and then each pixel in any one of these channels can be arranged in 255, so for all three channels we can arrange in 255*255* 255 combinations of colours known as 16 million colours [36]. These pixels are the building blocks of any image, combination of pixels in a specific pattern is called features of the images, and these features are very crucial for CNN to extract the underlying information of the image's dataset. The below image shows the intensity distribution of the colour channel present in the images [36, 37].

Figures 3 and 4 depict the range of each pixel value of the channels present in greyscale or RGB image, and by analysing the above-mentioned figures, we can mathematically compute the following information.

Similarly, if we deduce for images which consist of the same properties but instead of one channel, it is composed of three channels, where each channel, i.e. red, green and blue can have value in the range of 0–255 (inclusively), so mathematically we can calculate:

$$\text{pixel can have intensity} = 255 \quad (1)$$

$$\text{no. of channels in RGB image} = 3 \quad (2)$$

Fig. 3 Greyscale image

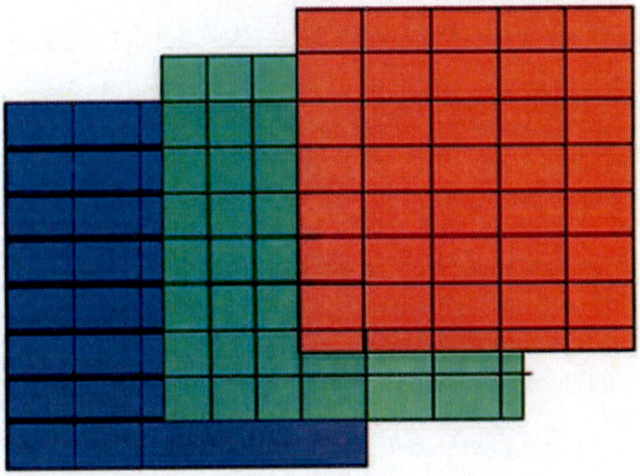

Fig. 4 RGB image

$$\text{no. of ways pixels can have intensity} = 16\,\text{millions} \tag{3}$$

Each channel of the image has specific features like some is more prone to noise as compared to other channels. Most importantly, pixel intensity is the building block of the image, and features emerge in the images if these pixels with definite quantitative value combine in a specific pattern [37]. Channels play a very crucial role when it comes to extracting the feature map using convolution layers. Convolution layer consists of filters which is just a matrix consisting of integers also known as weights. Forward propagation helps to calculate the loss propagated backward to adjust the weight of the filters present in the convolution layer of CNN to extract the underlying feature of the image [35, 37]. These underlying features are collection of the pixels present in all the channels in which image is composed of, and if we separate these channels, then we can be able to train the CNN to compute the similarity between these channels, and finally by aggregating it and calculating the sigmoid, we can compute the similarity percentage of the target image with respect to the query image.

3.3 Annotating the Dataset

In this paper, we have used channel-based CNN tower so images which were similar were labelled as 0 to annotate high similarity and images which are not similar, and they were labelled 1 to annotate high similarity to neural network tower. This paper removes the concept of labelling the dataset because it is based on the similarity in the target images with respect to the input image.

The dataset was labelled in such a way that the images which are similar were labelled as 0 and images which are dissimilar were labelled as 1 randomly, but it was ensured that image was paired with the subset of the images belonging to the classes which is defined by the numeric value of X.

3.4 Channel Extraction and Label Mapping

In section Image Channel, we have discussed the benefits of the channels present in the image; also in section Annotating the Dataset, we have annotated our dataset in such a way that the labelling procedure for unstructured image data is reduced to 90%, where labels consist of 0—dissimilar image and 1—stands for similar image. In this section, we will separate the channel and create new dataset from existing dataset by separating channel from the images.

After pre-processing, the input and target image of shape [32, 32, 3] will be separated into three consecutive images based on the channel of the respective image of size [32, 32, 1] and will be mapped to the respective label. Figure 5 summarizes the channel extracted from the RGB image.

Fig. 5 Channel from dataset

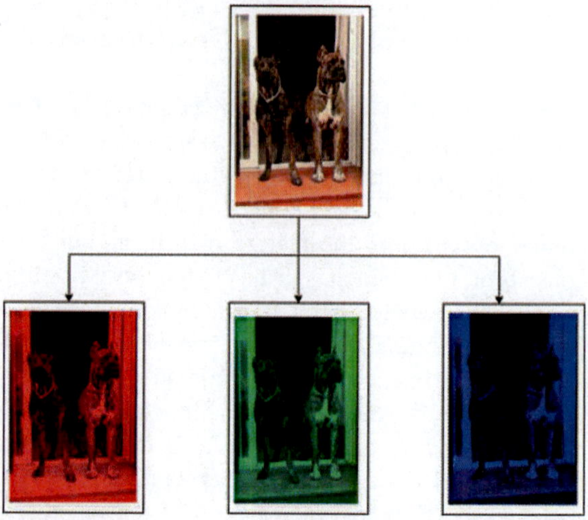

Table 1 Train dataset

Red channel * 2	Green channel * 2	Blue channel * 2	Labels
90,000,32,32,1	90,000,32,32,1	90,000,32,32,1	90,000,1
18,000,32,32,1	18,000,32,32,1	18,000,32,32,1	18,000,1

Table 2 Test dataset

Red channel * 2	Green channel * 2	Blue channel * 2	Labels
18,000,32,32,1	18,000,32,32,1	8000,32,32,1	18,000,1
5000,32,32,1	5000,32,32,1	5000,32,32,1	5000,32,32,1

After pre-processing and annotating the images, the dataset was organized in following way (Tables 1 and 2).

In the above table, first row is for Cifar-10 dataset and second is for Animal-10 dataset, respectively. The representation *X*, *Y*, *Z* and *C* denote number of samples, rows, columns, and channels, respectively, whereas *X* and *Y* denote number of samples and column, respectively.

3.5 Neural Network Architecture

In the above sections, we have discussed the importance of channels in the image. In image, it is a very high probability that one channel can have specific feature and another channel does not have the same feature because of the difference in the pixel intensity and combination of the pixel values leads to the formation of the features in the images. If we analyse image as a whole, i.e. combining all three channels red, green and blue by stacking them, there is a huge probability that one channel can overpower the intensity of that respective feature—group of pixel as a result, it can result in neglecting the feature as a whole. To counter this problem of neglecting the feature as a whole, we have developed similarity-based learning neural network based on channel-based CNN towers which take these separate channels as input and compute the similarity between these respective channels, and finally, it computes the softmax between these similarity indexes of each channel with the aggregated similarity index to compute the similarity in the target image with respect to the input image. The below image summarizes the neural network structure which we are going to discuss in this section.

In Fig. 6, we can see that the channel-based CNN tower basically consists of three layers—input layer, functional model and finally lambda layer.

We have discussed that colour image consists of three channels, where pixels value has range of 0–255 (inclusively), and also we have discussed that there is high probability of losing the spatial as well the features which are embedded into the respective channel. In Fig. 7, we have three pairs of input channels of each shape

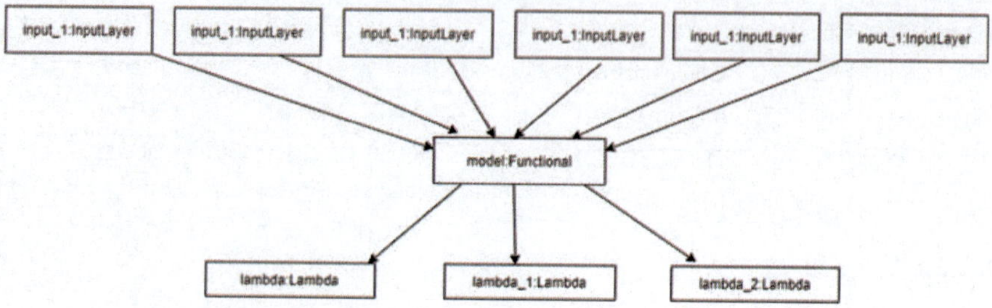

Fig. 6 Abstract view of the channel-based CNN tower

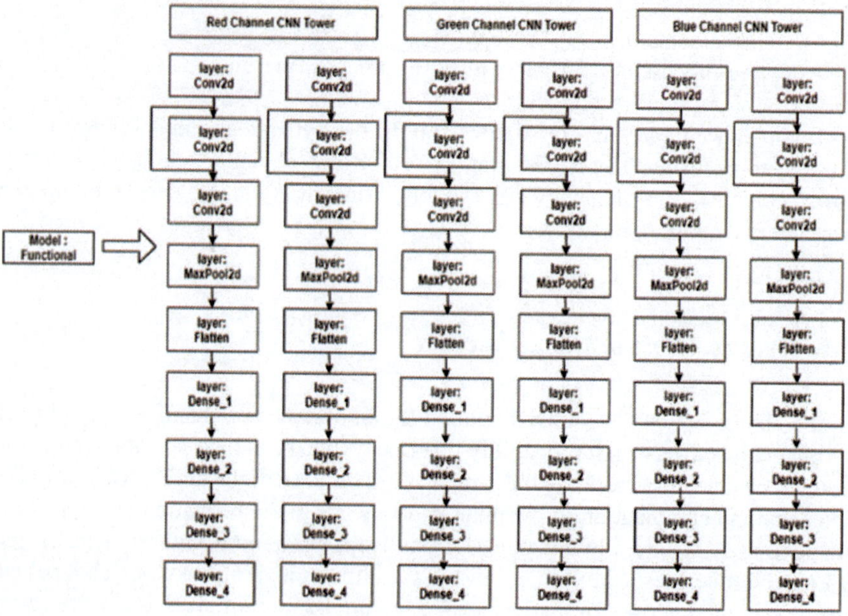

Fig. 7 Functional model

[32, 32, 1]. Pair of input channel is defined as in pair the same channel of input as well as the target image is queried and forward propagated for computing the similarity index between the respective channel like Input_red [32, 32, 1] will be paired with target_red [32, 32, 1], similarly for green and blue channel of the image to compute the similarity between the feature map of unstructured image dataset. Table 3 summarizes the input layer of channel-based CNN tower.

Model functional as shown in Fig. 7 is channel-based CNN tower, where input image and query channel tower have been paired which can be better understood by below image.

The above function model is composed of traditional residual CNN network which comprises traditional convolutional filters, Maxpooling2D layers, flatten layer and

Table 3 Input layer

Input layer name	Colour channel	Input shape
Input_2 and Input_3	Red channel	[32, 32, 1]
Input_4 and Input_5	Green channel	[32, 32, 1]
Input_6 and Input_7	Blue channel	[32, 32, 1]

dense layers. The CNN towers have been paired with respective of their channel, i.e. red, green and blue, from input and query images. It has a dense layer followed by three custom lambda functions which calculate the similarity between the channels.

In this paper, we have used channel-based CNN towers, where CNN towers have been paired according to their colour channel in input and query image. The 2D convolutional layers were incorporated because our dataset consists of images which are 2D matrices. We have mentioned two 2D convolutional layers which are employed to extract the feature maps from the image. A number of the filters which constitute the convolution layers and responsible for extracting the feature maps from the images are 2^9 and 2^8 filters, respectively. Though we have created the second convolutional layer with half the number of filters in the previous layers, it was compensated by concatenating the feature map obtained from the previous convolution layer. We have employed the concept of residual learning with 2D CNN as to make the deep neural network to learn faster [38, 39].

Maxpooling2D layer is employed to down sample the feature map by analysing the feature map produced when convolution filters slide over the images to extract the underlying features. In this paper, we employed Maxpooling layer of size 2×2 with stride of 1 followed by the flatten layer to transform the 2D tensor representing the sparse feature map of the input and query into 1D tensor. As shown in Fig. 7, our proposed model employs four dense layers with a number of neurons in each layer of 1000, 800, 200 and 20, respectively. Dense layer is employed in the network to transform the spatial feature extracted from the CNN into 1D tensor which we will use to calculate the similarity between input and query image.

In the proposed model, we have used two types of activation function to fire up the neuron present in CNN as well as the dense layer. In the CNN, we have employed Tanh activation function, and in the dense layer, we implemented the ReLU activation function. ReLU stands for rectifier linear unit. It is one of the most prominent used nonlinear activation function functions [38]. We have used ReLU because of its feature which does not allow all neurons to fire up simultaneously at the same point of time. ReLU has the advantage that it converges very quickly when compared to other activation functions [39] (Fig. 8).

The above graph shows that the ReLU activates the neurons when they have values greater than zero.

$$f(u) = \max(0, u) \tag{4}$$

The above (4) is a mathematical representation of the ReLU function which states that if the value of the neuron is greater than zero, it will be fired up as the input for

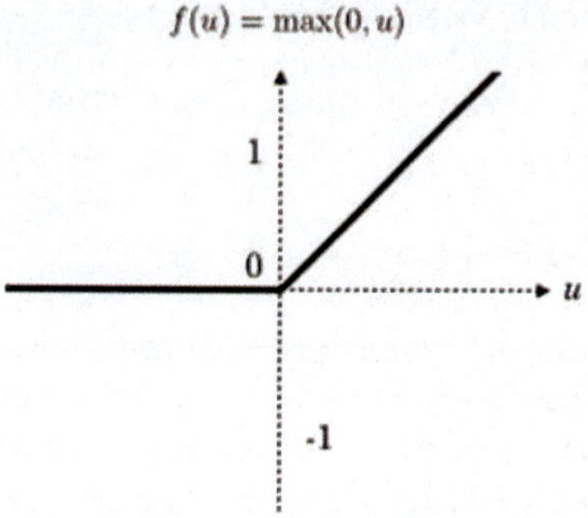

Fig. 8 Activation: ReLU

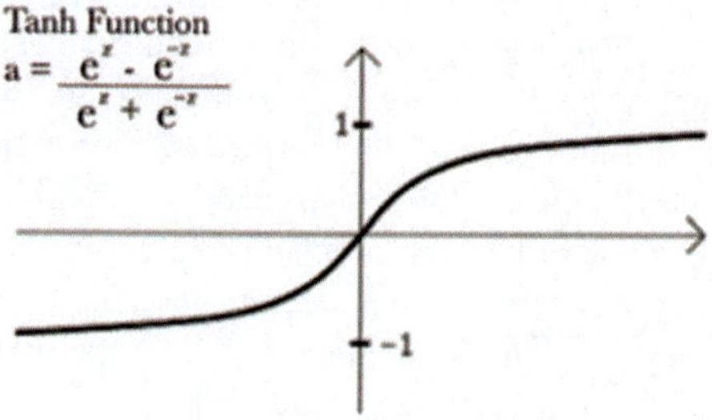

Fig. 9 Activation: Tanh

the next layer, else it will not be fired up and has range between 0 and + infinity [38, 39]. ReLU function ensures that all the neurons do not fire up at a same time and helps the model to converge as quickly as possible.

Tanh is known as the hyperbolic function. It has zero centre and a range between −1 and 1 inclusive. It is more often used in place of sigmoid because it delivers better training computational performance for deep neural network [40]. The main advantage of the Tanh is that it produces zero centred thus supports in backpropagation process while updating the weights. It also acts a dropout for certain layers because it produces some neurons as dead during computational process [41] (Fig. 9).

$$f(z) = \frac{e^z - e^{-z}}{e^z + e^{-z}}, \tag{5}$$

The above (5) is a mathematical representation of the Tanh function which states that if the value of the neuron has range between −1 and 1 inclusively.

In this paper, we have implemented Rmsprop as an optimizer to manipulate the neural network parameters for best outcome. Rmsprop utilizes the magnitude of the latest gradient to normalize the gradient along with the moving average over the root of the mean-squared gradient that is why it is called RMS by which we divide the current gradient [38, 42]. Below is the mathematical representation of Rmsprop:

$$r_t = (1 - \gamma) + * f'(\vartheta t)^2 + \gamma_{t-1}^{t} \tag{6}$$

$$\upsilon_{t+1} = \alpha/\sqrt{rt} * f'(\vartheta t)^2 \tag{7}$$

$$\vartheta_{t+1} = \vartheta t - \upsilon_{t+1} \tag{8}$$

where $f'(\vartheta t)$ is derivative of loss with respect to all the parameters in time steps t, α is step rate, and γ is the weight decay. In Rmsprop, the learning rate is adapted for each parameter. Rmsprop has shown good adaptation of learning rate in different applications [42]. Rmsprop can be seen as a generalization of Rmsprop and is capable of working with mini-batches as well opposed to only full batches.

Function lambda and similarity function are shown in Fig. 7, and our channel-based CNN tower consists of three pairs of the CNN towers, where each tower pair is specifically dedicated to the specific channel from the input images and the query image. These pairs of the CNN tower's sole purpose are to calculate the similarity between the specific channel feature map of the image the below formulae:

$$Rs = \sum(\mathrm{IR} - \mathrm{TR}) \tag{9}$$

$$Gs = \sum(\mathrm{IG} - \mathrm{TG}) \tag{10}$$

$$Bs = \sum(\mathrm{IB} - \mathrm{TB}) \tag{11}$$

where Rs, Gs and Bs represent the similarity of the each channel in the images, Ix and Tx represent the feature map image of the input image and query image in the vector form which has a range from 0 to 1, where 0 represents identical image and 1 represents no similarity, and x is the similar singleton set from the set R, G and B as shown in (9), (10) and (11).

Above equation was used to calculate the distance between the images feature map of the input image and the target image. Here we have used a modified form of the softmax activation function because the range of softmax is 0 to 1 [38, 39] (Fig. 10).

$$\mathrm{softmax} = \frac{e^{a_i}}{\sum e^{a_k}} \tag{12}$$

The range of the softmax function is 0 to 1, whereas the input is a continuous number, but in our case the input will have a range between 0 and 1 [31, 39]. If we analyse (9), (10) and (11), its range is between 0 and 1 inclusively. By considering the (9) which represents the softmax activation function by calculating the similarity with respect to the other channels present in the image and substituting it in (12), the range was exceeding 1 by two decimal number, but probability is not greater than 1, so to encounter this problem of softmax activation function becomes undefined, we

Fig. 10 Activation: softmax

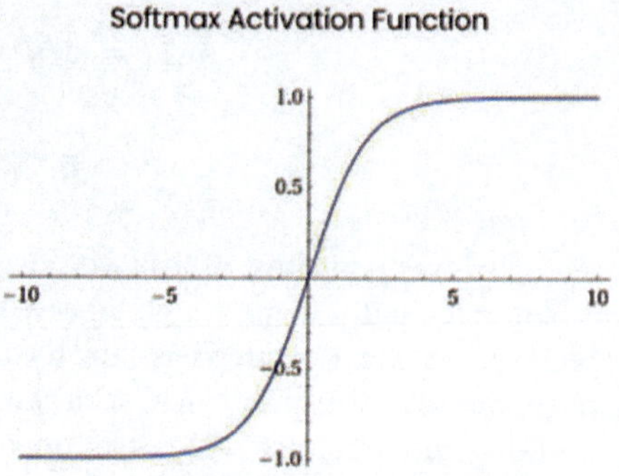

added epsilon ($\acute{\varepsilon}$)-very small number positive number to denominator and the new formula is

$$\text{modified softmax} = \frac{e^{a_i}}{\acute{\epsilon} + \sum e^{a_k}} \tag{13}$$

By using above (13), we can calculate the similarity of each channel with respect to other channels with the following equation:

$$\text{SR} = \frac{e^{R_s}}{\acute{\epsilon} + e^{Rs+Gs+Bs}} \tag{14}$$

$$\text{SG} = \frac{e^{Gs}}{\acute{\epsilon} + e^{Rs+Gs+Bs}}, \tag{15}$$

$$\text{SB} = \frac{e^{Bs}}{\acute{\epsilon} + e^{Rs+Gs+Bs}} \tag{16}$$

The above (14), (15) and (16) will give the similarity index of each channel with respect to other channels which has a range between 0 and 1. Since we need to combine these quantitative values which represents the similarity index of the channels with respect to each channel separately, we took average of the SR, SG and SB which bring down the quantitative value between 0 and 1 which is shown below:

$$\text{Cumulative Similarity} = \frac{(\text{SR} + \text{SG} + \text{SB})}{3} \tag{17}$$

The above (17) represents the similarity of the query image with respect to the input image in the range inclusive of 0 and 1, where 0 represents dissimilar image and 1 represents similar image. This cumulative similarity can be filtered out with the specific threshold value to get the similar kind of image based on the requirement and the features of the image.

4 Result and Discussion

As we have discussed in Sect. 3.1, the dataset consists of 1,20,000 images, where image channel was separated by the data processing. Additionally, we collected 400 random images from the Internet which was shuffled with the test set to calculate the efficiency of the tower-based neural network (Fig. 11).

The above graph explains the accuracy of the channel-based CNN tower between the training set and the test set over 20 epochs, in order to counter any effect of the overfitting, and the callbacks routines were used to roll back the model to the point where the difference between the test accuracy and the train accuracy are minimum. The final accuracy obtained on training set was 97.47% and on test set was 97.75% (Fig. 12).

Similarly, below graph explains the loss of the channel-based CNN tower between the training set and the test set over 20 epochs, in order to counter any effect of the overfitting and the callbacks routines were used to roll back the model to the point,

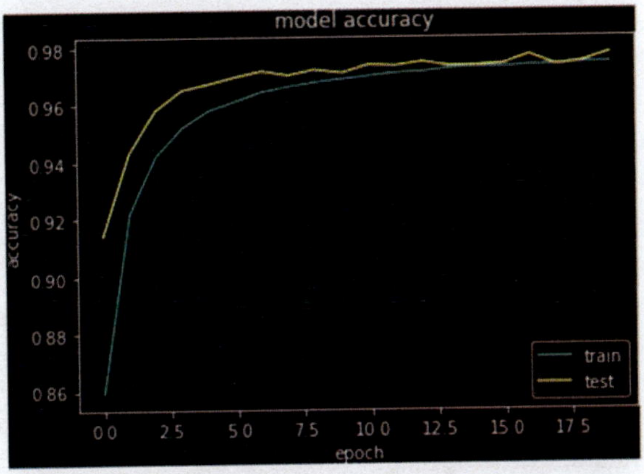

Fig. 11 Model accuracy

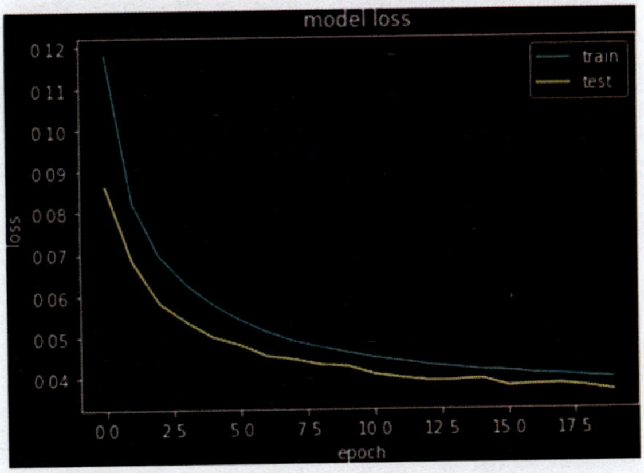

Fig. 12 Model loss

where the difference between the test loss and the train loss are minimum. The final accuracy obtained on training set was 03.95% and on test set was 03.65%.

5 Conclusion and Future Works

We developed a channel-based CNN tower model to compute the similarity of the two images target and the input image. This model used the structural as well as the physical dimension by carrying out CNN and the feature map of the images to compute the similarity between the combination of pixels present in the images. The constructed deep CNN generates numerous features maps by convoluting and Maxpooling on the images. The images are pre-processed in such a way that the channel present in the images was separated and the label was generated according to the similarity of the images. We trained the model on two separate datasets Cifar-10(mini) consisting of 60,000 images with ten objects and on Animals-10 dataset consisting of ten objects and 28,000 images. We pre-processed each image from these datasets as shown in Table 1, then propagated—forward and backward to the CNN as were fully connected layer using Rmsprop optimizer—variant of gradient descent algorithm and the distance between the feature map were calculated separately using channel-based CNN tower model. We obtained comparable similarity; hence, it can be applied to the real-world scenario.

In this paper, we used a supervised channel-based CNN tower to calculate the similarity percentage of the two images. There are existing methods which have achieved satisfactory performance in analysing the object by grouping them into clusters. This channel-based CNN tower model was trained using a supervised algorithm which requires labelled samples, but it can be used to label the unstructured image as well as because it calculates channel-based similarity of the input and target image. There are an obvious number of improvements, which can be made in the future, including improvement in the CNN tower and as well by training it using transfer learning as well as object detection to compute the channel-based similarity of the unstructured image data.

Acknowledgements I would like to thank Dr. P. Supraja and Mrs. A. Helen Victoria, who has helped in all the stages of my educational life. Without their help and guidance, it would have been a very difficult task for me. I would also like to thank Corrado Alessio for giving access to the Animal-10 dataset and Alex Krizhevsky for giving me access to the Cifar-10 dataset.

References

1. Petrovic VMP (2018) Artificial intelligence and virtual worlds toward human-level AI agents. IEEE. https://doi.org/10.1109/ACCESS.2018.2855970
2. Lpuridas P, Ebert (2016) Machine learning. IEEE software. https://doi.org/10.1109/MS.2016.114

3. Angra S, Ahuja S, Machine learning and its application. IEEE. https://doi.org/10.1109/ICBDACI.2017.8070809
4. Adadi A (2021) A survey on data-efficient algorithms in big data era. J Big Data 8:24. https://doi.org/10.1186/s40537-021-00419-9
5. Qi G, Luo J (2019) Small data challenges in big data era: a survey of recent progress on unsupervised and semi-supervised methods. arXiv preprint arXiv:1903.11260
6. Ojha U, Adhikari U, Singh DK (2017) Image annotation using deep learning: a review. In: 2017 International conference on intelligent computing and control (I2C2), pp 1–5. https://doi.org/10.1109/I2C2.2017.8321819
7. Aletti G, Micheletti A (2017) A clustering algorithm for multivariate data streams with correlated components. J Big Data 4:48. https://doi.org/10.1186/s40537-017-0109-0
8. Marginean S (2009) Advantages and disadvantages of clusters—a theoretical approach. Revista Econ 44:36–40
9. Koch G, Zemel R, Salakhutdinov R, Siamese neural networks for one-shot image recognition
10. Cano F, Cruz-Roa A (2020) An exploratory study of one-shot learning using Siamese convolutional neural network for histopathology image classification in breast cancer from few data examples. In: Proceedings of SPIE 11330, 15th international symposium on medical information processing and analysis, 113300A. https://doi.org/10.1117/12.2546488
11. Melamud O, Goldberger J, Dagan I, Riezler S, Goldberg Y (2016) context2vec: learning generic context embedding with bidirectional LSTM. In: Proceedings of the 20th SIGNLL conference on computational natural language learning, association for computational linguistics (ACL), Berlin, Germany, 11–12 Aug 2016, pp 51–61
12. Gothwal R, Gupta S, Gupta D, Dahiya AK (2014) Color image segmentation algorithm based on RGB channels. In: Proceedings of 3rd international conference on reliability, infocom technologies and optimization, pp 1–5. https://doi.org/10.1109/ICRITO.2014.7014669
13. He K, Zhang X, Ren S, Sun J, Deep residual learning for image recognition. arXiv:1512.03385
14. Deng J, Dong W, Socher R, Li L, Li K, Li F-F (2009) ImageNet: a large-scale hierarchical image database. In: 2009 IEEE conference on computer vision and pattern recognition, pp 248–255. https://doi.org/10.1109/CVPR.2009.5206848
15. Adnan K, Akbar R (2019) An analytical study of information extraction from unstructured and multidimensional big data. J Big Data 6:91. https://doi.org/10.1186/s40537-019-0254-8
16. Verma S, Jain K, Prakash C (2020) An unstructured to structured data conversion using machine learning algorithm in internet of things (IoT. In: Proceedings of the international conference on innovative computing & communications (ICICC) 2020
17. Huang X, Li J, Liang Y et al (2017) Spatial hidden Markov chain models for estimation of petroleum reservoir categorical variables. J Petrol Explor Prod Technol 7:11–22. https://doi.org/10.1007/s13202-016-0251-9
18. Blake A, Kohli P, Rother C (2011) Markov random fields for vision and image processing. The MIT Press, Cambridge, pp 11–22
19. Bindra K, Mishra A (2017) A detailed study of clustering algorithms. In: 2017 6th international conference on reliability, infocom technologies and optimization (trends and future directions) (ICRITO), pp 371–376. https://doi.org/10.1109/ICRITO.2017.8342454
20. Estivill-Castro V (2002) Why so many clustering algorithms: a position paper. ACM SIGKDD Explor Newsl 4:65–75
21. Xu D, Tian Y (2015) A comprehensive survey of clustering algorithms. Ann Data Sci 2:165–193. https://doi.org/10.1007/s40745-015-0040-1
22. Färber I, Günnemann S, Kriegel H, Kröger P, Müller E, Schubert E, Seidl T, Zimek A (2010) On using class-labels in evaluation of clusterings. In: MultiClust: 1st international workshop on discovering
23. Boley D, Gini M, Gross R, Han E, Hastings K, Karypis G, Kumar V, Mobasher B, Moore J (1999) Partitioning-based clustering for web document categorization. Decis Support Syst 27:329–341
24. Kleinberg J (2002) An impossibility theorem for clustering. In: Proceedings of 2002 conference advances in neural information processing systems, vol 15, pp 463–470

25. Shen C, Jin Z, Zhao Y, Fu Z (2017) Deep siamese network with multi-level simillarity perception for person re-identification. In: ACM conference. https://doi.org/10.1145/3123266.3123452
26. Varior RR, Haloi M, Wang G (2016) Gated siamese convolutional neural network architecture for human re-identifcation. In: European conference on computer vision. Springer, pp 791–808
27. Subramaniam A, Chatterjee M, Mittal A (2016) Deep neural networks with inexact matching for person re-identifcation. In: Advances in neural information processing systems, pp 2667–2675
28. Chen W, Chen X, Zhang J, Huang K (2016) A multitask deep network for person re-identifcation. arXiv preprint arXiv:1607.05369
29. Ahmed E, Jones M, Marks TK (2015) An improved deep learning architecture for person re-identifcation. In: Proceedings of the IEEE conference on computer vision and pattern recognition, pp 3908–3916
30. Li W, Zhao R, Xiao T, Wang X (2014) Deepreid: deep filter pairing neural network for person re-identifcation. In: Proceedings of the IEEE conference on computer vision and pattern recognition, pp 152–159
31. Liu J, Zha ZJ, Tian QI, Liu D, Yao T, Ling Q, Mei T (2016) Multi-scale triplet CNN for person re-identifcation. In: Proceedings of the 2016 ACM on multimedia conference. ACM, pp 192–196
32. Schroff F, Kalenichenko D, Philbin J (2015) Facenet: a unied embedding for face recognition and clustering. In: Proceedings of the IEEE conference on computer vision and pattern recognition, pp 815–823
33. Krizhevsky A, Sutskever I, Hinton GE (2012) ImageNet classification with deep convolutional neural networks. In: NIPS 2012: neural information processing systems, Lake Tahoe, Nevada
34. Song H, Kim M, Lee J (2019) SELFIE: refurbishing unclean samples for robust deep learning. In: Proceedings of 36th international conference on machine learning (ICML), Long Beach California
35. Wach HB, Dowski ER, Cathey WT (2000) Channel reduction and applications to image processing. Appl Opt 39(11):1794–1798
36. Ramírez Paulino I, Hounie I (2021) Image inpainting using patch consensus and DCT priors. In: Image processing on line, vol 11, pp 1–17. https://doi.org/10.5201/ipol.2021.286
37. Lisani JL, Petro AB (2021) Automatic 1D histogram segmentation and application to the computation of color palettes. In: Image processing on line, vol 11, pp 76–104. https://doi.org/10.5201/ipol.2021.344
38. Szegedy C, Liu W, Jia Y, Sermanet P, Reed S, Anguelov D, Erhan D, Vanhoucke V, Rabinovich A (2015) Going deeper with convolutions. In: Proceedings of the IEEE conference on computer vision and pattern recognition, pp 1–9
39. Krizhevsky A, Sutskever I, Hinton GE (2012) Imagenet classification with deep convolutional neural networks. In: Advances in neural information processing systems, pp 1097–1105
40. Marra S, Iachino MA, Morabito FC (2006) Tanh-like activation function implementation for high-performance digital neural systems. In: 2006 Ph.D. research in microelectronics and electronics, pp 237–240. https://doi.org/10.1109/RME.2006.1689940
41. Cong J, Xiao B (2014) Minimizing computation in convolutional neural networks. In: International conference on artificial neural networks. Springer, pp 281–290
42. Ruder S (2017) Insight centre for data analytics, NUI Galway Aylien Ltd., Dublin An overview of gradient descent optimisation algorithms arXiv:1609.04747v2 [cs.LG]. 15 Jun 2017

Design and Analysis of Adversarial Samples in Safety–Critical Environment: Disease Prediction System

Aruna Pavate and **Rajesh Bansode**

Abstract Deep learning is a part of machine learning applied in many applications, from object detection to disease prediction. In 2013, deep neural networks were vulnerable to perturbed samples during the testing/deployment stage. Deep learning becomes a significant issue when we apply such algorithms in designing safety–critical applications. Deep neural networks are fooled by adding some noise intentionally in the image such that the changes are sometimes not noticeable to the human eye and are known as an adversarial mechanism. The noise adds in such a direction that the difference between the original and perturbed images should be minimal. This work demonstrates the designing of adversarial images and analyzing them on the disease prediction using a chest X-ray prototype system. This work illustrates the design of adversarial samples using three different methods: an image augmentation scheme, filtering, and patches and analysis of how adversarial samples affect the result in real-world clinical settings. The change in the clinical setting not only affects the healthcare economy but also raises technical vulnerability. This work attempts to design and improve a more robust medical learning system. The analysis shows if patches applied in specific regions give expected results to the adversary. In this work, after the successful attack model shows 71% of confidence.

Keywords Deep neural network · Adversarial samples · Medical images · X-ray images · Data augmentation

1 Introduction

Machine learning algorithms such as deep neural networks are most extensively applicable in the medical domain, which helps the practitioner and expert concentrate on more essential tasks to predict the diseases early [1]. Recently, in 2018 US administration permitted the AI-based device for early detection of retinopathy which

A. Pavate (✉) · R. Bansode
Department of Information Technology, Thakur College of Engineering, Mumbai University, Mumbai, India
e-mail: arunaapavate@gmail.com; arunapavate@sfit.ac.in

M. Gupta et al. (eds.), *Artificial Intelligence on Medical Data*,
Lecture Notes in Computational Vision and Biomechanics 37,
https://doi.org/10.1007/978-981-19-0151-5_29

helps the patient to further evaluation and treatment in the early phases to avoid further complications [2]. There has been lots of work done in the medical field in integration with machine learning like a cancer diagnosis, based on morphological samples [3], diabetic retinopathy detection [4], heart disease prediction [5], skin disease prediction [6], Alzheimer's disease prediction using autoML tool [7], kidney disease prediction [8], and many more. For designing such applications, lots of training data are necessary. Appropriate training gives the correct results. Other than training, many factors need to be considered, like labeling data: manual task, expert advice, model selection, and many more. Convolution neural networks have given surprising results for designing such applications. Recently, in 2013 it has been observed that deep neural networks give expected results to the adversary by using adversarial samples during the training or deployment phase [9]. Adversarial samples are created by crafting the samples taken from the same domain with some modification. These samples are either impartible or sturdy [10] modification of samples [11] to make DNN give results other than the actual one. Adversarial samples are more harmful to safety–critical applications like self-driving cars, medical analysis, prediction, and robotics surgery. The researchers have tried handcrafted features in many applications [12, 13]. Adversarial mechanisms can be derivative-based optimization methods and non-derivative-based optimization methods that help to generate adversarial samples. Adversaries deliberately modify input samples by adding some noise to the original samples to make model mistakes. Adversarial attacks are possible during the testing and deployment phases without directly accessing the model, considering no knowledge about the model. There are many domains where researchers illustrated adversarial samples on the model. Recently, Samual et al. [14] demonstrated the attack of adversarial samples considering three clinical domains. Originally, adversarial samples applied and studied image classification and proved how adversarial samples change the output of the classifier [9, 10, 15]. Nowadays, these attacks are not limited to images but have enlarged to using other types of inputs such as speech recognition [16] and cyber security [17]. Adversarial attacks for safety–critical applications like disease prediction need to be analyzed as they break up human lives and result in an economic pull-down. Therefore, the contribution of this work includes:

- To design and analyze adversarial mechanisms for safety–critical applications.
- To demonstrate how slight changes in the input samples give different results than those applied to the prediction system.
- To create adversarial samples using the handcrafted method and using the data augmentation and filter method.
- To demonstrate generated adversarial samples on free medical tools available for medical practitioners.

This work proposed adversarial mechanisms considering deep learning environments and applied them to Chester X-ray prediction systems. This work concentrates on adding perturbation to the samples to design adversarial inputs and working on a deep neural network model by generating adversarial samples using data augmentation, filtering, and adversarial patch. The objective of this work is to make the machine learning model more robust.

The remaining sections of the chapter discuss related work, Sect. 2 describes reviews of design and analysis of adversarial samples. The related work mainly focuses on generating adversarial mechanisms and results after applying adversarial samples on various applications designed using deep neural networks. Section 3 describes the method for experimental design, and Sect. 4 discusses the experimental results and discussion; the final section concludes the work and future scope.

2 Related Work

This section highlights different approaches for generating adversarial attacks in the physical world only. In recent studies, there are many approaches described and defeated by countermeasures. The current approaches need to be studied, which improves the transferability and thus robustness of the neural network models. The related work mainly focuses on generating adversarial mechanisms and results after applying adversarial samples on various applications designed using deep neural networks.

Samuel et al. [14] applied adversarial samples on a medical system designed using deep neural networks. The author confirmed that medical systems designed using deep neural networks are vulnerable to adversarial samples. To demonstrate this, the authors discussed use cases in the real environment and the practical feasibility of adversarial samples in the healthcare domain.

Sharif et al. [18] developed a method to generate automatic attacks, which perceives through applying perturbation by injecting printed pairs of eyeglass frames in the physical world. In this work, the optimization function modifies by adding a penalty of non-printable score to enhance the printed adversarial images. Attack tried for both the targeted samples as well as for non-targeted samples. Authors successfully tried non-targeted attacks alongside face recognition system over 80% of the time and for targeted attack misguided the face recognition system with high success rate. The success rate depends on the target samples. Here universal perturbation attack generates using a set of input samples. Figure 1 shows the invisibility attack on the face recognition system. The middle image in Fig. 1 shows the attack by perturbing pixels. An adversary can make a machine learning system fail to detect a person's existence.

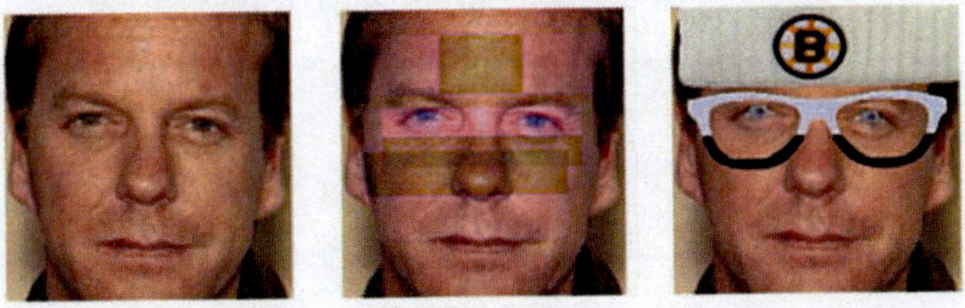

Fig. 1 Invisibility attack on face recognition system [18]

Papernot et al. [19] developed an attack against DNN classifiers. It crafts adversarial examples without the knowledge of the classifier training data or model. Author targeted model hosted by Amazon and Google. They misclassified adversarial examples at a rate of 96.19% and 88.94%. The substitute model is not trained to be robust to the small perturbations while defending against the model needs to apply infinite perturbation.

Evtimov et al. [20] proposed robust physical perturbations (RP2) to generate the perturbation by considering the images under different environments for real-world sign recognition. To generate perturbation (modifying the road sign like stop sign from speeding limit sign), two types of attacks proposed (1) Overlapping a perturbed road sign over a physical road sign (2) By pasting adversarial samples on an actual sign. To improve the printability score in the optimization objective function, the author included a non-printability score. The success of the attack varies under different physical conditions.

To mislead autonomous vehicles by modifying stop signs to traffic signs, an adversary can generate physical adversarial examples and easily confuse sign recognition systems [20, 21].

Xie et al. [22] algorithm dense adversary generation to cause deep neural networks to fail an image classification by removing the segmentation of pedestrians in object segmentation and detection systems.

Chen et al. [23] proposed a system that helps to solve more challenging problems to fool object detectors similar to faster R-CNN, as shown in Fig. 2. Expectation over transformation technique applied to improve the robustness of adversarial examples and successfully adapted to object detection.

Hokuto et al. [24] focused on three diseases—skin cancer, diabetic retinopathy, pneumonia classification, and vulnerability of the diseases analyzed on seven different neural network models. The attack tried for both the targeted and non-targeted classes with an 80% of success rate.

Review work discussed various approaches proposed for generating adversarial samples and applied in various domains. Considering deep neural network applications in safety–critical environment applications such as healthcare, the use of deep learning is truly inspiring. The use of machine learning models improves system

Fig. 2 Untargeted attacks that were never detected as a stop sign or something else [23]

performance and reduces costs. However, the use of a large scale of technologies leads to fraudulent practices that harm patients and influence the economy. This work outlined the generation of adversarial samples and discussed how the model's output changes with slight modification.

3 Methodology

The deep neural network is nonlinear and helps to relate the input and output, so minor changes also give the output more than the actual one. Adversarial mechanisms help create adversarial samples by slightly changing the input, e.g., changing some points of interest within a specified boundary, making a slight difference between the neighboring pixels, or resizing the images. Figure 3 illustrates how slight changes to the input sample $\times$ 0 given to the neural network model get a different result than the actual one.

Designing adversarial samples is considered an optimization problem. The optimization function helps train the neural network models $f(\theta)$ that generate the loss function and uses this loss function model to retrain to improve the performance. The aim is to minimize loss, where smaller values represent a better model than more significant values. Assume that we have a network parameterized by θ that maps a sample $\times$ 0 to real label Y^{true} and an adversary wants to function f to misclassify x' to Y^{false}.

The deep neural network uses the gradient for generating adversarial samples. For a given classifier $\left[Y^{\text{true}}|x0\right]$, input samples $\times$ 0 belong to the domain and target class. Here we need to edit the input $\times$ 0 that maximizes $\log(p[Y^{\text{true}}|x0]$ such that $\left|\left|x0 - x'\right|\right|_{\infty} \leq$. Traditional methods of generation of adversarial mechanisms require internal information about the model such as gradient information which is in real life actually not available. These methods gave good results for specific applications and the essential need to change the desired image. Some methods consider adversary knowledge as a black box—does not know the internal model details and white-box attack—it is considered adversary know the model details.

CNN is a popular network where a machine trains to classify images based on patterns in the images. Figure 4 shows the basic framework of the proposed system. Convolution neural networks are a special type of neural network where hidden

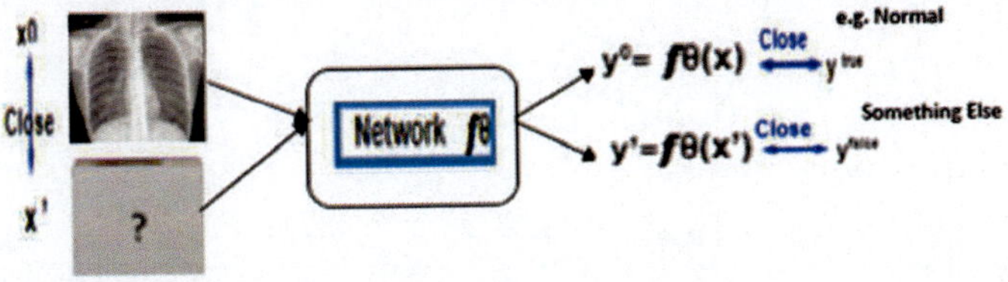

Fig. 3 An illustration of adversarial examples with a brief concept

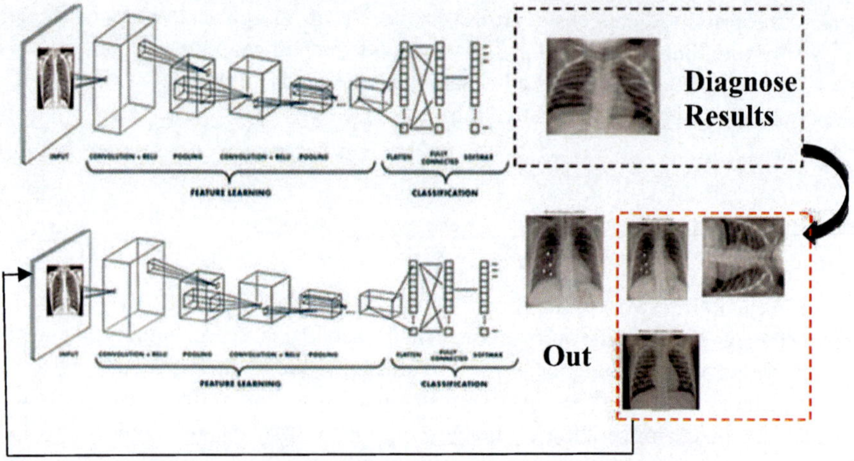

Fig. 4 Working of the proposed system

units connect local receptive fields. After training, CNN can be used to recognize objects in the images. Neural networks train using an optimization process that requires a loss function to calculate the model error. If the training data of the target network from the same domain is available, it can use it to train a proxy network to generate attacked objects and obtain input–output pairs from the target network. The adversarial samples generate to minimize the error loss and increase the error rate so that the attack will succeed. Generated adversarial samples apply to the original model during the testing/deployment phase.

Here, in this work, we have created adversarial samples using three different schemes: First, adding an adversarial patch, second augmenting the images (rotate by 90°, rotate by 180°, and rotate by 270°), third by applying filters (Neo filter, Mercury filter, Rouge Filter, and Sahara Filter). ResNet50 architecture trained using a variety of images by using the above schemes. Here, final output $\times$ 0 achieves using the variable mask (vm), rotating mask (rm), filtering mask (fm). We define adversarial sample corresponding to every transformation such that it applies the patch onto the image using element-wise binary operation:

$$\begin{aligned} Ap\left(x', y^{\text{true}}\right) = t\left[\left(vmoy^{\text{true}}\right) + [t(1 - vm)ox0]\middle|\left(rmoy^{\text{true}}\right)\right. \\ \left. + [t(1 - rm)ox0]\middle|\left(fmoy^{\text{true}}\right) + [t(1 - fm)ox0]\right] \end{aligned}$$

$$\left(rmoy^{true}\right) + [t(1 - rm)ox0]|$$

$$\left(fmoy^{true}\right) + [t(1 - fm)ox0]] \tag{1}$$

Here in Eq. 1, symbol o refers to the element-wise binary operation referred to as Schur product. The final result adversarial sample satisfies the criteria $x' = Ap\left(x', y^{\text{false}}\right)$ for the final patch.

In this experiment, generated adversarial samples apply to a proxy model. These samples are given input to the chester AI radiology assistant (prototype model) [26], specially designed as an expert system to assist and understand the chest X-ray diagnostics based on deep neural networks. This prototype helps the radiologist to predict 14 disease risk levels. To train the system, architecture DenseNet-121 and xrv-all-45rot15trans15scale using data collected from different sources. In this work, adversarial examples are produced using patches, augmentation schemes, and applying filters. Generated adversarial samples are analyzed by applying on a prototype and verifying the model's prediction changes after applying adversarial examples.

Model Setting and Considerations

In this work, for applying adversarial mechanisms, we consider adversarial attack setting as follows: the adversary's knowledge is white box (adversary knows the details of the architecture model, training set, etc.). Generated adversarial samples applied on non-targeted labels. Attack generalization is for individual attacks, and the attack frequency is one time. Network training model (victims model) used various data sets PadChest, NIH, RSNA Pneumonia, MIMIC-CXR, CheXpert [26] for training the model.

For our proposed system, i.e., adversary network COVID-19-Kaggle, NIH data sets [25] are used for training and testing the model. These data sets are different from the threat model but from the same domain. The data set contains an image of size 224 × 224 chest X-ray images. As for designing adversarial samples, we have used two data sets that are publicly available COVID-19 and NIH.

4 Experimental Results and Discussion

This section analyzed the outcome of the adversarial samples generated using three different methods and applied them to the proxy model for predicting 14 different diseases. Here, analyzing prediction risks of diseases from healthy to risky levels. The model's performance was analyzed using 368 samples (122 samples in each category). The samples were analyzed using an augmentation scheme-(rotate by 90°, rotate by 180°, and rotate by 270°), By adding patches to the samples and applying four different filters (Mercury, Sahara, Neo, Rouge). A detailed analysis of each result is as follows:

4.1 Performance of the Adversarial Samples Designed Using Augmentation Scheme

The generated adversarial samples applied on the chest X-ray disease prediction system [25] help predict the 14 independent diseases' pathological-related risk. Figure 5 represents the original sample, and Fig. 6 shows the adversarial samples generated using the augmentation scheme. The leftmost top corner, Fig. 6 represents sample generated rotating 90^0, the following two middle images are rotated by 180^0, and the last row represents samples rotated by 270^0.

Figure 6 shows how the risk of the disease changes after applying an augmentation scheme. The original research authors [26] applied a data augmentation scheme considering different angles such as 5°, 10°, 15°, 25°, 45°, 65°, 90°, 180°, and horizontal flip, and the result analysis was done for diseases. The model has responded very well for the augmentation scheme applied using 180° and 270° with 61% of images being too far from training distribution as the images are not cropped correctly and for 39% of images model has given results with some risk level of the diseases

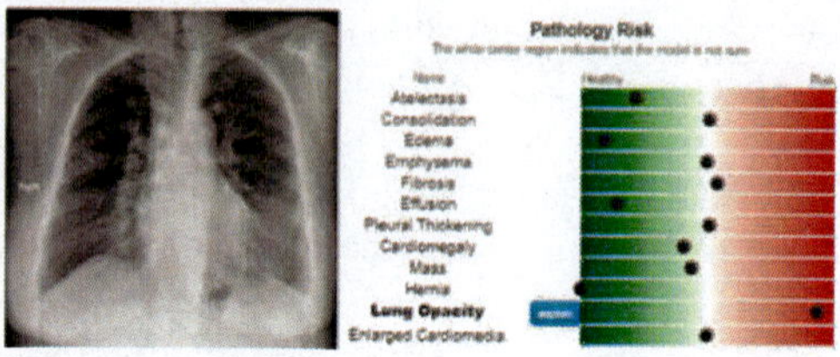

Fig. 5 Original image

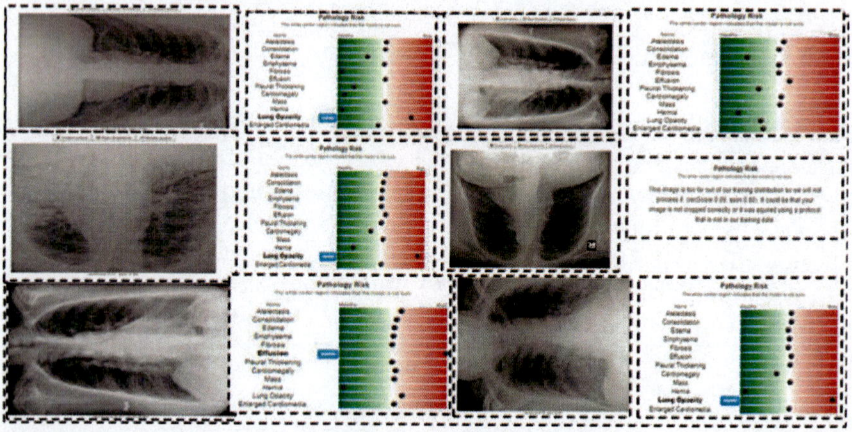

Fig. 6 Adversarial samples generated using an augmentation scheme. The first row represents the left rotation (90° rotated samples). The second row represents 180° rotated sample, and the third row represents the sample Right (rotated by 270°)

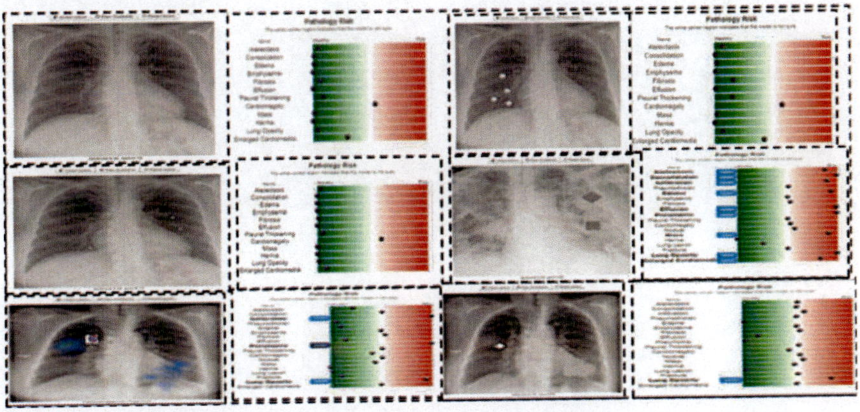

Fig. 7 Adversarial samples generated using adversarial patches. The first row represents the adversarial sample with four dots of dried ink. The second row represents the adversarial sample generated with a rec patch, and the third row represents only a single rec patch

so here adversary has a chance to access the model and to make able the system to give unexpected result.

4.2 Performance of the Adversarial Samples Designed Using Patches

4.3 Performance of the Adversarial Samples Designed Using Filter

Figure 8 shows adversarial samples generated using a filtering scheme. Here O1 represents the original sample, M1 represents the sample after applying Mercury filter, R1 represents rough filter, N1 represents Neo filter, and S1 represents Sahara filter. The observation shows that pathology risks associated with few diseases changes after applying this scheme. The analysis observes that the average confidence of the model after applying Neo 37%, Rouge 40%, Sahara 37%, and Mercury filter 42%.

The analysis of the sample of five diseases is shown in Fig. 9. This plot takes all the values mentioned in Table 1. The observation has been done for the pathology risk associated with each disease after applying adversarial samples observed. The model predicts the risk with Low risk, High risk, and Not Sure. The model gives output not sure for all the values greater than 0.5 and less than 0.7, the disease is at low risk for the confidence less than 0.5, and the risk of the disease is high when the model predicts the risk above 0.7 thresholds. For adversarial samples generated

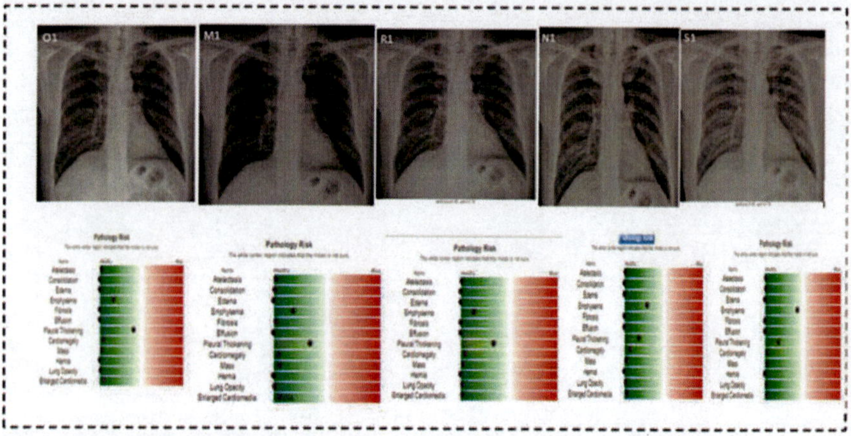

Fig. 8 Filter schemes had been considered for generating various adversarial samples—From left to right, the first row represents the O1: original sample, N1: Neo filter adversarial sample, M1: Mercury filter adversarial sample, R1: Rouge filter adversarial sample, S1: Sahara: filter adversarial sample

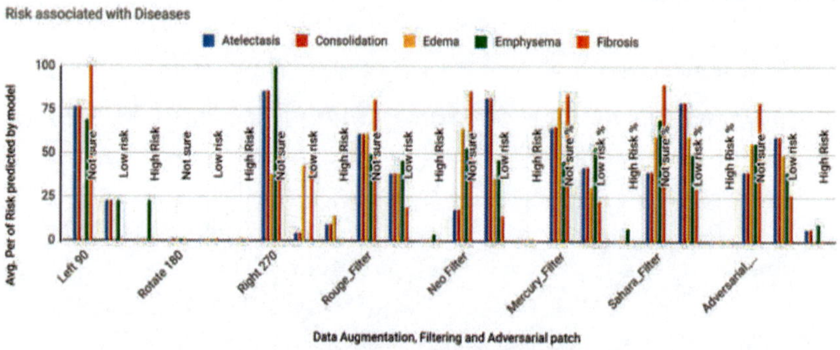

Fig. 9 Plot for the Risk associated with diseases: Low risk, High Risk, Not Sure (Five diseases Atelectasis, Consolidation, Edema, Emphysema, and Fibrosis)

using the rotate 180° augmentation scheme models show a high risk for Emphysema disease and for other diseases the model is not sure.

The original research [26] data augmentation scheme is applied while training the model, yet after applying our adversarial samples generated using three different schemes, the model predicted the samples belong to one of the classes with some confidence. Unexpectedly, after applying an adversarial samples model given remarkable success. The generated attack is non-targeted. In the future, targeted attacks are possible to design. The observation shows that the 0.2–0.79 model successfully fooled the disease prediction model. These results are possible under a white-box setting with an average success rate of 50%, considering very few changes not recognizable by humans. However, as the system [26] does not take the testing samples to

Table 1 Analysis of the performance (Confidence) of the model using adversarial samples generated and pre-trained model. The table represents the disease's name in short as: Atelectasis-A, Consolidation-C, Edema-Ed, Emphysema-Em, Fibrosis-F, Effusion-EF, Pleural Thickening-PT, Cardiomegaly C, Mass-M, Hernia-H, Lung Opacity-L, Enlarged Cardiomedia-En

	A	C	ED	EM	F	EF	PT	C	M	H	L	EN
Pre-trained model	0.83	0.82	0.81	0.81	0.75	0.68	0.62	0.89	0.88	0.77	0.59	0.65
No data augment	0.82	0.79	0.75	0.68	0.66	0.59	0.87	0.86	0.73	0.66	0.67	0.76
Horizontal flip	0.81	0.81	0.78	0.76	0.76	0.78	0.87	0.88	0.76	0.85	0.76	0.66
Left flip 90	0.50	0.54	0.42	0.51	0.37	0.52	0.38	0.29	0.55	0.06	0.79	0.48
Right flip 120	0.49	0.50	0.47	0.48	0.49	0.45	0.40	0.38	0.43	0.41	0.60	0.66
Rotate 180	0.22	15.0	0.21	0.11	0.10	0.15	0.12	0.16	0.17	0.34	0.23	0.42
Neo filter	0.28	0.46	0.35	0.47	0.30	0.28	0.31	0.23	0.52	0.03	0.72	0.46
Rouge filter	0.40	0.44	0.38	0.47	0.34	0.36	0.36	0.34	0.53	0.05	0.64	0.48
Sahara filter	0.33	0.51	0.37	0.48	0.31	0.31	0.29	0.21	0.50	0.02	0.72	0.43
Mercury filter	0.42	0.49	0.35	0.45	0.40	0.35	0.39	0.27	0.62	0.07	0.67	0.50
Adversarial patch	0.41	0.50	0.40	0.48	0.38	0.41	0.36	0.33	0.54	0.04	0.71	0.49

improve the model, it is difficult for the adversary to fool the model; in that case, the attack needs to consider the high-level feature maps between convolution layers.

5 Conclusion

The area of deep neural networks holds a vast prospective for discovering adversarial samples in today's exponentially increasing use of deep models in different areas. This work presents a category-wise compilation and an analysis of each scheme. Adversarial mechanisms are essential considerations, especially highlighted by considering safety–critical applications. Adversarial mechanisms are a significant concern for the security and privacy of models in the healthcare domain, especially biometric recognition, face recognition. The utilization of adversarial samples reduces the accuracy and hence robustness of the model. We demonstrated that using various adversarial schemes that attack is very effective.

Additionally, such attacks are transferable from one model to another model considering samples from the same domain. We demonstrated that using various adversarial schemes that attack is very effective. The purpose of this work is to focus the public, researchers, and medical professionals on the types of possibilities of attacks during testing/deploying the model in medical diagnostics and decision support systems so that the future direction is to build a more robust model.

References

1. Bohr A, Memarzadeh K (2020) The rise of artificial intelligence in healthcare applications. Artif Intell Healthc, pp 25–60. https://doi.org/10.1016/B978-0-12-818438-7.00002-2
2. 2018. Press Announcements, FDA permits marketing of artificial intelligence-based device to detect certain diabetes-related eye problems. https://www.fda.gov/news-events/press-announcements/fda-permits-marketing-artificial-intelligence-based-device-detect-certaindiabetes-related-eye
3. Junqueira LC, Carneiro J (2005) Basic histology text & atlas. McGraw-Hill Professional
4. Pavate A, Mistry J, Palve R, Nirav G (2020) Diabetic retinopathy detection-MobileNet binary classifier. Acta Sci Med Sci 4(12):86–91
5. Shankar V, Kumar V, Devagade U, Karanth V, Rohitaksha K (2020) Heart disease prediction using CNN algorithm. SN Comput Sci 1:170. https://doi.org/10.1007/s42979-020-0097-6
6. Jany Shabu SL, Nithin MS, Santhosh M, Roobini MS et al (2020) Skin disease prediction. J Comput Theoret Nanosci 17(8): 3458–3462(5). American Scientific Publishers. https://doi.org/10.1166/jctn.2020.9210
7. Francis CR., Lokhande U, Bamrah PK, D'costa A, Alzheimer's disease prediction using Fastai. In: Senjyu T, Mahalle PN, Perumal T, Joshi A (eds) Information and communication technology for intelligent systems. ICTIS 2020. Smart innovation, systems and technologies, vol 195. Springer, Singapore. https://doi.org/10.1007/978-981-15-7078-0_76
8. Kumar G, Naik DA, Jaiswal VK, Manollas M, Ankitha V (2019) Chronic kidney disease prediction. Int J Comput Sci Eng 7(4):1065–1069
9. Szegedy C, Zaremba W, Sutskever I, Bruna J, Erhan D, Goodfellow I, Fergus R (2014) Intriguing properties of neural networks. ICLR, abs/1312.6199
10. Goodfellow J, Shlens J, Szegedy C (2014) Explaining and harnessing adversarial examples. CoRR, abs/1412.6572
11. Furukawa Y, Curless B, Seitz SM, Szeliski R (2010) Towards Internet-scale multi-view stereo. In: 2010 IEEE computer society conference on computer vision and pattern recognition, pp 1434–1441. https://doi.org/10.1109/CVPR.2010.5539802
12. Khan H, Shah PM, Shah MA, ul Islam S, Rodrigues JJPC (2020) Cascading handcrafted features and convolutional neural network for IoT-enabled brain tumor segmentation. In: Computer communications, vol 153, pp 196–207. ISSN 0140-3664. https://doi.org/10.1016/j.comcom.2020.01.013
13. Lu J, Sibai H, Fabry E, Forsyth D (2017) No need to worry about adversarial examples in object detection in autonomous vehicles. arXiv preprint arXiv:1707.03501
14. Finlayson SG, Chung HW, Kohane IS, Beam AL (2018) Adversarial attacks against medical deep learning systems. CoRR, abs/1804.05296, 13 Aug 2018. https://dblp.org/rec/journals/corr/abs-1804-05296
15. Biggio B, Corona I, Maiorca D, Nelson B, Šrndić N, Laskov P, Giacinto G, Roli F (2013) Evasion attacks against machine Learning at test time. In: Proceedings of joint European conference on machine learning and knowledge discovery in databases, set.ECML PKDD
16. Carlini N, Wagner DA (2018) Audio adversarial examples: targeted attacks on speech-to-text. CoRR, vol.abs/1801.01944 [Online]. Available: http://arxiv.org/abs/1801.01944
17. Srndic N, Laskov P (2014) Practical evasion of a learning -based classifier. A case study. In: Proceedings of IEEE security and privacy symposium
18. Sharif M, Bhagavatula S, Bauer L, Reiter MK (2016) Accessorize to a crime: Real and stealthy attacks on state-of-the-art face recognition. In: Proceedings of the 2016 ACM SIGSAC conference on computer and communications security. ACM, pp 1528–1540

19. Papernot N, McDaniel O, Goodfellow I, Jha S, Berkay Celik Z, Swami A (2016) Practical black-box attacks against machine learning. In: Proceedings of the 2017 ACM Asia conference on computer and communications security, Abu Dhabi, UAE
20. Evtimov I, Eykholt K, Fernandes E, Kohno T, Li B, Prakash A, Rahmati A, Song D (2017) Robust physical-world attacks on deep learning models, vol 1. arXiv preprint arXiv:1707.08945
21. Kurakin A, Goodfellow I, Bengio S (2016) Adversarial examples in the physical world. arXiv preprint arXiv:1607.02533
22. Xie C, Wang J, Zhang Z, Zhou Y, Xie L, Yuille A (2017) Adversarial examples for semantic segmentation and object detection. In: International conference on computer vision. IEEE
23. Chen S-T, Cornelius C, Martin J, Chau DH (2018) ShapeShifter: robust physical adversarial attack on faster R-CNN object detector. In: Joint European conference on machine learning and knowledge discovery in databases, pp 52–68
24. Hirano H, Minagi A, Takemoto K (2021) Universal adversarial attacks on deep neural networks for medical image classification. BMC Med Imag 21(9). https://doi.org/10.1186/s12880-020-00530-y
25. https://www.kaggle.com/bachrr/covid-chest-xray
26. Cohen JP, Bertin P, Frappier V, Chester: a web delivered locally computed chest x-ray disease prediction system. arxiv, https://arxiv.org/abs/1901.11210

Brain Tumor Detection Using Fine-Tuned YOLO Model with Transfer Learning

Abhishek Dixit and Pooja Singh

Abstract The main motive of this research is to propose a transfer learning (TL) with fine-tuning for a deep learning model to detect brain tumor (BT) from MRI scans dataset. The brain tumor has obtrusive properties leading to a high demise rate but it is curable if the diagnosis is performed early. For this purpose computer aided diagnosis is used which is fast and reliable as compared to traditional method and this paper presents a YOLO based model for detection of tumor in MRI. The procedure is to use YOLO (version-4) model which is trained using a dataset which contain 3064 T1 weighted, contrast-enhanced (CE) MRI which were preprocessed and labeled using tool. By the help of TL we had the advantage of pre-trained loads of the COCO dataset which boosts the learning and provided the required features for detecting tumor in the brain. The work presented is prepared with the 29-layer YOLO Tiny and fine-tuned to work efficiently and perform task productively and accurately in most cases with solid execution. The outcome of the model is the highest like precision, recall and F1-score beating other previous results of earlier versions of YOLO and other studies like Fast R-CNN.

Keywords Brain tumor · Deep learning · Object detection · Computer aided diagnosis

1 Introduction

The nervous system (NS) of our body has more critical components as that of galaxy. Basically, NS has two main parts first is the central nervous system which contains two parts which are the brain and the spinal cord (SC) and second is the peripheral nervous system (PNS) and it is made of nerves which branch off from the SC and

A. Dixit (✉) · P. Singh
Amity School of Engineering & Technology, Amity University, Uttar Pradesh, Noida 201313, India
e-mail: abhishek.dixit@amity.edu

P. Singh
e-mail: psingh22@amity.edu

M. Gupta et al. (eds.), *Artificial Intelligence on Medical Data*,
Lecture Notes in Computational Vision and Biomechanics 37,
https://doi.org/10.1007/978-981-19-0151-5_30

reach out to all pieces of the body. The components of central nervous system (CNS) perform complex capacities that until the present time are not completely perceived, one of which is the development of cerebrum tumors. Universally, one of the major causes of death and disability is the brain tumor, as it affects the most important organ of human being. Brain tumor is caused due to the development of strange cellular bodies inside the brain. The phenomenon behind development of these cells is still unknown. The studies shows that the following tumors can be diagnosed, Glioma, Meningioma and Pituitary tumor. The most dangerous type of tumor is the cerebral glioma as it originates in the glial cell of brain. As stated by WHO in its disease report, brain tumor addresses under 2.00% of malignant growths in people; all things considered, genuine bleakness and challenges are accounted [4]. As of now, mind malignant growth is the tenth driving reason for death because of tumors for people [1]. Brain tumors can be dangerous fundamentally affecting the personal satisfaction and making a huge difference for an individual and their close ones. As tumor has obtrusive properties, leading to a high demise rate. But a motivational fact is that the disease is curable if the diagnosis is performed early. The Magnetic Resonance Imaging (MRI), Computed Tomography (CT), Positron Imaging Tomography (PET), X-ray Screening and Ultra-Sound Screening are the standard imaging technology used in the field of medical science for detecting tumor [2, 3].

In addition, the deep convolutional neural networks (DCNN) which is the outcome of machine learning (ML) and computer vision (CV), plays a very important role in performing critical tasks. DCNNs are cutting edge models that had tackled complex CAD issues as far as acknowledgment, arrangement, division and even location [4]. In any case, most existing CAD solutions which are based on DCNN do not run effectively on many platforms and they too required very high computational power. The lighter versions of DCNN are also available but the problem with them is that they are unable to point out the exact location of the tumor. The other model which are capable of locating the exact position of the tumor is the segmentation model but it requires high computation cost. The main focus of this work is to present a recent deep learning model which is lighter and has better performance than the existing ones. Such a solution has many advantages in medical field, like the diagnosis can be done in a faster manner.

2 Literature Review

The author in [5] introduced an overview on brain tumor detection utilizing image processing. The focus of the paper is to summarize the techniques used in each step for detection of BT in the MRI. The general procedure for identifying the BT in MRI are: Preprocessing, Filtering, Segmentation and Post-preparing. In [6], the author started the utilization of the Content-based-Image-Retrieval (CBIR) method from a vigorous dataset to segment brain tumor. The work presented by them used a new framework which augmented the tumor area to act as the region of inters, which are further divided into subregion by acknowledging the intensity with the help of adaptive spatial division. They achieved a mAP of 94.68% by using the Fisher kernel which help in accumulating all the regions and the resultant is image level signature. Many started to investigate deep learning to detect brain tumors from MRIs. Additionally, DCNN turned out to be notable because of its performance in classification of classes as compared to other techniques. As crafted by Swati et al. [7], by the help of transfer learning they figured out how to utilize a pre-prepared DCNN similar to VGG19. Transfer learning help in the way that it gives initial training which can be used for their classifier to use image recognition features right away. Deepak et al. [8], performed multi-class classification for CAD with the help of transfer learning. Their work included the patient-level five-fold cross-validation to categorize different brain tumor from the similar source. They achieved an accuracy of 98% by utilizing the pre-trained DCNN GoogleNet and their training approach. Their work proved that brain tumor classification can be achieved by transfer learning and fine-tuning. The research work of Rehman et al. [9], they produced synthetic data by using image processing to improve the performance of their model, this step helped their DCNNs like AlexNet by Alex Krizhevsky other is GoogleNet by Google and last one is VGG16 to deliver more features. 97.39, 98.04, and 98.69% are the accuracies of their classifier. DCNNs [10] performed multi-class classification by using a custom-build CNN model and it is performed on the same set of brain tumors.

In the proposed CNN the overfitting is controlled by some activation functions, dropout layers, pooling and normalization. Their experiments achieved a remarkable performance with an accuracy of 97.7% which is best among the other state-of-the-art techniques, which clearly proves the importance of model tuning on image classification. Noreen et al. the author in [11] used DenseNet201 and InceptionV3, to diagnose cerebrum tumors, they developed a multi-stage feature extraction, and the results were outstanding, achieved 99.51% for DenseNet201 and 99.34% for InceptionV3. Many used an object detection model for the purpose of detecting brain tumor, Bhanothu et al. [12] Faster R-CNN. The used model was able to detect the location of the tumor by using bounding box, as compared to the DCNN, Faster R-CNN achieved mAP of 77.60%.

Many have worked on brain tumor diagnosis and few of them are mentioned in the Table 1. Many of the work achieved a good result with classification and segmentation. The author Al-masni et al. [13], used "You Only Look Once" (YOLO), an object detection model, for detection of breast masses in mammograms. During the training and testing stage the YOLO model takes the whole input image into consideration rather than dealing with the images in patches, it increased the speed of the model as compared to the other models like Fast R-CNN [14] and the window-sliding method [15]. As they achieved an accuracy score of 99.70%, they concluded that the YOLO model can lead to the new levels in the field of CAD. The effectiveness of YOLO model is proved by another research done by Ünver et al. [16], they used a YOLO version 3 model which was trained on 2000 labeled pictures (without synthetic data) to classify skin disease. They also used a segmentation layer which is GrabCut and attain an accuracy of 93.4% outperforming others which uses U-Net and ResNet

Table 1 List of algorithms used for detecting brain tumor

Author	Model	Approach	Performance (%)	Year
Cheng et al.	Novel-segmentation Model	Segmentation	mAP of 94.68	2016
Bhanothu et al.	Fast R-CNN	Detection	mAP 77.60	2020
Swati et al.	VGG19	Classification	Accuracy of 94.82	2019
Deepak et al.	GoogleNet	Classification	Accuracy of 98	2019
Rehman et al.	Alex Net, GoogleNet, VGG16	Classification	Accuracy of 97.39, 98.04, and 98.69, respectively	2019
Sultan et al.	Customer CNN	Classification	Accuracy of 98.7	2019
Noreen et al.	DenseNet201 and InceptionV3	Classification	Accuracy of 99.34 and 99.51	2020

model. It proves that YOLO can be helpful in solving the medical problems. Few previous research claims that a high computational power is required and an ordinary performance is shown by the YOLO model leading it to be trusted less for the future arrangements and adaptations. By this the use of YOLO decreased for medical purposes. But this silence was broken when the YOLO version 3 and 4 made a significant improvement in its performance as compared to similar models and the computational cost is also less.

The presented work tests the performance of YOLOv4 model in term of training and to produce a model which is automated and capable of diagnosing brain tumor. The work also aims to develop a model which is less bulky and require less computational power which helps in deploying the model in any platform. By going through the studies on the YOLO we can say that the models can be improved for the automated detection of brain tumor. None of the research work include version 4 of YOLO for this purpose.

3 Methodology

The models which are available for the brain tumor detection are based on DCNN and their performance is not so good on many platforms and they also need a high computational cost. There are the lighter versions of DCNN but they are unable to point out the exact location of the tumor. The model which is capable of locating the exact location of the tumor is the segmentation model but it requires a high computational cost. This work used an existing DL model to produce efficient solution with excellent performance to detect brain tumor from MRIs.

Figure 1 introduced that the version of YOLO used is YOLOv4 Tiny which uses less computational power [17]. To gain the features of image recognition the model was pre-trained with the COCO dataset [18]. The tiny version of YOLO version 4 does not immediately work up to the mark because of the hyper parameter settings

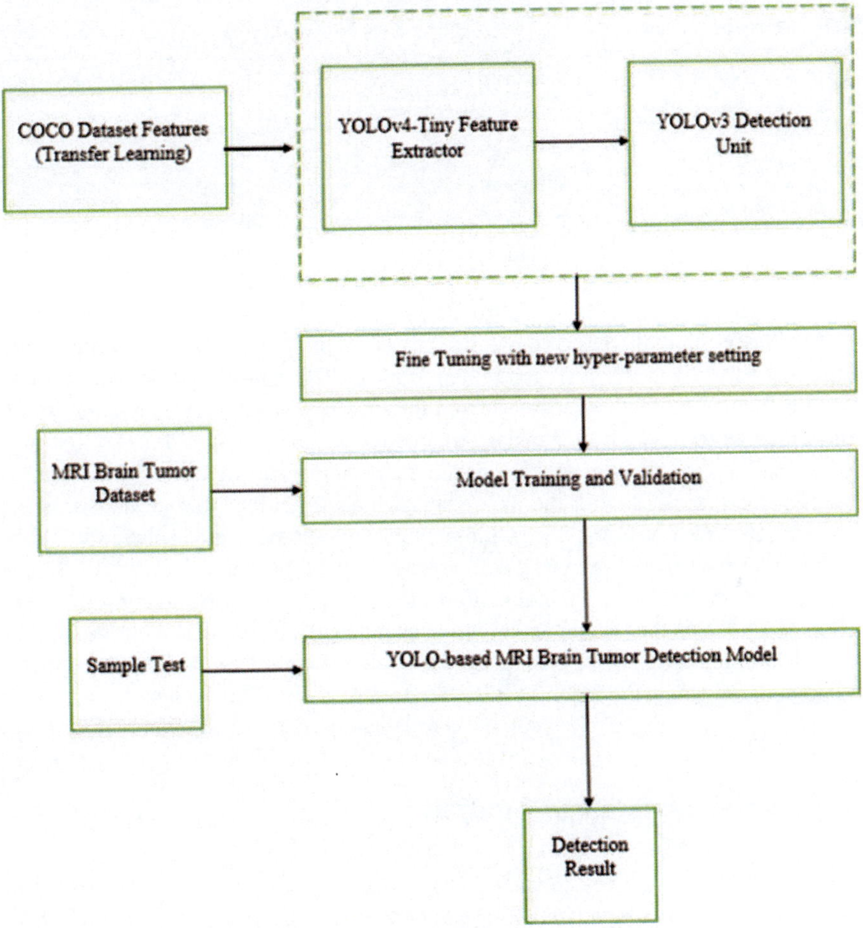

Fig. 1 The methodology adapted for brain tumor detection

and the features from the COCO dataset. The work further fine-tuned and redevelop the model to detect the object of interest. As the training and the fine-tuning is done, the model is again trained on the MRI dataset to gain new weights. Hence, the resultant model can classify different tumors like meningioma, glioma and pituitary tumors from MRIs.

3.1 TL, Model Training and Fine-Tuning

The deep leaning depends upon the data, using inappropriate data can lead to incorrect performance. The transfer learning provide a way of training models and attaining

Table 2 Hyper-parameters taken

Hyper-parameters	Value
Batch-size	64
Sub-divisions	8
Learning-rate	0.0000261
Momentum	0.9
Decay	0.0005
Iterations	6000

good results without the use of large dataset. The work presented also used transfer learning and utilized the weights gained from COCO dataset for detecting brain tumor in MRI. The features learned from the COCO dataset adds the capability of image recognition. To advance the pre-trained model fine-tuning is done for optimal resource allocation and to take care of the memory loss during testing and training [19]. For the proposed work the parameters which are tailor fitted are cluster-size, learning-rate, sub-divisions, decay, momentum and iterations as displayed in Table 2. The fine-tuned model prepared with a group size of 64, a development of 8, and emphases of 6000. The decay, momentum and learning-rate for the preparation interaction depended on the current resource and changed ideally with upsides of 0.00261, 0.9 and 0.0005. Moreover, to give introductory aftereffects of the presentation during the preparation stage, the loads serialized naturally every 1000 cycles.

4 Experimentation and Analysis

Table 3 depicts the average precision (AP) which is based on the number of true positive (TP) and false positive (FP) of each model. As shown in the table the YOLO version 4 with dimension 415 × 415 has the largest AP for classifying meningioma tumor. For the dimensions of 607 × 607 has an average precision of 98.50% and

Table 3 The total detection and average precision

Model	Meningioma-0			Glioma-1			Pituitary-2		
	TP	FP	AP (%)	TP	FP	AP (%)	TP	FP	AP (%)
YOLOv4 tiny (607 × 607)	138	8	98.50	220	57	83.30	174	9	96.20
YOLOv4 tiny (415 × 415)	138	6	98.70	231	39	84.60	174	13	96.00
YOLOv4 tiny (319 × 319)	137	9	97.80	234	42	86.10	168	12	94.70

319 × 319 has 97.80%. If we see for the Glioma, the 319 × 319 has the highest AP which is 86.10% with 607 × 607 at 83.30% and 415 × 415 at 84.60%. The category which is the, 607 × 607 has the highest AP which is 96.20% followed by 415 × 415 with 96.00% and 319 × 319 with 94.70%. The results obtained from the experiments are excellent, each model has performed differently and has different result in classifying tumors, and so it is very difficult to point out the best one. So one more evaluation set is prepared which help us to determine which is the best model for the classifying tumor.

Table 4 presents the impact of provided dimensions on the execution. A mAP of 93.10% is obtained from the 415 × 415 input dimension, the 319 × 319 is with 92.80% but the 607 × 607 has no improvement and had the lowest mAP of 92.80%. Notwithstanding, it had a little expansion in IoU with 71.70% and 319 × 319 just had 71.50%. If we see the overall performance the dimension 416 × 416 has the best mAP, IoU being 72.50, RC of 88.60 and F1-score being 89.50. The dimension 607 × 607 has the lowest MR which is 1.10% but 415 × 415 has 2.20% and 319 × 319 has 1.80%. The largest FPPI goes with 607 × 607 of 12.20%, 319 × 319 with 10.80% and 415 × 415 with 9.70%.

Table 5 contains the research which imply the similar strategies and a comparison is given in the table. As presented in the table our work got the highest mAP of 93.10% with YOLO version 4 model. The YOLO version 4 tiny model outperforms all existing models which can detect brain tumor.

Table 4 Detection in various input dimensions

Model	mAP	IoU	PR	RC	F1	MR	FPPI
YOLOv4 tiny (607 × 607)	92.80	71.70	88.00	86.90	87.40	1.10	12.20
YOLOv4 tiny (415 × 415)	93.10	72.50	90.30	88.60	89.50	2.20	9.70
YOLOv4 tiny (319 × 319)	92.80	71.50	89.50	87.90	88.70	1.80	10.80

Table 5 Comparison with existing approach

Method	Dataset size	Classes	Dataset	Result (%)
YOLOv4 tiny	3064	3	[22]	93.10
YOLOv3 tiny	3064	3	[22]	91.98
YOLOv2 tiny	3064	3	[22]	79.90
Fast R-CNN [20]	3064	3	[22]	91.66
Fast R-CNN [21]	2406	3	[22]	77.60

5 Conclusion

This work introduced the productivity of utilizing TL and calibrating a new YOLO model to perform recognition of some categories of brain tumors in MRIs. The preprocessing of data is done by using some techniques like labeling, file-conversion, normalizing pixel contrast. The main part of the model is the YOLO version 4 Tiny which extracted the features for the detector. The proposed model utilized both the weights of COCO dataset and the features extracted from the MRI. The result of the assessment shows that the tiny version of YOLO4 gives 93.10% mAP with 415 × 415 as dimension and can detect tumor correctly.

The presented work proves that if an object detection model is pre-trained and fine-tuned then it will be able to detect brain tumor. If we compare this method with the classification one then this work proves to be the most efficient and effective, it also outperforms the segmentation methods in term of computation and cost. Besides that training a YOLO and a similar model is quite lengthy process as if requires labeling of the dataset. The results of the YOLO model depend on how well the model is trained and how much data is used to train the model, it requires a large dataset. Further research and work can improve the mentioned concerns. The labeling.

6 Future Work

As the YOLOv4 performs well for the detection of brain tumor, in the future we will be working on the methods and parameters which will enhance the performance of the model. There is a new version of YOLO which is version 5, in the future we will also be working on the YOLOv5 for the task of tumor detection and will compare the performance of both the models.

References

1. Schapira ML, Dale MW (2021) Brain tumor: statistics. https://www.cancer.net/cancer-types/brain-tumor/statistics Accessed 10 Apr 2021
2. Hossain A, Islam MT, Chowdhury MEH, Samsuzzaman M (2020) A grounded coplanar waveguide-based slotted inverted delta-shaped wideband antenna for microwave head imaging
3. Mohammed BJ, Abbosh AM, Mustafa S, Ireland D (2014) Microwave system for head imaging. IEEE Trans Instrum Meas 63(1):117–123
4. Stewart BW, Wild CP (2014) World cancer report 2014. IARC. IARC Nonserial Publ, Lyon, France, p 630
5. Kapoor L, Thakur S (2017) A survey on brain tumor detection using image processing techniques. In: 7th international conference on cloud computing, data science & engineering
6. Cheng J, Yang W, Huang M, Huang W, Jiang J et al (2016) Retrieval of brain tumors by adaptive spatial pooling and fisher vector representation. PLoS ONE 11(6):e0157112. https://doi.org/10.1371/journal.pone.0157112

7. Zhao Q, Ali F, Kabir M, Ali Z, Ahmed S, Lu J, Swati ZNK (2019) Brain tumor classification for MR images using TL and fine-tuning. Comput Med Imag Graph, Article (Link)
8. Deepak S, Ameer PM (2019) Brain-tumor classification using deep-CNN features via TL. Comput Biol Med, Article (Link)
9. Naz S, Rehman A, Akram F, Razzak MI, Imran M (2019) A deep learning-based framework for automatic brain tumors classification using TL. Circuits Syst Signal Process, Article (Link)
10. Salem NM, Sultan HH, Al Atabany W (2019) Multi classification of brain-tumor images using deep neural network. IEEE Access, Article (Link)
11. Sarwinda D, Bustamam A (2016) Detection of Alzheimer's disease using advanced local binary pattern from hippocampus and whole brain of MR images. In: 2016, international joint conference neural networks (IJCNN) IEEE. (pp 5051–5056)
12. Kamalakannan A, Bhanothu Y, Rajamanickam G (2020) Detection and classification of brain-tumor in MRI images using deep convolutional network. In: Proceedings of 2020 6th ICACCS, Article (Link)
13. Kumar PSJ, Saha MA Digital image processing based detection of brain abnormality for Alzheimer's disease
14. Yao HM, Jiang L, Sha WEI (2020) Enhanced deep learning approach based on the deep convolutional encoder–decoder architecture for electromagnetic inverse scattering problems. IEEE Antennas Wirel Propag Lett 19(7):1211–1215
15. Gerazov B, Conceicao RC (2017) Deep learning for tumour classification in homogeneous breast tissue in medical microwave imaging. In: Proceedings IEEE 17th international conference smart technology (EUROCON), pp 564–569
16. Khoshdel V, Asefi M, Ashraf A, LoVetri J (2020) Full 3D microwave breast imaging using a deep-learning technique. J Imag 6(8):80
17. Bochkovskiy A, Wang CY, Mark Liao HY (2004) YOLOv4: optimal speed and accuracy of object detection, arXiv:2004.10934, Article (Link)
18. Lin TY, Maire M, Belongie S, Hays J, Perona P, Ramanan D, Dollár P, Zitnick CL (2014) Microsoft COCO: Common objects in context. In: Proceedings European conference on computer vision, pp 740–755, Article (Link)
19. Ferrari V, Henderson P (2017) End-to-end training of object class detectors for mean-average precision. In: Proceedings of ACCV, Article (Link)
20. Palaniappan S, Noreen N, Ahmad I, Qayyum A, Shoaib M, Imran M (2020) A deep learning model based on concatenation technique for diagnosis of brain tumor. IEEE Access 8:55135–55144, Article (Link)
21. Nayak DR, Dash R, Majhi B (2015) Classification of brain MR images using discrete-wavelet-transform and random forests. In: 2015 fifth national conference computer vision, pattern recognition, image processing and graphics (NCVPRIPG), IEEE. (pp 1–4)
22. Goswami S, Bhaiya LKP (2013) Brain tumour detection using unsupervised learning based neural network. In: 2013 international conference communication systems and network technologies (CSNT), IEEE. (pp 573–577)

Computational Image Analysis of Painful and Pain-Free Intervertebral Disc

Jerina Tiwari, Siddhi Raj Sharma, Sukirti Chauhan, Mike Adams, and Polly Lama

Abstract Disc pain is associated with reinnervations and revascularization, and these ingrowing vessels are confined to (a) the peripheral 2 mm of annulus and (b) vicinity of fissures and clefts deeper in the matrix. We hypothesize that 'pain-free' aged intervertebral discs even though structurally intact show certain histological changes similar to the pathological 'painful' disc removed at surgery matched thorough grade of degeneration; however, blood vessels ingrowths are a distinct feature of 'painful disc'. To analyses the tissues differences with accuracy, confocal system and volocity image analysis tool was used for assessments of images, to develop a predictive tool compatible to be used with macroscopic images. Sixty intervertebral disc tissue samples were collected from lower lumbar regions, and 30 discs tissue were from the patient's undergoing surgeries due to 'pain' while 30 discs were from non-degenerated 'pain-free' structurally intact disc samples. Histological and confocal analysis was used to score the images, and the confirmatory changes was matched with grades of degeneration. The results indicate blood vessels and inflammatory cells into the discs are confined to fissures and disrupted regions in 'painful disc' in the oaf and the iaf regions, while 'pain-free' disc shows fissures devoid of any vessels. Age-related changes were mainly in the form of cell clusters and loss of

J. Tiwari · S. Chauhan · P. Lama (✉)
Sikkim Manipal Institute of Medical Sciences, Sikkim Manipal University, 5th Mile Tadong, Sikkim 737102, India
e-mail: pollylama@smims.smu.edu.in

J. Tiwari
e-mail: jerina.t@smims.smu.edu.in

S. Chauhan
e-mail: sukirti@smims.smu.edu.in

S. R. Sharma
North Bengal Neuro Centre, Mahananda Sarni, Siliguri, West Bengal, India

M. Adams
Centre for Clinical Anatomy, University of Bristol, Bristol, UK
e-mail: mike.adam@bris.ac.uk

M. Gupta et al. (eds.), *Artificial Intelligence on Medical Data*,
Lecture Notes in Computational Vision and Biomechanics 37,
https://doi.org/10.1007/978-981-19-0151-5_31

proteoglycans. The source of pain thus seems to be focal and correspond to fissured or delaminated outer and inner annulus regions of the intervertebral discs.

Keywords Intervertebral discs · Image analysis · Disc degeneration · Pain · Fissures

1 Introduction

As age advances, the cartilaginous intervertebral discs located between the vertebral bodies of the spine undergoes immense morphological and microscopic histochemical changes [1, 2]. The nutritional supply to the disc in particular changes and influence the process of disc metabolism which makes the core of the nucleus pulposus, inner and outer annulus fibrosus regions of the intervertebral disc relatively toxic for the survival of the disc cells [7]. Therefore, the amount of matrix metalloproteases (MMP's) a degradative enzyme increases with age, injury and pathogenesis [3, 11, 38], which facilities the breakdown of extracellular matrix leading to defects or alterations in the orientation of the collagen fibres [23]. These changes impair the healing process and complicates transport of the nutrients especially to the centre of the disc allowing cell death and loss in maintenance of structural strength and integrity [23]. However, it is known that age-related degeneration of the intervertebral discs may not always lead to pathogenesis [10], and pain can be accelerated by environmental factor, genetics, obesity, poor physical conditions, trauma and osteoarthritis [1, 8, 12].

Several studies have been conducted to distinguish the changes behind pathological and age-related disc degeneration [1, 7, 25–27, 43]; however, less is known about aged non-degenerated disc with features similar to painful human discs. The similarities in human disc between the 'painful' pathological and aged-matched 'pain-free' subgroups is relatively unexplored especially from a histological perspective [13].

Therefore, the aim of the study was to compare the pathological surgically removed painful discs with the non-degenerated cadaveric pain-free intervertebral discs by confocal and volocity image analysis software through the process of grading the disc tissues with well-established degenerative histo-pathological scoring system [7, 25]. The hypothesis is that the non-degenerated cadaveric 'pain-free' intervertebral discs even though structurally intact show certain histological changes similar to the pathological 'painful' disc removed at surgery matched thorough grade of degeneration. Further, the study analyses volocity image analysis tool for assessments of histological image data as a process for development of a predictive tool compatible to be used with macroscopic image analysis.

2 Materials and Methods

The study was approved by the Sikkim Manipal Institute of Medical Sciences, Sikkim Manipal University Institution Ethics committee, (SMIMS/IEC Registration No IEC/462/16-070) India. Sixty intervertebral disc tissue samples were collected from L1 to S1 regions, out of which 30 discs tissue were from the patient's undergoing surgeries for painful disc degeneration, disc herniation, nerve root pain and spondylis-thesis. Thirty non-degenerated, structurally intact, pain-free cadaveric disc samples were collected from the Department of Anatomy. Cadaveric discs samples were scored thorough gross examination based on Thompson's grade of disc degeneration [41].

An anonymous clinical data sheet was collected from the pathological 'painful disc', and those discs were graded according to the Pfirmann grade of disc degeneration [33]. Disc tissues were collected from 41 male to 19 female patients for this purpose, and age range were between 34 and 72 years.

2.1 *Tissue Characterization for Histological Assessment*

Disc tissue were cut into small 8mm^3 cubes, snap-frozen before sectioning, and embedded in optimal cutting tissue (oct) medium. 5 μm thin serial sections were cut using Leica cryostat, and the tissue sections were fixed in 10% neutral buffer formalin (nbf). Ehrlich haematoxylin and eosin (h and e) stain were used to reveal cells and the features of the matrix, and toluidine blue stain was used to assess loss of proteoglycan especially around fissures and tears.

Three tissue types were identified histologically as nucleus pulposus (np), inner annulus fibrosus (iaf) and outer annulus fibrosus (oaf). Oaf regions comprised of highly orientated and lamellar type I collagen fibres with fibroblast-like cells running parallel to the collagen fibre in each successive lamella. The collagen fibres have high affinity for eosin stain and are birefringent when viewed under polarized light microscope (Fig. 1a). Iaf comprised of wider and less organized collagen lamellae especially around cell clusters, collagen fibres showed diminished birefringence and eosin staining (Fig. 1b). Np had an amorphous collagen-poor but glycosaminoglycans-rich matrix that did not exhibit birefringence, with a few rounded chondrocyte-like cells (Fig. 1c) in clusters or pairs [9, 14].

2.2 *Histological Scoring System*

All stained sections were observed under a LEICA light microscope with 20 × and 40 × objectives and graded for histological changes based on a modified Boos scale [7] shown in Table 1.

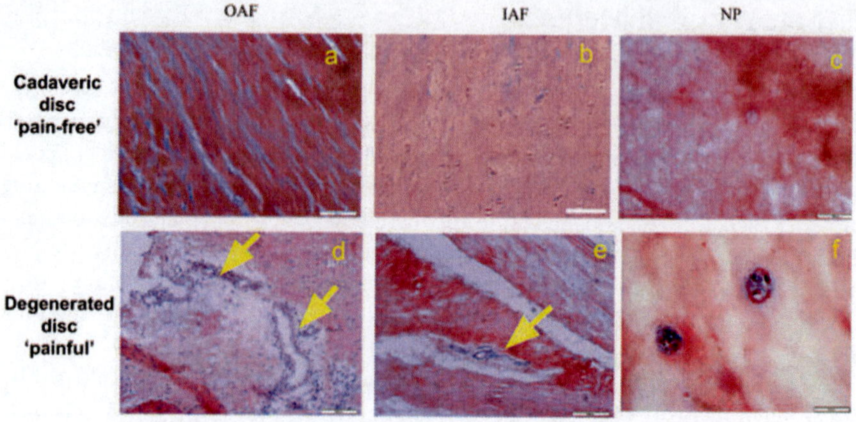

Fig. 1 Comparisons of three regions: (**a** and **d**) outer annulus fibrosus (oaf), (**b** and **e**) inner annulus (iaf) and (**c** and **f**) nucleus pulpous (np) regions in 'pain-free and painful disc'. Arrow shows blood vessels in the iaf and oaf regions of painful disc. Scale bar 100 μm

Table 1 Histological assessment of degenerative features, modified Boss et al. [7]

Variables of histological assessment	Grades
Neovascularity/blood vessels Newly formed blood vessels with reparative alterations	0 = absent 1 = rarely present 2 = present in intermediate amount 3 = abundant
Tears/fissures Splitting of collagen fibre bundles in the annulus fibrosus and nucleus regions creating a gap or break within the tissue	0 = absent 1 = rarely present 2 = present in intermediate amount 3 = abundant
GAG depletion Cystic, oval or irregular areas with loss in sulphated glycosaminosoglycans staining. (GAG rich matrix usually appears dark blue to purple with toluidine blue staining)	0 = absent 1 = rarely present 2 = present in intermediate amount 3 = abundant
Inflammatory cells Immunocytes with small irregular nuclei, often in swarms along the edges of/surfaces of disc tissue	0 = absent 1 = rarely present 2 = present in intermediate amount 3 = abundant
Chondrocyte clustering Multiple chondrocyte in small rounded groups or clusters sharply demarcated by a rim of territorial matrix	0 = none 1 = increase in single chondrocytes 2 = paired chondrocytes 3 = clusters of 3–7 cells 4 = maximum clusters of 8–15 cells 5 = clusters of > 15 cells

2.3 Immunofluorescence

In brief, 30 μm thick sections were fixed in acetone cooled at −20 °C for 10 min and rinsed in phosphate buffer saline (PBS). Sections were treated with 20% donkey serum (Invitrogen, UK) in PBS for 60 min at 4 °C. Sections were washed in three changes of PBS and were then treated with mouse monoclonal primary antibodies at 1:25 dilutions in PBS for PGP 9.5 (general cytoplasmic neuronal marker, ABD Sorotec UK), 1:20 dilution for CD 31 (endothelial cell marker, Dako, UK) and 1:500 dilutions for sub-P (nociceptive neuropeptides marker, ABCAM, UK). Secondary donkey anti-mouse antibody was conjugated with alexa-fluorophore (Invitrogen, UK) and used at 1:200 dilutions in PBS. To quench auto fluorescence, sections were immersed in 0.1% Sudan black B for 5 min [5]. Sections were then washed in PBS solution, counter stained and mounted with DAPI (Vector laboratories UK). The excitation/emission spectrum for the secondary antibody was at 550/607 nm. For, DAPI the excitation was at 405 nm and emission at 488 nm. Primary antibodies were omitted during the incubation stage in the case of negative controls, and autofluorescence was quenched using 0.1% Sudan black B [5].

2.4 Computational and Confocal Analysis of Degenerative Features

To *quantitatively* determine the extent of ingrowths, identification of fissures and delaminated annulus was made with a confocal microscope. Disrupted and fissured regions were filled with inflammatory cells, blood vessels within them. The areas occupied by distinct blood vessels and nerves, and their distance were then measured from the nearest fissure based on whether np, iaf, oaf regions was being visualized.

To take such measurements 'Volocity' 3D (Perkin Elmer) image analysis software was used. A built-in freehand region of interest (ROI) was used to outline and trace the path of blood vessels in each field of view. The area occupied by each such vessel was automatically calculated in μm^2 (after adjusting the settings of the software). 'Line' tool in the software was used to measure ingrowths from the nearest fissure axis: a straight line was made by clicking the line tool to create two anchor points; by continuing to click, a path could be created and the length of the path was automatically calculated in μm (after adjusting the software settings). Once a vessel or nerve had been identified in the image data a unique coloured-coded number was assigned to it. Counts of blood vessels were then made together with their total cross-sectional areas in mm^2.Ten 30 μm thick sections were examined from each disc specimen, taken from superficial and deep zones. Depending upon the sections, three fields of view were taken each of the nucleus pulposus (np), inner annulus fibrosus (iaf) and outer annulus fibrosus (oaf) regions per section. Immunoreactive blood vessels were observed with 20 × magnification primarily, but for quality images 40 × objective magnification was also used. Digital images were obtained with a Leica DMIRBE argon laser confocal scanning inverted microscope. All sections were sequentially scanned to prevent crosstalk between different fluorophores, and to prevent photobleaching a four-line average was used for each image. Line average scans each line that corresponds to the section and was useful in improving signal-to-noise ratio. For each tissue type, and patient group, values were expressed as mean ± standard deviation.

2.5 *Statistical Analysis*

One-way ANOVA was used to compare mean values of histological, immunohistochemical and radiological grades between (a) the three regions of interest (np, iaf and oaf) and (b) different patient groups. A two-way ANOVA was used to examine interactions between the three regions of interest and the two patient groups. Associations between histological variables were analysed using Spearman's rank correlation. All statistical tests were executed using SPSS software version 18. All data were expressed as mean ± standard deviation of the mean. A p value < 0.05 was considered as statistical significance.

3 Results

3.1 *Histological Characterization of Three Tissue Region in Painful and Pain-Free Disc*

Oaf regions were highly disorganized with evident displacement of collagen fibres especially in painful disc patient groups. Disrupted areas were infiltrated with inflammatory cells, blood vessels and fibroblasts (Fig. 1d–e) in the oaf region. Large, moderate and small-sized cell clusters were abundant around fissured/disrupted iaf regions in painful disc tissue. Clusters were abundant within np regions, and a few cells showed characteristic features of apoptosis (Fig. 1g). A few cells in the np regions showed lysed cell membrane under high magnification in few severely degenerated disc tissues. Large-sized cell clusters in the np regions existed around fissures. In contrast, degenerative features such as blood vessels and inflammatory cells were minimal or completely absent in cadaveric 'pain-free' discs (Fig. 1a–f) and were the hallmark features of disrupted discs. Changes in cellular apoptotic feature were similar in the painful and pain-free groups, and large and medium sized cell clusters were frequently observed in both the groups. Blood vessels were always near or contained within fissures in the painful disc, but all fissured or delaminated regions did not show blood vessels in growths. Especially, in cadaveric disc splits and tears that resembled fissures were observed to be artefacts introduced during tissue sectioning, thus the presence of vessels and other cells lining a fissure acted as a reliable indicator for identification of a natural fissure and allowed distinction from sectioning artefacts.

3.2 *Other Histologically Determined Degenerative Changes*

Toluidine blue staining showed that proteoglycan/GAG loss varied significantly between pain-free and painful patient groups in the three tissue type analysed, np, iaf and oaf regions. Proteoglycan staining was always depleted in severely disrupted parts of iaf and oaf regions, and more intense loss was scored in the np regions. In general, painful disc had the highest scored values for loss of proteoglycans ($p < 0.003$), which correlated with evident developments of tears and inflammatory cells ($p < 0.004$), and an increase in blood vessels score ($p < 0.003$). These changes occurred largely in the iaf and oaf regions, Table 2.

3.3 *Blood Vessels Quantified with Confocal Analysis*

The density of blood vessels was greatest in the superficial part of the tissue regions, thus it could be a phenomenon seen after herniation and fissuring of the disc. Serial 5 μm sections were analysed with 'volocity' image analysis software, and the mean number of blood vessels count, the area occupied by them, and the extent of penetrations were colour coded and numbered to create a distinct set of variable, which were compared in the two painful and pain-free groups. Figure 2a–i shows gross

Table 2 Comparison of histological variables between three tissue types: NP, IAF and OAF in pain-free and painful discs

Tissue type	Patients groups (mean ± SD)					
	Pain-free disc ($n = 30$)			Painful disc ($n = 30$)		
	NP	IAF	OAF	NP	IAF	OAF
Tears (0–3)	0.7 (0.5)	1.1 (0.3)	0.9 (0.4)	1.8 (0.9)	2.3 (0.7)	2.2 (0.4)
GAG loss (0–3)	1.0 (0.6)	1.3 (0.5)	0.7 (0.6)	1.3 (0.8)	2.1 (0.3)	1.8 (1.0)
Cell clusters (0–5)	1.1 (0.3)**	1.2 (0.9)**	0**	1.4 (0.5)**	2.3 (1.2)**	0**
Blood vessels (0–3)	0	0	0	0**	1.5 (0.7)**	1.5 (1.1)**
Inflammatory cells (0–3)	0	0	0.4 (0.5)	0*	1.2 (1.0)*	1.4 (1.3)*

The 'n' values refer to the number of patients in each group. Significant differences between the three tissue types for various histological variables in each patient group are shown in bold, with significance levels denoted as: $*P < 0.05$) or $**P < 0.01$

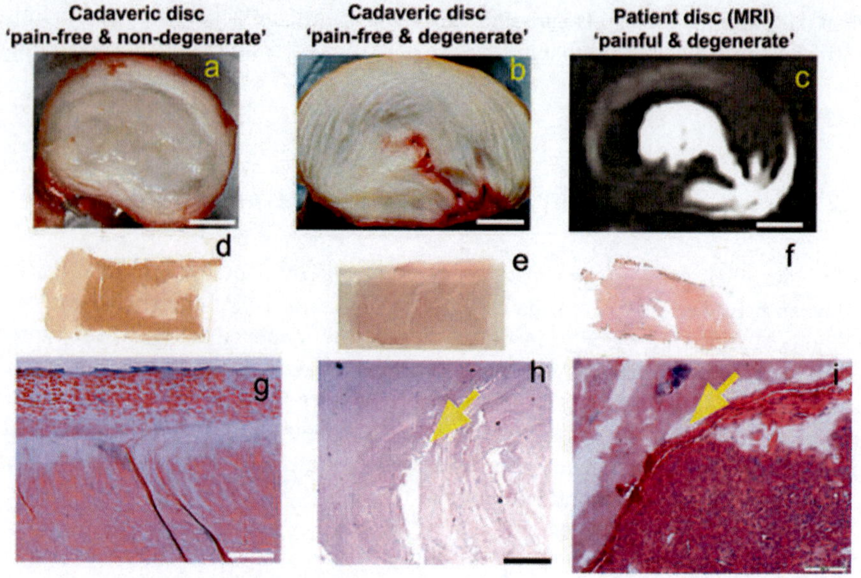

Fig. 2 Macroscopic and microscopic comparisons between 'pain-free and painful disc'. Arrow shows blood vessels in the iaf and oaf regions of painful disc. **a-c** shows gross morphological differences between pain-free and painful discs, **d-f** histological sections on slide, **g-i** shows assessment of section with evident displacement of collagen fibres and presence of inflammatory cells in disrupted region (arrows). Scale bar 100 μm

morphological differences between pain-free and painful discs, histological sections on slide and (g–i) shows assessment of section with evident displacement of collagen fibres and presence of inflammatory cells in disrupted region.

Painful and pain-free groups differences were recorded for blood vessels ingrowths ($p < 0.001$), and the area occupied by blood vessels ($p < 0.002$), Fig. 3a–e.

3.4 Blood Vessels Colour Code and Grade of Degeneration

The colour-coded blood vessels were converted into a binary data and compared with MRI grade of degeneration (Fig. 3a–c). The data sequence and identifiable image code were matched with the pain-free tissue types to analyses if a predictive model could be made with the code-based identification of the structure in MRI images. The accuracy level was recoded at 99% as preliminary data findings.

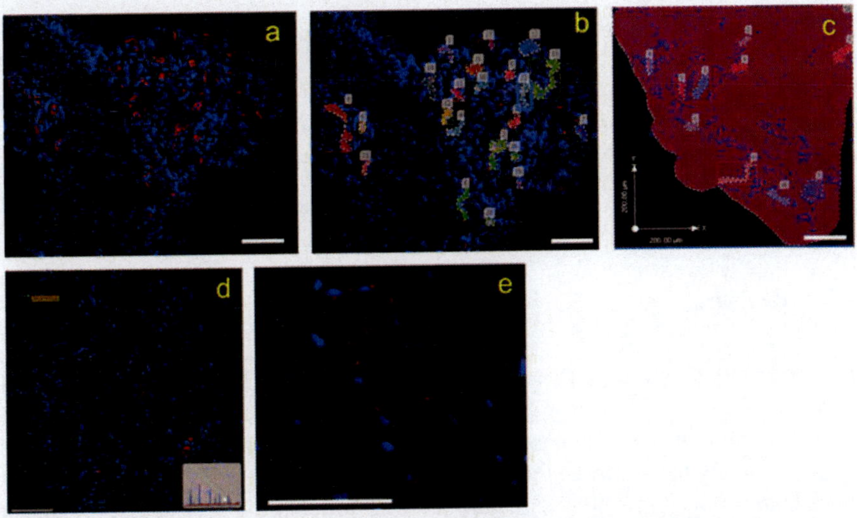

Fig. 3 Confocal analysis of tiny blood vessels in 'painful disc'. **a** Blue indicates DAPI stained cell-nuclei, and red are CD-31 immunoreactive blood vessels. **b–d** Quantification and colour coding of vessels with 'volocity' image analysis software, and **e** magnified view of CD-31 positive vessel. scale bar 70 and 50 μm

4 Discussion

4.1 Summary of Findings

This study reports blood vessels and inflammatory cells into the discs are confined to fissures and disrupted regions in painful disc in the oaf and the iaf regions. Blood vessels correlates with pfirmann grades of degeneration, presence of tears or fissures and loss of GAG staining which were absent in the age-matched cadaveric 'pain-free' disc. Thus, this is the first study to identify features of pain in age matched 'human pain-free and painful discs'. Cell clusters of different sizes were a feature of painful and pain-free disc and exhibited similar features of apoptosis, with increase in density of cells around fissures. The images used for grading the tissue samples from both the group were colour coded to map a predictive stimulation for degree of degeneration which correlated with the pfirmann grades of disc degeneration in the painful tissue group.

4.2 Study Strengths

The use of surgically removed tissues from the patients group allows the data to be applicable to human subjects with pain and who has not undergone surgery, thus the results from this study are applicable to human populations. The age-matched cadaveric pain-free disc tissue confirms localized changes such as inflammatory cell invasion, and blood vessels ingrowth are a feature of painful disc, and changes in cellularity are a common feature of pain-free disc and painful disc. The study also shows that colour coding of images can be developed to deliver a predictable software technology.

4.3 Weaknesses

The study has only focused on histological images, and therefore identification of nerves in the painful tissue group is not taken into consideration, though previous published data our own studies had shown the presence of nerves in painful discs [25]. This study has explored the potential of colour coding and image analysis, but these findings need to be further confirmed with patient trails with the development of a robust software.

4.4 Relationship to Other Studies

It has been suggested that the neuropathological basis of pain is the nerve endings or nociceptors present within the lumbar intervertebral disc [3]. Stimulation of nociceptors causes the release of neuropeptides such as sub-P, CGRP and VIP [4, 8, 11, 12]. Neuropeptides have been suggested to play an important role in initiating the process of neurogenic inflammation capable of vasodilatation of blood vessels [16] resulting in infiltration of the discs tissue with inflammatory cells, release of various inflammatory mediator such as phospholipase A2, prostaglandin E2, nitric oxide and cytokines [17, 18, 21, 22, 36, 37]. While cytokines have been implicated to play a role in neuropathic pain response, [38] it has been suggested that the cells of the nucleus pulposus also secrete pro-inflammatory mediators such as TNF-α,IL-1 andIL-6 that can induces pathological changes when in contact with nerve roots, inducing nerve oedema, demyelination, activation of macrophages and radicular pain (Saal et al., 1990). Increased cytokines production by NP cells in degenerated discs has also been associated with activation of nerve growth factors and its receptors [10, 20]. It is also been suggested that increased expression of pro-inflammatory mediators such as IL-1β and TNF-α may stimulate degenerative changes that increases production of matrix metalloproteinases, degradation of collagen fibres and loss of GAG's [26, 34, 39]. Tissue disruptions/tears could lead to loss of compressive stress resisting properties of nucleus pulposus [2] allowing swelling of disc tissue in order to balance the internal stress, followed by GAG loss, promoting ingrowths of blood vessels and nerves [27] as well as sensitization of nerves and vasodilatation of vessels by inflammatory mediators with release of substance P [4, 8] giving rise to elevated levels of pain. It has been suggested that inflammatory cells synthesize, store and secrete nerve growth factors, which may be considered as an inducer to promote nerve ingrowths into inner layers of the discs [17, 32]. Such infiltrating inflammatory cells were always observed to be most abundant amongst the patients with painful disc.

SP, CGRP and VIP immunoreactivity have been described to occur in the superficial layers of the annulus fibrosus, not exceeding the outer 3 mm [4]. Such neovascularization was localized in painful patients especially around the superficial zones and were less frequent within deeper zones. This suggest that increased presence of vessels could have been a surface phenomenon which could have occurred after the disc tissue lost its structural integrity and came in contact with surrounding structures like dorsal root ganglia (DRG) or nerve roots [29] resulting in mechanical compression of the nerve roots, inflammation and pain [28].

4.5 *Explanation of Results*

Structurally intact or minimally disrupted discs tissue from pain-free disc showed no vessels, had fewer fissures, and showed minimal proteoglycan loss while severely degenerated painful discs showed abundant vascular proliferation, loss of proteoglycans, cell clustering and fissures. Stress profilometry studies show that severely degenerated discs with radial fissures and delaminated annulus have highly irregular stress profiles that aids an overall reduction in compressive stress resisting properties of the discs [2]. Such loss of compressive stress correlates with loss of proteoglycans/GAG's and water content of the discs. These changes increase load on the annulus resulting in delamination and fissures, focal swelling and proteoglycan loss leaving behind a scaffold of GAG depleted collagen lamellae which provide an ideal substrate for blood vessels and nerve ingrowths [1, 2, 40]. Therefore, it can be suggested that fissures could prompt blood vessels and nerve ingrowths. Nociceptive neuropeptides sense changes in the intra-discal pressure upon tissue damage [2], promoting peripherally located blood vessel vasodilatation [8] and inflammatory cell invasion, thus facilitating further ingrowths into structurally weakened and damaged regions of degenerated [14, 15] and herniated discs tissue [30].

In this study GAG's loss, inflammatory cell infiltration and fissures were mostly observed within iaf and oaf regions, while in np and iaf/oaf junctions large cell cluster formations were frequent. It is likely that degenerative changes such as large cell cluster formations with gradual loss in GAG's synthesis could alter the hydrostatic pressure making the discs tissue weak and vulnerable to injury even with minor mechanical stimuli, resulting in fissures and herniation of disc, and thus sensitization of blood vessels and nerves with elevated expressions of cytokines and substance P could be rapid process. One of the potential observations suggesting this view is the association of blood vessels and nerves within GAG depleted fissured regions and its correlations with pain.

Technically, this study shows that use tissue sections and image coding make it possible to indicate with confidence the painful changes are most localized, the results obtained from this study thus support our aim, and shows that blood vessels into the annulus is near fissures, and predominates in painful disc.

4.6 Future Questions and Research

In agreement with previous published studies, we observed that GAG depleted, and fissured areas served as an ambient environment for blood vessels ingrowths, and on the contrary proteoglycan rich areas show absence of blood vessels with presence of cell clusters in pain-free discs. Cell cluster formations are possibly due to loss in collagen restraint and inevitable loss of GAG's particularly with age are unlikely to cause pain. Focal swelling is perhaps a feature of painful disc that weaken the binding of cells with the surrounding matrix, promoting inferior matrix synthesis, which alters the stability of the discs tissue and allow it to herniate even with minor mechanical stimuli, thus vessels and nerves could lead to synthesis and release of substance P, a pain transmitter. The source of pain could be focal and correspond to fissured or delaminated iaf and oaf regions of the disc that are heavily vascularized. Therefore, the focus for further studies should be why and how does intervertebral discs tissue develop fissures and delamination, and how does the unusual environment of annulus fissure affect disc cell metabolism.

4.7 Clinical Relevance

Attempts to treat low back pain by denervation should address pathologies and regions where innervations are most concentrated.

References

1. Adams MA, Bogduk N, Burton K, Dolan P (2012) The biomechanics of back pain, 3rd edn. Churchill Livingstone, Edinburgh
2. Adams MA, McNally DS, Dolan P (1996) 'Stress' distributions inside intervertebral discs: the effects of age and degeneration. J Bone Joint Surg Br 78:965–972
3. Aoki Y, Ohtori S, Takahashi K, Ino H, Takahashi Y, Chiba T, Moriya H (2004) Innervation of the lumbar intervertebral disc by nerve growth factor-dependent neurons related to inflammatory pain. Spine 29:1077–1081
4. Ashton IK, Walsh DA, Polak JM, Eisenstein SM (1994) Substance P in intervertebral discs: binding sites on vascular endothelium of the human annulus fibrosus. Acta Orthop Scand 65:635–639
5. Baschong W, Suetterlin R, Laeng RH (2001) Control of autofluorescence of archival formaldehyde-fixed, paraffin-embedded tissue in confocal laser scanning microscopy (CLSM). J Histochem Cytochem 49:1565–1571

6. Bogduk N, Tynan W, Wilson AS (1981) The nerve supply to the human lumbar intervertebral discs. J Anat 132:39–56
7. Boos N, Weissbach S, Rohrbach H, Weiler C, Spratt KF, Nerlich AG (2002) Classification of age-related changes in lumbar intervertebral discs: 2002 Volvo Award in basic science. Spine 27:2631–2644
8. Brown MF, Hukkanen MV, McCarthy ID, Redfern DR, Batten JJ, Crock HV, Hughes SP, Polak JM (1997) Sensory and sympathetic innervation of the vertebral endplate in patients with degenerative disc disease. J Bone Joint Surg Br 79:147–153
9. Bruehlmann SB, Rattner JB, Matyas JR, Duncan NA (2002) Regional variations in the cellular matrix of the annulus fibrosus of the intervertebral disc. J Anat 201:159–171
10. Burke JG, Watson RWG, McCormack D, Dowling FE, Walsh MG, Fitzpatrick JM (2002) Intervertebral discs which cause low back pain secrete high levels of proinflammatory mediators. J Bone Joint Surg Br 84:196–201
11. Coppes MH, Marani E, Thomeer RTWM, Oudega M, Groen GJ (1990) Innervation of annulus fibrosis in low back pain. Lancet 336:189–190
12. Edgar MA (2007) The nerve supply of the lumbar intervertebral disc. J Bone Joint Surg Br 89-B:1135–1139
13. Fagan A, Moore R, Vernon Roberts B, Blumbergs P, Fraser R (2003) ISSLS prize winner: the innervation of the intervertebral disc: a quantitative analysis. Spine 28:2570–2576
14. Freemont AJ, Peacock TE, Goupille P, Hoyland JA, O'Brien J, Jayson MI (1997) Nerve ingrowth into diseased intervertebral disc in chronic back pain. Lancet 350:178–181
15. Freemont AJ, Watkins A, Le Maitre C, Baird P, Jeziorska M, Knight MT, Ross ER, O'Brien JP, Hoyland JA (2002) Nerve growth factor expression and innervation of the painful intervertebral disc. J Pathol 197:286–292
16. Goupille P, Jayson MIV, Valat J-P, Freemont AJ (1998) The role of inflammation in disk herniation-associated radiculopathy. Semin Arthritis Rheum 28:60–71
17. Grönblad M, Virri J, Tolonen J, Seitsalo S, Kääpä E, Kankare J, Myllynen P, Karaharju EO (1994) A controlled immunohistochemical study of inflammatory cells in disc herniation tissue. Spine 19:2744–2751
18. Habtemariam A, Virri J, Grönblad M, Holm S, Kaigle A, Karaharju E (1998) Inflammatory cells in full-thickness anulus injury in pigs: an experimental disc herniation animal model. Spine 23:524–529
19. Johnson WEBP, Caterson BP, Eisenstein SMPF, Roberts SP (2005) Human intervertebral disc aggrecan inhibits endothelial cell adhesion and cell migration in vitro. Spine 30:1139–1147
20. Johnson WEBP, Sivan SP, Wright KTB, Eisenstein SMPF, Maroudas AP, Roberts SP (2006) Human intervertebral disc cells promote nerve growth over substrata of human intervertebral disc aggrecan. Spine 31:1187–1193
21. Kang JD, Georgescu HI, McIntyre-Larkin L, Stefanovic-Racic M, Donaldson WFI, Evans CH (1996) Herniated lumbar intervertebral discs spontaneously produce matrix metalloproteinases, nitric oxide, interleukin-6, and prostaglandin E2. Spine 21:271–277
22. Kang JDMD, Stefanovic-Racic MMD, McIntyre LABS, Georgescu HIBS, Evans CHP (1997) Toward a biochemical understanding of human intervertebral disc degeneration and herniation: contributions of nitric oxide, interleukins, prostaglandin e2, and matrix metalloproteinases. [miscellaneous article]. Spine 22:1065–1073
23. Lama P, Claireaux H, Flower L, Harding IJ, Dolan T, Le Maitre CL, Adams MA (2019) Physical disruption of intervertebral disc promotes cell clustering and a degenerative phenotype. Cell Death Discov 5:154
24. Lama P, Le Maitre CL, Dolan P, Tarlton JF, Harding IJ, Adams MA (2013) Do intervertebral discs degenerate before they herniate, or after? Bone Joint J 95-B(8):1127–1133
25. Lama P, Le Maitre CL, Harding IJ, Dolan P, Adams MA (2018) Nerves and blood vessels in degenerated intervertebral discs are confined to physically disrupted tissue. J Anat 233(1):86–97
26. Le Maitre CL, Freemont AJ, Hoyland JA (2005) The role of interleukin-1 in the pathogenesis of human intervertebral disc degeneration. Arthritis Res Ther 7:R732-745

27. Melrose J, Roberts S, Smith S, Menage J, Ghosh P (2002) Increased nerve and blood vessel ingrowth associated with proteoglycan depletion in an ovine anular lesion model of experimental disc degeneration. Spine 27:1278–1285
28. Olmarker K, Larsson K (1998) Tumor necrosis factor [alpha] and nucleus-pulposus-induced nerve root injury. Spine 23:2538–2544
29. Omarker K, Myers RR (1998) Pathogenesis of sciatic pain: role of herniated nucleus pulposus and deformation of spinal nerve root and dorsal root ganglion. Pain 78:99–105
30. Ozaki S, Muro T, Ito S, Mizushima M (1999) Neovascularization of the outermost area of herniated lumbar intervertebral discs. J Orthop Sci 4:286–292
31. Palmgren T, Grönblad M, Virri J, Kääpä E, Karaharju E (1999) An immunohistochemical study of nerve structures in the anulus fibrosus of human normal lumbar intervertebral discs. Spine 24:2075
32. Peng B, Wu W, Hou S, Li P, Zhang C, Yang Y (2005) The pathogenesis of discogenic low back pain. J Bone Joint Surg Br 87:62–67
33. Pfirrmann CW, Metzdorf A, Zanetti M, Hodler J, Boos N (2001) Magnetic resonance classification of lumbar intervertebral disc degeneration. Spine 26:1873–1878
34. Purmessur D, Freemont AJ, Hoyland JA (2008) Expression and regulation of neurotrophins in the nondegenerate and degenerate human intervertebral disc. Arthritis Res Ther 10:R99
35. Roberts S, Eisenstein SM, Menage J, Evans EH, Ashton IK (1995) Mechanoreceptors in intervertebral discs: morphology, distribution, and neuropeptides. Spine 20:2645–2651
36. Saal JS (1995) The role of inflammation in lumbar pain. Spine 20:1821–1827
37. Saal JS, Franson RC, Dobrow R, Saal JA, White AH, Goldthwaite N (1990) High levels of inflammatory phospholipase A2 activity in lumbar disc herniations. Spine (Phila Pa 1976) 15:674–678
38. Séguin CA, Pilliar RM, Roughley PJ, Kandel RA (2005) Tumor necrosis factorα modulates matrix production and catabolism in nucleus pulposus tissue. Spine 30:1940–1948
39. Smith LJ, Chiaro JA, Nerurkar NL, Cortes DH, Horava SD, Hebela NM, Mauck RL, Dodge GR, Elliott DM (2011) Nucleus pulposus cells synthesize a functional extracellular matrix and respond to inflammatory cytokine challenge following long-term agarose culture. Eur Cell Mater 22:291–301
40. Stefanakis M, Al- M, Harding I, Pollintine P, Dolan P, Tarlton J, Adams MA (2012) Annulus fissures are mechanically and chemically conducive to the ingrowth of nerves and blood vessels. Spine 37:1883–1891
41. Thompson JP, Pearce RH, Schechter MT, Adams ME, Tsang IK, Bishop PB (1990) Preliminary evaluation of a scheme for grading the gross morphology of the human intervertebral disc. Spine 15(5):411–415
42. Tolofari S, Richardson S, Freemont A, Hoyland J (2010) Expression of semaphorin 3A and its receptors in the human intervertebral disc: potential role in regulating neural ingrowth in the degenerate intervertebral disc. Arthritis Res Ther 12:R1
43. Weiler C, Lopez-Ramos M, Mayer H, Korge A, Siepe C, Wuertz K, Weiler V, Boos N, Nerlich A (2011) Histological analysis of surgical lumbar intervertebral disc tissue provides evidence for an association between disc degeneration and increased body mass index. BMC Res Notes 4:497

Imaging Informatics: Next Generation and Innovations

A Deep Learning Technique for Bi-Fold Grading of an Eye Disorder DR-Diabetic Retinopathy

Padmanayana and Anoop B K

Abstract Nowadays, with increasing cases of diabetes, one should control the blood sugar as well as perform regular examination of eyes to prevent oneself from blindness. Any person having diabetes is likely to develop diabetic retinopathy (DR). DR is triggered by high blood sugar due to diabetes. After some time, having excessive amount of sugar in blood, can damage retina. When sugar jams the tiny blood vessels the eyes are damaged and this will affect the blood vessels and result in leakage of fluid. Millions of working aged adults suffers from loss of sight due to diabetic retinopathy. DR cannot be treated completely, but early detection of DR prevents the person from vision loss. We proposed a deep learning model for detection of diabetic retinopathy. Detection of DR is a slow process. Physical detection of DR involves a trained clinician to study and estimate the color fundus photographs of the retina. Normal process of identification takes a minimum of two days. In our paper, convolutional neural network architecture has been used to classify images into two classes which is no-diabetic retinopathy and with diabetic retinopathy. APTOS-2019 blindness detection dataset has been used from Kaggle which contains high-resolution retinal images. Those images are used to train the model. Web-based interface has been created for easy interaction with the model.

Keywords Contrast limited adaptive histogram equalization (CLAHE) · Convolution neural network (CNN) · Deep learning · Diabetic retinopathy (DR) · Gaussian-blur filter

Padmanayana (✉)
Computer Science Department, SUIET, Mangaluru 574146, India
e-mail: Padmanayana10@sitmng.ac.in

Anoop B K
HOD,AI & ML Department, SIT, Mangaluru 574143, India

M. Gupta et al. (eds.), *Artificial Intelligence on Medical Data*,
Lecture Notes in Computational Vision and Biomechanics 37,
https://doi.org/10.1007/978-981-19-0151-5_32

1 Introduction

Diabetic retinopathy is a diabetes problem that affects eyes. It is affected due to the damage of blood vessels to photosensitive tissues in retina. Some of the symptoms are blurred vision, vision loss, dark or empty area of vision or dark strings floating in one's vision.

Timely detection of diabetic retinopathy precludes the person from vision loss [1]. Manual detection of diabetic retinopathy takes more time and requires a trained clinician for detection. Convolutional neural network architecture model has been used to train the model. High-resolution retinal images are used to train the model.

Web-based interface has been used as the front end of the model, which interacts with the model. The web page takes high-resolution retinal images as input. The image is then pre-processed and enhanced, [2] and then result is predicted. The predicted result is displayed in the web page. In our paper, we used convolutional neural network architecture to train the model.

The trained model predicts the result as no-diabetic retinopathy and with diabetic retinopathy. The predicted result is displayed in the web page. Figure 1 displays the retina image which shows the normal image as well as the retina image having diabetic retinopathy, where we can see the swelling and blood clot in the blood vessels.

In our paper, we used 3600 high-resolution retinal image from APTOS_2019 blindness detection dataset to train the model of Kaggle. These images taken to train are pre-processed and enhanced before training the model. The enhanced images are used for classification of the image as no-diabetic retinopathy and with diabetic retinopathy. In pre-processing, high-resolution images are cropped, resized, and converted to gray scale image. Gray scale images are blurred using Gaussian-blur filter. CLAHE is implemented to enhance the image and to remove noise and increase contrast of the image.

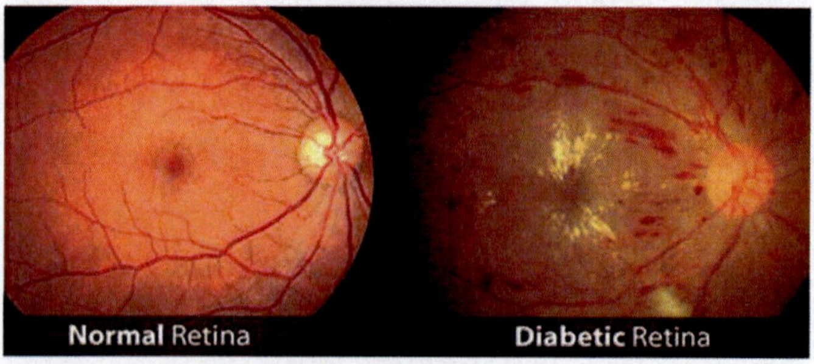

Fig. 1 Retina image before and after diabetic retinopathy

2 Literature Review

Revathy et al. [3] proposed method which detects hemorrhages, exudates, and micro aneurysms. Dilation and erosion operators are performed. The three-classifier used to classify the images are support vector machine, k-nearest neighbor, and random forest, and these classifiers classifies the image as no-diabetic retinopathy and with diabetic retinopathy. In Zeng et al. [4], the paper proposes CNN model with Siamese-like architecture. The model takes binocular fundus images as inputs. In Borys et al. [5] proposed an ensemble model for final solution. Three CNN architectures are used for transfer learning such as EfficientNet-B4, EfficientNet-B5, SE-ResNeXt50. In Biran et al. [6], the model used many image processing techniques such as circular Hough transform (CTH), contrast limited adaptive histogram equalization (CLAHE), Gabor filter, and thresholding. Support vector machine is used to classify the image. The model. Pao et al. [7] proposed bichannel-convolutional neural network architecture used to train the model. The model is trained in this manner, which accepts the inputs of both entropy images and the green component from unsharp masking. Kumaran et al. [8] in their paper have used artificial neural network and used high-resolution retinal image for training the model.

3 Proposed Method

3.1 Preprocessing

In our paper, we have trained the model with high-resolution retina images. Figure 2 shows the block diagram of the model. Color fundus retinal images are used for the

Fig. 2 Block diagram

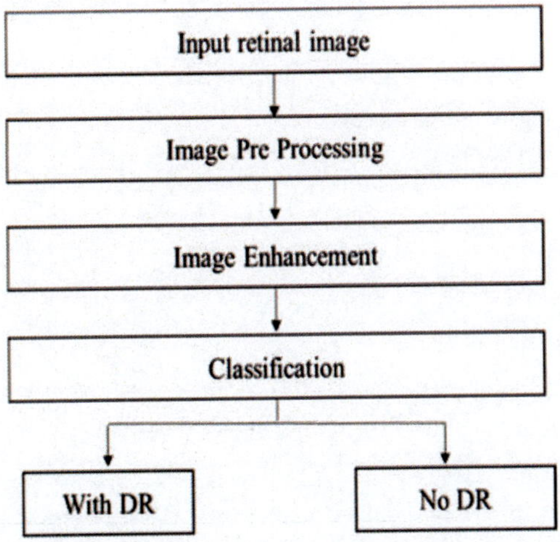

DR detection. These retinal images are pre-processed, enhanced and classifies it as No DR or with DR.

3.2 Enhancement

In this process, the contrast of the image is increased by using contrast limited adaptive histogram equalization (CLAHE) filter. This filter is used to increase the contrast of the image, and noise amplification problem is decreased.

CLAHE does equalization of histogram [10] in small patches or in small tiles with high accuracy and contrast limiting. The function add weighted from cv2 Python library is used to unsharp masking and high boost filtering of the image. In this function, the images are blurred and smoothened then the smoothened image is subtracted from the original image making it unsharp masking. This mask is then added to the original image enhancing the high-frequency components. Figure 3 exhibits the original input image, and Fig. 4 exhibits the input image after pre-processing and enhancement.

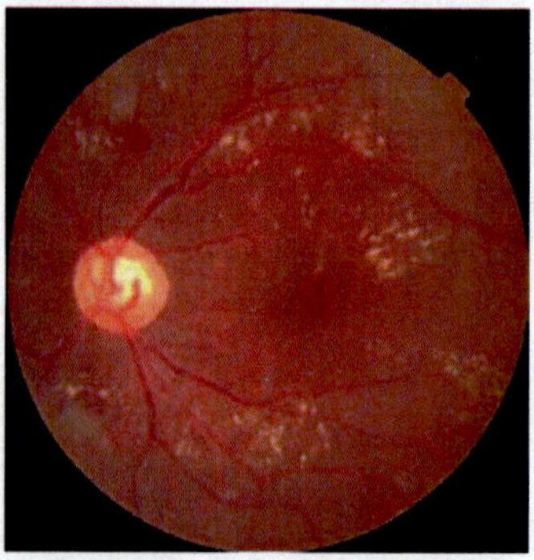

Fig. 3 Original image

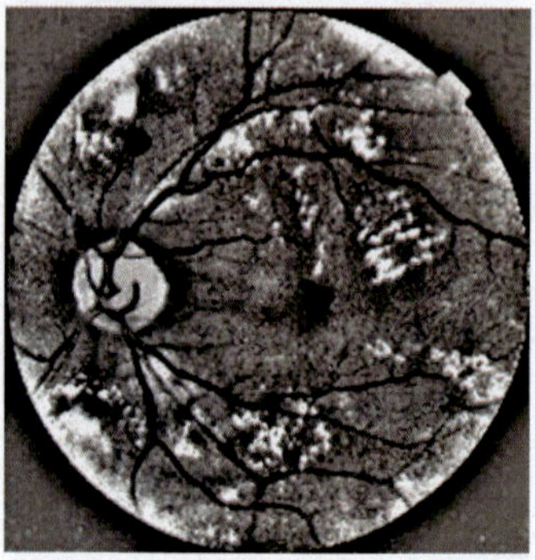

Fig. 4 Image processing and enhancement

Table 1 Summary of CNN layers used in this work

Layer type	Output shape	Parameters
Conv2D	(None, 222, 222, 32)	320
Batch_normalization	(None, 222, 222, 32)	128
Conv2D	(None, 220, 220, 32)	9248
Batch_normalization	(None, 220, 220, 32)	128
MaxPooling	(None, 110, 110, 32)	0
Droupot	(None, 110, 110, 32)	0
Conv2D	(None, 108, 108, 64)	18,496
Batch_normalization	(None, 108, 108, 64)	256
Conv2D	(None, 106, 106, 128)	73,856
Batch_normalization	(None, 106, 106, 128)	512
MaxPooling	(None, 53, 53, 128)	0
Droupot	(None, 53, 53, 128)	0
Dense	(None, 512)	184,091,136
Batch_normalization	(None, 512)	2048
Droupot	(None, 512)	0
Dense	(None, 128)	65,664

3.3 Architecture

In our paper, we proposed a convolutional neural network model which is a type of neural network model. The model is trained using convolutional neural network architecture. The core building block of convolutional neural network is convolutional layer. Implementation of different layers of CNN such as convolutional layer, pooling layer, and fully connected layer which has rectified linear units (ReLU) correction layer.

CNN model used in this work is sequential with the input image size 224 × 224 × 1.

Total parameters: 184,262,562 Trainable parameters: 184,260,770 Non-trainable parameters: 1792. Table 1 shows the summary of CNN layers used in this work.

3.4 Training

The model is trained with total of 3600 high-resolution retinal images [9]. The model is trained separately. The images are randomly split into 30% of total images as testing image and remaining as training image. Further, the dataset is again randomly split into 40% of total images as validation testing images and remaining as validation training images.

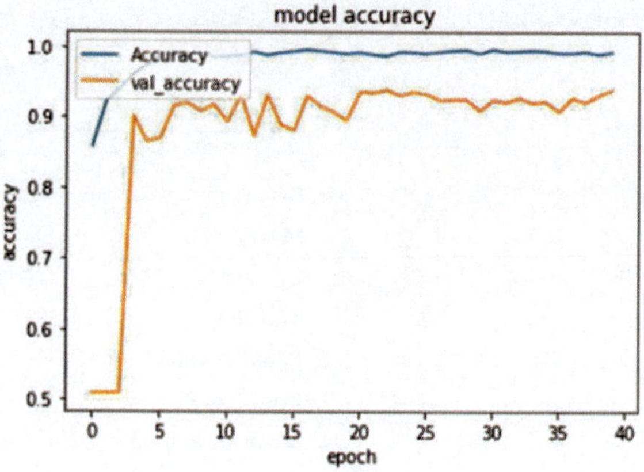

Fig. 5 Model accuracy graph

Figure 5 shows the model accuracy graph for 40 epoch having label as accuracy and epoch. We can see the validation accuracy increases as epoch increases. The trained model is saved and connected to web interface. Web-based interface has been created for easy interaction with the user. The web page will accept the high-resolution retinal image as input. This retinal image is pre-processed. In pre-processing, the images are cropped, resized, and blurred using Gaussian-blur filter and converted to gray scale. Then the images are enhanced by CLAHE filter [5] which improve the contrast of the image. The model then predicts the result as 0 or 1. The predicted result is sent back, and the classification result is displayed in web page as No DR or with DR.

3.5 WEB Application

The Web-based interface has been created for easy interaction with the model. The web page is created using the FLASK web frame work which is written using Python. The first page gives the details or knowledge about the diabetic retinopathy. In second page, user can upload the image to check for the result. After a few seconds, the web page displays the result as No DR or with DR, classifying the retinal image.

4 Results

In this paper, an efficient deep learning model has been proposed to detect DR-diabetic retinopathy using convolutional neural network. To train the model, architecture convolutional neural network has been used. The web-based interface has been created which connects the model for better interaction. High-resolution images are given as input from the web page which then given input to the model. The input image is pre-processed and enhanced. The model then predicts the result as 0 or 1. The classification is then displayed in the web page as No DR or with DR.

True positive (TP) is the condition where the images are with DR and are predicted as with DR. True negative (TN) is the condition where the images are No DR and are predicted as No DR. False positive (FP) is the case where retinal images are No DR and got predicted as with DR. False negative (FN) is the case where retinal images are with DR and got predicted as No DR. Figure 6 shows the confusion matrix which will summarize the performance of a classification algorithm. Confusion matrix will summarize the true positive, true negative, false positive, and false negative of the trained model.

The model has the sensitivity of 86% and specificity of 96%. Sensitivity is the measurement of true positive rate which means it predicts very few negative results or false positive rates. Sensitivity formula can be defined as Sensitivity = TP/(TP + FN). Specificity is the measurement of true negative rate which means it predicts very few negative results or false negative rates. Specificity formula can be defined as Specificity = TN/(TN + FP).

Table 2 shows the performance metric for the binary classifier used in this work.

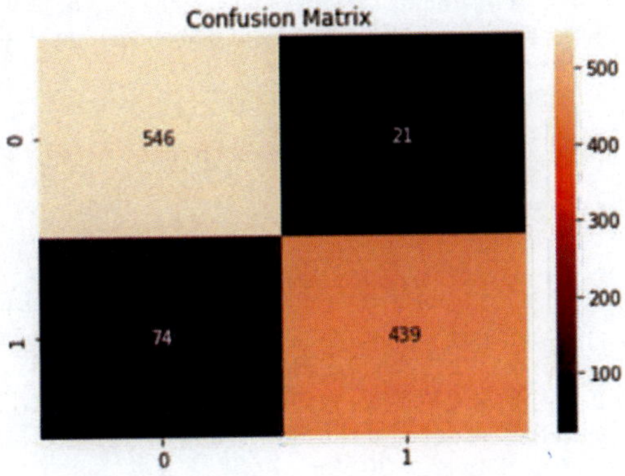

Fig. 6 Confusion matrix

Table 2 Performance metric for binary classification

Parameters	Precision	Recall	F1-score	Support
No DR	0.88	0.96	0.92	567
DR	0.95	0.86	0.90	513
Accuracy	0.95	0.86	0.91	1080
Macro average	0.92	0.91	0.91	1080
Weighted average	0.92	0.91	0.91	1080

5 Conclusion

In our paper, we have created a binary classification of diabetic retinopathy detection using web interface for easy interaction. In this paper, we have considered the convolutional neural network architecture to train the model. Web-based interface has been used for accepting the input image and for predicting the result. The given input images are pre- processed, enhanced, and predicted as 0 or 1. The web page displays the predicted result as No DR or with DR. The model shows 99% of training accuracy and 93% of validation accuracy, and gives the testing accuracy of 91%. The model has the sensitivity and specificity of 86% and 96%, respectively. Future work may include in improving the performance of the model and increasing the accuracy of the model. Training the model with real dataset from the hospitals. Training the model with equal number of datasets of each classification. Using high-resolution retinal images for training and testing the model. Classifying the model with different types of diabetic retinopathy such as mild, moderate, severe, and proliferative. Using the equal number of datasets containing high-resolution of the retinal images to train the model.

Acknowledgements We acknowledge the help of Dr. Govindraj for providing us the useful inputs for the detection of the presence of diabetic retinopathy-DR in the early stage.

References

1. Tian C, Fang T, Fan Y, Wu W (2020) Multi-path convolutional neural network in fundus segmentation of blood vessels. Biocybern Biomed Eng 40(2):583–95. http://doi.org/https://doi.org/10.1016/j.bbe.2020.01.011
2. Wu Y, Xia Y, Song Y, Zhang Y, Cai W (2020) NFN+: a novel network followed network for retinal vessel segmentation. Neural Netw 126:153–162. https://doi.org/10.1016/j.neunet.2020.02.018 Epub 2020 Mar 4 PMID: 32222424
3. Revathy R, Nithya BS, Reshma JJ, Ragendhu SS, Sumithra MD (2020) Diabetic retinopathy detection using machine learning. Int J Eng Res Technol
4. Zeng X, Chen H, Luo Y, Ye W (2020) Deep learning approach to diabetic retinopathy detection. IJRTE
5. Borys T, Marchenko P, Spodarets D (2020) Deep learning approach to diabetic retinopathy detection. IJRTE
6. Biran A, Sobhe Bidari P, Almazroe A, Lakshminarayanan V, Raahemifar K (2016) Automatic detection and classification of diabetic retinopathy using retinal fundus images. IEEE
7. Pao SI, Lin HZ, Chien KH, Tai MC (2020) Detection of diabetic retinopathy using bichannel convolutional neural network. IEEE
8. Kumaran Y, Patil CM (2018) A brief review of the detection of diabetic retinopathy in human eyes using pre-processing & segmentation techniques. IJRTE
9. https://www.kaggle.com/ratthachat/aptos-eye-preprocessing-in-diabetic-retinopathy?scriptVersionId=20340219
10. Wang J, Luo J, Liu B, Feng R, Lu L, Zou H (2020) Automated diabetic retinopathy grading and lesion detection based on the modified R-FCN object-detection algorithm. IET Comput Vis 14:1–8. https://doi.org/10.1049/iet-cvi.2018.5508

SARM-Net: A Spatial Attention-Based Residual M-Net for Polyp Segmentation

Debapriya Banik and Debotosh Bhattacharjee

Abstract Polyp segmentation is a crucial step for the early diagnosis of colorectal cancer. However, the heterogeneous nature of polyps poses a significant challenge in the segmentation task, and it is still an unsolved problem. So in this study, we have proposed a deep learning network, namely "a spatial attention-based residual M-Net for polyp segmentation" (SARM-Net). The network is inspired by the M-Net architecture, where some additional modules are added to the existing architecture to improve the segmentation performance. We have employed residual connections in the M-Net architecture to preserve gradient information during backpropagation, facilitating optimal gradient flow. Meanwhile, unlike M-Net, where the contextual information is directly fed from the encoder to the decoder by skip connections, a spatial attention block (SAB) is introduced in our proposed network to focus on the relevant significant features and ignore the redundant features in the spatial dimension prior to the concatenation which will facilitate better optimization of the network. The segmentation performance was evaluated on the "Kvasir-SEG" database. The experimental results reflect the segmentation performance improvement compared to the traditional deep learning models and a recent state-of-the-art method.

Keywords Colorectal cancer · Polyp · Colonoscopy · Segmentation · Deep learning

1 Introduction

Colorectal cancer (CRC) is the third most commonly occurring cancer worldwide. However, in terms of mortality rate due to cancer, it ranked second, and in India,

D. Banik (✉) · D. Bhattacharjee
Department of Computer Science and Engineering, Jadavpur University, Kolkata-32, India
e-mail: debu.cse88@gmail.com

D. Bhattacharjee
e-mail: debotosh@ieee.org

M. Gupta et al. (eds.), *Artificial Intelligence on Medical Data*,
Lecture Notes in Computational Vision and Biomechanics 37,
https://doi.org/10.1007/978-981-19-0151-5_33

it ranks in the top five among all cancers [1]. So these ranks alarm the foreseeable consequences of CRC. CRC is the most frequent form of malignancy in the gastrointestinal (GI) tract. CRC initiates from the protrusions known as colorectal polyp(s) that develops in the colorectum [2]. These precursor lesions are benign in their initial stage, and there is no significant symptom of the disease in its early stage which makes it difficult to diagnose the disease. CRC has a survival rate of more than 90% if the prognosis is done in its initial stage and the rate drastically falls below 10% in its advanced stage [3]. Colonoscopy examination is considered a gold standard for early diagnosis of CRC where direct visualization of the colorectum is available, and polyps can be diagnosed with high sensitivity and specificity [4]. However, many advanced non-invasive modalities are available, such as wireless capsule endoscopy (WCE) and CT colonography. But these examinations are very time consuming and suffer due to the residual fluid or stool, which results in the obscuration of the polyp region.

During the colonoscopy examination, once the polyp is detected, it undergoes resection for biopsy analysis to evaluate the degree of malignancy and may undergo resection based on the malignancy report. This procedure is time consuming and depends on the expertise of the endoscopists, leading to a favorable risk of polyp miss rates and resection of benign polyps degrades the system's performance [5]. To overcome such limitations, computer-aided polyp segmentation plays a crucial role in the early diagnosis of CRC. Once the polyp is accurately segmented, it undergoes in-vivo optical biopsy analysis based on the polyp patterns, and polyp(s) may be resected accordingly, increasing the efficiency of the diagnosis system. The procedure is cost effective and consumes less time, which may assist the endoscopists in planning for the following treatment sequence.

In the last few years, there has been a significant improvement in developing deep learning (DL)-based techniques for the segmentation of polyp for early diagnosis of CRC. The DL-based techniques seemed to provide a powerful solution for segmenting the polyp regions as it does not require domain-based knowledge to extract relevant features. It surpassed the handcrafted based techniques where features are extracted manually based on texture, color, shape and fed into a machine learning (ML) classifier to segment the polyp regions [6]. This strategy often fails to deal with the high intraclass variation and low interclass variation of the polyps in the colonoscopy frames, which significantly increases the miss-detection rate and suffers from poor generalization ability.

For instance, [2] have proposed an attention gated U-Net architecture, namely focus U-Net. This architecture uses a focus gate that provides spatial and channel attention and introduces focal parameters to give attention to the salient regions in the image and suppress the irrelevant background region. In another study, [7] proposed a shuffle efficient channel attention network (sECANet). The improved channel attention mechanism is employed to each of the convolutional blocks of mask R-CNN, which explicitly models the interdependencies between the channels of the convolutional blocks. In [8], authors have proposed a parallel reverse attention network (PraNet) to segment the polyp region. The authors have employed a partial parallel decoder to aggregate the features from different high-level layers for course segmentation. The parallel reverse attention mechanism can precisely segment by tuning the existing approximated polyp regions from high-level features. In [9], authors have introduced different modules to extract robust information from different levels, such as pyramid vision transformer (PVT) encoder module, cascaded fusion module (CFM), camouflage identification module (CIM), and similarity aggregation module (SAM). These modules address different concerns in the polyp frame and precisely segment the polyp region. A ResUNet++ architecture with conditional random field (CRF) and test time augmentation (TTA) is proposed in the works of [10]. Authors have employed SE block in the encoder and attention module in the decoder with ASPP in the bridge between encoder and decoder. These modules can extract multi-scale features and highlight relevant semantic information and filter our irrelevant information. A multimodel fusion-based segmentation method, namely polyp-Net, is proposed in the works of [11]. Authors have proposed a CNN with a pooling mechanism based on DTCWT and a level set method, namely LGWe-LSM. The pixel-level fusion of those models significantly improves the segmentation map with the reduction of the false-positive regions. In [12], authors have proposed a network PolypSegNet with a backbone of U-Net. Authors have incorporated different blocks in the network, namely DDI module-based D-Unit layer, DFSM block, and DRM block. These blocks improve the performance as these modules extract features from different receptive fields and reduce the semantic gap between the encoder and decoder.

Regardless of such significant progress, the computer-aided polyp segmentation task is still a challenging problem due to its heterogeneous appearances in different stages of its development in size, color, shape, and texture. Moreover [7], there is no sharp boundary between the polyp region and the mucosa, making the segmentation task very difficult, and still, it is an unresolved problem. This study has proposed a deep learning network, namely "a spatial attention-based residual M-Net for polyp segmentation" (SARM-Net). In our proposed network, residual connections are employed in the M-Net architecture to preserve the gradient information during backpropagation, facilitating optimal gradient flow. Meanwhile, unlike M-Net, where the contextual information is directly fed from the encoder to the decoder by skip connections, a spatial attention block (SAB) is introduced in our network to focus on the relevant significant features and ignore the redundant features in spatial dimension prior to the concatenation which will facilitate better optimization of the network.

The rest of the paper is organized as follows. Section 2 describes our proposed methodology and gives a detailed description of the architecture. Section 3 presents the experimental results to evaluate the performance and compare it with traditional deep learning models and a SOTA method. Finally, the conclusion is presented in Sect. 4.

2 Proposed Methodology

This section presents a detailed explanation of our proposed network employed to segment polyp(s) in colonoscopy frames. Figure 1 shows some diversified forms of polyps in colonoscopy frames.

Let Ω be the set of intensity values of the input colonoscopy frame and the input image has the functionality $P : \Omega \rightarrow \Re^{+}$ and $\Omega \subset \Re^{3}$. A grayscale image (P_g) is constructed from P such that $P_g : \Omega \rightarrow \Re^{2}$. $\mathbb{Q} \in Q$ be the ground truth binary mask, and $\mathcal{M} \in M$ be the predicted segmented mask. Each pixel $\varkappa$in the ground truth mask $\mathcal{M} = \{m_{\varkappa}, \varkappa = 1,2, \ldots, n\}$and the predicted mask $\mathbb{Q} = \{q_{\varkappa}, \varkappa = 1,2, \ldots, n\}$takes a value from the set {0, 1} where label 1 represents the polyp regions, and label 0 represents the background regions. For a given input image P_g, the goal of this work is to predict either polyp or background labels for each pixel accurately. Figure 2 represents our proposed SARM-Net architecture. Each of the components used in this architecture is explained in the following subsections.

2.1 Proposed SARM-Net Architecture

The proposed SARM-Net is inspired by the M-Net architecture [13], which is a variant of U-Net [14]. The network has an encoding and decoding branch. Each level of the encoding branch has two cascaded 2D convolution layers (same padding) with

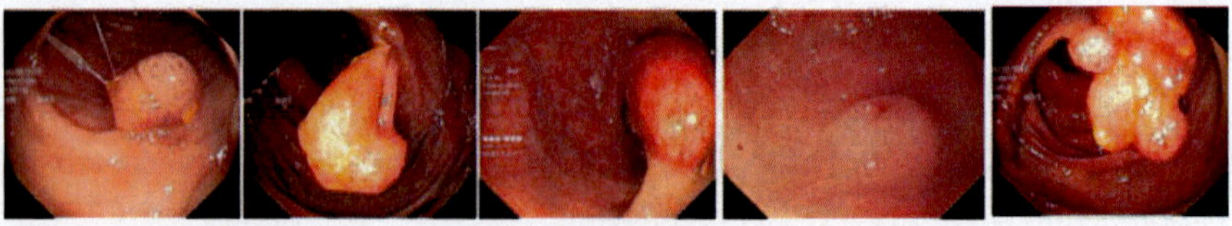

Fig. 1 Different variations of polyps in terms of shape, size, color, and texture

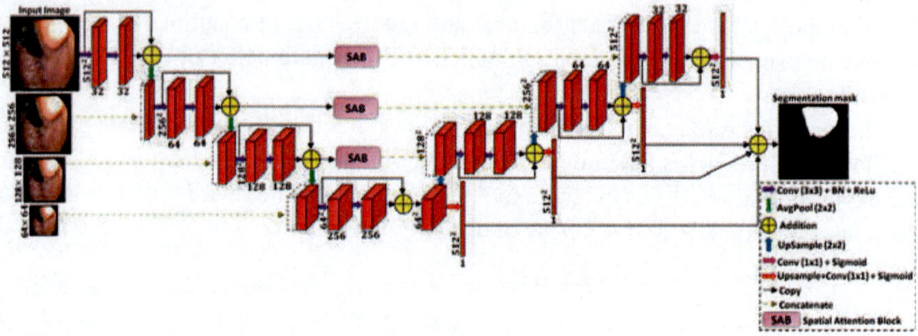

Fig. 2 Proposed SARM-Net architecture

filters size (3 × 3) and a residual connection to facilitate better gradient flow during backpropagation. The convolutional operations in the encoder branch are responsible for extracting minute, complex features in the image. The average pooling (2 × 2) operation is used to reduce the size of the feature maps, which enables the network to capture fine contextual information in the feature map. The network uses multiscale inputs as the input image is down-sampled at a rate of (2, 4, and 8) and concatenated into the corresponding encoder levels to extract distinctive information. The decoder branch is a mirror structure of the encoder branch with slight modification where Up-sampling with size (2×2) is used to increase the size of the feature map and restore the size of the output image. The final feature map from each decoder level is up-sampled to the output image size except the last decoder level as it is already up-sampled to the size of the output image, and point-wise convolution with filters size (1 × 1) is performed at each level to provide multi-layer deep supervision. Finally, all the generated feature maps by the point-wise convolutions are fused to generate the final segmentation mask. All the convolution layers used in the network are passed through a non-linear activation function ReLu. In contrast, the final point-wise convolution layers are passed through the sigmoid activation function, and batch normalization is applied to all the convolutional layers except for the point-wise convolutions to prevent overfitting.

In M-Net, skip connections are provided from each encoder level to the corresponding decoder level to propagate spatial information lost during downsampling. These skip connections may convey redundant low-level insignificant information. So in our proposed network, three numbers of spatial attention blocks (SAB) [15] are added where each SAB takes the output from each level of the encoder and generates the spatial attention map by exploiting the inter-spatial relationship of features in the spatial dimension. This attention map focuses on where the significant information is located. The output of each SAB is concatenated with the corresponding level of the decoder by skip connections. This mechanism highlights the significant features relevant to the context of the image, which overcomes the problem of propagation of irrelevant information. In Fig. 3, we have given the pictorial representation of the spatial attention block (SAB) used in our proposed network.

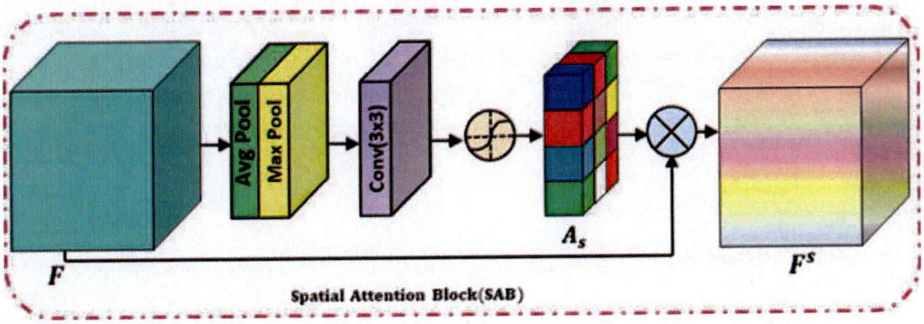

Fig. 3 Spatial attention block

Let us consider a feature map from a particular level in the encoder branch $F \in \Re^{H \times W \times C}$ where H and W represent the height, and the width, and C represents the channels in the F. To compute spatial attention, average pool and max pool operations are done at each pixel location across the channel axis which produces D_{max} and $D_{\text{avg}} \in \Re^{H \times W \times 1}$. The generated two-channel feature descriptor is concatenated and passed through a convolution layer with filter size (3×3) to produce the spatial attention map (A_s) where $A_S \in \Re^{H \times W \times 1}$. A_S signifies which location needs to be emphasized or suppress. Finally, A_s is concatenated with each channel C of F by element-wise multiplication, which results in F^s such that $F^s = A_s \otimes F$. This F^s is then concatenated with the corresponding level in the decoder branch by a skip connection.

2.2 Training Strategy and Implementation Details

The proposed model is optimized using the intersection over union (IoU) loss function [16], which may be defined as in (1)

$$\arg \min_{w} \mathcal{L}_{IoU} = 1 - IoU (1)$$

where w is the set of parameters used in the network.

The proposed network was trained on a workstation with Intel Xeon Processor powered by NVIDIA Quadro P5000 GPU with 16 GB RAM. The training process of the model was done using an Adam optimizer with a learning rate of 0.0001. The model was trained for 300 epochs with a batch size of 15.

3 Experimental Results and Discussion

In this section, we have elaborated on the experiments performed to judge the performance of the proposed method. A detailed description of the dataset used in this work is given in Sect. 3.1. The quantitative and qualitative evaluation performed to judge the efficacy of the proposed method is elaborated in Sect 3.2.

3.1 Dataset Description

The proposed study is evaluated on a bench-mark open-access dataset "Kvasir-SEG," [10] containing gastrointestinal polyp images. The dataset comprises 1000 colonoscopy frames containing polyp images, and the corresponding polyp mask manually annotated by expert gastroenterologists from Oslo University Hospital (Norway). These masks serve as the ground truth for the evaluation of our proposed method. The resolution of the images ranges from 332× 487 to 1920 × 1072. Prior to the training process of our proposed network, we have resized the images to a fixed resolution of 512 × 512. We have randomly split the dataset into 70–20–10%, where 70% images are devoted for training the model, 20% for validation, and the remaining 10% to test the model.

3.2 *Quantitative and Qualitative Evaluation*

Firstly, we have statistically shown the degree of similarity between our predicted segmentation mask and the corresponding ground truth (g_t) mask by the Bland–Altman plot. From Fig. 4, it can be perceived that in most colonoscopy frames, the polyp regions have a good agreement with the ground truth as it belongs within the agreement limit marked in the blue dotted line (95% confidence level).

We have compared our proposed method (SARM-Net) with some of the traditional deep learning methods (M-Net [13] and attention U-Net [17]) generally employed for the segmentation task and also with a SOTA method [11]. We have considered some of the standard measures such as dice similarity coefficient (DSC), precision (PRE), recall (REC), and F1-score (F1) [18] to quantify the performance of the segmentation results. From Table 1, it can be seen that SARM-Net outperforms the other two DL methods in terms of all the considered measures. Moreover, there is

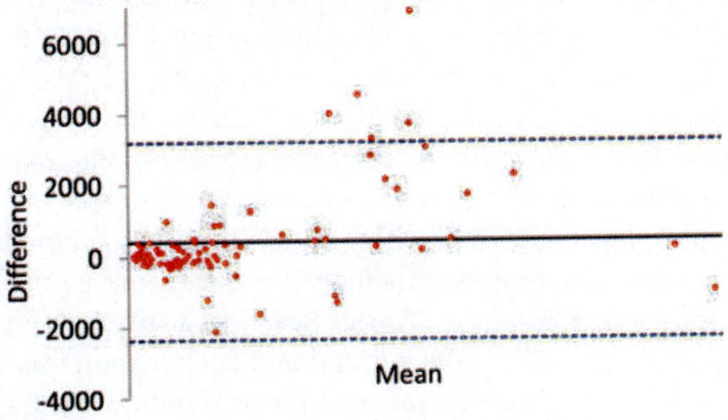

Fig. 4 Bland-Altman plot to statically evaluate the predicted segmentation mask with the g_t

Table 1 Comparison of the segmentation performance with traditional deep learning models and a SOTA method

MODELS	DSC ↑	PRE ↑	REC ↑	F1 ↑
M-Net	0.787	0.723	0.651	0.685
Attention U-Net	0.811	0.781	0.693	0.734
[11]	0.819	0.734	0.847	0.786
SARM-Net	0.837	0.793	0.833	0.812

Red color for best performing results

a tradeoff between the precision and recall of our proposed method. A low recall may lead to a miss rate, and low precision may lead to reexamine the colorectum. It can also be observed that SARM-Net surpasses [11] in DSC, but the REC of [11] is more compared to SARM-Net. But there is no tradeoff between PRE and REC in [11], which is clinically insignificant.

In Fig. 6, we have done a pictorial representation to highlight the degree of overlap between our predicted segmentation mask and the corresponding ground truth mask. We have done an overlay analysis to highlight different regions to differentiate the segmentation result. The pink color signifies the false-positive region, and the green color signifies the false-negative region. As can be seen from Fig. 6, there is a small amount of pink region and a negligible amount of green region, which justifies that our proposed method can segment the polyp mask efficiently. In Fig. 6, we have done a qualitative analysis of the segmentation performance of our proposed method with M-Net [13] and attention U-Net [17]. We have considered some colonoscopy frames with complex shapes and sizes of polyps. The results are represented in the corresponding rows. It can be seen that the segmentation mask generated by our proposed method has a good agreement with the ground truth mask in comparison to other DL models where there is a significant amount of false-positive and false-negative regions (Fig. 5).

4 Conclusion

Accurate segmentation of precursor polyp is very crucial for the early diagnosis of CRC. This study has proposed a deep learning framework named "SARM-Net" to segment polyp in colonoscopy frames. The network is motivated from M-Net with residual connections to enhance the optimization capability of the network. Further, spatial attention block is added in the proposed network to focus on where the significant information is located and ignore the irrelevant information, which improves the overall representation power of the network. The model also provides multi-layer deep supervision and multiscale learning to extract discriminative information. The proposed method is evaluated on an open-access gastrointestinal (GI) tract image database "Kvasir-SEG" to justify its efficacy. The proposed method is also compared with two traditional deep learning methods such as attention U-Net and M-Net. The experimental result signifies that the additional modules incorporated in our SARM-Net improve the segmentation performance compared to the DL methods. Furthermore, we have also compared our model with a SOTA method. In the future, we will study the generalization proficiency of the proposed network by exploring other medical imaging datasets to justify its robustness and clinical acceptance. We believe that this study will motivate more minds to enhance and resolve the polyp segmentation task.

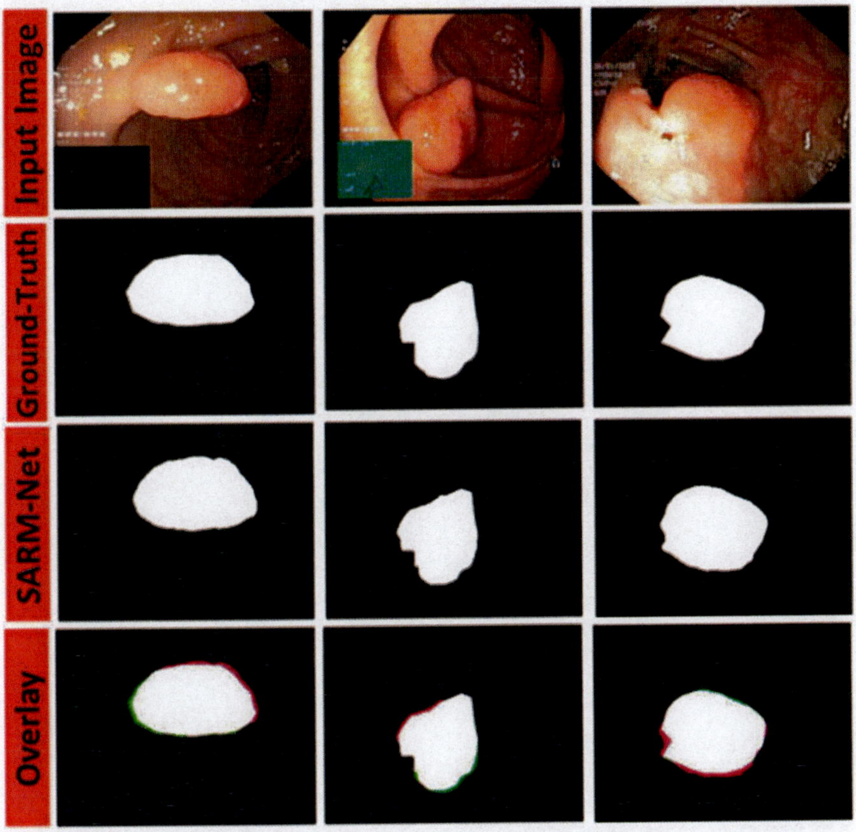

Fig. 5 Comparison of the segmentation result with an overlay measure

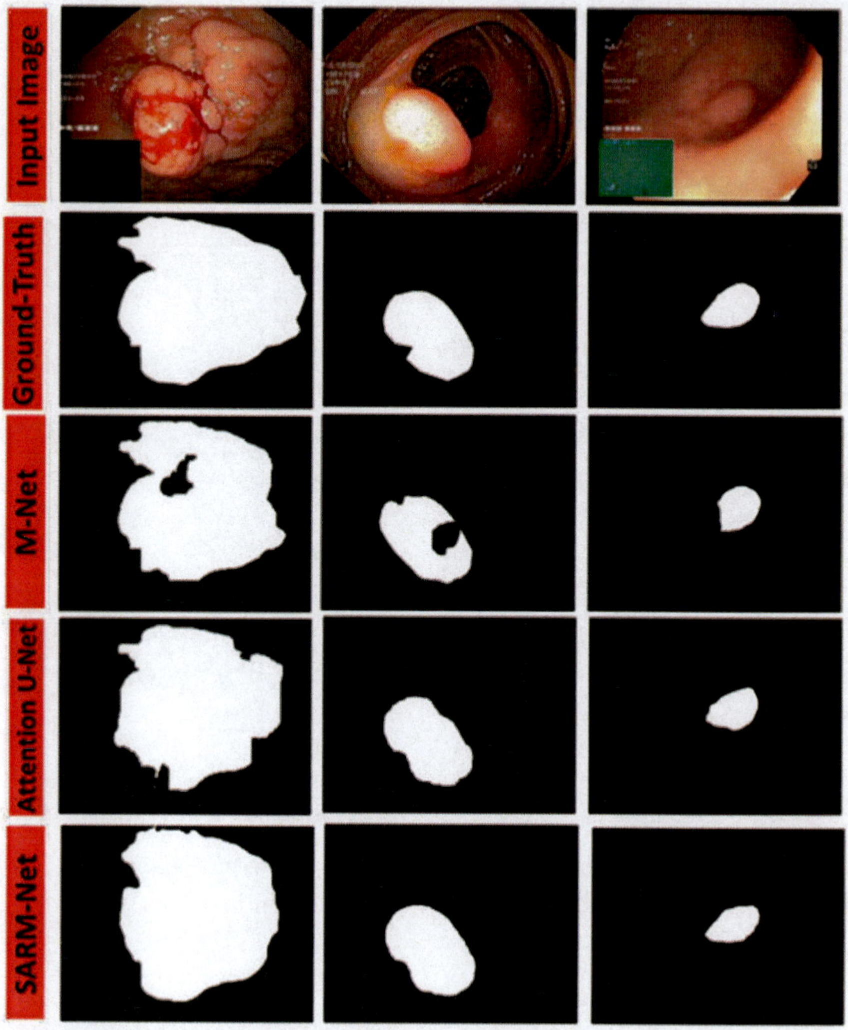

Fig. 6 Comparison of the segmentation result of SARM-Net with other deep learning models

References

1. Rawla P, Sunkara T, Barsouk A (2019) Epidemiology of colorectal cancer: incidence, mortality, survival, and risk factors. Przeglad Gastroenterologiczny 14(2):89
2. Yeung M, Sala E, Schönlieb CB, Rundo L (2021) Focus U-Net: a novel dual attention-gated CNN for polyp segmentation during colonoscopy. Comput Biol Med 104815
3. Jia X, Mai X, Cui Y, Yuan Y, Xing X, Seo H, Xing L, Meng MQH (2020) Automatic polyp recognition in colonoscopy images using deep learning and two-stage pyramidal feature prediction. IEEE Trans Autom Sci Eng 17(3):1570–1584
4. Banik D, Bhattacharjee D, Nasipuri M (221) Recent advances in intelligent imaging systems

for early prediction of colorectal cancer: a perspective. Enabling Mach Learn Appl Data Sci 39–61
5. Ahmad OF, Lovat LB (2019) Artificial intelligence for colorectal polyp detection: are we ready for prime time? J Med Artif Intell 2
6. Wang Wi, Tian J, Zhang C, Luo Y, Wang X, Li J (2020) An improved deep learning approach and its applications on colonic polyp images detection. BMC Med Imaging 20(1):1–14
7. Yang K, Chang S, Tian Z, Cong Gao YuDu, Zhang X, Liu K, Meng J, Xue L (2022) Automatic polyp detection and segmentation using shuffle efficient channel attention network. Alex Eng J 61(1):917–926
8. Fan DP, Ji GP, Zhou T, Chen G, Fu H, Shen J, Shao L (2020) Pranet: parallel reverse attention network for polyp segmentation. In: International conference on medical image computing and computer-assisted intervention, pp 263–273. Springer
9. Dong B, Wang W, Fan DP, Li J, Fu H, Shao L (2021) Polyp-PVT: polyp segmentation with pyramid vision transformers. arXiv preprint arXiv:2108.06932
10. Jha D, Smedsrud PH, Johansen D, de Lange T, Johansen HD, Halvorsen P, Riegler MA (2021) A comprehensive study on colorectal polyp segmentation with ResUNet++, conditional random field and test-time augmentation. IEEE J Biomed Health Inform 25(6):2029–2040
11. Banik D, Roy K, Bhattacharjee D, Nasipuri M, Krejcar O (2020) Polyp-Net: a multimodel fusion network for polyp segmentation. IEEE Trans Instrum Meas 70:1–12
12. Mahmud T, Paul B, Anowarul Fattah S (2021) PolypSegNet: a modified encoder-decoder architecture for automated polyp segmentation from colonoscopy images." Comput Biol Med 128:104119
13. Mehta R, Sivaswamy J (2017) M-net: a convolutional neural network for deep brain structure segmentation. In: 2017 IEEE 14th international symposium on biomedical imaging (ISBI 2017), pp 437–440. IEEE
14. Ronneberger O, Fischer P, Brox T (2015) U-net: convolutional networks for biomedical image segmentation. In: International conference on medical image computing and computer-assisted intervention, pp 234–241. Springer
15. Liu W, Chen C, KYK Wong, Su Z, Han J (2016) STAR-Net: a spatial attention residue network for scene text recognition. BMVC 2(7)
16. Rahman MA, Wang Y (2016) Optimizing intersection-over-union in deep neural networks for image segmentation. In: International symposium on visual computing, pp 234–244. Springer
17. Oktay O, Schlemper J, Le Folgoc L, Lee M, Heinrich M, Misawa K, Mori K et al (2018) Attention u-net: learning where to look for the pancreas. *arXiv preprint* arXiv:1804.03999
18. Banik D, Bhattacharjee D (2021) Mitigating data imbalance issues in medical image analysis. In: Data preprocessing, active learning, and cost perceptive approaches for resolving data imbalance, pp 66–89. IGI Global

Domain-Specific Cues for the Usability of Marker-Controlled Watershed Algorithm and U-Net for Medical Image Segmentation

Kaushiki Roy, Debotosh Bhattacharjee, Masuma Khatun, and Anindita Dutta

Abstract In medical image analysis, segmentation of the region-of-interest is the crucial phase for proper diagnosis. However, this task is very challenging due to missing or diffuse organ/tissue boundaries. However, deep learning frameworks based on the U-Net backbone have gained immense popularity since those can deal with image inconsistencies. This study compared U-Net's performance with a marker-controlled watershed algorithm, a traditional segmentation algorithm. We tried to find which method is more efficient for polyp and overlapping cell segmentation. These two segmentation tasks were chosen since they are the most challenging task in medical history. The study of cell morphology is of great importance since it helps to understand the structure and conditions of cells, which helps us understand our underlying health conditions. A necessary before understanding cell morphology is cell segmentation. Polyps are protrusions from the inner lining of the colon or rectum which turns into colorectal cancer if left untreated. Thus, accurate segmentation of the polyp region is essential to help the physician properly diagnose the disease. In this work, we have used the KVASIR dataset to report the performance of the U-Net and watershed algorithm for polyp segmentation. A private dataset cultured on HeLa cell line was used for cell segmentation. Watershed algorithm obtained higher accuracy of 91.34% over U-Net which obtained accuracy 80.46% for cell segmentation. On the contrary, U-Net performed better for polyp segmentation and obtained accuracy 91.29% compared to watershed with reported accuracy of 89.19%.

Keywords HeLa cell line · Polyp segmentation · U-NET · Watershed algorithm · KVASIR dataset

K. Roy (✉) · D. Bhattacharjee · M. Khatun · A. Dutta
Department of Computer Science and Engineering, Jadavpur University, Kolkata 700032, India
e-mail: kroy.cse.rs@jadavpuruniversity.in

D. Bhattacharjee
e-mail: debotosh@ieee.org

M. Gupta et al. (eds.), *Artificial Intelligence on Medical Data*,
Lecture Notes in Computational Vision and Biomechanics 37,
https://doi.org/10.1007/978-981-19-0151-5_34

1 Introduction

Colorectal cancer is the most commonly occurring cancer in women and men in the United States (US). According to the American cancer society (ACS) [1], nearly, 1,04,270 new cases of colon cancer and 45,230 new cases of rectal cancer will be diagnosed in 2021. Luckily, colorectal cancer commonly develops gradually over many years. The disease can be prevented if the polyps are detected and removed before they turn into cancer. Therefore, numerous research has been conducted by scientists to develop an automated system for polyp segmentation. Yu et al. [2] developed a novel offline and online 3D deep learning integration framework for polyp segmentation in the ASU-Mayo dataset. Zhang et al. [3] developed FCN with novel hybrid classification for polyp detection in the CVC-ColonDB dataset. Bardhi et al. [4] used convolutional encoder–decoder model for polyp segmentation in CVC-ColonDB, CVC-ClinicDB, and ETIS-LARIB datasets. Xiao et al. [5] used deep lab and LSTM network for polyp identification in the CVC-ClinicDB dataset.

Our second area of study in this manuscript is cell segmentation. Cell segmentation is a fundamental step to understand cell behaviors and their response to changes in the local environment. Detection and segmentation of microscopic cells are required for drug development, searching for new drugs, disease diagnosis, and treatment. Segmentation is a crucial problem in many analysis approaches, especially those that follow individual cells over time. However, automated segmentation of cell images is typically very challenging due to high density, low contrast, deformable cell shapes, and occlusion of one cell by another leading to missing or diffused boundaries. Many traditional approaches could not overcome these challenges effectively since traditional approaches rely on seed detection for segmentation. However, for cells with missing separating boundaries among these, seeds are not appropriately detected for individual cells. This leads to failure in accurately segmenting cells by traditional approaches. In [6], the authors proposed a bounding box-based convolutional neural network for cell segmentation. The network was tested on three datasets, namely Fluo-N2DL-HeLa, PhC-HeLa, and Hist-BM. In [7], a hybrid network was developed by integrating convolutional long short-term memory (C-LSTM) and U-Net for live-cell segmentation. Authors in [8] developed a deep learning-based framework for cell detection and segmentation in fluorescence microscopy images containing channels for nuclei and cells.

2 Proposed Methodology

This section will describe the marker-controlled watershed algorithm and U-Net architecture used to segment cells in the HeLa cell line and polyp. Figure 1a, b shows an input microscopy frame of the HeLa cell line's corresponding mask, respectively. Figure 2a, b shows a colorectal frame with a polyp and its corresponding ground truth, respectively.

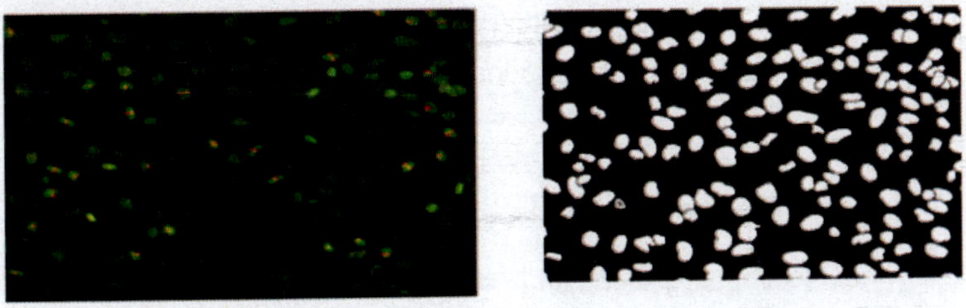

(a) (b)

Fig. 1 **a** Input microscopy frame cultured from HeLa cell line and **b** its corresponding mask

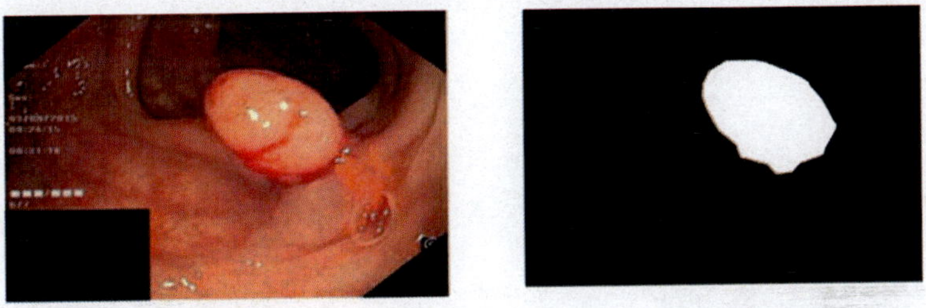

(a) (b)

Fig. 2 **a** Input colonoscopy frame with a polyp and **b** its corresponding mask

2.1 Marker-Controlled Watershed Algorithm

The marker-controlled watershed segmentation [9] method has shown to be robust and flexible for the segmentation of objects with closed contours and where the boundaries are expressed as ridges. This algorithm considers the input image a topographic surface and simulates its flooding from specific seed points or markers. The accuracy and efficiency of this algorithm are dependent on the accurate generation of markers. To generate the markers, we first convert the input RGB frames I_f to grayscale I_g. Next, we generate the distance transform (DT) map D_{map} from I_g. The DT value for a pixel $p \in I_g$ is represented by Eq. 1.

$$D_{map}(p) = \min\{(p, b)|b \in I'_g \tag{1}$$

D_{map} is the DT image for I_g, and I'_g represents the set of all background pixels. Finally, we apply H-maxima transform and extended maxima transform [10] on D_{map} to generate the final set of markers. The H-maxima transform [10] suppresses all maxima whose depth f is less than or equal to a given threshold h. This is performed by dilation of f from $f - h$, as shown in Eq. 2.

$$H_{max}(f) = R^d_f(f - h) \tag{2}$$

Here, R_f^d represents reconstruction by dilation function, and H_{max} represents the H-maxima transformation. Extended maxima transform, E_{max}, is defined as the regional maxima (R_{max}) of H-maxima transformation and is defined in Eq. 3.

$$E_{max}(f) = R_{max}(H_{max}(f)) \tag{3}$$

Figure 3 shows input images and their corresponding distance-transformed images, and the seed generated after applying extended maxima transform on the distance-transformed image. Each of the generated seeds is then used as markers for segmentation by the watershed algorithm. For polyp segmentation, multiple spurious regions are generated after segmentation. These regions are eliminated using a circularity check (Eq. 4) on all the objects.

$$Cir_{index} = 4 \times \pi \times \text{Area}/\text{Perimeter}^2 \tag{4}$$

Only, those objects which are highly circular ($Cir_{index} > 0.7$) are retained, while the others are eliminated, as shown in Fig. 4.

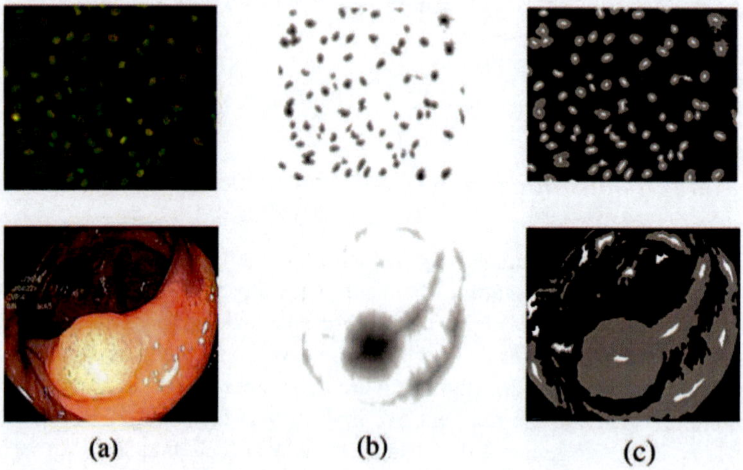

Fig. 3 **a** Represent input frames, **b** distance transformed image, and **c** seeds generated by extended maxima transform

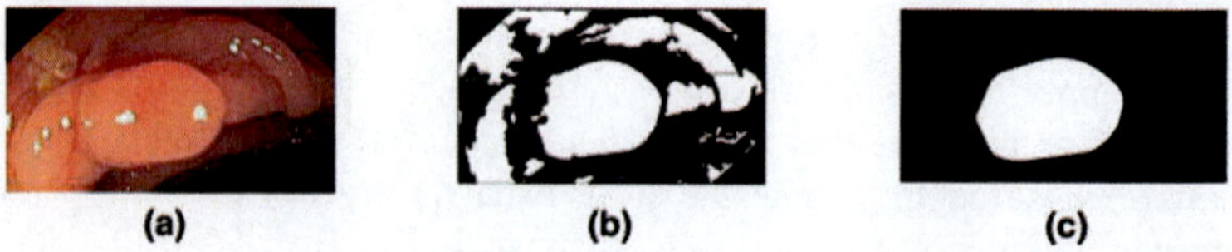

Fig. 4 **a** input image, **b** image after watershed segmentation, and **c** final image obtained after circularity check

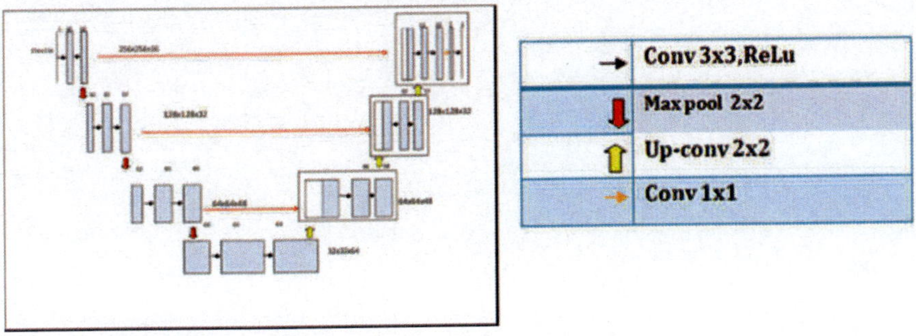

Fig. 5 U-Net architecture

2.2 *U-Net Architecture*

A typical U-Net architecture [11] has two parts: encoder and decoder. The encoder is generally constituted by the convolutional blocks followed by max-pooling to encode the input image. In the second half, the decoder part is generally constituted by transposed 2D convolutional layers (a type of upsampling technique). In Fig. 5, we have shown the architecture of the U-Net framework.

3 Experimental Results

3.1 *Dataset Description*

The "Kvasir" dataset [12] consists of 1000 gastrointestinal tract images with a polyp. Each of the images in the dataset was annotated by expert gastroenterologists. A private dataset was collected from Gordon's lab, Oncology department in University of Alberta for cell segmentation. This dataset consists of 53 microscopy frames captured from cultured HeLa cell lines annotated by expert oncologists from Dr. Gordon's lab.

3.2 *Performance Evaluation of Watershed Versus U-Net for Cell and Polyp Segmentation*

Various discrepancy evaluation methods were used to compare the performance of watershed versus U-Net for segmentation. We constructed a precision-recall plot in Fig. 6 to report watershed performance for segmenting each microscopy frame in the cell segmentation dataset.

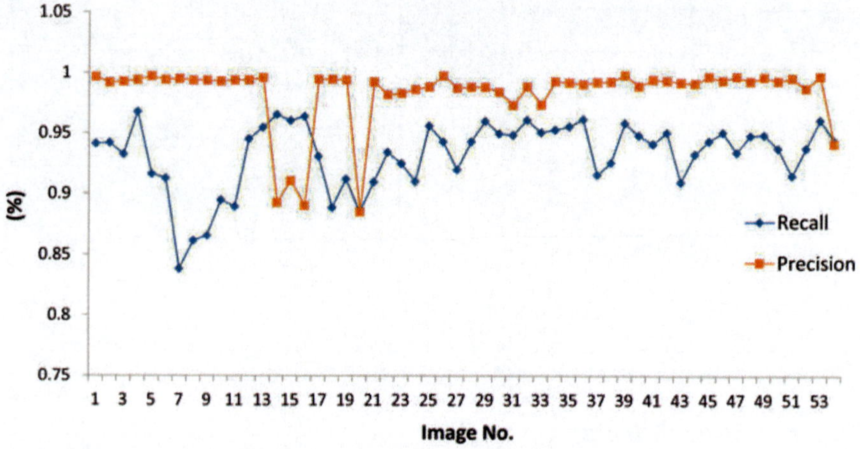

Fig. 6 Recall and precision plot for watershed

In Fig. 7, we report the Jaccard value obtained by watershed for segmenting each input frame.

The average recall, precision, F1-score, accuracy, and Jaccard index obtained by the watershed algorithm are 91.91%, 91.53%, 90.80%, 91.34%, and 83.87%, respectively. However, U-Net underperforms watershed for cell segmentation and achieves recall, precision, F1-score, accuracy, and Jaccard index of 89.24%, 68.83%, 79.10%, 80.46%, and 40.35%, respectively.

Next, we report the performance of these two algorithms for polyp segmentation. The watershed algorithm achieves accuracy, Jaccard, recall, precision, and F1-score of 50.85%, 49.78%, 50.86%, 89.19%, and 62.37%, respectively, whereas the value of these evaluation metrics obtained by U-Net is 81.93%, 73.45%, 63.6%, 91.29%, and 67.93%, respectively. A precise inspection of these metrics helps us conclude that the watershed performs better than U-Net for cell segmentation, whereas the reverse is true for polyp segmentation.

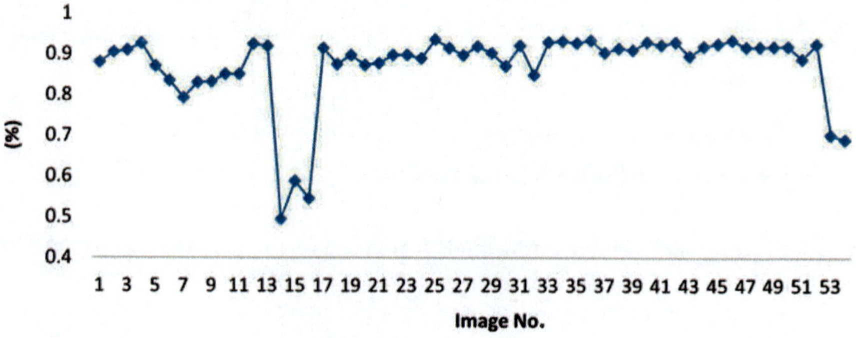

Fig. 7 Jaccard index for cell segmentation by watershed

Table 1 Comparative analysis of U-Net and marker-controlled watershed with state-of-the-art for polyp segmentation

Method	Accuracy	Pre	Rec	Specificity	F1-score
Yu et al. [2]	-	0.78	0.53	-	0.63
Zhang et al. [3]	**0.97**	-	**0.75**	**0.98**	-
U-net	0.81	**0.91**	0.63	-	**0.67**
Marker-controlled watershed	0.51	0.89	0.50	-	0.62

Red for best-performing methods

3.3 *Comparison with State-Of-The-Art*

In this section, we compare the performance of U-Net and watershed for polyp segmentation with some of the state-of-the-art methods in Table 1.

It can be seen from Table 1 that the performance of U-Net is way better than watershed for polyp segmentation. U-net achieves the highest performance value for two metrics, namely precision and F1-score. However, we could not provide a comparative analysis between U-Net and watershed with state-of-the-art methods for cell segmentation since the HeLa dataset used for the work is our private dataset, and no work has been reported in the state-of-the-art on this dataset.

3.4 *Qualitative Evaluation*

In Figs. 8 and 9, we show the output of watershed and U-Net for cell segmentation. In Figs. 10 and 11, we show the output of watershed and U-Net for polyp segmentation.

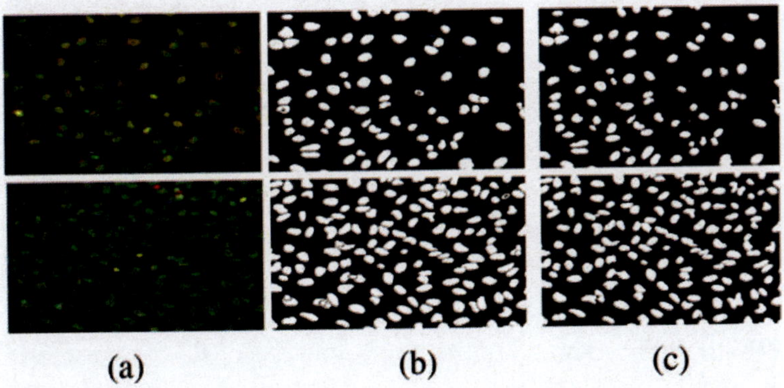

Fig. 8 **a** Represent input image, **b** represent annotated ground truth and **c** represent watershed output

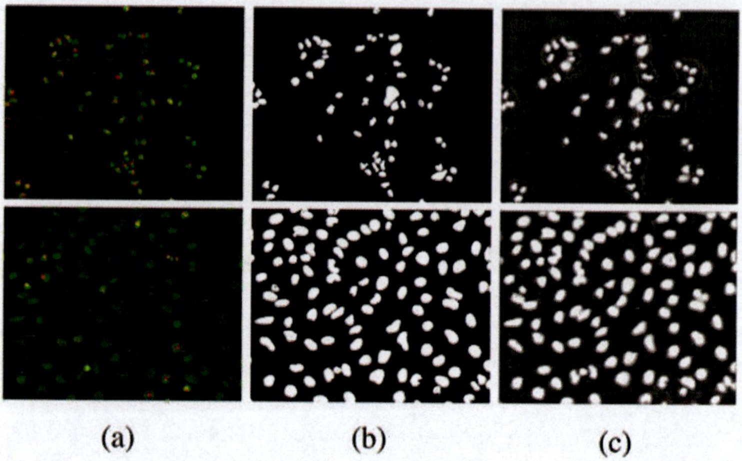

Fig. 9 **a** Represent input image, **b** represent annotated ground truth, and **c** represent U-net segmentation map

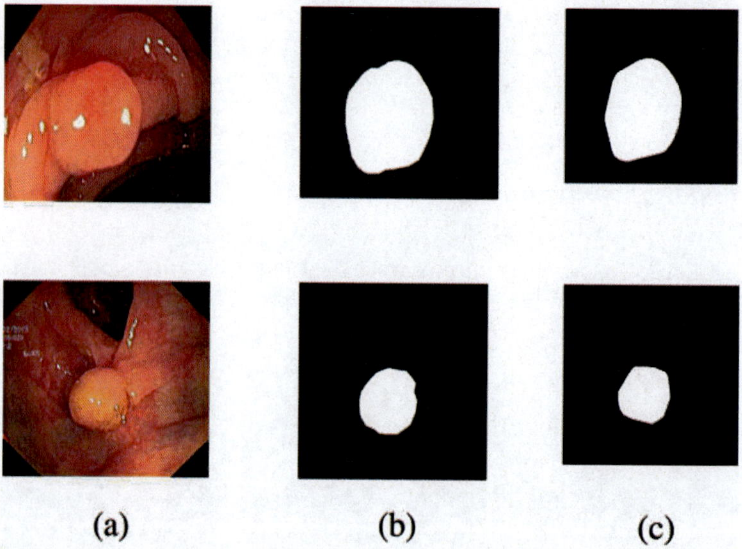

Fig. 10 **a** Represent input image, **b** represent annotated ground truth, and **c** represent watershed output

4 Conclusion

In this study, we compared the performance of the marker-controlled watershed algorithm with U-Net for overlapping cell and polyp segmentation. It was observed in the study that for cell segmentation, the marker-controlled watershed algorithm

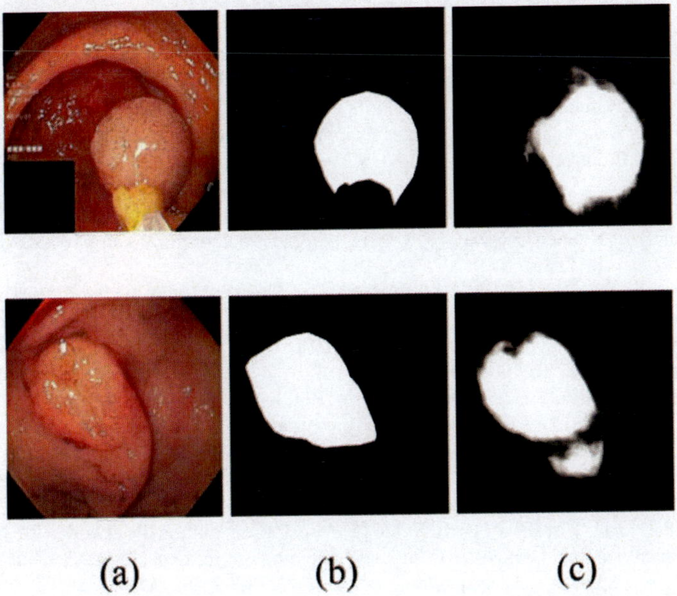

Fig. 11 **a** Represent input image, **b** represent annotated ground truth, and **c** represent U-Net segmentation map

performs better than U-net, and just, the reverse observation was made for polyp segmentation. In cell segmentation, U-Net failed when there were missing tissue boundaries, and the degree of overlap was high. However, in colorectal frames, most of the polyps had a well-distinguished boundary from the surrounding. Thus, U-Net was able to segment polyp regions efficiently and performed better than a watershed.

Acknowledgements The first author is thankful to the DST INSPIRE fellowship program (IF170366).

References

1. Colorectal Cancer Statistics | How Common Is Colorectal Cancer? https://www.cancer.org/cancer/colon-rectal-cancer/about/key-statistics.html. Accessed 31 Aug 2021
2. Yu L (2019) Towards ai-empowered healthcare: medical image analysis with data-efficient deep learning. The Chinese University of Hong Kong (Hong Kong)
3. Zhang L, Dolwani S, Ye X (2017) Automated polyp segmentation in colonoscopy frames using fully convolutional neural network and textons. Commun Comput Inf Sci 723:707–717. https://doi.org/10.1007/978-3-319-60964-5_62
4. Bardhi O, Sierra-Sosa D, Garcia-Zapirain B, Elmaghraby A (2018) Automatic colon polyp detection using convolutional encoder-decoder model. IEEE Int Symp Signal Process Inf Technol ISSPIT 2017:445–448. https://doi.org/10.1109/ISSPIT.2017.8388684

5. Xiao WT, Chang LJ, Liu WM (2018) Semantic segmentation of colorectal polyps with deeplab and LSTM networks. IEEE Int Conf Consum Electron ICCE-TW 2018. doi: https://doi.org/10.1109/ICCE-CHINA.2018.8448568
6. Akram SU, Kannala J, Eklund L, Heikkilä J (2016) Cell segmentation proposal network for microscopy image analysis. Lect Notes Comput Sci (including Subser Lect Notes Artif Intell Lect Notes Bioinformatics), 10008 LNCS, pp 21–29, doi: https://doi.org/10.1007/978-3-319-46976-8_3
7. Arbelle A, Raviv TR (2019) Microscopy cell segmentation via convolutional LSTM networks. Proc Int Symp Biomed Imaging, 1008–1012, Apr 2019. doi: https://doi.org/10.1109/ISBI.2019.8759447
8. Korfhage N, Mühling M, Ringshandl S, Becker A, Schmeck B, Freisleben B (2020) Detection and segmentation of morphologically complex eukaryotic cells in fluorescence microscopy images via feature pyramid fusion. PLOS Comput Biol 16(9):e1008179. https://doi.org/10.1371/JOURNAL.PCBI.1008179
9. Veta M, Huisman A, Viergever MA, Van Diest PJ, Pluim JPW (2011) Marker-controlled watershed segmentation of nuclei in H&E stained breast cancer biopsy images. Proc Int Symp Biomed Imaging 618–621. doi: https://doi.org/10.1109/ISBI.2011.5872483
10. Qin Y, Wang W, Liu W, Yuan N (2013) Extended-maxima transform watershed segmentation algorithm for touching corn kernels. https://doi.org/10.1155/2013/268046
11. Ronneberger O, Fischer P, Brox T (2015) U-net: Convolutional networks for biomedical image segmentation. Lect Notes Comp Sci (Including Subser Lect Notes Art Intell Lect Notes Bioinform) 9351:234–241. https://doi.org/10.1007/978-3-319-24574-4_28
12. Jha D et al (2020) Kvasir-SEG: a segmented polyp dataset. Lect Notes Comput Sci (Including Subser Lect Notes Artif Intell Lect Notes Bioinform), vol 11962:451–462. doi: https://doi.org/10.1007/978-3-030-37734-2_37

Hybrid CNN-SVM Model for Face Mask Detector to Protect from COVID-19

Charu Agarwal, Inderjeet Kaur, and Sunita Yadav

Abstract The 2019 coronavirus outbreak (COVID-19) has had a huge impact on humanity. By May 2021, nearly, 172 million people worldwide were affected by the infectious spread of COVID-19. While the distribution of vaccines has already begun, mass distribution around the world has yet to take place. According to the World Health Organization (WHO), wearing a face mask can significantly reduce the spread of the COVID-19 virus. However, even improper wearing of face mask can prevent the purposes and lead to the spread of the virus. Under the influence of public health and the global economy, an effective Covid-19 pandemic strategy requires a lot of attention of humanity. To prevent the spread of such deadly virus, intelligent techniques are required. In the proposed work, an intelligent face mask detector framework is proposed based on deep learning concept which can classify the person who wear mask from those who are not wearing mask. In the proposed work, a hybrid model of convolution neural network with support vector machine is used for designing the mask detector. The performance of the proposed method is evaluated on real-world masked face recognition dataset (RMFD) and medical mask dataset (MDD). When implemented, it has been found that the proposed method can achieve high accuracy (99.11%). The excellent performance of the proposed model is very suitable for video surveillance equipment also.

Keywords Covid-19 · Face mask · Deep learning · Convolution neural network · Accuracy · Precision

---

C. Agarwal (✉) · I. Kaur · S. Yadav
Department of Computer Science and Engineering, Ajay Kumar Garg Engineering College, Ghaziabad, India
e-mail: agarwalcharu2@gmail.com

M. Gupta et al. (eds.), *Artificial Intelligence on Medical Data*,
Lecture Notes in Computational Vision and Biomechanics 37,
https://doi.org/10.1007/978-981-19-0151-5_35

1 Introduction

The world has been hit by the coronavirus, also known as COVID-19, in the past two years. According to the World Health Organization (WHO) [1], as of September 18, 2021, there were a total of 226,844,344confirmed cases of COVID-19, of which 4,666,334 people had died.

According to the World Health Organization [2], this is a virus that shows the following symptoms in the people such as cold, fever, or nausea and causes serious problems such as severe acute respiratory syndrome (SARS). The virus was first detected in Wuhan, China in December 2019 and is slowly spreading around the world, affecting countries around the world. As it can cause serious problems with the disease, everyone must be protected against it. According to the guidelines of the World Health Organization, everyone must follow appropriate behaviors to protect themselves from COVID-19, such as wearing face mask, maintaining social distancing, and regular hand washing. Many researchers are also concerned about the importance of wearing masks.

According to Howard [3], when the density is high, wearing mask in public places can reduce the spread of infections. According to Dowd [4], until there are no vaccines and specific treatments, the twenty-first century pandemic will be a treated with the strategy starting in the fourteenth century. Around the world, masks have become an unstoppable part of the public technology to fight the current coronavirus pandemic. Holger et al. [5] discussed in detail the advantages and disadvantages of using face masks. In the end, they concluded that masks are useful for both the general population and healthcare professionals. He thought about the early ideas about the use of masks and continued to participate in the subsequent political discussions.

The above requirement has drawn the attention of many researchers toward the development of face mask detector strategies. Ge et al. [6] used the LLE-CNN model to identify masks and obtained an accuracy of 76.1. Ejaz et al. [7] have used principal component analysis (PCA) technology and reached 70% accuracy.

In addition to preventing the coronavirus, wearing masks have also become standard due to the increase in pollution and contamination. Therefore, it is necessary to distinguish between those who wear masks and those who do not. With this in mind, intelligent technologies should be developed to identify and identify people with or without masks.

In the proposed work, a novel intelligent face mask detector using the deep neural network method is proposed. This model is based on hybrid approach in convolution neural network (CNN) is used along with support vector machine (SVM). The model is executed on real-world masked face recognition dataset (RMFD) and medical mask dataset (MDD), compiled our results, and compared them with other studies done in the same domain.

2 Preliminaries

- Deep Neural Network (DNN)

In the proposed method, to design the face mask detector, a most popular deep neural network technique known as convolution neural network (CNN) is used. CNN is best in classifying images into different classes. CNN is a deep learning process that consists of many layers overlapping each other, and these layers use a close correlation between values. CNN is used as a deep learning model, especially when the dataset is represented as images, as it can perform feature extraction and classification tasks. There is no need to manually add functions to the classifier. It has the ability to extract elements from images using weave layers. A CNN model mainly consists of four layers such as:

The first layer in CNN is known as the convolution layers. As the name suggests, this layer convolves between the image pixel window and the filter. This layer is capable in extracting the features of an image. The convolution operation (marked with *) is used to extract features from an image and produces feature map. You can think of this as a mathematical formulation given by Eq. (1) with an input image of A and a B kernel (also known as a feature detector).

$$FM(i, j) = (A * B)(i, j) \tag{1}$$

The second layer in CNN is ReLU layer (Rectified linear unit). ReLU is an activation function. This function takes as input the output generated by the convolution layer known as feature map. ReLU function converts negative values into zero, and all the positive values remain same. After ReLU layer, the next layer in CNN is pooling layer. The main function of this layer is to reduce the size by using a pool of values, from which this function keeps only the maximum value from that pool and discarding other values. The next layer is called the fully connected layer and will eventually classify the merged feature maps. The most popular classification function is softmax. These layers can be staked over one another to achieve high degree of accuracy.

- Support Vector Machine (SVM)

Support vector machine (SVM) can be used to solve the problems related with classification as well as regression. It can solve both linearly separable problems as well as nonlinearly separable problems. To deal with nonlinearly separable problems, it uses kernel functions. The SVM hypothesis is given in Eq. (2).

$$h_\theta(z) = \begin{cases} 1, & \text{if } \theta^T z \geq 0 \\ 0, & \text{otherwise} \end{cases} \tag{2}$$

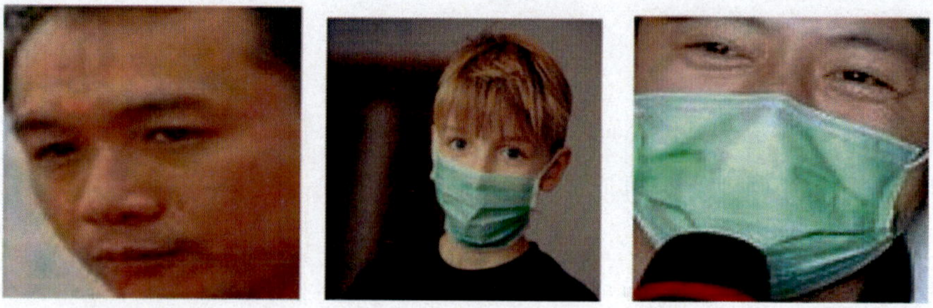

Fig. 1 Few images from RMFD dataset

Fig. 2 Few images from MMD dataset

3 Dataset

Real-world masked face recognition dataset (RMFD) is the most popular dataset among researchers for face mask detection problem [8]. This dataset is one the largest dataset in which total 525 people are involved. This dataset has 5000 images of wearing mask and 90,000 images in which person is wearing mask. This dataset is very useful as it contains the images of the same persons with mask and without mask. Few images from this dataset are given in Fig. 1.

The second dataset used is MMD dataset published by Mikolaj Witkowski [9]. This dataset consists of 682 pictures out of which it consists of 3000 images in which person is wearing medical mask with over 3 k medical masked faces wearing masks. Few images from this dataset are given in Fig. 2.

4 Proposed Approach

In the current situation, we are all forced to wear masks to protect ourselves from the deadly coronavirus. People also wear masks to protect themselves from dust, pollution, and other diseases. Therefore, in order to control the spread of the virus, it is imperative to develop a smart mask detector. In this research paper, we propose the basis of a smart mask detector based on hybrid CNN-SVM model. As it is already

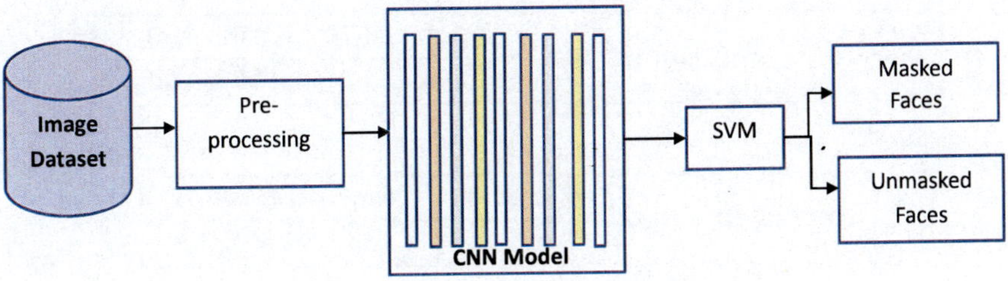

Fig. 3 Proposed model

proved in the previous research works that SVM is one of the best classifiers, so, in this model, the feature map is computed by CNN, whereas the classification task is done by SVM instead of fully connected layer of CNN. The proposed model can be depicted from Fig. 3. The CNN architecture used in the proposed model can be depicted form Fig. 4.

The feature map extracted from the above proposed CNN model is then passed to SVM classifier using with regularizer l2 and hinge loss.

5 Result Analysis and Discussion

The present research work presents an approach to classify the faces as masked faces or unmasked faces. To design such a classifier, a novel hybrid model is utilized known as CNN-SVM model in which CNN is used for feature extraction and SVM is used for classification. The proposed scheme is carried out in Python 3.7 in Google Colaboratory. The proposed scheme is implemented on RMFD dataset. Table 1 depicts the validation accuracy, precision, recall, and F1-score values when the model is trained with dataset, and Table 2 depicts the testing accuracy, precision, recall, and F1-score values obtained from testing the model on dataset.

Results tabulated in Tables 1 and 2 show that the proposed model reports good results for RMFD dataset in comparison to MMD dataset. For RMFD dataset, it achieves more than 99% testing accuracy, whereas for MMD dataset, it gives 95.16% accuracy.

To further analyze the performance of the proposed model, we compared our results with other similar research works in the same domain. Table 3 tabulates the results for comparison purpose of our model and other similar schemes in terms of testing accuracy.

From Table 3, it is observed that the proposed scheme outperformed the other schemes and shows very good results which indicates that to reduce the spread of CORONA virus the proposed scheme is very effective. It can be implemented on real-time scale also to identify the people who are not wearing the mask.

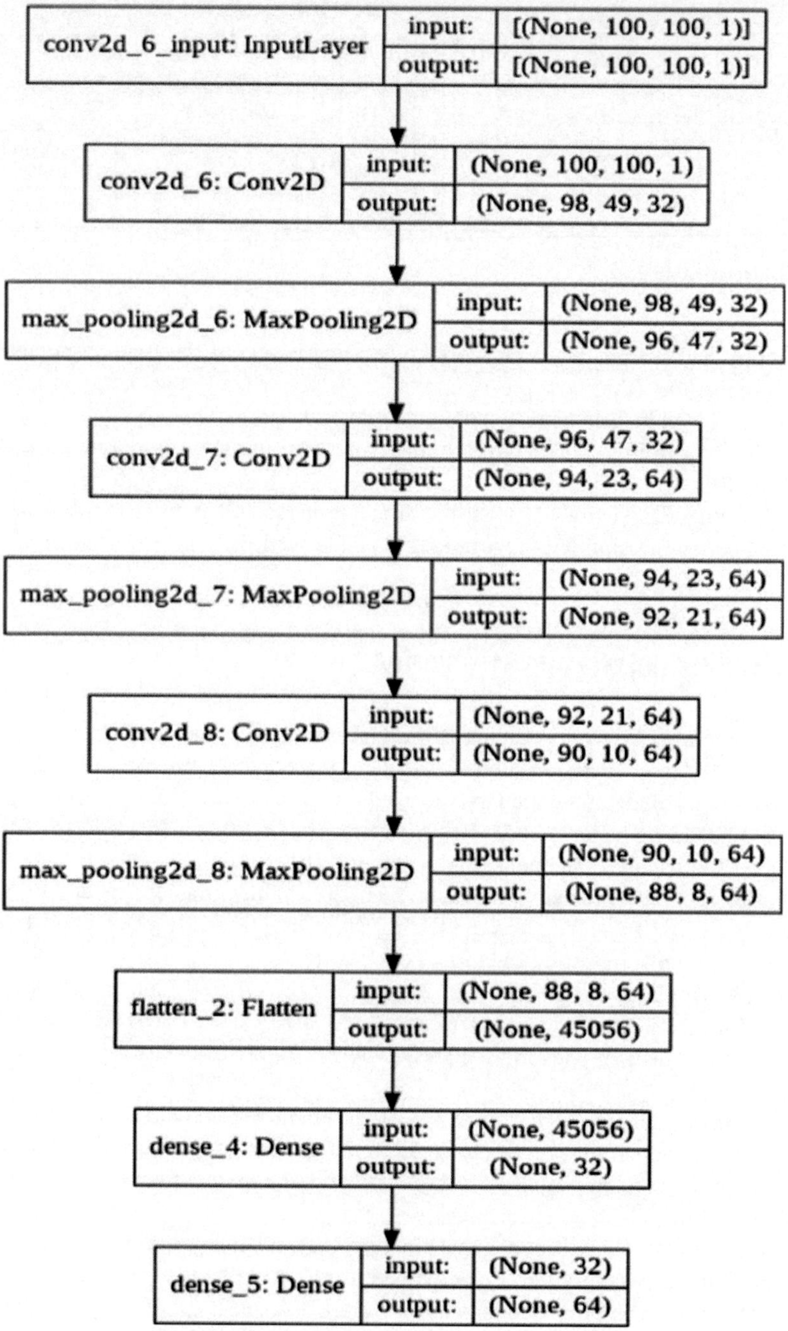

Fig. 4 Proposed CNN architecture

Table 1 Performance metrics outcomes (training phase)

Dataset	Testing accuracy	Precision	Recall	F1-score
MMD	95.03	95.86	97.91	95.78
RFMD	98.92	98.5	99.3	98.5

Table 2 Performance metrics outcomes (testing phase)

Dataset	Testing accuracy	Precision	Recall	F1-score
MMD	95.16	95.44	94.85	95.14
RFMD	99.033	98.11	1	99.04

Table 3 Testing accuracy comparison

Dataset	Testing accuracy (%)
Principle component analysis (Ejaz et al. [6])	70
LLE-CNNs (Ge et al. [7])	76.1
Proposed model (CNN-SVM)	99.03

6 Conclusion

Proposed research is being conducted to develop a new framework for face mask image recognition problems in response to the current COVID-19 pandemic. Hybrid model of CNN-SVM has been successfully used for classification tasks. The proposed shows good accuracy results and is also compared with other similar work. It is found that the proposed work outperforms the other schemes in comparison. This is because, in our scheme, CNN is used for feature extraction from the images, and SVM is used for classification. Therefore, we summarize the development and offer a successful research framework based mainly on training and testing of deep learning methods based on CNN models. As a future method, this work can be extended to test a newly created real-image dataset that contains thousands of available image objects.

References

1. WHO Coronavirus Disease (COVID-19) Dashboard. https://COVID-19.who.int/
2. WHO Coronavirus disease (COVID-19) advice for the public: when and how to use masks (2020) https://www.who.int/emergencies/diseases/novel-coronavirus-2019/advice-for-public/when-and-how-to-use-masks
3. Howard J, Huang A, Li Z, Tufekci Z, Zdimal V, van der Westhuizen H, Tang V (2020) Face masks against COVID-19: an evidence review.
4. Vnv O'Dowd K, Nair K, Forouzandeh P, Mathew S, Grant J, Moran R, Pillai S (2020) Face masks and respirators in the fight against the COVID-19pandemic: a review of current materials, advances and future perspectives. Materials 13(15):3363

5. Holger JS, Akl EA, Chou R, Chu DK, Loeb M, Lotfi T, Mustafa RA, Neumann I, Saxinger L, Sultan S, Mertz D. Use of facemasks during the COVID-19pandemic. https://doi.org/10.1016/S2213-2600(20)30359-3
6. Ejaz MS, Islam MR, Sifatullah M, Sarker A (2019) Implementation of principal component analysis on masked and non-masked face recognition. In: 2019 1st international conference on advances in science, engineering and robotics technology (ICASERT), pp 1–5
7. Ge S, Li J, Ye Q, Luo Z (2017) Detecting masked faces in the wild with LLECNNs. IEEE Conf Comp Vision Pattern Recogn (CVPR) 2017:426–434. https://doi.org/10.1109/CVPR.2017.53
8. Medical Mask Dataset (2021) Accessed 17 Jun 2021. [Online]. Available: https://www.kaggle.com/shreyashwaghe/medical-mask-dataset
9. Real World Masked Face Recognition Dataset (2021) Accessed 17 Jun 2021. [Online]. Available: https://github.com/X-zhangyang/Real-WorldMasked-Face-Dataset

Enhancing ECG Signal in Noisy Environment: A Review

Ganta Naga Sandhya Devi and Vinay Kumar Mittal

Abstract Today, for treatment of any disease, the patient needs to undergo many tests. The extraction of biomedical signal is done initially, and then, depending on the report, the diagnosis or treatment can be preferred. There are many biomedical signals which we extract for accurate analysis and treatment. The biomedical signal that deals with heart diseases is ECG signal. This review paper focuses on the ECG signal analysis. While recording the ECG signal, there may be many types of noise present. Even the heat generated in the electronic device also cause noise. So, this noise must be eliminated in order to extract the exact ECG signal so that correct treatment can be done. For this purpose, filtering techniques can be used. Depending on the noise that has been present in the ECG signal, the filter has to be chosen accordingly. This paper reviews databases, features, filtering techniques, signal processing methods, and results obtained in different studies of ECG signal analysis.

Keywords Noise · Databases · Signal processing methods · Signal-to-noise ratio

1 Introduction

There is a lot of information in biomedical signal related to human health. This information is very important since based on this medication is given to the patient. Therefore, biomedical signals play a vital role. The different types of biomedical signals [1] present are electrocardiogram (ECG), electrooculogram (EoG), electroretinogram (ERG), electroencephalogram (EEG) [2], electromyogram (EMG), electrogastrogram (EGG). Electrocardiogram signal determines the electrical functioning of the heart. The three main components of ECG wave P-wave, QRS complex [3], and the T-wave represent the depolarization of the atria, the depolarization, and

G. Naga Sandhya Devi (✉)
KL University, Vijayawada, India
e-mail: 2002041040@kluniversity.in

V. K. Mittal
GITAM (Deemed to Be University), Hyderabad, India
e-mail: vmittal@gitam.edu

M. Gupta et al. (eds.), *Artificial Intelligence on Medical Data*,
Lecture Notes in Computational Vision and Biomechanics 37,
https://doi.org/10.1007/978-981-19-0151-5_36

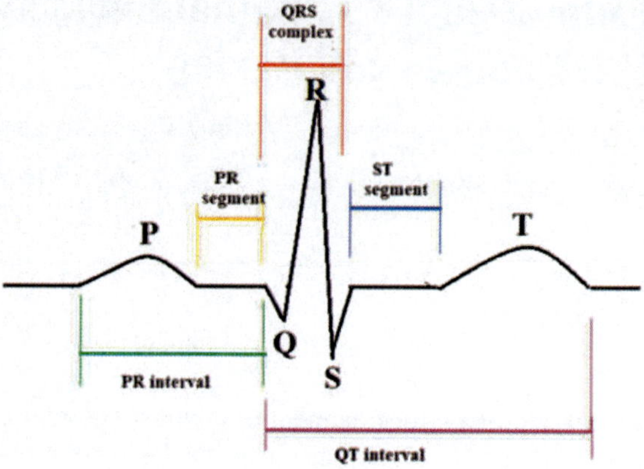

Fig. 1 ECG of a heart in normal sinus rhythm

repolarization of the ventricles [4], respectively. Figure 1 shows the ECG signal of a heart. Some of the noises present in ECG signal are baseline wander, powerline interference, electromyography noise [5].

The focus of this paper is removing the noise that is present in the ECG signal. The objective of this review paper is to discuss various filtering techniques for removing the noise in the ECG signal. The scope of this paper is to define which filtering technique can be applied exactly for that particular noise in the ECG signal and which performs better in giving a clear ECG signal.

The databases of biomedical signals used in different studies are MIT-BIH database, PhysioNet database, MITDB, MIT-BIH arrhythmia, MIT-BIH noise stress database [6]. These databases provide ECG signal records and can be used for processing of ECG signal.

The various signal processing methods used in different studies are complete ensemble empirical decomposition with adaptive noise (CEEMDAN), adaptive noise canceller, subtractive procedure, normalized least mean squares algorithm, bionic wavelet transform. The various filtering techniques [7] used in different studies are adaptive IIR filter, second-order Q-varying IIR notch filter, discrete Fourier transform (DFT), and mean square error approximation.

In this paper, a review of filtering techniques for noise removal in ECG signals is presented. This review paper consists of the databases, signal processing methods used in different studies, and the corresponding results obtained.

This paper is organized as follows. Section 1 consists of Introduction. Section 2 comprises of noises that are present in ECG signal. Section 3 comprises of databases used in previous studies. Section 4 comprises of signal processing methods used and the results in previous studies. Section 6 comprises of applications. Section 7 comprises of Analysis and scope for future work. Section 8 comprises of Summary and Conclusions.

2 Noises in Biomedical Signals

In this section, different types of noises in ECG signal are presented. In any signal, while recording or transmitting, there will be noise added to the signal. Similarly, in biomedical signals, also, there will be noise [8]. The noises present in ECG signal are listed in Table 1. The noises present in biomedical signals such as ECG are baseline wander noise, powerline interference, electromyography noise, channel noise, electrode contact noise, and motion artifacts.

Electromyography noise: If muscles present beside the heart contract, then this noise called EMG noise will exist [9]. This also depends on the probe quality. The ECG noise frequency is about −150 Hz.

Channel noise: If ECG signal is transmitted through the channel, then this noise is introduced [10]. This noise is called channel noise. It is similar to the additive white Gaussian noise (AWGN).

Baseline wander noise: This noise is because of patient's respiration and body movement. It does not allow us to detect correct peaks in the ECG signal. That is QRS complex cannot be detected correctly.

Powerline Interference: If grounding of the ECG equipment or patient is not correctly done, then this powerline interference occurs. The 50 Hz signals can be induced from other electrical equipment. Some may induce 60 Hz powerline interference [11]. Figure 2 shows the baseline wandering and powerline interference.

Table 1 (b) Noise in ECG signal and (c) factor responsible for that noise

(a) Sl. No.	(b) Noise in ECG signal	(c) Factor responsible for the noise	(d) Ref. papers
1	Baseline wander noise	Because of the patient's respiration and body movement	[5, 8, 9]
2	Powerline interference	If the ECG equipment or patient grounding is not correct	[5, 8, 9]
3	Electromyography noise	If the muscles present beside the heart contract	[5, 8, 9]
4	Channel noise	When ECG signal is sent over the channel	[8, 9]
5	Electrode contact noise	If the contact between electrode and skin is lost	[8, 9]
6	Motion artifacts	Because of change in electrode–skin impedance	[8, 9]

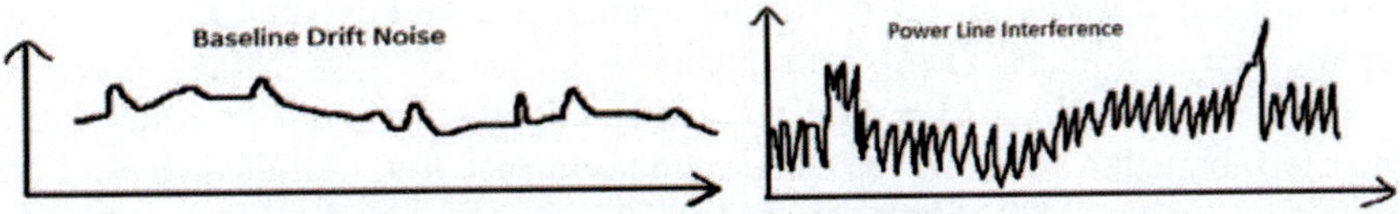

Fig. 2 Baseline wandering and powerline interference in ECG signal

Table 2 (b) Databases used in different studies and (c) purpose of the database

(a) Sl. no.	(b) Database	(c) Purpose	(d) Ref. papers
1	MIT-BIH arrhythmia	This consists of 48 half-hour excerpts of two-channel ambulatory ECG recordings	[5, 9–11]
2	PhysioNet	It consists of all the biomedical databases	[6]
3	MIT-BIH stress database	3 half-hour recordings of noise in ambulatory ECG recordings. The clean ECG signal is added with noise	[6]
4	MITDB database	It provides ECG signals for studying cardiac diseases	[7]

Electrode contact noise: This noise occurs when the contact between the electrodes and skin is lost. The duration of this noise is about 1 s.

Motion artifacts: These are due to the change in electrode–skin impedance.

3 Databases

The database represents collection of signals. The biomedical ECG signal database used in some of the studies is MIT-BIH database, PhysioNet database,

MIT database, MIT-BIH noise stress database. The PhysioNet and MIT-BIH database provides various ECG signals. The MIT-BIH noise stress database provides different types of noise such as AWGN, and they can be added to the ECG signal and can be analyzed accordingly. The purpose of the databases used in biomedical is listed in Table 2.

4 Signal Processing Methods Used

In 1983, Mark Yelderman proposed an adaptive noise canceller [12] to reduce the electrosurgical interference. An improved SNR of 110 dB has been achieved. In 1985, J. A. Van Alste has used a FIR filter [13] for removing baseline wander noise and powerline interference with a smaller number of taps. For an ECG signal having sampling rate of about 250 Hz and a filter impulse length of duration 1 s, the no. of multiplications is reduced to 26 per sample.

In 1995, Y. Ziyalder, M. Cem Saki, and H. AlperGiiger proposed a line interference subtraction filter for reducing line interference [14]. All the signals were sampled at a rate of 1000 samples/sec and digitized using a 12-bit A/D converter. The result obtained is for 200 beats; the level of noise reducing to 15% of the original value is 98.82%. In 1995, SooChang Pei proposed a high-order notch filter [15] for suppressing the 50 Hz powerline (AC) interference. The proposed technique uses the vector projection which removes the transient states.

In 1998, N. Kumaravel proposed a genetic algorithm [16] for removing sinusoidal interference. Using this algorithm, for frequency interference of 51.5 Hz, an SNR of 84.5 dB, RMS output error = 0 mV, whereas for adaptive filter SNR = 40.6 dB, FIR filter, SNR = 17.6 dB. In 2002, Shahriar Iravanian proposed a filter residue method [17] for cardiac biosignal filtering. The filter residue method has achieved high SNR of 5.2 dB when compared to linear and median filter. In 2004, I. P. Mitov proposed a discrete Fourier transform (DFT) and minimum-squared-error (MSE) approximation [18]. For input SNR more than 25 dB, this algorithm has achieved logarithmic-relative error which is less than 32 dB, and the error-to-signal ratio is below 24 dB (Table 3).

In 2005, Chavdar Levkov proposed a subtractive procedure [19] to remove the unacceptable transient response time that exists with adaptive filtering. With this subtraction procedure, a clean ECG signal has been obtained with ripple frequency of 60 Hz and sampling frequency of 250 Hz. Jacek Piskorowski proposed a second-order Q-varying IIR notch filter which removes 15 Hz sinusoidal interference. In 2006, Omid Sayadi proposed a bionic wavelet transform [20]. For an input SNR of −5 dB, an improvement of 6 dB is achieved.

In 2010, D. JeyaRani proposed removal of noises in ECG signal by using different filters [21] such as band stop filter, lowpass filter, highpass filter, and average digital filter. The three noises such as baseline wander noise below 1 Hz, 50/60 Hz powerline interference, and electromyography noise above 100 Hz are removed using these filters [22].

Zia-ul-Haque proposed LMS, normalized least mean square (NLMS), and log-LMS algorithms [23]. Out of these, NLMS algorithm has given SNR of 22.17, and MSE is approximately zero. In 2019, N. M. Verulkar has proposed an adaptive IIR filter and compared the results with non-adaptive IIR filter for removing the powerline interference. The results indicated that mean square error (MSE) is 5 dB [8] for a frequency of 48.5 Hz and about 9 dB for frequency of 51 Hz for adaptive IIR filter, whereas for non-adaptive filter, MSE is from 48 to 50 dB. With this, the signal is enhanced [24].

In 2020, Krishna Teja proposed a denoising signal processing techniques such as empirical mode decomposition (EMD), ensemble empirical mode decomposition (EEMD), and CEEMDAN [25, 26] for removing the baseline noise. CEEDMAN has outperformed of all the three methods with kurtosis = 20.5177, RMS value = 0.1740, Crest factor = 7.0475, skewness = 3.6722 (Table 4).

Table 3 (b) Noise, (c) signal processing methods, and (d) results in different studies

(a) Sl. No.	(b) Noise	(c) Signal processing methods	(d) Results	(e) Ref. papers
1	Baseline noise	CEEMDAN Highpass filter	CEEDMAN has outperformed of all the three methods with kurtosis = 20.5177, RMS value = 0.1740, crest factor = 7.0475, skewness = 3.6722 This has been used, and the noise below 1 Hz has been removed	[25] [21]
2	Powerline interference	Adaptive IIR filter DFT and MSE Band stop filter Line interference subtraction (LIS) filter High-order notch filter	The results indicated that mean square error (MSE) is 5 dB for a frequency of 48.5 Hz adaptive IIR filter, whereas for non-adaptive filter, MSE is from 48 to 50 dB; for SNR more than 25 dB, this has achieved logarithmic relative error <32 dB, and the error-to-signal ratio is also below 24 dB For frequency interference of 51.5 Hz, an SNR of 84.5 dB, RMS output error = 0 mV. Whereas for adaptive filter, SNR = 40.6 dB, FIR filter, SNR = 17.6 dB Using this 50/60 Hz, powerline interference has been removed By using this, for an average of 200 samples, the noise level reducing to 15% is about 98.82%. The 50/60 Hz powerline interference has been removed	[8] [18] [32] [14] [15]
3	Electrosurgical interference	Adaptive noise canceller	This noise canceller has given improved SNR of 110 dB	[12]

(continued)

Table 3 (continued)

(a) Sl. No.	(b) Noise	(c) Signal processing methods	(d) Results	(e) Ref. papers
4	Unacceptable transient response time	Subtractive procedure	Clear ECG signal	[9]
5	15 Hz sinusoidal interference	Q-varying IIR notch filter of second-order	Clear ECG signal	[10]
6	Baseline wander and powerline	FIR filter	Multiplications have been reduced to 26 per sample	[10]
7	High frequency noise	NLMS algorithm	For given SNR of 22.17, MSE is approximately zero	[23]
8	Noises in ECG	Bionic wavelet transform	Soft WT shows better performance for low input SNRs. But, for high input SNRs, the BWT performs better than WT	[20]
9	Low frequency noise	Filter residue method	Filter residue method has achieved high SNR of 5.2 dB when compared to linear and median filters	[17]
10	Electromyography noise	Lowpass filter	The noise above 100 Hz has been removed	[21]

Table 4 (b) Databases (c) features (d) classifiers used and (e) results obtained in previous studies

(a) S.no	(b) Database used	(c) Features used	(d) Classifiers used	(e) Results	(f) Ref. papers
1	MIT-BIH arrhythmia database	DFT	Grow and learn, Kohonen neural networks	GAL proved to be better. Time is 4 and 7 ms for GAL and Kohonen	[27]
2	MIT-BIH arrhythmia database	Wavelet transform	Independent component analysis	An accuracy of 97.5% has been achieved	[30]
3	MIT-BIH database	Dyadic wavelet transform	Artificial neural networks	An accuracy of 97.77% has been achieved	[32]
4	MIT-BIH database	DFT and dyadic wavelet transform	Neural network trained by genetic algorithm	An accuracy of 99.4% and 92.2% with wavelet transform and DFT, respectively	[31]

5 Features and Classifiers Used for Improving Accuracy of ECG Waveform

In 1996, for detecting the ECG waveform [27], the artificial neural networks such as grow and learn (GAL) and Kohonen are used. Among these two neural networks, the GAL network has fast learning and also better classification performance. In 1998, adaptive filtering [28] technique is for determining the limits of P and T waves so that the searched point corresponds to a less error.

In 2006, the ECG is analyzed through the hidden Markov models (HMMs) [29] because these models have the ability of beat detection, segmentation, and classification. The two-channel QT database is used for evaluating the performance of HMM, and this method has given high performance of beat detection. With this method, the sensitivity and positive predictivity obtained are 99.79% and 99.96%, respectively. In 2008, the method uses the feature extraction techniques such as independent component analysis (ICA) and wavelet transform [30]. This is done on ECG samples taken from MIT-BIH arrhythmia database, and 97.5% accuracy has been achieved.

In 1999, by using the feature extraction methods such as DFT and DWT and the classifier used is neural network trained by genetic algorithm [31] tested on ten different waveforms from MIT-BIH database, Wavelet transform and DFT have given accuracy of 99.4% and 92.2% respectively.

In 2000, ECG signal filtering has been proposed using adaptive neural network and a wavelet transform-based QRS detection method [32]. In this mixed method, detection rate failures are comparatively less with wavelet transform. For ECG diagnosis, feature extraction is done using the dyadic wavelet transform (DWT) and artificial neural network as classifier on MIT-BIH database; the accuracy achieved is 97.77%.

6 Applications

As today, heart diseases are becoming more and more, detecting them in advance saves human lives. So, this ECG signal monitoring is important, and hence, we have to obtain clear ECG signal. The applications of noise removal in ECG [33, 34] are as follows to monitor the health condition of the patient, to detect if there is a chance of heart attack and medicate the patient accordingly so that he can survive.

- To identify whether pattern of contraction of heart is accurate
- To monitor the condition of heart muscle
- To determine whether the heart is normal, enlarged, or is damaged
- To detect irregularities in hearts rhythm called arrhythmia
- To detect the extent of recovery possible for the patient undergoing treatment of heart.

7 Scope and Analysis of Future Work

The scope of this paper can be extended to any type of noise present in the ECG signal. This study helps to remove the noise in ECG signal [22, 35] so that the condition of the patient can be monitored correctly and the treatment can be given accordingly. The future work in this area includes the enhancement of the ECG signal by eliminating the low-frequency and high-frequency noises.

8 Summary and Conclusions

This study mainly helps to identify the features extracted [36] and the signal processing methods [37] used for the noise removal in ECG signals. Different signal processing methods and their performance are studied and evaluated based on the noise present. The results are evaluated in terms of factors such as SNR, signal-to error ratio [38], RMS error, kurtosis, crest factor, and skewness. Depending on the noise present, the filtering techniques must be used to reduce the noise in the ECG signal. This helps in obtaining a clear ECG signal in the end.

References

1. Sharma S, Ahmad Ansari I (2010) A review on biomedical systems and biomedical signal processing techniques. In: International conference on biomedical engineering and assistive technologies, NIT Jalandhar, Punjab, India, December, 2010
2. Mahesh Kumar J, Mittal VK (2021) EEG data acquisition system and analysis of EEG signals. In: 2021 2nd international conference for emerging technology (INCET), Belgaum, India. 21–23 May 2021
3. Thuy Nguyen NV, Duc Tran L, Van Huynh T (2017) Detect QRS complex in ECG. In: 12th IEEE conference on industrial electronics and applications, June 2017
4. Inderbir Kaur R (2013) Electrocardiogram signal analysis—an overview. Int J Comp Appl (0975–8887) 84(7)
5. Sudheer G, Sai Durga Prasad G, Naveen Kumar E, Srinivasulu A (2017) Noise removal from ECG signal by using FIR filter with different window techniques. Int J Adv Res Sci Eng 6(8)
6. Mitraa S, Mitrab M, Chaudhuria BB (2004) Generation of digital time database from paper ECG records and Fourier transform-based analysis for disease identification. Comp Biol Med 34(1):551–560
7. Nayak S, Soni MK, Dipali B (2012) Filtering techniques for ECG signal processing. Int J Res Eng Appl Sci 2(2):671–679
8. Verulkar NM, Zope PH, Suralkar SR (2012) Filtering techniques for reduction of power line interference in electrocardiogram signals. Int J Eng Res Technol (IJERT) 1(9)
9. Velayudhan A, Peter S (2016) Noise analysis and different denoising techniques of ECG signal—a survey. IOSR J Electron Commun Eng (IOSR-JECE) 40–44
10. Piskorowski J (2010) Digital Q-varying notch IIR filter with transient suppression. IEEE Trans Instrum Measur 59(4)
11. Karagiannis A (2009) Noise components identification in biomedical signals based on empirical mode decomposition. In: 9th international conference on information technology and applications in biomedicine, (ITAB), December 2009
12. Yelderman M, Widrow B, Cioffi JM, Hesler E, Leddy JA (1983) A ECG enhancement by adaptive cancellation of electrosurgical interference. IEEE Trans Biomed Eng BME-30(7)
13. Van Alste JA, Schilder TS (1985) Removal of base-line wander and power-line interference from the ECG by an efficient FIR filter with a reduced number of taps. IEEE Trans Biomed Eng BME-32(12)
14. Ider YZ, Saki MC, Gcer HA (1995) Removal of power line interference in signal-averaged electrocardiography systems. IEEE Trans Biomed Eng 42(7):731–735
15. Pei S-C, Tseng C-C (1995) Elimination of AC interference in electrocardiogram using IIR notch filter with transient suppression. IEEE Trans Biomed Eng 42(11):1128–1132
16. Kumaravel N, Nithiyanandam N (1998) Genetic-algorithm cancellation of sinusoidal powerline interference in electrocardiograms. Med Biol Eng Comput
17. Iravanian S, Tung L (2002) A novel algorithm for cardiac biosignal filtering based on filtered residue method. IEEE Trans Biomed Eng 49(11)
18. Mitov IP (2004) A method for reduction of power line interference in the ECG. Med Eng Phys 879–887
19. Levkov C, Mihov G, Ivanov R, Daskalov I, Christov I, Dotsinsky I (2005) Removal of power-line interference from the ECG: a review of the subtraction procedure. Biomed Eng Online
20. Sayadi O, Shamsollahi MB (2006) ECG denoising with adaptive bionic wavelet transform. In: Proceedings of the 28th IEEE EMBS annual international conference, New York City, USA, August 2006
21. Jeyarani D, Jaya Singh T (2010) Analysis of noise reduction techniques on QRS ECG waveform—by applying different filters. In: IEEE conference on recent advances in space technology services and climate change (RSTSCC), Chennai, 2010
22. Sandhya Devi GN, Mittal VK (2021) Enhancing image in noisy environment: a review. In: International conference on signal processing and communication engineering systems (SPACES 2021), KL University, Vijayawada, 11–12 June 2021 (in press)

23. Rizwan Qureshi Z, Nawaz M, Yar Khuhawar F, Tunio N, Uzair M (2019) Analysis of ECG signal processing and filtering algorithms. Int J Adv Comp Sci Appl (IJACSA) 10(3)
24. Sandhya Devi GN, Mittal VK (2021) Enhancing signal in noisy environment: a review. In: Fifth international conference on information and communication technology for intelligent systems (ICTIS-2021), April 2021, Ahmedabad, in press
25. Teja K, Tiwari R, Mohanty S (2020) Adaptive denoising of ECG using EMD, EEMD and CEEMDAN signal processing techniques. In: First international conference on advances in physical sciences and materials: journal of physics
26. Mittal S, Mittal VK (2019) Biomedical requirements for human machine interface towards building a humanoid: a review. In: Proceedings 16th international IEEE India conference (INDICON 2019), Marwadi University, Rajkot, Gujrat, India, 13–15 Dec 2019
27. Dokur Z, Olmez T, Korurek M, Yazgan E (1996) Detection of ECG waveforms by using artificial neural networks. In: 18th annual international conference of the IEEE engineering in medicine and biology society, Amsterdam, 1996
28. Soria-Olivas E, Martinez-Sober M, Calpe-Maravilla J, Francisco J, Chorro J (1998) Application of adaptive signal processing for determining the limits of P and T waves in an ECG. IEEE transactions on biomedical engineering, vol 45, no 8, August 1998
29. Andreão RV, Dorizzi B, Boudy J (2006) ECG signal analysis through hidden markov models. IEEE transactions on biomedical engineering, vol 53, no 8, August, 2006
30. Yong Z, Wenxue H, Yonghong X, Jianxin C (2008) ECG beats classification based on ensemble feature composed of independent components and QRS complex width. In: 2008 international conference on computer science and software engineering
31. Ziimray DOH, Olmez T, Yazgan E (1999) ECG waveform classification using the neural network and wavelet transform. In: Proceedings of the first joint BMEWEMBS conference serving humanity, advancing technology 1316, 1999. Atlanta, GA, USA
32. Sziliigyi SM, Sziliigyi U (2000) Wavelet transform and neural-network-based adaptive filtering for QRS detection. In: Proceedings of the 22nd annual EMBS international conference, July 23–28, 2000, Chicago
33. Devi Bhavani K, Mittal VK (2020) Studies on paralinguistic speech sounds. In: Proceedings of the 17th international IEEE India conference (INDICON 2020), IEEE, 2020, pp 1–6
34. Baumert M, Porta A, Cichocki A (2016) Biomedical signal processing: from a conceptual framework to clinical applications. Proc IEEE 104(2):220–222
35. Naga Sandhya Devi G, Mittal VK (2021) Enhancing speech in noisy environment: a review. In: IEEE region 10 symposium (TENSYMP), Grand Hyatt Jeju, Republic of Korea, 23–25 August, 2021
36. Banu U, Patil GM, Fatima R (2016) A survey on sources of noise and advanced noise removal techniques of biosignals. Int J Emerg Technol (Special Issue on ICRIET-2016) 7(2):8–13
37. Mohanta A, Mukherjee P, Mittal VK (2020) Prosody features characterization of autism speech for automated detection and classification. In: Proceedings of the 26th national conference on communications (NCC 2020), IIT Kharagpur, India, 21–23 Feb 2020
38. Radha Krishna G, Krishnan R, Mittal VK (2019) Non-native accent partitioning for speakers of indian regional languages. In: Proceedings of the 16th international conference on natural language processing (ICON 2019), IIIT Hyderabad, India, 18–21 Dec 2019

AI and TB: A New Insight in Digital Chest Radiography

Karma G. Dolma, Alok K. Paul, Mohammed Rahmatullah, Maria de Lourdes Pereira, Christophe Wiart, Priyanka Shankarishan, Veeranoot Nissapatorn, and Bidita Khandelwal

Abstract With reports of 9.9 million people being infected with tuberculosis by WHO, there is a dire need to curtail the spread of tuberculosis. In spite of having faced many impediments like lack of certified radiologist and chest radiography hardware which are expensive, diagnosis of tuberculosis still remains undetected at early stage. Chest radiography is one of the earliest method of detection used and is an asset for diagnosis of TB especially in early stages of infection, in a resource limited setting as well as for differential diagnosis. In the times of artificial intelligence (AI), we can see many modern platforms for the development of Computer-aided detection (CAD) through machine learning (ML) and deep learning (DL) and there are data coming forth indicating their utilization to the maximum. These approaches involve in hospital settings for examining the diseases through clinical aetiology as well as X-ray images of the patient. Presently, efforts and strategies are being framed and articulated to bring more accuracy adopting the use of the AI and machine learning approaches for the diagnosis of TB. This survey provides an insight to the application and use of CAD for the diagnosis of TB.

K. G. Dolma
Department of Microbiology, Sikkim Manipal Institute of Medical Sciences, Sikkim Manipal University, Gangtok, Sikkim 737102, India
e-mail: karma.d@smims.smu.edu.in

A. K. Paul
School of Pharmacy and Pharmacology, University of Tasmania, Hobart, TAS 7001, Australia
e-mail: alok.paul@utas.edu.au

M. Rahmatullah
Department of Biotechnology & Genetic Engineering, University of Development Alternative, Dhanmondi, Dhaka 1207, Bangladesh

M. de Lourdes Pereira
Department of Medical Sciences, CICECO-Aveiro Institute of Materials &, University of Aveiro, Aveiro, Portugal
e-mail: mlourdespereira@ua.pt

C. Wiart
School of Pharmacy, University of Nottingham Malaysia Campus, Selangor, Malaysia
e-mail: Christophe.Wiart@nottingham.edu.my

M. Gupta et al. (eds.), *Artificial Intelligence on Medical Data*,
Lecture Notes in Computational Vision and Biomechanics 37,
https://doi.org/10.1007/978-981-19-0151-5_37

Keywords Artificial intelligence · Chest radiography · Computer-aided detection · Deep learning · Tuberculosis

1 Introduction

Tuberculosis (TB), which accounts for the increased incidence related to deaths globally for any bacterial pathogenic disease, is caused by *Mycobacterium tuberculosis* (*M.tb*). There is a dire urgency for the discovery of innovative diagnosis and treatment approaches [1]. Since the discovery of *M.tb* in 1882 by Robert Koch [2], the Bacillus-Calmette Guerin (BCG) in 1908 [3, 4] and the anti-mycobacterial drugs in 1943 [5], many huge steps were taken for control of tuberculosis. However, with the onset of AIDS [6] pandemic in the early 90's, the emergence of resistant strains [7] further plummeted the progress made over the times. TB is differentiated into two principal categories, the pulmonary and the extra-pulmonary infection (EP). In the pulmonary infection, there is the lung involvement and the EP infection basically depends upon the organ involved causing military TB, tubercular meningitis, tubercular lymphadenopathy, etc. There are several other factors that has a significant dynamic function in consolidating the spread of the illness. The perilous blend of AIDS and TB [8], Diabetes and TB [9] makes the immunity of the host more susceptible to TB infection. The other risk factors that cause TB are silicosis, smoking, crowding, prisons, alcoholism, end-stage renal diseases substance abuse and malnutrition [10–12]. The bacteria transmits via aerosol, when a person sneezes, coughs or speaks in the form of respiratory droplets (Fig. 1). Once infected, the spectrum of manifestations includes cough $\geq$2 weeks, fever and chills, loss of appetite and weight, night sweats, weakness and so on. If the person does not show active symptoms, the bacteria can remain dormant for many years to come and finally spring to action when the host's immunity fades. The problematic emergence of multi drug resistance (MDR) tuberculosis has further supplemented due to inadequate usage and non-compliance of anti-TB medicines, poor quality drugs, spread of resistant strains of *M.tb* and initiation of drug efflux pumps [13]. Information technology

P. Shankarishan
Departemnt of Applied Biology, University of Sciences and Technology Meghalaya (USTM), Ri-Bhoi, India
e-mail: p.shankarishan@rediffmail.com

V. Nissapatorn (✉)
School of Allied Health Sciences, World Union for Herbal Drug Discovery (WUHeDD), and Research Excellence Center for Innovation and Health Products (RECIHP), Walailak University, Nakhon Si Thammarat, Thailand
e-mail: nisspat@gmail.com

B. Khandelwal (✉)
Department of General Medicine, Sikkim Manipal Institute of Medical Sciences, Sikkim Manipal University, Gangtok, Sikkim 737102, India
e-mail: bidita.k@smims.smu.edu.in

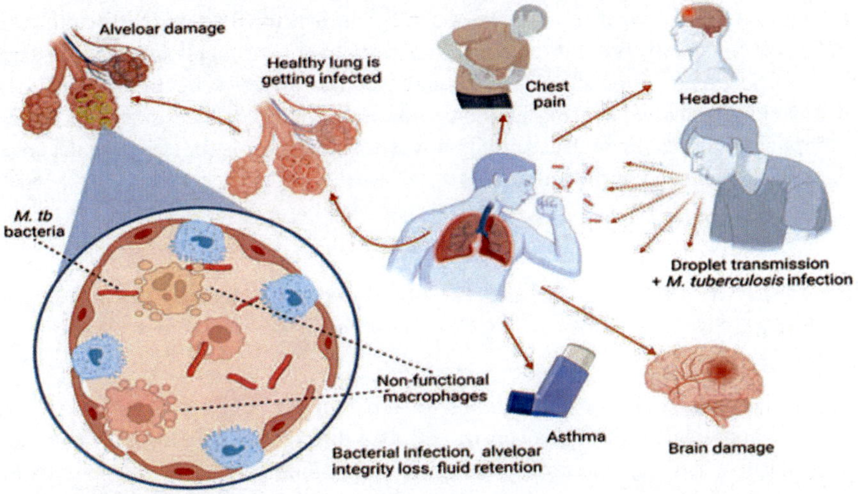

Fig. 1 Pathophysiology and symptoms of tuberculosis. The Figure was made with www.biorender.com (accessed on: 29 October 2021)

(IT) enabled adherence to therapy has led to major success in tackling the menace of TB and health sector is optimistic about the use of AI in the diagnostic aspect also Till date, there are scanty data on AI-TB diagnosis reported in India, therefore, this review timely aimed to highlight the significant contribution of AI in TB diagnosis, especially in this modern era, to benefit affected patients as far as the clinical management is concerned.

In the year 2021, the World Health Organization (WHO) estimated that 9.9 million people were affected with TB in 2020. Further, WHO recorded a global decline in the number of new fresh TB cases. Reports of decline in cases ranged from 7.1 million in 2019 to 5.8 million in 2020, with 16 countries accounting for 93% of this reduction [17]. In India, The National TB Elimination Programme (NTEP) identifies the patients by the systematic screening and surveillance system through which the patients present themselves to medicare facilities with suspected TB symptoms. The national lockdown imposed by the government in March–April 2020 resulted in downhill notification by 38% (44% private notification) [15]. An unprecedented derailing in the momentum gained by the TB projects has been caused due to COVID-19. A large-scale active TB case finding was launched at the community level in all parts of the country with the objective of massive screening and testing, engaging the health and community workers, surveillance of symptoms within the household, contacting the residents through phone or video apps, etc. In India, the active case finding has been executed in closely associated of sputum smear positive TB patients, mainly via suspected symptoms screening. These suspected TB cases would undergo chest radiography and routine laboratory examinations in practice. A CXR can detect the lesions suggestive of TB, identify the locations, type, state and complexity of the lesions. WHO has recommended that CXR screening can be adopted in case finding of TB [16]. Over times there is a stiff competition amongst the health care bodies to constantly progress and advance the diagnostic application and lessen the disease transmission by efficient diagnosis and effective treatment [17]. The overwhelming challenge facing TB control in India remains delayed diagnosis, hence prompt diagnosis using high sensitivity novel diagnostic tools for early and accurate diagnosis is the need of the hour.

2 The Role of CXR in TB Diagnosis—From Past to Present Scenario

One of the primary diagnostic methods for tuberculosis is Chest X-Ray (CXR). It is a well-appreciated tool to categorize TB as a differential diagnosis for patients, particularly when any abnormal findings are identified related to TB. One of its main advantage is that it is easy to operate and has a low running cost, which is an added bonus for low-income countries. Furthermore, it can be useful in many other modalities apart from tuberculosis diagnosis. However, the CXR aberrations that are persistent with PTB have also been observed in various pathologies of lung infection, making it a challenging diagnosis, as a result of which CXR has a poor specificity. Furthermore, there exist a substantial intra- and inter-observer discrepancy in the interpretation of CXRs. Hence, to be dependent only on CXR for TB diagnosis can lead to over diagnosis or under diagnosis [18]. In addition, reports of lack of well-qualified and experienced radiologists, expensive hardware of CXR are some of the drawback of CXR [19, 20] (Table 1).

Over the years, the utilization of CXR in examining and testing algorithms has been endorsed as a beneficial diagnostic mechanism. With the advent of artificial intelligence (AI), newer machineries and expertise have been developed for the CXR evaluations. Recently, the most noted one is the Computer-aided detection (CAD)

Table 1 Clinical presentation of PTB in an X-ray [21]

Forms	Mode of action	Findings
Primary tuberculosis	Infection effects are observed immediately	Integrated pleural effusion, Lymphadenopathy, and miliary nodules
Secondary tuberculosis	Reactivation of Latent tuberculosis	Cavities, consolidations, and centrilobular nodules involving posterior segments of the upper lobe, or the superior segment of the lower lobe
Miliary tuberculosis	Haematogenous dissemination	Miliary lung nodules
Old tuberculosis	–	Fibro-nodular opacities and variation in the apical and upper lung zones

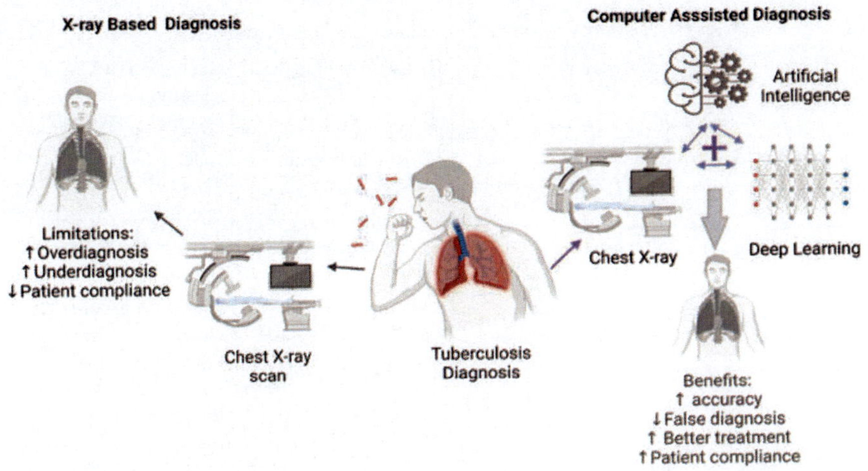

Fig. 2 Application of CAD in TB infection. The Figure was made with www.biorender.com (accessed on: 29 October 2021)

software that can analyse digital CXR images for abnormalities and the likelihood of TB being present. This new development in technology forces us to reconsider the use of CXR in screening and diagnostic algorithms [22]. Till 2016, WHO had provided no approvals on the usage of CAD for TB. However, in 2020, the WHO expert panel on reviewing the performance of CAD endorses the application of CAD as a substitute to elucidate CXR for testing and screening of PTB in individuals aged 15 [23–25] (Fig. 2). There are various diseases like pleural effusion, neoplastic disease, cardiac disease, pneumonia and occupational lung diseases which can exhibit similar radiological patterns like tuberculosis. Hence, a differential diagnosis of the disease needs to be taken into consideration when evaluating such a case. Even an experienced radiologist can succumb to perceptual as well as cognitive biases leading to errors [26]. Hence, in the CXR interpretation a particular extent of higher and lower diagnosis is pertinent as at times it is problematic to discriminate analogous lesions, or one may come across ambiguous nodules [27] (Table 2).

3 Computer-Aided Detection (CAD) System—A New Paradigm

The developments and enhancements observed in the automated computer image recognition have been substantial over the years. Machine learning (ML) and deep learning (DL) are apparently the two extensively utilized AI methods for constructing CAD systems able of analysing Chest X-rays [28]. Their use in the diagnosis of various other diseases has been proficient as observed in neurological disorders [29], bone suppression and skin lesions [30] and various cancer related diseases [31].

Table 2 Comparison of various CAD systems for diagnosing tuberculosis

Author and year	Software used	Dataset/database used	Accuracy (AUC, sensitivity and specificity, AUROC)
Melendez et al. [45]	Combining the score of CAD4TB algorithm	12 clinical features findings	AUC: 0.84 Sensitivity: 49% Specificity: 95%
Murphy et al. [46]	Deep learning CAD4TB model	500 labelled cohort images	Sensitivity—90% Specificity—98%
Lakhani et al. [47]	GoogleNet and AlexNet	1007 images	Untrained version (AUC): • AlexNet—0.90 • GoogleNet—0.88 Pre-trained version (AUC): • AlexNet—0.98 • GoogleNet—0.97
Zaidi et al. [48]	Retrospective analysing the CAD4TB and Xpert MTB/RIF testing	6845 individuals from two private TB diagnostic centres in Karachi	CAD4TB: Sensitivity: (range 85.8–97.3%) Negative predictive value (NPV): (range 96.1–98.4%) The area under the ROC curve (AUC): 0.79
Pasa et al. [49]	Convolutional neural network (CNN)	National Institutes of Health (NIH)—138 CXRs Belarus tuberculosis portal—304 CXRs	Accuracy: • NIH—79.0% • Belarus—84.4% • Combined—86.2% AUC: • NIH—0.811 • Belarus—0.900 • Combined—0.925
Khan et al. [50]	Artificial neural network (ANN)	12,638 TB records	Test and validation: ≥93% Accuracy: ≥94
Zhi Zhen Qin et al. [41]	CAD4TB Lunit INSIGHT qXR	1196 cases (515 from Nepal and 681 from Cameroon)	AUC: CAD4TB (0.92, 95% CI: 0.90–0.95) Lunit (0.94, 95% CI: 0.93–0.96) qXR (0.94, 95% CI:0.92–0.97)

(continued)

Table 2 (continued)

Author and year	Software used	Dataset/database used	Accuracy (AUC, sensitivity and specificity, AUROC)
Sathitratanacheewin et al. [51]	Deep convolutional neural network (DCNN) model using a tuberculosis (TB)-specific chest x-ray (CXR) dataset DAC4TB	National library of medicine Shenzhen No. 3 hospital National Institute of Health Chest X-ray 8 database	AUC: DCCN model: 0.9845 For detecting TB: 0.8502 Supervised DCNN model in the Chest X-ray 8 dataset was:0.7054 Sensitivity—72% Specificity—82%, 36.51% of abnormal Radiographs in the final DCNN model estimated that the ChestX-ray8 dataset were related to TB
Zhi Zhen Qin et al. [52]	• CAD4TB • InferRead DR • Lunit INSIGHT CXR • JF CXR-1 • qXR	23,954 individuals	Areas under the receiver operating characteristic curve (AUROC): qXR: 90·81% (95% CI 90·33–91·29) CAD4TB:90·34% (89·81–90·87) Lunit INSIGHT CXR: 88·61% (88·03–89·20) InferRead DR:84·90% (84·27–85·54) JF CXR:1–84·89% (84·26–85·53) The CAD4TB (72·9% specificity [72·3–73·5] and qXR: (74·3% specificity [95% CI 73·3–74·9]) exhibited the WHO's Target Product Profile (TPP) of triage tests (≥90% sensitivity and ≥70% specificity

Artificial intelligence is a branch of computer science that deals with developing systems to facilitate tasks that generally needs human skills and intelligence [32]. ML is a branch of artificial intelligence that employs computers algorithms to decide what variables are important [33] and are executed to carry out tasks without definite programming [32].

In machine learning, the expert distinguishes and prepare qualities that appear as characteristic in the data. Then different statistical techniques set up or manage the data based on these features [32]. Deep learning belongs to the subgroup of ML, where the read features are compositional or hierarchical [16, 32]. Machine learning (ML) techniques are efficaciously utilized for constructing predictive classification models including biological compound identification [34, 35], side-effect predictions [36, 37], gene prediction for novel disease [38], microarray data analysis [39] and drug discovery against TB [40, 41]. Several studies have proved that with the usage of CAD, we can identify and confirm a TB case on CXR. The AUC of ROC curves around 0.80–0.95 suggest that CAD solutions can read a CXR as well as an experienced radiologist [42–44].

Melendez et al. mixed the score of CAD4TB algorithm with 12 clinical features, which gave the algorithm an AUC of 0.84, sensitivity of 49% and specificity of 95% [45]. In 2019, Murphy et al. [46] builded a deep learning CAD4TB model, from a cohort of 500 CXRs from Pakistan. This model had a sensitivity—90% and specificity—98%, respectively. In 2017, Lakhani and Sundaram [47] developed the GoogleNet and AlexNet models (pre-trained and untrained) using around1, 007images and 68% in algorithm training. The findings of the study exhibited the AUC of untrained version of AlexNet and GoogleNet with 0.90 and 0.88, respectively, however, on the other hand the pre-trained versions of AlexNet and GoogleNet confirmed AUC values of 0.98 and 0.97, respectively. Zaidi and colleagues [48], performed a retrospective analysis of 6,845 cases using four different models of CAD4TB and Xpert MTB/Rif in two private TB diagnostic centres of Karachi, Pakistan. They observed an NPV (range 96.1–98.4%) and a greater sensitivity (range 85.8–97.3%) and the AUROC curve for CAD4TB was 0.79. When the demographic variables were taken in consideration, an AUC of 0.83 was expressed exhibiting good analytical analysis. Pasa et al. [49] developed an effective deep neural network model using two different public databases: National Institutes of Health (NIH)—138 CXRs and Belarus Tuberculosis portal—304 CXRs. The accuracy and area under the curve (AUC) for NIH, Belarus Tuberculosis portal and combined was observed as 79%,84.4%, 86.2%, 0.811, 0.900 and 0.925, respectively. Using artificial neural network (ANN), Khan et al. [50], in 2019 created a model using trained dataset of 12,638 TB records. These were obtained from the Tuberculosis Laboratory, Khyber Pakhtunkhwa, Pakistan from 2016 to 2017. The model established an overall accuracy of 94% and above and test and validation accuracy of 93% and above. Qin et al. in 2019 [41] led a contemplative assessment of 3 DL systems (CAD4TB, Lunit INSIGHT and qXR) from 1196 patients from Nepal and Cameroon, after receiving the Xpert MTB/Rif assay and independent chest reading from 2 groups of radiologist. They found that the use of DL system could significantly decrease the use of Xpert MTB/RIF assay by 66% and increase the sensitivity by ≥95%. Sathitratanacheewin et al. [51], 2020 established a deep convolutional neural network (DCNN) model comparing National Library of Medicine Shenzhen No.3 Hospital (TB specific CXR) and National Institute of Health Clinical Centres (non-TB-specific CXR). An AUC of 0.9845 and 0.8502, respectively, was observed. Another study published in 2021 by Zhi Zhen Qin et al. [52] undertook in 3 Tuberculosis centres in Dhaka, Bangladesh from 2014 to 2016, registering ≥15 years and above. The CXR was analysed by five commercial AI algorithms: JF CXR-1 (version 2), qXR (version 3), Lunit INSIGHT CXR (version 4.9.0), InferRead DR (version 2) and CAD4TB (version 7) and further read by a team of three registered radiologists. The analysis was that all the 5 algorithms could significantly lessen the use of Xpert MTB/RIF assay by 50% however, a high sensitivity range of 90% was maintained and the five AI algorithms significantly outperformed the radiologists.

4 Conclusion

In order to realize the global dream of TB elimination, early and accurate diagnosis is of paramount importance. Linking the diagnostic accuracy of a standard CXR with the technology of AI will facilitate the prompt and accurate diagnosis of PTB and selective referral of presumptive cases to more sensitive, but expensive, molecular methods for confirmation. Recently, there has been an uprise in the application of artificial intelligence (AI) that have led to eminent progress in automated computer image identification. There is much room for further exploration on CAD-based TB identification utilizing profound learning. Further, it is guaranteed that TB can be diagnosed using the AI-based CAD systems and improve active case finding in TB detection in the coming times. In this evidence-based survey, apart from CAD4TB [53, 54] there are various software programmes like Lunit INSIGHT and qXR that have accomplished the same diagnostic accuracy as an expert. In this context, approaches like Identification Tree (IDT), General Regression Neural Network (GRNN), and other Multi-Layer Neural Networks (MLNNs) and Genetic Algorithm systems may prove to be essential tools in utilizing AI for diagnosing TBs and 'reading' chest X-rays. CAD systems used AI to screen X-ray images to categorize aberrations and resolve the paucity of radiologists, mainly in developing countries [55]. The CAD software programmes can be especially utilized in a country like India, where the load of PTB is maximum and the proportion of radiologists to patient is low [56]. The survey on the application of CAD for TB diagnosis in this paper emphasizes the need for further research on CAD systems that utilize CXR for TB diagnosis. With WHO recommendation, these newer solutions can be incorporated into the TB programmes with operational research (ORs) to elevate effect. However, other significant categories like HIV, smear status, demographic association and clinical data should be explored to attain customized risk scores, bringing enhanced performance.

Contributors: All authors contributed to and approved the final manuscript.

Declaration of interests: We declare no competing interests.

Acknowledgements The authors thank Professor Karma Sonam Sherpa, Registrar, Sikkim Manipal University for his support and expertise.

References

1. Jamal S, Khubaib M, Gangwar R, Grover S, Grover A, Hasnain SE (2020) Artifcial Intelligence and Machine learning based prediction of resistant and susceptible mutations in Mycobacterium tuberculosis. Sci Rep 10:5487. https://doi.org/10.1038/s41598-020-62368-2
2. Murray JF (2004) Mycobacterium tuberculosis and the cause of consumption: from discovery to the fact. Am J Respir Crit Care Med 169(10):1086–1088
3. Vargová L, Vymazalová K, Horávcková L (2017) A brief history of tuberculosis in the Czech Lands. Tuberculosis (9)
4. Luca S, Mihaescu T (2013) History of BCG vaccine. Maedica (Buchar) 8(1): 53–58
5. Fogel N (2015) Tuberculosis: a disease without boundaries. Tuberculosis 95(5):527–531
6. Manosuthi W, Wiboonchutikul S, Sungkanuparph S (2016) Integrated therapy for HIV and tuberculosis. AIDS Res Therapy 13(1):22
7. Pawlowski A, Jansson M, Sköld M, Rottenberg ME, Källenius G (2012) Tuberculosis and HIV co-infection. PLoS Pathog 8(2):e1002464
8. Nantha S (2014) A review of tuberculosis research in Malaysia. Med J Malaysia (69):88
9. Baghaei P, Marjani M, Javanmard P, Tabarsi P, Masjedi MR (2013) Diabetes mellitus and tuberculosis facts and controversies. J Diabetes Metab Disord 12(1):58–65
10. Lamberti M, Muoio MR, Westermann C, Nienhaus A, Arnese A, Sobrinho AP, Giuseppe G, Garzillo EM, Crispino V, Coppola N (2017) Prevalence and associated risk factors of latent tuberculosis infection among undergraduate and postgraduate dental students: a retrospective study. Arch Environ Occup Health 72(2):99–105
11. Tudor C, Van der Walt ML, Margot B, Dorman SE, Pan WK, Yenokyan G, Farley JE (2016) Occupational risk factors for tuberculosis among healthcare workers in KwaZulu-Natal, South Africa. Clin Infect Dis 62(3):S255–S261
12. Gupta S, Shenoy VP, Bairy I, Srinivasa H, Mukhopadhyay C (2011) Diabetes mellitus and HIV as co-morbidities in tuberculosis patients of rural south India. J Infect Public Health 4(3):140–144
13. Siddiqi N, Das R, Hasnain SE (2004) Mycobacterium tuberculosis isolate with a distinct genomic identity overexpresses a tap-like efflux pump. Infection 32:109–111
14. WHO, GLOBAL TUBERCULOSIS REPORT 2021, www.who.int/tb/data
15. India TB Report 2021, Central TB Division, Ministry of Health and Family Welfare. www.tbcindia.gov.in
16. Cao XF, Li Y, Xin HN, Zhang HR, Pai M, Gao L (2021) Application of artificial intelligence in digital chest radiography reading for pulmonary tuberculosis screening Chronic Dis Transl Med 7:35–40
17. Asgharzadeh M, Kafil HS, Ghorghanlu S, Pourostadi M (2015) Laboratory cross-contamination of Mycobacterium tuberculosis in Northwest of Iran. Egypt J Chest Dis Tuberc 64:665–669. https://doi.org/10.1016/j.ejcdt.2015.03.018
18. Koppaka R, Bock N (2004) How reliable is chest radiography? In: Frieden T (ed) Toman's tuberculosis: case detection, treatment, and monitoring. Questions and answers, second edition. Geneva: World Health Organization, pp 51–60. (WHO/HTM/TB/2004.344; http://apps.who.int/iris/bitstream/10665/42701/1/9241546034.pdf)
19. Qin C, Yao D, Shi Y, Song Z (2018) Computer-aided detection in chest radiography based on artificial intelligence: a survey. Biomed Eng Online 17:113. https://doi.org/ https://doi.org/10.1186/s12938-018-0544-y
20. Pande T, Pai M, Khan FA, Denkinger CM (2015) Use of chest radiographyin the 22 highest tuberculosis burden countries. Eur Respir J 46:1816e1819. https://doi.org/10.1183/13993003.01064-2015
21. Nachiappan AC, Rahbar K, Shi X, Guy ES, Eduardo J, Barbosa M, Shroff GS, Ocazionez D, Schlesinger AE, Katz SI, Hammer MM (2017) Pulmonary tuberculosis: role of radiology in diagnosis and management. Radiographics 37:52e72. https://doi.org/10. 1148/rg.2017160032
22. CAD for TB screening and triaging: Foundation for Innovative New Diagnostics; 2021 [Available from: https://www.finddx.org/tb/cad-tb/]

23. Rapid communication on systematic screening for tuberculosis [Internet]. World Health Organization. 2020. [Available from: https://extranet.who.int/iris/restricted/bitstream/handle/10665/337372/9789240016552-eng.pdf]
24. WHO consolidated guidelines on tuberculosis (2021) Module 2: screening—systematic screening for tuberculosis disease. World Health Organization, WHO, Geneva
25. WHO operational handbook on tuberculosis (2021) Module 2: screening—systematic screening for tuberculosis disease. World Health Organization, WHO, Geneva
26. Donovan T, Litchfield D (2013) Looking for cancer: expertise related differences in searching and decision making. Appl Cogn Psychol 27:43e49
27. Kulkarni S, Jha S (2020) Artificial intelligence, radiology, and tuberculosis: a review. Acad Radiol 27:71e75. https://doi.org/10.1016/j.acra.2019.10.003
28. Wernick M, Yang Y, Brankov J, Yourganov G, Strother S (2010) Machine learning in medical imaging. IEEE Signal Process Mag 27:25–38
29. Valliani AAA, Ranti D, Oermann EK (2019) Deep learning and neurology: a systematic review. Neurol Ther 8:351–65
30. Bakkouri I, Afdel K (2019) Computer-aided diagnosis (CAD) system based on multi-layer feature fusion network for skin lesion recognition in dermoscopy images. Multimed Tools Appl 79:20483–20518
31. Firmino M, Angelo G, Morais H, Dantas MR, Valentim R (2016) Computer-aided detection (CADe) and diagnosis (CADx) system for lung cancer with likelihood of malignancy. Biomed Eng Online 15:1–17
32. Chartrand G, Cheng MP, Vorontsov E, Drozdzal M, Turcotte S, Pal CJ, Kadoury S, Tang A (2017) Deep learning: a primer for radiologists. Radio Graph 37:2113–2131. https://doi.org/10.1148/rg.2017170077
33. Beam AL, Kohane IS (2018) Big data and machine learning in healthcare. J Am Med Assoc 319:1317e1318. https://doi.org/10.1001/jama.2017.18391
34. Jamal S, Goyal S, Shanker A, Grover A (2015) Checking the STEP-associated trafcking and internalization of glutamate receptors for reduced cognitive deficits: a machine learning approach-based cheminformatics study and its application for drug repurposing. PLoS One 10:e0129370
35. Tiwari K, Jamal S, Grover S, Goyal S, Singh A, Grover A (2016) Cheminformatics based machine learning approaches for assessing glycolytic pathway antagonists of mycobacterium tuberculosis. Comb Chem High Troughput Screen 19:667–675
36. Jamal S, Goyal S, Shanker A, Grover A (2017) Predicting neurological adverse drug reactions based on biological, chemical and phenotypic properties of drugs using machine learning models. Sci Rep 7:872
37. Liu M, Wu Y, Chen Y, Sun J, Zhao Z, Matheny ME, Xu H (2012) Large-scale prediction of adverse drug reactions using chemical, biological, and phenotypic properties of drugs. J Am Med Inform Assoc 19:e28–35. doi:https://doi.org/10.1136/amiajnl-2011-000699
38. Jamal S, Goyal S, Shanker A, Grover A (2016) Integrating network, sequence and functional features using machine learning approaches towards identifcation of novel Alzheimer genes. BMC Genomics 17:807
39. Pirooznia M, Yang JY, Yang MQ, Deng Y (2008) A comparative study of diferent machine learning methods on microarray gene expression data. BMC Genomics 9(Suppl 1):S13
40. Lane T, Russo DP, Zorn KM, Clark AM, Korotcov A, Tkachenko V, Reynolds RC, Perryman AL (2018) Comparing and validating machine learning models for mycobacterium tuberculosis drug discovery. Mol Pharm 15:4346–4360
41. Usmani SS, Bhalla S, Aghava GPS (2018) Prediction of antitubercular peptides from sequence information using ensemble classifer and hybrid features. Front Pharmacol 9:954
42. Qin ZZ, Sander MS, Rai B, Titahong CN, Sudrungrot S, Laah SN, Adhikari LM, Carter EJ, Puri L, Codlin AJ, Creswell J (2019) Using artificial intelligence to read chest radiographs for tuberculosis detection: a multi-site evaluation of the diagnostic accuracy of three deep learning systems. Sci Rep 9(1):1–10

43. Murphy K, Habib SS, Zaidi SMA, Khowaja S, Khan A, Melendez J (2019) Computer aided detection of tuberculosis on chest radiographs: an evaluation of the CAD4TB v6 system. arXiv preprint arXiv:190303349
44. Nash M, Kadavigere R, Andrade J, Sukumar CA, Chawla K, Shenoy VP et al (2020) Deep learning, computer-aided radiography reading for tuberculosis: a diagnostic accuracy study from a tertiary hospital in India. Sci Rep 10(1):1–10
45. Melendez J, Sánchez CI, Philipsen RH, Maduskar P, Dawson R, Theron G, Van Ginneken B (2016) An automated tuberculosis screening strategy combining X-ray-based computer-aided detection and clinical information. Sci Rep 6:25265
46. Murphy K, Habib SS, Zaidi SMA, Khowaja S, Khan A, Melendez J, Scholten ET, Farhan A, Schalekam S, Verhagen M, Rick H, Philipsen MA, Ginneken BVHM (2020) Computer aided detection of tuberculosis on chest radiographs: an evaluation of the CAD4TB v6 system. Sci Rep 10:5492. https://doi.org/10.1038/s41598-020-62148-y
47. Lakhani P, Sundaram B (2017) Deep learning at chest radiography: automated classification of pulmonary tuberculosis by using convolutional neural networks. Radiology 284(2):574–582
48. Zaidi SMA, Habib SS, Ginneken BV, Ferrand RA, Creswell J, Khowaja, Khan AE (2018) Valuation of the diagnostic accuracy of computer-aided detection of tuberculosis on chest radiography among private sector patients in Pakistan. Sci Rep 8:12339. doi:https://doi.org/10.1038/s41598-018-30810-1
49. Pasa F, Golkov V, Pfeiffer F, Cremers D, Pfeiffer D (2019) Efficient deep network architectures for fast chest X-ray tuberculosis screening and visualization. Sci Rep 9(1):1–9
50. Khan MT, Kaushik AC, Ji L, Malik SI, Ali S, Wei DQ (2019) Artificial neural networks for prediction of tuberculosis disease. Front Microbiol 10:395
51. Sathitratanacheewin S, Sunanta P, Pongpirul K (2020) Deep learning for automated classification of tuberculosis-related chest X-Ray: dataset distribution shift limits diagnostic performance generalizability. Heliyon 6:e04614
52. Qin ZZ, Ahmed S, Sarker MS, Paul K, Adel ASS, Naheyan T, Barrett R, Banu S, Creswell J (2021) Tuberculosis detection from chest x-rays for triaging in a high tuberculosis-burden setting: an evaluation of five artificial intelligence algorithms. Lancet Digit Health 3:e543–54
53. Philipsen RHHM, Sanchez C, Maduskar P, Melen DEZJ, Bax LP, Peter JG, Dawson R, Theron G, Dheda K, Ginneka BV (2015) Automated chestradiography as a triage for Xpert testing in resourceconstrained settings: a prospective study of diagnostic accuracy and costs. Sci Rep 5:12215
54. Breuninger M, Ginneken BV, Philipsen RHM, Mhimbira F, Hella JJ, Lwilla F, Hombergh JVD, Wagner D, Reither K (2014) Diagnostic accuracy of computer-aided detection of pulmonary tuberculosis in chest radiographs: a validation study from sub-saharan Africa. PLOS One 9:e106381
55. Raghavendra U, Acharya UR, Adeli H (2019) Artificial intelligence techniques for automated diagnosis of neurological disorders. Eur Neurol 43210:1–24
56. Arora R (2014) The training and practice of radiology in India: current trends. Quant Imaging Med Surg 4:449–450. https://doi.org/10.3978/j.issn.2223-4292.2014.11.04

Thermophoretic and Chemical Reaction Effects on Unsteady Dissipative MHD Convection Flow-Stream Past an Upright Permeable Plate Accompanying Radiation and Absorption of Heat

Bhrigu Kumar Kalita and Rita Choudhury

Abstract The essence of this study revolves around investigating the impact of thermophoresis, radiation as well as absorption of heat, and cataclysmal chemical reaction on thermal dispensation and mobility traits of an unsteady flow-stream passing by an upright permeable plate. The fluid is conductive of electricity and is subjected to viscous and ohmic dissipation. A magnetic field persistent with static intensity is applied normal to the plate. The dimensional governing equations and corresponding initial as well as frontier conditions are transformed into non-dimensional ones incorporating appropriate dimensionless variables. The dimensionless equation system is then solved numerically utilizing numerical computing environment of MATLAB. The impacts of concomitant flow dictating parameters on dimensionless flow attributes are depicted graphically and with the help of these pictorial visualizations, inferences pertaining to the problem are attempted. Magnetohydrodynamics is crucial in various biomedical utilizations viz. magnetic resonance imaging, target specific drug delivery, etc. MATLAB, with its image processing tools and built-in image segmentation algorithms, is of great utility for medical image analysis.

Keywords Chemical reaction · Heat absorption · MHD · Ohmic dissipation · Thermal radiation · Thermophoresis

Nomenclature

u^*, v^*	Velocity constituents along x^* and y^* directions, respectively, ms^{-1}
t^*	Time variable, s
y^*	Dimensional ordinate, m
U^*_∞	Free stream velocity parallel to the plate, ms^{-1}
m^*	Frequency of asymptotic suction, s^{-1}
g	Gravitational acceleration, ms^{-2}

B. K. Kalita (✉) · R. Choudhury
Department of Mathematics, Gauhati University, Guwahati, Assam 781014, India
e-mail: kumar90bhrigu@gmail.com

M. Gupta et al. (eds.), *Artificial Intelligence on Medical Data*,
Lecture Notes in Computational Vision and Biomechanics 37,
https://doi.org/10.1007/978-981-19-0151-5_38

T^*	Variable fluid temperature, K
T^*_∞	Environment temperature, K
C^*	Variable fluid concentration, Kg m^{-3}
C^*_∞	Concentration away from plate, Kg m^{-3}
B_0	Magnetic field strength, Am^{-1}
k^*	Porosity of the plate, m^2
C_p	Specific heat capacity, JKg^{-1} K^{-1}
Q_r	Radiative heat flux, Wm^{-2}
k_r	Mean absorption coefficient, m^{-1}
Q_s	Heat sink coefficient, Wm^{-3} K^{-1}
D	Coefficient of mass diffusion, m^{2}s^{-1}
K_c	Rate of chemical reaction, s^{-1}
V_t	Thermophoretic velocity, ms^{-1}
K_t	Thermophoretic coefficient
T_r	Mean reference temperature, K
U^*_p	Velocity of initial plate motion, ms^{-1}
y	Dimensionless ordinate
t	Dimensionless time
u	Dimensionless fluid velocity
G_t	Thermal Grashof number
G_c	Solutal Grashof number
M	Magnetic parameter
k	Porosity parameter
R	Radiation parameter
P_r	Prandtl number
S	Heat absorption parameter
E_c	Eckert number
S_c	Schmidt number
S_f	Skin friction coefficient
N_u	Nusselt number
S_h	Sherwood number

Greek Symbols

ρ	Fluid density, Kg m^{-3}
ν	Kinematic viscosity of fluid, m^{2}s^{-1}
λ_t	Thermal expansion coefficient, K^{-1}
λ_c	Mass expansion coefficient, Kg^{-1}
σ_e	Electrical conductivity, S m^{-1}
κ	Thermal conductivity, Wm^{-1} K^{-1}
σ_r	Stefan-Boltzmann constant, Wm^{-2} K^{-4}
θ	Dimensionless temperature
ϕ	Dimensionless concentration

γ	Chemical reaction parameter
δ	Thermophoretic parameter

1 Introduction

Eventuality of natural convection and forced convection have been greatly enthralling for researchers. Convection flow, natural or forced or mixed, past a permeable plate has been endorsed in numerous industrial implementations. Energy transport, thermal insulation, oil extraction procedure, dying of porous solids, nuclear reactor cooling, and filtering device manufacturing implement convection mechanism in large extent. Thermal and spatial propagation in perforated medium in contribution of hydromagnetic convection has been captivated by researchers' fascination. This paradigm of heat as well as mass propagation has been very effective in edaphology, designing machineries like MHD generators and accelerators, designing subterranean energy storage arrangement, nuclear power plant and so on. The genre of moving conducting fluids, known as magnetohydrodynamics (MHD), has grown into a subject matter of exploration since its inception due to its implementation in scientific endeavors viz. experiments regarding earth's magnetic field in geophysics, earthquake monitoring mechanism in seismology, study of solar wind in astrophysics, etc. In the field of industrial engineering, MHD has been amply utilized in plasma captivation, thermonuclear reactor cooling and metal casting. In medical science, MHD is employed in magnetic resonance imaging, computed tomography, magnetic drug targeting, etc. By virtue of diversified academic and experimental prominence, traits of aforementioned flows have been inspected by a great many researchers viz. Magyari et al. [1], Muthukumaraswamy and Ganesan [2], Chamkha [3], Satar and Kalim [4], Das et al. [5], Takhar et al. [6] and Ahmed et al. [7]. Noteworthy scrutiny has been put forth by researchers like Kim [8], Ahmed et al. [9], Mbeledogu et al. [10] and Elbashbeshy [11], etc.

Heat transfer mechanism of fluid flow in various physical situations, more often or not, gets escorted with thermal radiation. Thermal radiation can disseminate even in absence of mediator and contributes in heat propagation process. However, heat sink or source is an exigency in several allied processes of combined heat-mass transposition environments such as nuclear reactor, consumer electronic goods, soldering of electronic materials. Explorations of Das et al. [12], Makinde [13], Siddiqa and Hossain [14] and Barik et al. [15] have been contributory in assessing influence of heat ejection or assimilation, and thermal radiation on disparate type of flows.

In furtherance of chemical kinematics, heat dissemination and mass transposition, in view of boundary layer concept have been beneficial. Numerous contemplations have been prompted, in recent years, around combined thermal and spatial mobilization of fluid flow processes involving chemical reaction. A whole lot of procedures viz. separation of chemical components, liquid extraction, cooling towers in industries, drying, and membrane filtration, capitalizes on chemical reactive component of integrated heat-mass transposition mechanism. Deviand Kandasamy [16, 17], and Das et al. [18] have scrutinized effect of chemical reaction on flow past flat plates. Chambre and Young [19] have inspected diffusion of species that encounter chemical reaction. Significant upshots of chemical reaction accompanied with other physiological traits have been put forth in inquisitions of Reddy et al. [20, 25], Muthucumaraswamy [21] and Kandasamy et al. [22]. Muthucumaraswamy and Amutha [23] have probed thermal diffusion consequences of hydromagnetic convection influenced under chemical interaction.

The action of viscous force on adjoining fluid flow layers bring about heat cultivation. This process, being irreversible, is dissipative and hence is coined as viscous dissipation. Likewise, hydromagnetic convection of electrically conducting fluid inflicts ohmic dissipation owing to electric mobility. Recently, ample curiosity about effect of viscous as well as ohmic dissipation on thermal energy of fluid flow has been developed. Raju et al. [24] have inspected heat transfer effect, accompanying viscous dissipation, on fluid flow past an erect plate. Ohmic heating, in hydromagnetic flow, in view of thermo-diffusion and chemical effects has been scrutinized in report of Reddy et al. [25]. Combined influence of viscous as well as ohmic dissipation on MHD convection flow has been explored by authors viz. Ibrahim [26], Jamalabadi and Park [27], Ibrahim and Suneetha [28], Babu and Reddy [29] and Chen [30], etc.

Movement of dangling minuscule particles in a fluid due to pre-existent temperature gradient, known in literatures as thermophoresis, possesses applicability in numerous chemical engineering and bio-engineering practices. Procedures like separation of different type of particles from a mixture, removing impure ions from semiconductor wafer, and separation of polymers in field flow fractionation utilizes thermophoresis to a large extent. Moreover, thermophoresis owns promising potential in the field of drug discovery and maneuvering bio-molecules such as DNA and viruses. Effect of thermophoresis on convective flow past different sorts of physical configurations and geometries, along with other flow parameters, has been assessed by Goren [31], Selim et al. [32], Chamkha and Pop [33, 34], Alam et al. [35] and Chamkha and Camille [36], etc.

The prospect of this study has been to probe the thermophoretic and chemical reaction effect on MHD convection time dependent fluid flow past an upright perforated plate accompanied with viscous and ohmic dissipation, thermal radiation, and heat sink. After suitable transformation of dimensional mathematical system equations into non-dimensional ones, numerical technique has been implemented for obtaining solutions. It is indispensable to report that the mathematical model, elaborated in the upcoming section, considers unidirectional incompressible laminar flow. For numerous physical world problems, this is a limitation of categorization. Moreover, negligence of induced magnetic field is a limitation in implementation level. Presuming specific form of stream velocity is another limitation regarding the model.

2 Mathematical Model

We reckon with time-reliant convection flow of a viscous and electrically conducting fluid. The flow geometry is such that x^*-axis is placed along the semi-infinite perforated vertical plate which is set into sudden motion and y^*-axis is normal to it. Magnetic field of unaltered strength is inflicted along y^*-axis. Since the plate is of sizable length, owing to classical boundary layer theory, order of physical quantities are very small as compared to that of length of the plate, and hence physical quantities, in this scenario, do not depend on x^*. The outer stream velocity U_∞, according to Stuart [37], is presumably of the form $U_\infty = U_0[1 + g(t)]$. Chemical reaction of order one, with its rate of reaction being directly proportional to species concentration, is incorporated. The governing equations, gleaned from above premise, are given by the following:

$$\frac{\partial v^*}{\partial y^*} = 0 \tag{1}$$

$$\begin{aligned}\frac{\partial u^*}{\partial t^*} + v^*\frac{\partial u^*}{\partial y^*} &= \frac{\mathrm{d}U_\infty^*}{\mathrm{d}t^*} + \nu\frac{\partial^2 u^*}{\partial y^{*2}} + g\lambda_t\left(T^* - T_\infty^*\right)\\ &+ g\lambda_c\left(C^* - C_\infty^*\right) + \left(\frac{\sigma B_0^2}{\rho} + \frac{\nu}{k^*}\right)\left(U_\infty^* - u^*\right)\end{aligned} \tag{2}$$

$$\begin{aligned}\frac{\partial T^*}{\partial t^*} + v^*\frac{\partial T^*}{\partial y^*} &= \frac{\kappa}{\rho C_p}\frac{\partial^2 T^*}{\partial y^{*2}} - \frac{1}{\rho C_p}\frac{\partial Q_r}{\partial y^*}\\ &- \frac{Q_a}{\rho C_p}\left(T^* - T_\infty^*\right) + \frac{\nu}{C_p}\left(\frac{\partial u^*}{\partial y^*}\right)^2 + \frac{\sigma B_0^2}{\rho C_p}(u^*)^2\end{aligned} \tag{3}$$

$$\begin{aligned}\frac{\partial C^*}{\partial t^*} + v^*\frac{\partial C^*}{\partial y^*} &= D\frac{\partial^2 C^*}{\partial y^{*2}} - K_c\left(C^* - C_\infty^*\right)\\ &- \frac{\partial}{\partial y^*}\left\{V_t\left(C^* - C_\infty^*\right)\right\}\end{aligned} \tag{4}$$

Inceptive and frontier conditions on motility, temperature and mass distribution of fluid are

$$t^* = 0{:}u^* = 0,\ T^* = T_\infty^*,\ C^* = C_\infty^*\ \forall y^* \tag{5}$$

$$\begin{aligned}&t^* > 0{:}u^* = u_p^*,\ T^* = T_w^* + \varepsilon\left(T_w^* - T_\infty^*\right)\ \exp\ (m^* t^*),\\ &C^* = C_w^* + \varepsilon\left(C_w^* - C_\infty^*\right)\ \exp(m^* t^*)\ \text{at}\ y^* = 0\end{aligned} \tag{6}$$

$$\begin{aligned}&u^* \to U_\infty^* = u_0\left(1 + \varepsilon\ \exp(m^* t^*)\right),\\ &T^* \to T_\infty^*,\ C^* \to C_\infty^*\ \text{as}\ y^* \to \infty\end{aligned} \tag{7}$$

where $\varepsilon < 1$ is a non-dimensional positive constant.

Equation (1) suggests that the transverse velocity component is dependent on time variable only. In compliance with nature of outer flow, it is legitimate to assume the suction to confiscate the form:

$$v^* = -v_0\left(1 + \varepsilon A \exp(m^* t^*)\right) \tag{8}$$

where A is dimensionless positive constant such that $A << 1$ and consequently, $\varepsilon A << 1$.

Adopting Roseland approximation, heat flux due to radiation is given by

$$Q_r = -\frac{4\sigma_r}{3k_r}\frac{\partial T^{*4}}{\partial y^*} \tag{9}$$

With the presumption that the thermal variations within the flow are considerably small, the term T^{*4} can be linearized by using Taylor series. Expressing T^{*4} about the free stream temperature T_∞^* and ignoring higher order terms, we arrive at the following approximation:

$$T^{*4} \approx 4T_\infty^{*3}T^* - 3T_\infty^{*4} \tag{10}$$

Thermophoretic velocity V_t is proportional to the temperature gradient and is given by

$$V_t = -\frac{\nu K_t}{T_r}\frac{\partial T^*}{\partial y^*} \tag{11}$$

Plugging in v^* from Eq. (8) in Eqs. (2)–(4); incorporating expressions (9) and (10) in Eq. (3) and using Eq. (11) in Eq. (4), the governing equations transfigure into the forms below:

$$\frac{\partial u^*}{\partial t^*} - v_0\left(1 + \varepsilon A \exp(m^* t^*)\right)\frac{\partial u^*}{\partial y^*} = \frac{dU_\infty^*}{dt^*} + \nu\frac{\partial^2 u^*}{\partial y^{*2}} + g\lambda_t\left(T^* - T_\infty^*\right) + g\lambda_c\left(C^* - C_\infty^*\right) + \left(\frac{\sigma_e B_0^2}{\rho} + \frac{\nu}{k^*}\right)\left(U_\infty^* - u^*\right) \tag{12}$$

$$\frac{\partial T^*}{\partial t^*} - v_0\left(1 + \varepsilon A \exp(m^* t^*)\right)\frac{\partial T^*}{\partial y^*} = \frac{1}{\rho C_p}\left(\kappa + \frac{16}{3}\frac{\sigma_r T_\infty^{*3}}{k_r}\right)\frac{\partial^2 T^*}{\partial y^{*2}} - \frac{Q_a}{\rho C_p}\left(T^* - T_\infty^*\right) + \frac{\nu}{C_p}\left(\frac{\partial u^*}{\partial y^*}\right)^2 + \frac{\sigma_e B_0^2}{\rho C_p}\left(u^*\right)^2 \tag{13}$$

$$\frac{\partial C^*}{\partial t^*} - v_0\left(1 + \varepsilon A \exp(m^* t^*)\right)\frac{\partial C^*}{\partial y^*} = D\frac{\partial^2 C^*}{\partial y^{*2}} - K_c\left(C^* - C_\infty^*\right)$$

$$-\left(-\frac{\nu K_t}{T_r}\right)\frac{\partial}{\partial y^*}\left\{(C^* - C^*_\infty)\frac{\partial T^*}{\partial y^*}\right\} \tag{14}$$

At this point it is imperative to introduce the following non-dimensional quantities

$$\begin{aligned}
y &= \frac{v_0 y^*}{\nu}, t = \frac{v_0^2 t^*}{\nu}, m = \frac{m^* \nu}{v_0^2}, u = \frac{u^*}{u_0},\\
U_\infty &= \frac{U^*_\infty}{u_0}, \theta = \frac{T^* - T^*_\infty}{T^*_w - T^*_\infty}, \phi = \frac{C^* - C^*_\infty}{C^*_w - C^*_\infty},\\
G_t &= \frac{g\lambda_t(T^*_w - T^*_\infty)\nu}{u_0 v_0^2}, G_c = \frac{g\lambda_c(C^*_w - C^*_\infty)\nu}{u_0 v_0^2},\\
M &= \frac{\sigma_e B_0^2 \nu}{\rho v_0^2}, k = \frac{v_0^2 k^*}{\nu^2}, P_r = \frac{\nu\rho C_p}{\kappa},\\
R &= \frac{4\sigma_r T^{*3}_\infty}{k_r}, S = \frac{Q_s \nu}{\rho C_p v_0^2}, E_c = \frac{u_0^2}{C_p(T^*_w - T^*_\infty)}, S_c = \frac{\nu}{D},\\
\gamma &= \frac{K_c \nu}{v_0^2}, \delta = -\frac{K_t(T^*_w - T^*_\infty)}{T_r}
\end{aligned} \tag{15}$$

Equations (12), (13) and (14), in view of non-dimensional quantities given by (15), transform into the following dimensionless form

$$\begin{aligned}
\frac{\partial u}{\partial t} - (1 + \varepsilon A\exp(mt))\frac{\partial u}{\partial y} &= \frac{\partial^2 u}{\partial y^2} + G_t\theta\\
&+ G_c\phi + \left(M + \frac{1}{k}\right)(U_\infty - u) + m(U_\infty - 1)
\end{aligned} \tag{16}$$

$$\begin{aligned}
\frac{\partial \theta}{\partial t} - (1 + \varepsilon A\exp(mt))\frac{\partial \theta}{\partial y} &= \left(\frac{3 + 4R}{3P_r}\right)\\
&\frac{\partial^2 \theta}{\partial y^2} - S\theta + E_c\left(\frac{\partial u}{\partial y}\right)^2 + E_c M u^2
\end{aligned} \tag{17}$$

$$\frac{\partial \phi}{\partial t} - (1 + \varepsilon A\exp(mt))\frac{\partial \phi}{\partial y} = \frac{1}{S_c}\frac{\partial^2 \phi}{\partial y^2} - \gamma\phi - \delta\frac{\partial}{\partial y}\left(\phi\frac{\partial \theta}{\partial y}\right) \tag{18}$$

The apropos inceptive and frontier conditions in non-dimensional form are given by

$$t = 0\colon u = 0, \theta = 0, \phi = 0\ \forall\, y \tag{19}$$

$$t > 0:\ u = U_p, \theta = 1 + \varepsilon\exp(mt), \phi = 1 + \varepsilon\exp(mt)\ \text{at}\ y = 0 \tag{20}$$

$$u \to U_\infty = 1 + \varepsilon \exp(mt), \theta \to 0, \phi \to 0 \text{ as } y \to \infty \qquad (21)$$

The assemblage of Eqs. (16)–(18) together with the conditions (19)–(21), represents the present problem.

3 Solution of the Problem

The pdepe solver incorporated in MATLAB works quite well for parabolic and elliptic partial differential equations (PDEs). In this particular problem, the set of Eqs. (16)–(18), with prerequisite conditions given by Eqs. (19)–(21), have been solved numerically using pdepe. The non-dimensional velocity, heat and mass distribution of the fluid flow along with shear rate, rate of heat transfer, rate of mass transfer at the plate in terms of skin friction coefficient, Nusselt number and Sherwood number, respectively, have been depicted graphically.

4 Results and Discussion

In the process of analyzing the results, the graphical delineations of flow attributes on account of alteration of germane parameters have been put forth. The most sought after procedure to visualize the impacts of flow dictating parameters is to vary one particular parameter while keeping others constant, and the same has been on the cards for this problem. This study has been executed with the numeric values endorsed as $M = 0.3$, $k = 2.6$, $G_t = 1.8$, $G_c = 2$, $R = 0.3$, $S_c = 0.4$, $S = 0.3$, $E_c = 0.01$, $P_r = 2.36$, $\gamma = 0.2$, $\delta = 0.16$, $\varepsilon = 0.1$, $A = 0.001$, $m = 0.1$, $U_p = 0.4$. These numeric endorsements are in implementation in all graphical depictions unless otherwise indicated at specific instances.

Effect of magnetic parameter, M and porosity parameter, k on fluid locomotion have been evinced in Figs. 2 and 3, sequentially. It is spotted that corresponding intensification of M and k inflict opposite type of impacts on fluid velocity. While an upsurge in M has retarding effect on fluid locomotion, rise in k encourages growth of the same. Exertion of magnetic field orthogonally to the flow forges a drag known as Lorentzforce which restrains the flow. Contrarily, increase in porosity parameter suggests more escaping aperture for the flow and consequently incites fluid velocity. Figures 4 and 5 delineate the impact of thermal radiation, R and heat absorption parameter, S on flow velocity, respectively. It is seen that escalation in R invigorates fluid motion, but a surge in S tails off fluid velocity. Subsiding upshot of chemical reaction parameter, γ and thermophoretic parameter, δ on mobility has been demonstrated in Figs. 6 and 7. Escalation in γ and δ suggest growth in pendulous particles due to rise in calamitous reaction and thermophoretic movement, respectively, which hinders the fluid mobility.

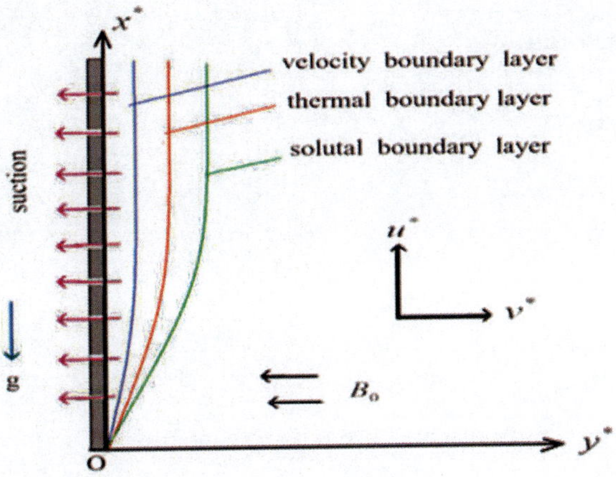

Fig. 1 Schematic flow configuration

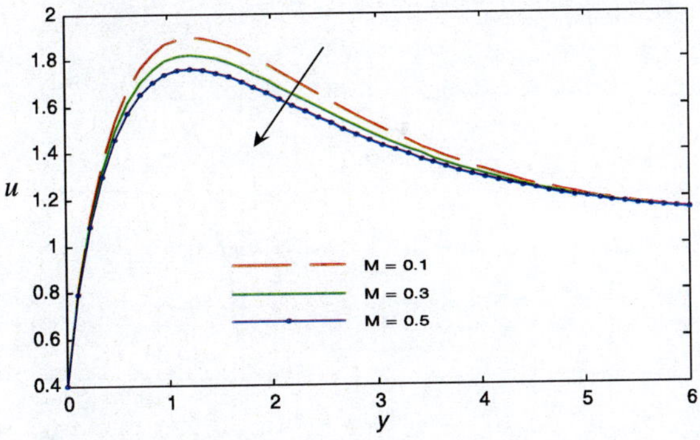

Fig. 2 Velocity for dissimilar M

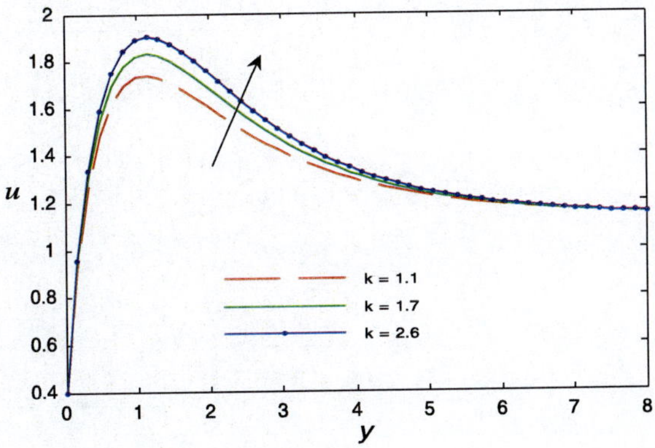

Fig. 3 Velocity for dissimilar k

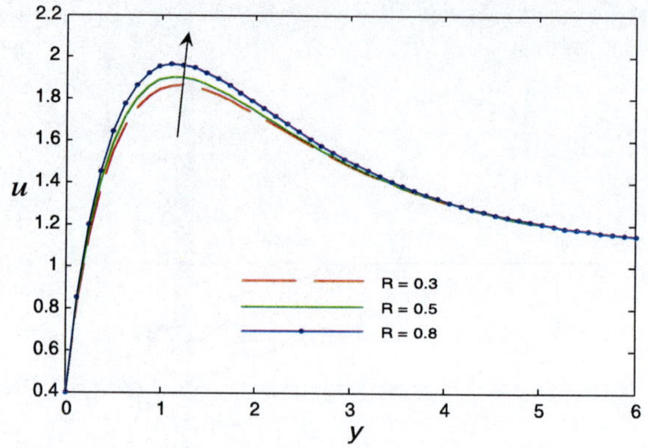

Fig. 4 Velocity for dissimilar R

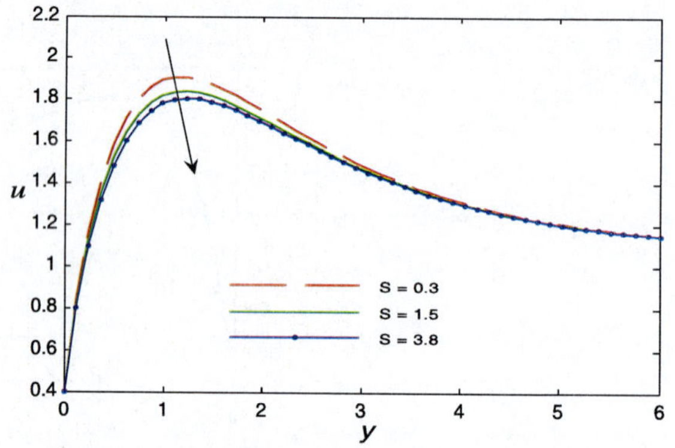

Fig. 5 Velocity for dissimilar S

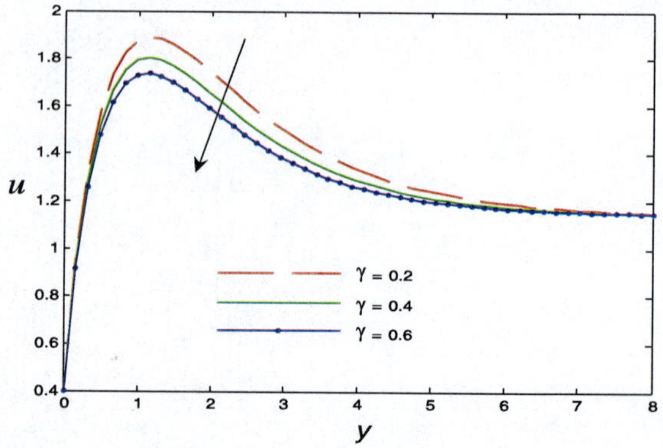

Fig. 6 Velocity for dissimilar γ

Figures 8, 9, 10, 11, 12, and 13 outline the effect of disparate parameters on local shear rate at the plate. Augmentation in k and R intensifies skin friction coefficient as indicated in Figs. 8 and 9, respectively. The reason for this is that surge in porosity and thermal radiation induce growth in velocity boundary layer resulting into more

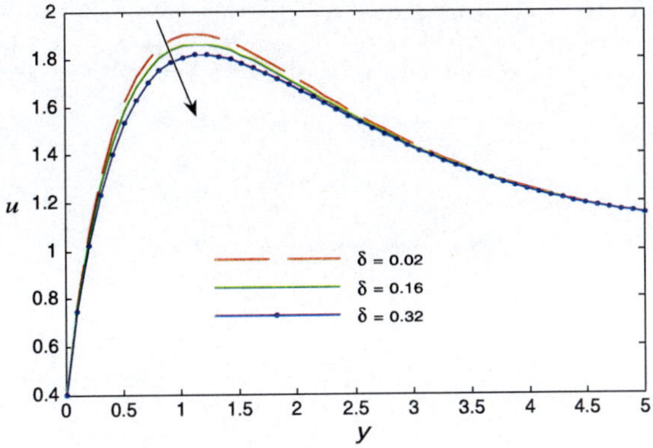

Fig. 7 Velocity for dissimilar δ

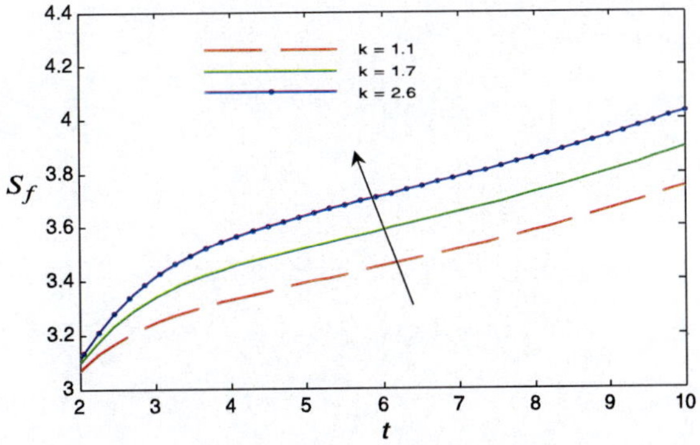

Fig. 8 Skin friction for varying k

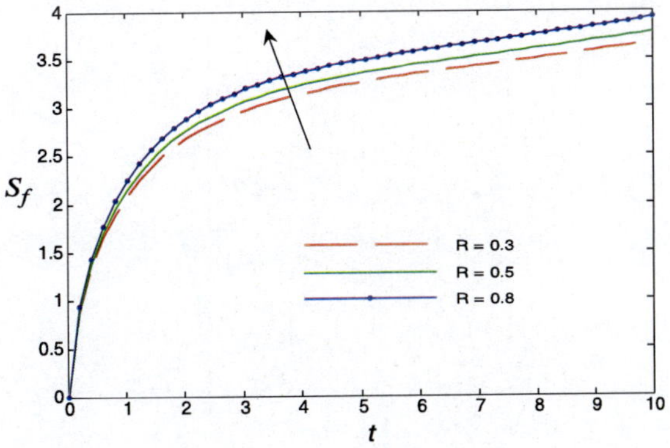

Fig. 9 Skin friction for varying R

inter-layer friction of fluid. However, Figs. 10, 11, 12, and 13 demonstrate that upsurge in M, S, γ, and δ, respectively, brings down the shear rate on the plate surface. The resistive force that arises due to application of magnetic field retards velocity boundary layer to impart impeding effect on skin friction. Heat absorption,

Fig. 10 Skin friction for varying M

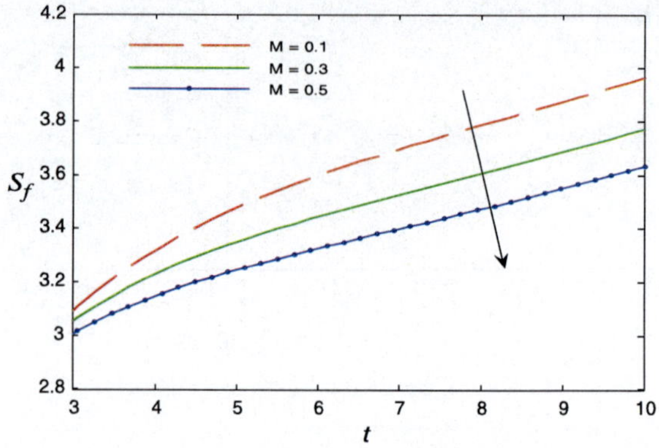

Fig. 11 Skin friction for varying S

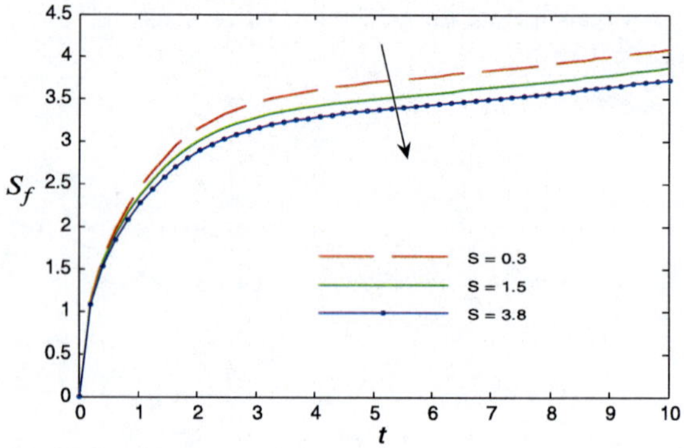

Fig. 12 Skin friction for varying γ

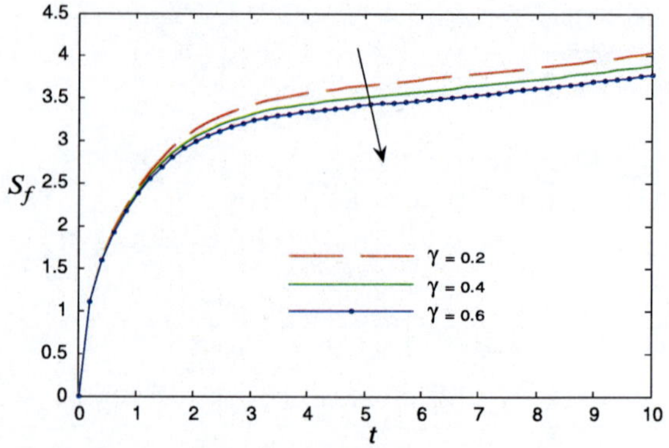

thermophoresis and endothermic chemical reaction with their coolant trait hinders locomotion of fluid particles to reduce inter-layer friction and thus causes depletion in shear rate on the plate.

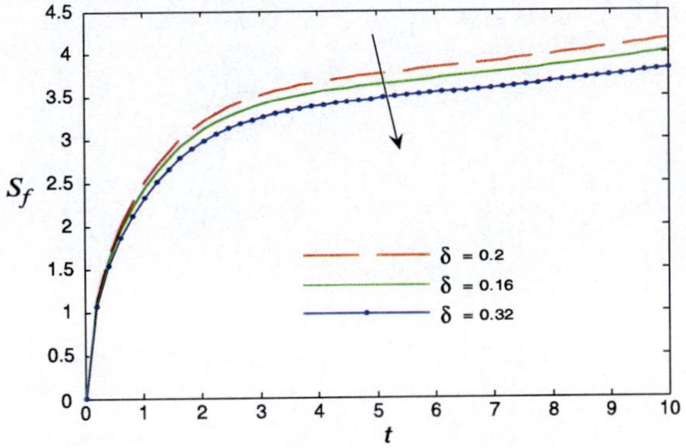

Fig. 13 Skin friction for varying δ

Impact of E_c, R, P_r, S, and δ on heat dispersal mechanism has been delineated in Figs. 14, 15, 16, 17, and 18. Inflating values of E_c and R intensifies fluid temperature and hence heat transfer process dwindles as revealed in Figs. 14 and 15. While Eckert number represents concern about the self-warming up of fluid owing to internal

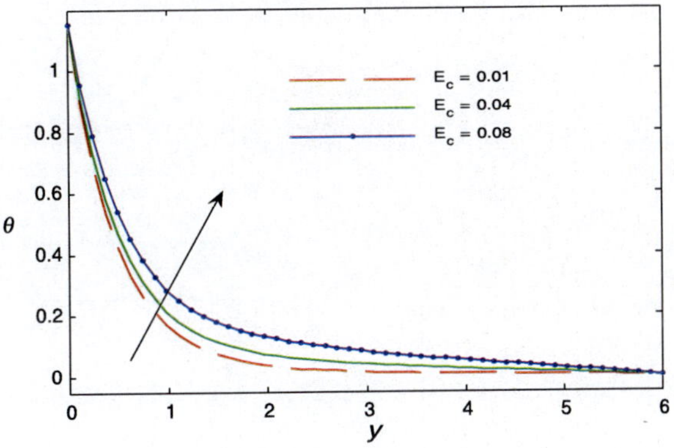

Fig. 14 Temperature for varying E

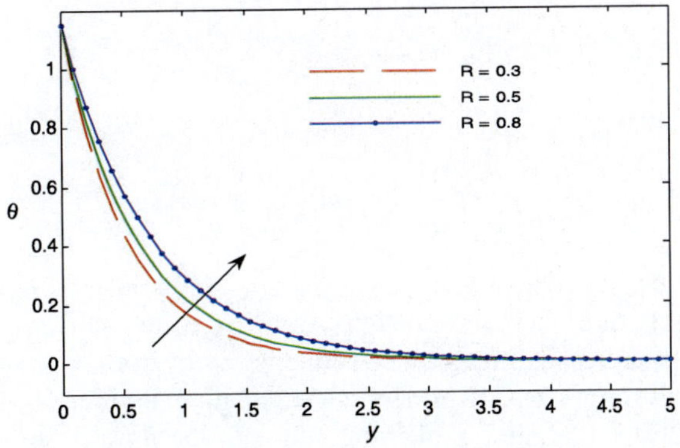

Fig. 15 Temperature for varying R

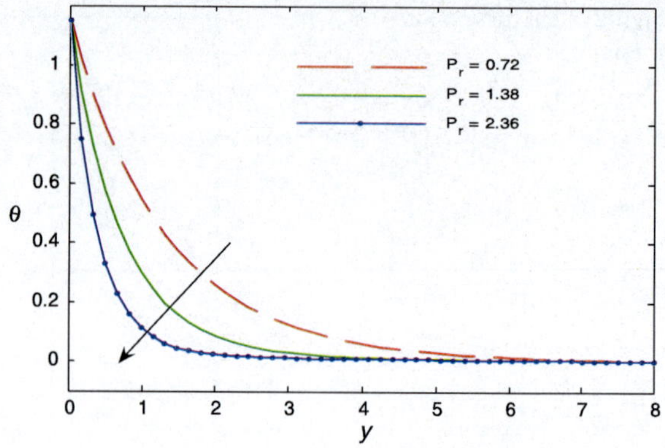

Fig. 16 Temperature for varying P_r

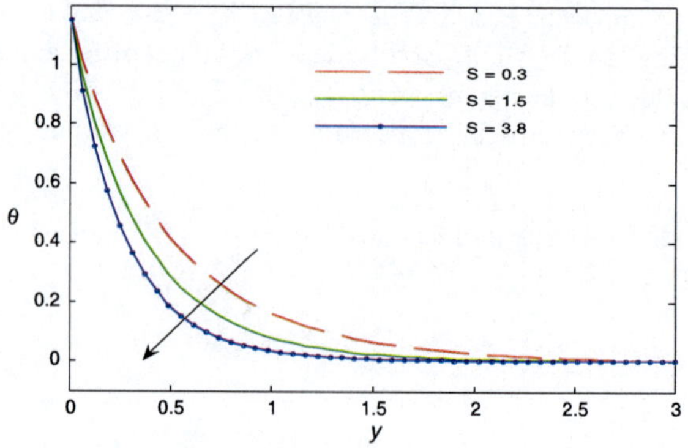

Fig. 17 Temperature for varying S

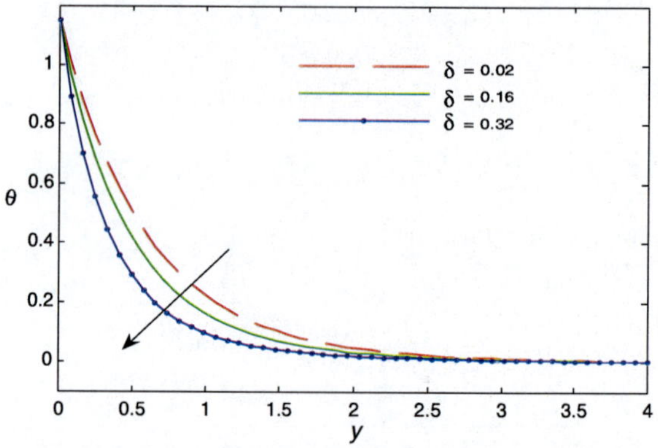

Fig. 18 Temperature for varying δ

friction, thermal radiation parameter is an external cause. Strengthening of Eckert number signifies growth in kinetic energy in comparison with the enthalpy of the flow system, which in turn inflicts dissipation. Figures 16, 17, and 18 explicate the impact of P_r, S and δ on thermal affairs of the flow. A surge in Prandtl number indicates

Fig. 19 Nusselt number for different R

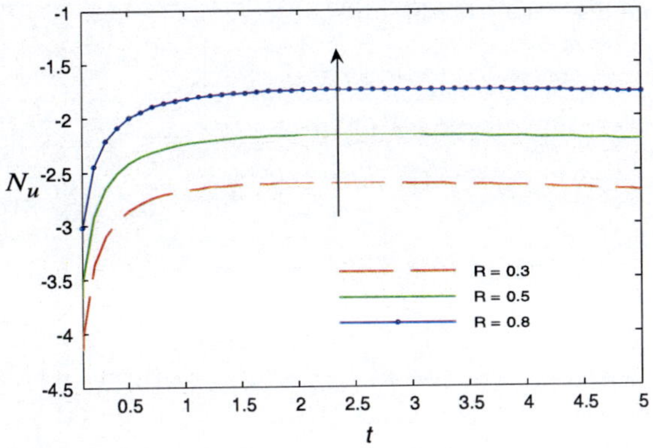

subjugation of thermal diffusivity under momentum diffusivity which manifests in downturn in fluid temperature as delineated in Fig. 16. Strengthening of heat absorption, as displayed in Fig. 17, righteously soaks up the heat from the flow. Figure 18 elucidates the impact of δ on fluid temperature. It is noticed that fluid temperature plunges with elevation of δ. Aggravation of thermophoresis demands more heat from the flow for motility of particles and hence the diminution of temperature occurs.

Influence of sss on rate of heat transmission on plate surface, indicated by Nusselt number, has been depicted in Figs. 19, 20, 21, and 22, respectively. As evinced in these figures, upsurge in thermal radiation and dissipative heat magnifies rate of heat shift at the plate. However, heat absorption and thermophoresis hamper the development of thermal gradient at the plate surface.

Upshot of Schmidt number, S_c, on species concentration and rate of mass transfer at the plate has been illustrated in Figs. 23 and 24, respectively. Improvement in Schmidt number signifies galvanization of momentum diffusion as compared to mass diffusion. This necessarily instigates rapid reduction of fluid concentration and lowers the rate of mass transfer. The upshots of chemical reaction parameter, γ, on fluid mass and Sherwood number has been delineated in Figs. 25 and 26. It is spotted

Fig. 20 Nusselt number for different S

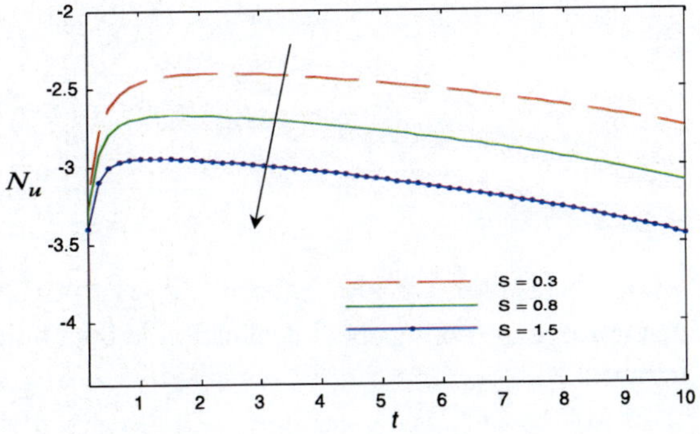

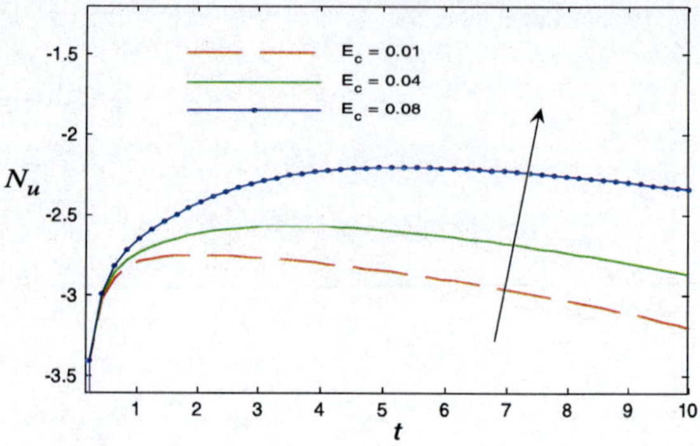

Fig. 21 Nusselt number for different *E*

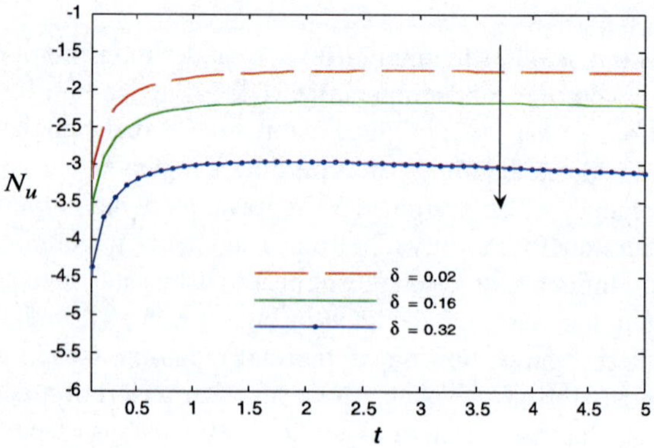

Fig. 22 Nusselt number for different δ

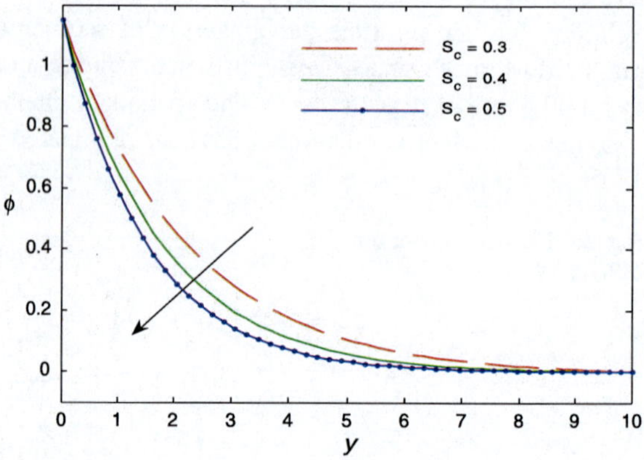

Fig. 23 Concentration for distinct S_c

that upsurge in γ bring about declination in both fluid concentration and Sherwood number.

Fig. 24 Sherwood number for distinct S_c

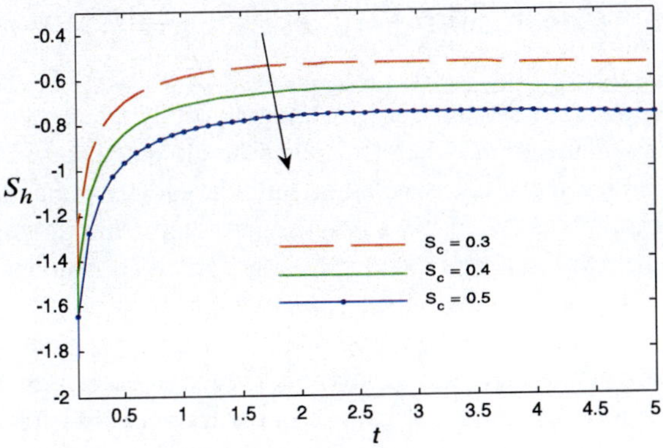

Fig. 25 Concentration for distinct γ

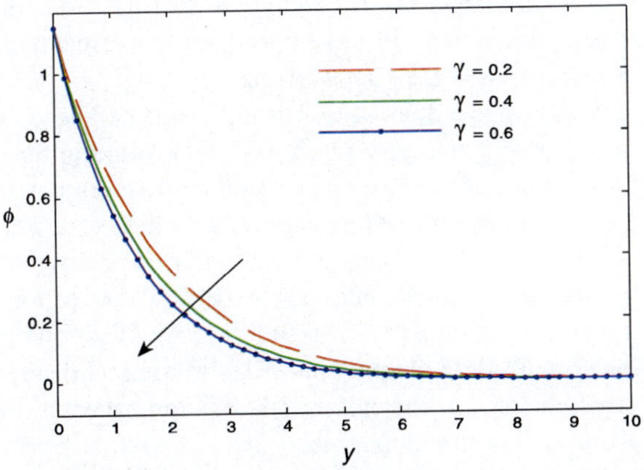

Fig. 26 Sherwood for distinct γ

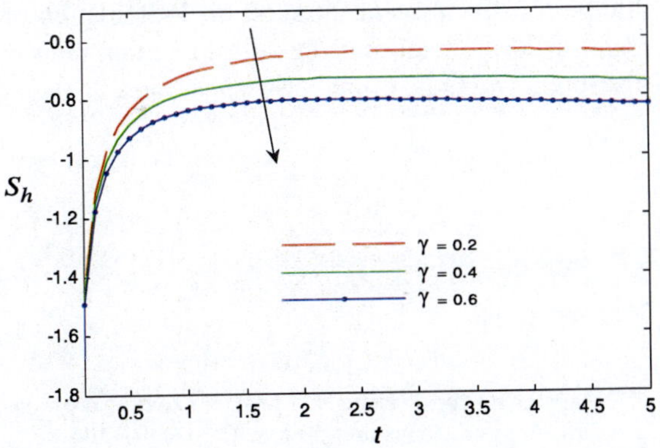

5 Conclusion

In this study, unsteady magnetohydrodynamic flow, in view of effects of thermophoresis and chemical reaction subjected to viscous dissipation, Ohmic dissipation, thermal assimilation and heat radiation, has been investigated. Visual illustrations in the form of graphs, contrived from numerical computations, have been instrumental in perceiving the upshots of concomitant flow parameters on the fluid mobility, heat distribution and mass disposition.

Representative dimensionless fluid velocity tends to speed up with intensification of porosity or thermal radiation. However, escalation in magnetic field intensity, heat absorption, chemical reaction or thermophoresis leads to retardation of fluid mobility. Shear rate at the plate drops with the upsurge in chemical reaction or thermophoresis. Growth in magnetic field strength or thermal radiation has depleting impact on skin friction coefficient. However, porosity or thermal radiation inflicts increment in rate of local shear at the plate surface.

Fluid gets warmer with surge in Eckert number. Mass transposition due to advection rises in comparison to heat enthalpy and in turn dissipation increases to raise temperature of the fluid. Augmentation in thermal radiation contributes in warming the fluid. Heat absorption, necessarily, checks growth of fluid temperature. Encouragement in dominance of momentum diffusion over thermal diffusion, as suggested by growth in Prandtl number, expedites convection in comparison with conduction, resulting into quicker heat loss. Amplification of thermophoresis contributes in lowering of fluid temperature. Rate of thermal dissemination at the plate rises with growth in dissipation or thermal radiation, whereas it declines with elevation in heat absorption or thermophoresis.

Inciting superiority of momentum diffusion over molecular diffusion results in dilution of species concentration, and retardation of mass transposition rate at the plate surface. Cataclysmic chemical reaction induces retarding effect on both mass distribution and rate of mass transposition.

Acknowledgements This work is supported financially by University Grants Commission, India.

References

1. Magyari E, Pop I, Keller B (2004) Analytical solutions for unsteady free convection flow through a porous medium. J Engg Math 48:93–104
2. Muthukumaraswamy R, Ganesan P (1998) Unsteady flow past an impulsively started vertical plate with heat and mass transfer. Heat Mass Transfer 34:187–193
3. Chamkha AJ (2004) Unsteady MHD convective heat and mass transfer past a semi-infinite vertical permeable moving plate with heat absorption. Int J Engg Sci 42:217–230
4. Abdus Satar MD, Hamid Kalim MD (1996) Unsteady free convection interaction with thermal radiation in a boundary layer flow past a vertical porous plate. J Math Phys Sci 30(1):25–37

5. Das SS, Sahoo SK, Dash GC (2006) Numerical solution of mass transfer effects on unsteady flow past an accelerated vertical porous plate with suction. Bulletin Malay Math Sci Soc 29(1):33–42
6. Takhar HS, Chamkha AJ, Nath G (1999) Unsteady flow and heat transfer on a semi-infinite flat plate with an aligned magnetic field. Int J Engg Sci 37:1723–1736
7. Ahmed SF, Das MK, Ali LE (2015) Analytical study on unsteady MHD free convection and mass transfer flow past a vertical porous plate. American J Appl Math 3(2):64–74
8. Kim YJ (2000) Unsteady MHD convective heat transfer past a semi-infinite vertical porous moving plate with variable suction. Int J Engg Sci 38:833–445
9. Ahmed SF, Mondal S, Roy A (2013) Numerical studies on MHD free convection and mass transfer flow past a vertical flat plate. IOSR J Eng 3(5):41–47
10. Mbeledogu IU, Amakiri ARC, Ogulu A (2007) Unsteady MHD free convection flow of a compressible fluid past a moving vertical plate in the presence of radioactive heat transfer. Int J Heat Mass Transfer 50:1668–1674
11. Elbashbeshy EMA (1997) Heat and mass transfer along a vertical plate with variable surface tension and concentration in the presence of magnetic field. Int J Engg Sci 35(5):515–522
12. Das SS, Satapathy A, Das JK, Panda JP (2009) Mass transfer effects on MHD flow and heat transfer past a vertical porous plate through porous medium under oscillatory suction and heat source. Int J Heat Mass Transfer 52(25–26):5962–5969
13. Makinde OD (2005) Free convection flow with thermal radiation and mass transfer past a moving vertical porous plate. Int Commun Heat Mass Transfer 32:1411–1419
14. Siddiqa S, Hossain MA (2012) Mixed convection boundary layer flow over a vertical flat plate with radiative heat transfer. Appl Math 3:7005–7016
15. Barik RN, Das GC, Kar M (2014) Unsteady free convective MHD flow and mass transfer through porous medium in a rotating system with fluctuating heat source/sink and chemical reaction. J Appl Anal Comput 4:113–123
16. Anjali Devi SP, Kandasamy R (2000) Effects of chemical reaction, heat and mass transfer on MHD flow past a semi-infinite plate. Z Angew Math Mech 80(10):697–701
17. Anjali Devi SP, Kandasamy R (1999) Effects of chemical reaction, heat and mass transfer on laminar flow along a semi-infinite horizontal plate and its application. Heat Mass Transfer 35:465–467
18. Das UN, Deka R, Soundalgekar VM (1994) Effects of mass transfer on flow past an impulsively started infinite vertical plate with constant heat flux and chemical reaction. J Forsch Ingenieurwesen Eng Res 60:284–287
19. Chambre PL, Young JD (1958) On the diffusion of a chemically reactive species in a laminar boundary layer flow. Phys Fluids 1:48–54
20. Ramana Reddy JV, Sugunamma V, Sandeep N, Sulochana C (2016) Influence of chemical reaction, radiation and rotation on MHD nanofluid flow past a permeable flat plate in porous medium. J Nigerian Math Soc 35:48–65
21. Muthucumaraswamy R (2002) Effects of suction on heat and mass transfer along a moving vertical surface in the presence of chemical reaction. Forschungim Ingenieurwesen 67(3):129–132
22. Kandasamy R, Periasamy K, Prabhu KK (2005) Effects of chemical reaction, heat and mass transfer along a wedge with heat source and concentration in the presence of suction or injection. Int J Heat Mass Transfer 48(7):1388–1394
23. Muthucumaraswamy R, Amutha K (2012) Thermal diffusion effects on MHD flow past an oscillating vertical plate with chemical reaction of first order. Int J Math Arch 3(12):4989–4996
24. Raju MC, Varma SVK, Seshaiah B (2015) Heat transfer effects on a viscous dissipative fluid flow past a vertical plate in presence of induced magnetic field. Ain Shams Engg J 6:333–339
25. Reddy NA, Varma SVK, Raju MC (2009) Thermo-diffusion and chemical effects with simultaneous thermal and mass diffusion in MHD mixed convection flow with ohmic heating. J Naval Arch Mar Engg 6:84–93
26. Ibrahim M (2013) Effects of mass transfer, radiation, Joule heating and viscous dissipation on steady MHD marangoni convection flow over a flat surface with suction and injection. Int J Engg Math 4:1–9

27. Jamalabadi MYA, Park JH (2014) Thermal radiation, Joule heating and viscous dissipation effects on MHD forced convection flow with uniform surface temperature. J Fluid Dyn 4:125–132
28. Ibrahim SM, Suneetha K (2016) Heat source and chemical effects on MHD convection flow embedded in a porous medium with Soret, viscous and Joules dissipation. Ain Shams Engg J 7(2):811–818
29. Babu VS, Reddy GVR (2011) Mass transfer effects on MHD mixed convective flow from a vertical surface with ohmic heating and viscous dissipation. Adv Appl Sci Res 2(4):138–146
30. Chen CH (2004) Combined heat and mass transfer in MHD free convection from a vertical surface with ohmic heating and viscous dissipation. Int J Engg Sci 42(7):699–413
31. Goren SL (1977) Thermophoresis of aerosol particles in laminar boundary layer on flat plate. J Colloid Interface Sci 61:77–85
32. Selim A, Hossain MA, Rees DAS (2003) The effect of surface mass transfer on mixed convection flow past a heated vertical flat permeable plate with thermophoresis. Int J Therm Sci 42:973–982
33. Chamkha AJ, Pop I (2004) Effects of thermophoresis particle deposition in free convection boundary layer from a vertical flat plate embedded in a porous medium. Int Commun Heat Mass Transfer 31:421–430
34. Chamkha AJ, Pop I (2006) Effects of heat generation or absorption on thermophoretic free convection boundary layer from a vertical flat plate embedded in a porous medium. Int Commun Heat Mass Transfer 33:1096–1102
35. Alam MS, Rahaman MM, Satar MA (2007) Similarity solutions for hydromagnetic free convective heat and mass transfer flow along a semi-infinite permeable inclined flat plate with heat generation and thermophoresis. Non-linear Anal Model Control 12(4):433–445
36. Chamkha AJ, Camille I (2000) Effects of heat generation/absorption and thermophoresis on hydromagnetic flow with heat and mass transfer flow over a flat surface. Int J Numer Methods Heat Fluid Flow 10:432–448
37. Stuart JT (1955) A solution of the Navier-Stokes and energy equations illustrating the response of skin friction and temperature of an infinite plate thermometer to fluctuations in the stream velocity. In: Proceedings of the royal society of London, series A, mathematical and physical sciences, 1955, vol 231, pp 116–130

Motion Detection of Human on Video: State of the Art

C. Jeba Nega Cheltha and Chirag Sharma

Abstract In current scenario, almost everyone is using camera in smart phones; in all buildings, CCTV is fixed for surveillance. Despite of all these, crime rate is increasing more and more every second. Human motion detection is one of the recent topics in computer vision. Human motion detection in video is used in many fields like medicine, military, cybercrime, road traffic control system, etc. The videos which are recorded in CCTV may or may not be clear due to various reasons. Hence, human motion detection is very much needed to identify the motion of human in the video which is recorded by CCTV or any other sources. This human motion detection will help to detect the crime happening in the world. There are many algorithms and techniques introduced by various authors. In this paper, I am going to analyse the various techniques introduced by various authors.

Keywords Human motion · Computer vision · Object classification · Motion segmentation · Tracking algorithm

1 Introduction

As we see in the current scenario, crime rate is increasing day by day even though CCTV is fixed in almost every building. Sometimes, it is difficult to identify the motion of human in videos due to various reason. Some of the reasons include flickering street light, low light intensity, shadow, illumination variations, human hidden behind any other object, etc. So it is very much necessary to focus on motion detection of human. In the recorded video, many objects and many background images will be available, in which identifying human and motion of the human body from other object in the video is the study of human motion analysis. Although many research papers are published in this area, there is some advantage and drawback in those papers. In human motion segmentation, object classification and tracking play very

C. J. N. Cheltha (✉) · C. Sharma
Department of Computer Science and Engineering, Lovely Professional University, Phagwara, India
e-mail: chelnega8@gmail.com; jeba.25111@lpu.co.in

M. Gupta et al. (eds.), *Artificial Intelligence on Medical Data*,
Lecture Notes in Computational Vision and Biomechanics 37,
https://doi.org/10.1007/978-981-19-0151-5_39

important role. Motion segmentation means finding autonomously moving pixels from the video and extricating them from the background motion. There are few algorithms related to motion segmentation. Some of them are background subtraction, statistical methods, temporal differencing, optical flow [1] and spatiotemporal filter [2, 3].

Background subtraction: It is one of the approaches which is used in motion segmentation. This methodology [4, 5] is utilized for tracking down the moving articles by discovering the contrast between the present frame [6, 7] and the reference outline pixel by pixel. Even though in paper [8] and [9] have taken many challenges in their paper related to background substitution, still the work should be expanded if the night is noisy. Statistical methods: It is one of the methods used in motion segmentation. This method is much useful to use in dynamic background model. In this model, each pixel in the current edge will be secret into frontal area or foundation by comparing the measurements of the current foundation model. Temporal differencing: To detect the moving region in the video, more than two consecutive frames difference is calculated in this method for motion segmentation. But this method is not good for extracting the complete appropriate feature pixels. But this method is used for human detection [10] in video. Optical flow: This is another method which is used for motion segmentation in the video. This method is used for scheming the motion of the image intensities. It is used to extract the moving regions in a video based on optical flow [11].

Now after motion segmentation from the video, it is necessary to find whether the object identified in the video is human or not. If the object is moving, then we can identify it as human, but still there are many other objects like animals, birds and vehicles also move. To detect whether the moving object is human or any other object, object classification is used. It is used in motion detection in videos. The classification of object is based on shape [3, 12], size, colour, motion [13, 14], texture based [15–18], etc. Various classifications of object are used in various papers.

Subsequent to recognizing if the moving item is human, next we really wanted to follow the stances of human body. This is finished by different human movement tracking algorithms [19, 20]. In this paper, I am dissecting different papers.

2 Literature Survey

2.1 *Vandit Gajjar et al.*

Gajjar et al. said in their paper [21] that the cameras utilized for taking video are RGB cameras, on the grounds that RGB camera is useful to take quality video during faint light. For identifying individuals in any casing, HOG elements to prepare a help vector machine are utilized. In the wake of preparing remarkable quality, window pictures are utilized. This decreased computational time in this paper for distinguishing human in any video. Additionally, *K* implies calculation which is utilized to recognize the movement of human. In HOG models' best case, the calculation time taken was 1768.29 s. In saliency model, it required 23.0047 s. In saliency model, any place human is distinguished, and it features with focused energy. Future work of this paper is to add movement following strategy for video reconnaissance by optical stream related to visual saliency windowing.

2.2 *Manikandan et al.*

Manikandan et al. explained in their paper [22] entitled "Prediction of Human Motion Detection in Video Surveillance Environment using Tensor Flow" deep learning method for detecting the motion of human being in video surveillance systems. In this paper, as video clip is taken long range, video human appears with different resolutions, and fixed, unmoving objects and also removes shaking, blurring and out of focus data. For detecting moving object, 2D convolution operation is used. AdaBoost classifier is applied for human detection algorithm. CNN is used for image-based classification in this paper. Neural network approaches like GoogleNet and Google Tensor are used for prediction and decision-making. The quality of the work is 88% accuracy.

2.3 *Nouar AlDahoul et al.*

In [1] paper, said about, human discovery in recordings assumes a significant part in different genuine applications. In this paper, moving camera is utilized. Edge of the video is given as info. This paper distinguishes human and non-human articles. In this paper, three profound models are utilized: regulated CNN, pretrained CNN and HELM. In this paper, foundation adjustment is finished by optical stream technique, which gauges the speed and heading of the movement between outlines. The upside of this paper is that it identifies human picture naturally, likewise it is hearty against different exercises. The principle disadvantage of this paper is that it relies upon the nature of optical stream handling.

2.4 *BirMohan et al.*

In paper [8], new algorithm was introduced for detecting the motion. In this paper, background substitution is used for motion segmentation and classification, and tracking of human motion a new algorithm was implemented called motion detection (MODE). This calculation is free of enlightenment varieties, bootstrapping, dynamic varieties and clamour issues.

2.5 Sapana et al.

Mishra et al. [23] explained about various segmentation, object classification and motion detection in static camera, and tracking in their work. Sapna et al. said [24] in their paper "Human Motion Detection and Video Surveillance using MATLAB". In this paper for detecting the object, eigen object is computed from three frames of the video. Shape-based object classification is used for human. To remove noise and unwanted objects from the video frame, and also to detect moving region, background substitution [25] is used in this paper. Spatiotemporal differencing is used in video to detect moving object. Background elimination is done by temporal differencing in this paper. Human is detected by shape in this paper. This paper works well to detect human in various challenges. But vertical shadow cannot be eliminated in this paper.

2.6 Geetanjali et al.

In paper [26], Kale et al. explained about human motion detection. In this paper, kinematic approach is used. For classifying whether the moving object detected is human or something else, few models are used. They are humanoid body model, humanoid image model, trajectory-based recognition, state model-based approach and hierarchical synthetic approach. Few limitations of this paper are in humanoid body model, and 3D object is difficult to track. Humanoid image model is sensitive if view and size vary. Trajectory-based recognition gives speed invariant recognition. In intention reasoning, if object hit during motion detection cannot be identified.

2.7 Qiang et al.

Qiang et al. said in their paper, named "Catch of 3D Human Motion Pose in Virtual Reality Based on Video Recognition" [27, 28], about movement present innovation in 3D activity. For versatile change, fuzzy logic is utilized in proposed work of this paper to perceive the movement state. Human movement model, two-stage Kalman channel plan which depends on fuzzy logic, is utilized in this paper. Managed learning is utilized in this paper to catch the human movement act catch for catching and 3D human movement signals.

2.8 Husien et al.

Husein et al. said in their paper [29], entitled as "Motion detect application with frame difference method on a surveillance camera", frame difference method to detect an object [30]. It helps to detect every motion of an object. This method compares two frames and finds its difference [31]. Along with frame differencing, background subtraction method is also used in this paper to find the reference on foreground objective. Background subtraction is mainly used to detect motion on static camera. It can also reduce the noise. Research gap in this paper is that paper wrong motions are captured due to lighting effect in surveillance camera.

2.9 *Wafae Mrabti et al.*

Mrabti et al., said in their paper about human motion tracking [32] about few visual tracking methods such as compressive tracking, online multiple instance algorithm, FragTrack tracking algorithm and tracking learning detection framework. Compressive tracking algorithm is online visual tracking algorithm. Two principles are used in compressive tracking. First rule is in the wake of choosing the example physically; set of windows will be produced close to the situation of chose tests which is likewise called as sure examples. Second guideline is that select the examples which is away from the middle. Next following strategy utilized in this paper is online different instance learning tracking algorithm which utilizes recognition procedure by utilizing numerous occasions learning for following. Next calculation looked at in this paper is FragTrack tracking calculation which track part astute. Following median flow tracker is utilized in this paper, which could be ready to anticipate [33] the area of the articles across continuous edges. Indicator part in this strategy assists with filtering the total picture and right the tracker. This load of following calculations is exact and powerful. Exploration hole is that speed changes, occultation of the objective ought to be improved.

2.10 *J. Josue Hernandez et al.*

Hernandez et al. [34] in their paper explained human motion detection. Using the frames, motion vectors are estimated, and this will help to reduce distortion and illumination. To identify whether the motion is relevant or irrelevant, motion vector estimation is used in this paper. To classify the object, trajectory estimation is used in this paper. At last, the development of the article is followed by getting movement vectors and by following individual movement vectors in each casing. In this paper, human development is distinguished from inside to outside of the confined zone.

2.11 K. Sheng-Yi Chiu et al.

Sheng-Yi Chiu et al. [35] said in their paper foundation deduction calculation dependent on class entropy investigation which powerfully makes shading classifications for every pixel in the pictures and additionally colour classification entropy to conquer dynamic entropy. This paper overcomes many challenges in moving object by using joint category entropy and EBBE algorithm.

2.12 Hariri et al.

Hariri et al. [36] examined random forests (RF), linear discriminant analysis (LDA) classifiers and support vector machines (SVMs). Furthermore, edge location division approach has been used in this examination, hence separated the highlights utilizing morphological tasks. The reproduction result shows the LDA and SVM give better execution. Notwithstanding, the dataset as older should be involved.

3 Comparison of Various Methods and Papers

Table 1 describes the survey of various segmentation methods used in human motion detection. Table 2 describes the future enhancement identified from various papers, and Table 3 explains about various object classifications. Figure 1 describes the accuracy of motion segmentation; Fig. 2 describes accuracy of various object classifications, and Fig. 3 describes overall performance of few authors.

Table 1 Motion segmentation method

Motion segmentation method	Description	Accuracy
Background subtraction	This method is used to distinguish moving objects in a categorization of frames from motionless cameras Accuracy is high when used with temporal differencing and hierarchical background model	High
Optical flow	It is the gesture of objects between successive frames of sequence, caused by the comparative movement among the object and camera	Medium to high
Spatiotemporal filter	It is used to recognize the objects existing in the scene. This method processes two successive frames at a time	Medium to high

Table 2 Overall analysis of work done in previous years

Previous paper	Object classification/detection method used	Tracking algorithm	Limitation/future enhancement
Sapana K Mishra et al.	Shape-based frame differencing	–	Vertical shadow elimination is not successful and also in moving camera human detection is not successful
Manikandan et al.	2D convolution operation and AdaBoost classifier	–	Average accuracy of this method is 88%, so accuracy can be improved further in future work
Nouar et al.	–	Supervised CNN, pretrained CNN, hierarchical extreme learning machine	Human detection depends on the optical flow processing stage
Josue Hernandez et al.		Trajectory estimation	
AM Husein et al.	Frame differencing	–	Need to improve the effectiveness in low light-based conditions
Birmohan et al.	MODE	–	Need to improve the processing speed and evaluation
Vandit et al.	HOG + SVM classifier	K-means algorithm	Detection precision is 83.11, and it can be improved in future work
Wafae et al.	Particle filter	Predictive learning tracking frame	Need to improve on more challenges such as velocity changes and occultation
Itasaso et al.	Key frame-based approach	Handcrafted motion, MPDMM, ELM	Deep learning is best, and it should be improved more

Table 3 Object classification method

Object classification method	Description	Accuracy
Shape based	This method identifies moving object based on shape	Medium
Texture based	This method is to allocate an unidentified sample image to one of a set of recognized texture classes	High
Motion based	This method identifies moving object based on motion	Medium

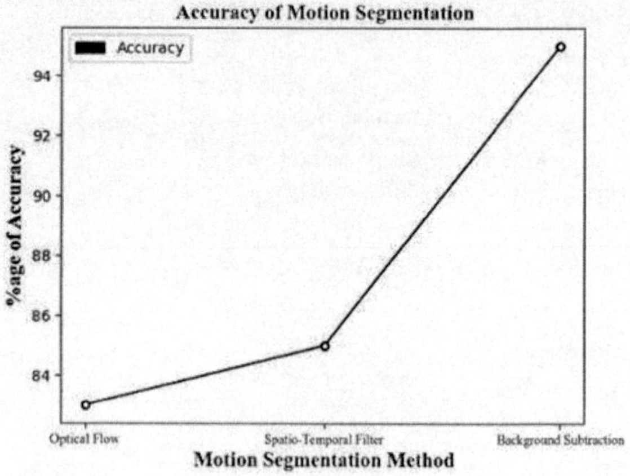

Fig. 1 Accuracy of object classification method

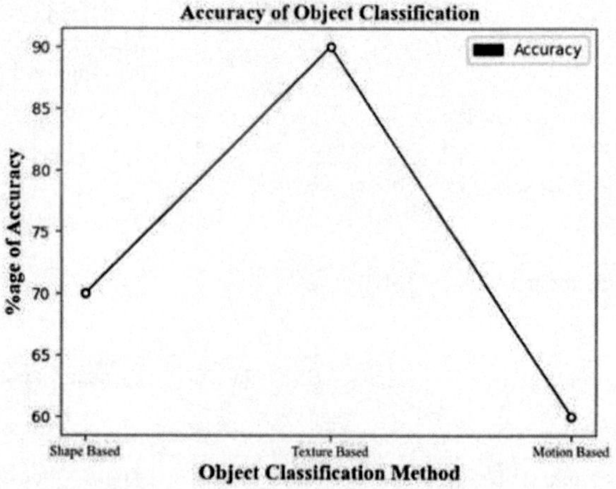

Fig. 2 Accuracy of object classification

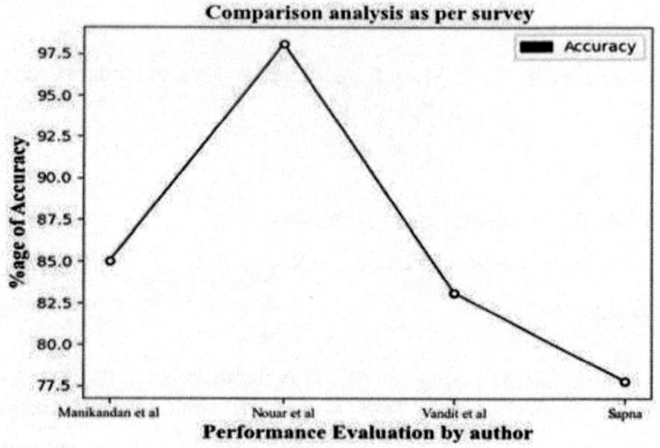

Fig. 3 Overall performance as per evaluation done by authors

4 Conclusion

Human motion detection is one of the recent topics in the digitized world, which mainly helps to reduce crime level around the world. It is also used in various other fields like medical science, sports, military, etc. This field of study contributes much more in societal issues. In this paper, motion segmentation, object classification and object tracking of various papers were analysed, and comparison table is also made. And also, it identified various papers limitation and future enhancement. As per the limitation and future enhancement of various papers analysed, human motion detection study can be enhanced in future.

References

1. AlDahoul N, Md Sabri AQ, Mansoor AM (2018) Real-time human detection for aerial captured video sequences via deep models. Hindawi
2. Piroddi R, Vlachos T (2006) A simple framework for spatio-temporal video segmentation and delayering using dense motion fields. IEEE Signal Process Lett 13(7):421
3. Candamo J, Shreve M, Goldgof DB, Sapper DB, Kasturi R (2010) Understanding transit scenes: a survey on human behavior-recognition algorithms. IEEE Trans Intell Transp Syst 11(1):206–224
4. Vo GD, Park C (2018) Robust regression for image binarization under heavy noise and nonuniform background. Pattern Recognit 81:224–239
5. Choudhury SK, Sa PK, Bakshi S, Majhi B (2016) An evaluation of background subtraction for object detection vis-a-vis mitigating challenging scenarios. IEEE Access 4:6133–6150
6. Barron JL, Fleet DJ, Beauchemin SS, Burkitt TA (1992) Performance of optical flow technique. In: IEEE conference on computer vision and pattern recognition (CVPR '92). Piscataway: IEEE, pp 236–242
7. Paul M, Haque SM, Chakraborty S (2013) Human detection in surveillance videos and its applications—a review. EURASIP J Adv Signal Process 2013:176
8. Singh B, Singh D, Singh G, Sharma N, Sibbal V (2014) Motion detection for video survilance. In: ICSPCT, IEEE
9. Maddalena L, Petrosino A (2008) A self-organizing approach to background subtraction for visual surveillance applications. IEEE Trans Image Process 17(7):1168–1177
10. Lipton AJ, Fujiyoshi H, Patil RS (1998) Moving target classification and tracking from real-time video. In: Proceedings fourth IEEE workshop application computer vision WACV98 Cat No98EX201, vol 98, no 2, pp 8–14
11. Zhong H, Shi J, Visontai M (2004) Detecting unusual activity in video. In: 2004 IEEE computer society conference on computer vision and pattern recognition (CVPR 2004). Piscataway: IEEE, pp 819–826
12. Eishita FZ, Rahman A, Azad SA, Rahman A (2013) Occlusion handling in object detection. In: Multidisciplinary computational intelligence techniques: applications in business, engineering, and medicine. IGI global. https://doi.org/10.4018/978-1-4666-1830-5.ch005
13. Bobick AF, Davis JW (2001) The recognition of human movement using temporal templates. IEEE Trans Pattern Anal Mach Intell 23(3):257–267
14. Cutler R, Davis LS (2000) Robust real-time periodic motion detection, analysis, and applications. IEEE Trans Pattern Anal Mach Intell 22(8):781–796
15. Ojala T, Pietikinen M, Maenpaa T (2002) Multi-resolution grayscale and rotation invariant texture classification with local binary patterns. PAMI 24(7):971–987

16. Zhang L, Li SZ, Yuan X, Xiang S (2007) Real-time object classification in video surveillance based on appearance learning. In IEEE computer society conference on computer vision and pattern recognition 2007 (CVPR 2007). Piscataway: IEEE, pp 1–8
17. Zhu Q, Yeh MC, Cheng KT, Avidan S (2006) Fast human detection using a cascade of histograms of oriented gradients. In: IEEE computer society conference on computer vision and pattern recognition 2006 (CVPR '06). Piscataway: IEEE, pp 1491–1498
18. Moctezuma D, Conde C, de Diego IM, Cabello E (2011) Person detection in surveillance environment with HoGG: gabor filters and histogram of oriented gradient. In: ICCV workshops. Piscataway: IEEE, pp 1793–1800
19. Rodríguez-Moreno I, Martínez-Otzeta JM, Sierra B, Rodriguez I, Jauregi E (2019) Video activity recognition: state-of-the-art. Sensors 19(3160) https://doi.org/10.3390/s19143160
20. Tarchoun B, Khalifa AB, Dhifallah S, Jegham I, Mahjoub MA (2020) Hand-crafted features vs deep learning for pedestrian detection in moving camera. Traitement du Signal 37(2):209–216
21. Gajjar V, Gurnani A, Khandhediya Y (2017) Human detection and tracking for video surveillance: a cognitive science approach. In: Proceedings of the IEEE international conference on computer vision workshops
22. Manikandan S, Chinnadurai M, Thiruvenkatesuresh MP, Sivakumar M (2020) Prediction of human motion detection in video surveillance environment using tensor flow. Int J Adv Sci Technol 29(05):2791–2798
23. Hernandez J, Morita H, Nakano-Miytake M, Perez-Meana H (2009) Motion detection and tracking using video frames. CIARP 2009, LNCS 5856:1054–1061
24. Mishra SK, Bhagat KS (2015) Human motion detection and video surveillance using MATLAB. Int J Sci Eng Res 3(7)
25. Haritaoglu I, Harwood D, Davis LS (2000) W4: real-time surveillance of people and their activities. IEEE Trans Pattern Anal Mach Intell 22(8):809–830
26. Kale GV, Patil VH (2016) A study of vision based human motion recognition and analysis. Int J Ambient Comput Intell 7(2)
27. Geng B, Pham N, Xue Q et al (2020) A three-dimensional vocal fold posturing model based on muscle mechanics and magnetic resonance imaging of a canine larynx. J Acoust Soc Am 147(4):2597–2608
28. Husein AM, Halim D, Leo R (2019) Motion detect application with frame difference method on a surveillance camera. J Phys Conf Ser 1230(2019):012017. https://doi.org/10.1088/1742-6596/1230/1/012017
29. Singla N (20141) Motion detection based on frame difference method. Int J Inf Comput Technol 4(15)
30. Guo J, Wang J, Bai R, Zhang Y, Li Y (2017) A new moving object detection method based on frame-difference and background subtraction. In: IOP conference series: materials science and engineering, vol 242
31. Mrabti W, Baibai K, Bellach B, Thami RO, Tairi H (2018) Human motion tracking: a comparative study. ICDS
32. Kalal Z, Mikolajczyk K, Matas J (2010) Forward-backward error: automatic detection of tracking failures, pp 2756–2759
33. Mishra MS, Bhagat KS (2015) A survey on human motion detection and surveillance. Int J Adv Res Electron Commun Eng 4(4)
34. Hu W, Tan T, Wang L, Maybank S (2004) A survey on visual surveillance of object motion and behaviors. IEEE Trans Syst Man Cybern Part C Appl Rev 34(3):334–352
35. Chiu SY, Chiu CC, Xu SS A (2018) Background substraction algorithm environments based on category entropy analysis. Appl Sci 8:885 https://doi.org/10.3390/app8060885
36. Elhariri E, El-Bendary N, Hassanien AE, Snasel V (2015) An assistive object recognition system for enhancing seniors quality of life. Procedia Comput Sci 65:691–700
37. Barnich O, Van Droogenbroeck M (2009) ViBe: a powerful random technique to estimate the background in video sequences. In: IEEE International conference on acoustics, speech and signal processing, pp 945–948

38. Bergen JR, Burt PJ, Hingorani R, Peleg S (1992) A three-frame algorithm for estimating two-component image motion. IEEE Trans Pattern Anal Mach Intell 9:886–896
39. Huynh-The T, Lee S, Hua CH (2017) ADM-HIPaR: an efficient background subtraction approach. In: Proceedings of the 2017 14th IEEE international conference on advanced video and signal based surveillance AVSS 2017, Lecce, Italy
40. Nagendran A, Dheivasenathipathy N, Nair RV, Sharma V 2014 Recognition and tracking moving objects using moving camera in complex scenes. Int J Comput Sci Eng Appl 4:31–36

Author Index

M. Gupta et al. (eds.), *Artificial Intelligence on Medical Data*,
Lecture Notes in Computational Vision and Biomechanics 37,
https://doi.org/10.1007/978-981-19-0151-5